Global Issues in Water Policy

Volume 24

More information about this series at http://www.springer.com/series/8877

Jean-Daniel Rinaudo • Cameron Holley
Steve Barnett • Marielle Montginoul
Editors

Sustainable Groundwater Management

A Comparative Analysis of French and Australian Policies and Implications to Other Countries

Springer

Editors
Jean-Daniel Rinaudo
BRGM
Montpellier University
Montpellier, France

Cameron Holley
Faculty of Law
University of New South Wales Sydney
Sydney, NSW, Australia

Steve Barnett
Adelaide, SA, Australia

Marielle Montginoul
INRAE – UMR G-Eau, Montpellier
University, Montpellier, France

ISSN 2211-0631 ISSN 2211-0658 (electronic)
Global Issues in Water Policy
ISBN 978-3-030-32765-1 ISBN 978-3-030-32766-8 (eBook)
https://doi.org/10.1007/978-3-030-32766-8

This Springer imprint is published by the registered company Springer Nature Switzerland AG.
The registered company address is: Gewerbestrasse 11, 6330 Cham, Switzerland

Preface

The idea for this book was born in November 2015 during a professional visit to Australia by Jean-Daniel Rinaudo, who had the opportunity to meet with Cameron Holley and Steve Barnett, as well as many of their colleagues in Adelaide, Melbourne, Canberra and Sydney to discuss many aspects of groundwater management. During the enthusiastic and productive exchanges that took place during this visit, it soon became clear to us that management policies and planning tools in both France and Australia were based on similar foundations and that continuing to share knowledge and experience would be mutually beneficial.

A year later after the IAH International Congress in Montpellier (France) in September 2016, we met again for a two-day workshop that brought together 30 French and 13 Australian experts, all directly involved in the management and planning of groundwater resources. For many participants, this workshop offered them the opportunity to share the results of several decades of personal experience for the first time and to engage with their peers from the other side of the world. One of the highlights of this meeting was the moment when each delegation reported their views of each other's management model in a game called 'report of bewilderment'. The main finding was that, reassuringly, French and Australian water managers employ similar approaches to solve similar problems using similar technology. But it was also the realisation of the existence of fundamental philosophical differences, of a 'clash of civilizations'. This was mostly apparent on the subject of ownership of water use rights where the French rejected the idea of water markets, while the Australians expressed a polite perplexity regarding the collective management of water allocation to existing water users.

At the end of the workshop, many participants agreed to contribute to an edited book so that the management approaches and techniques discussed therein may inform and benefit their peers, groundwater managers from other countries and future generations. A collective work project was thus submitted to Ariel Dinar, who strongly encouraged us to pursue the project. The group was extended to include several academic and professional experts from Australia, France and other nations in order to meet the requirements of an academic publication and to extend the coverage of the book. Although the editorial efforts were a collective endeavor, a significant amount of the work was shouldered by Rinaudo, whose leadership and diligence drove this collection. *Merci de nous avoir menés, Jean-Daniel.* Eighteen months later, the collection was completed. Our hope is that the comparative insights from the completed book will assist groundwater managers and scholars across the globe and, by doing so, help contribute to the efforts of the UN High Level Panel on Water's *Agenda for Water Action (2018)*, which calls for efforts 'catalyzing change, building partnerships & international cooperation at the global level'.

This journey has benefited from many supports that we wish to thank here. The initial work visit carried out in 2015 was financed by the National Research Agency (ANR) and Brgm as part of the Arena Groundwater project. The French Embassy in Australia also supported the short mobility visit of Cameron Holley to Montpellier in 2016 (Scientific Mobilisation Grant 2016), and Holley's work on the book benefited from an Australian Research Council Discovery Early Career Researcher Awards (DE140101216) and an Australian Research Council Discovery Grant (DP170100281). The Rhône Méditerranée Corse Water Agency subsequently funded the organisation of the workshop in Montpellier in September 2016. This could not have happened without the tremendous motivation of the Australian experts who convinced their institutions of the value of this exchange or who personally financed their travel to France. Brgm actively financed the translation or the English editing of most French chapters. Our translators, in particular Isis Olivier, must also be thanked for the quality of their work, as well as Emilie Lenoir and Marie-Adélaide Ethève for editing the manuscript. Finally, this book was only made possible by the dedication and hard work of the chapter authors, and we are extremely grateful to all of them for their willingness to collaborate on this project.

Montpellier, France — Jean-Daniel Rinaudo
Sydney, Australia — Cameron Holley
Adelaide, Australia — Steve Barnett
Montpellier, France — Marielle Montginoul

About the Book

The book comprises 27 chapters, covering four main topical areas. Chapters 1 to 10 provides background information on the French and Australian groundwater resources and policy context at federal and national levels, as well as at river basin level, where groundwater policy implementation and long-term planning actually takes place. Contributors describe how groundwater policies have progressively structured over the last 25 years, using primary information accumulated during their career, with the support of academic authors providing conceptual frameworks for policy analysis. The confrontation of the Australian and French approaches reveals fundamental political and philosophical values in relation to the property of water and to the role that users' communities should play in allocation.

Chapters 11 to 16 deal with conceptual approaches and operational tools used to assess sustainable abstraction limits. Contributions highlight the differences between conceptual approaches prevailing in the two countries. While French policy makers assert that sustainable should be defined based on scientific evidence, their Australian counterpart acknowledge that such limits must be acceptable, thus negotiated between stakeholders. This part of the book also provides a good overview of the tools and models used to estimate extraction limits at different geographic and time scales, considering climate variability and uncertainties about future changes. Some chapters also look at this issue from a political economy perspective, highlighting that extraction limits result from a negotiation where scientific evidence only plays a limited role.

Chapters 17 to 24 focus on approaches implemented to reduce groundwater entitlements in over-exploited aquifers. A number of case studies illustrate the different policy approaches that can be used to restore long-term sustainability. Issues addressed in these chapters include that of financial compensation for cutbacks in entitlements (either through buy-back programmes or the development of substitution water resources), possible differentiated treatment of used and sleeping water rights, the unbundling of water entitlements, and allocations. Two chapters discuss compliance and enforcement problems, the intensity of which increases as water scarcity rises. The part ends with a discussion of the linkages between groundwater management and agricultural policies.

The last two chapters develop an international perspective of the issues addressed in the book through contributions from California and Chile. A concluding chapter draws lessons from French, Australian and international experiences, highlighting common features observed in the long pathways taken by various countries to shift from open access to sustainable groundwater abstraction regimes.

Contents

1 **Sustainable Groundwater Management in France and Australia: Setting Extraction Limits, Allocating Rights and Reallocation** 1
Cameron Holley, Jean-Daniel Rinaudo, Steve Barnett, and Marielle Montginoul

2 **Groundwater in France: Resources, Use and Management Issues** 17
Jean-Christophe Maréchal and Josselin Rouillard

3 **Groundwater Policy in France: From Private to Collective Management** 47
Jean-Daniel Rinaudo

4 **Groundwater Management Planning at the River Basin District Level: Comparative Analysis of the Adour-Garonne and Loire-Bretagne River Basins** 67
Jean-Daniel Rinaudo, Pierre Marchet, and Pascal Billault

5 **Lessons from Twenty Years of Local Volumetric Groundwater Management: The Case of the Beauce Aquifer, Central France** 93
Frederic Verley

6 **Groundwater in Australia: Occurrence and Management Issues** . . . 109
Steve Barnett, Nikki Harrington, Peter Cook, and Craig T. Simmons

7 **The Evolution of Groundwater Management Policy in the States of Australia** 129
Rebecca Nelson, Steve Barnett, and Ann Kollmorgen

8 Developing a Coordinated Groundwater Management Plan for the Interstate Murray-Darling Basin 143
Glen Walker, Steve Barnett, and Stuart Richardson

9 Information Systems for Sustainable Management of Groundwater Extraction in France and Australia 163
John Sharples, Elisabetta Carrara, Lindsay Preece, Laurence Chery, Benjamin Lopez, and Jean-Daniel Rinaudo

10 The Challenge of Making Groundwater Visible: A Review of Communication Approaches and Tools in France 191
Audrey Richard-Ferroudji and Gaïa Lassaube

11 Conceptual Approaches, Methods and Models Used to Assess Abstraction Limits for Unconfined Aquifers in France 211
Luc Arnaud

12 Setting Sustainable Abstraction Limits in Confined Aquifers: Example from Deep Confined Aquifers in the Bordeaux Region, France 229
Frédéric Lapuyade, Marc Saltel, and Bruno de Grissac

13 A Tool to Determine Annual Ground-Water Allocations in the Tarn-et-Garonne Alluvial Aquifer (France) 253
Pierre Le Cointe, Vorlette Nuttinck, and Jean-Daniel Rinaudo

14 Conceptual Approaches, Methods and Models Used to Assess Extraction Limits in Australia: From Sustainable to Acceptable Yield 275
Daniel Pierce and Peter Cook

15 Case Study: An Integrated Approach to Determining Sustainable Abstraction Limits in Perth's North West Urban Growth Corridor 291
Mal McGivern and Clive Hampton

16 Using Resource Condition Limits to Define Groundwater Management Objectives in the Barossa Valley, South Australia 299
Daniel Pierce, Roger Cranswick, and Megan Hancock Lane

17 Reducing Entitlements When Groundwater Has Been Over-Allocated: Policy Issues and Options 315
Stefanie Schulte and Gabriela Cuadrado Quesada

18 Developing Substitution Resources as Compensation for Reduced Groundwater Entitlements: The Case of the Poitou Marshes (France) 333
Olivier Douez, Jean Eudes du Peuty, Daniel Lepercq, and Marielle Montginoul

19 New Approaches for Allocation Reductions and Groundwater Salinity Management in South Australia 355
Steve Barnett and David Williamson

20 Reducing Groundwater Entitlements in the Lower Murrumbidgee Groundwater Management Area 365
Ken Schuster, Amanda Kennedy, and Cameron Holley

21 Development of Groundwater Markets in Australia: Insights from Victoria in the Murray Darling Basin 385
Julia De Luca and Darren Sinclair

22 Groundwater Regulation, Compliance and Enforcement: Insights on Regulators, Regulated Actors and Frameworks in New South Wales, Australia 411
Cameron Holley, Tariro Mutongwizo, Susan Pucci, Juan Castilla-Rho, and Darren Sinclair

23 Compliance and Enforcement: The Achilles Heel of French Water Policy 435
Marielle Montginoul, Jean-Daniel Rinaudo, and Charlotte Alcouffe

24 Tracing the Impact of Agricultural Policies on Irrigation Water Demand and Groundwater Extraction in France 461
Josselin Rouillard

25 Groundwater Management Lessons from Chile 481
Guillermo Donoso, Elisabeth Lictevout, and Jean-Daniel Rinaudo

26 California's 2014 Sustainable Groundwater Management Act – From the Back Seat to the Driver Seat in the (Inter)National Groundwater Sustainability Movement 511
Thomas Harter

27 Changing from Unrestricted Access to Sustainable Abstraction Management Regimes: Lessons Learnt from France and Australia 537
Jean-Daniel Rinaudo, Steve Barnett, and Cameron Holley

About the Editors

Steve Barnett is a Principal Hydrogeologist at the Water Science and Monitoring Branch of the Department for Environment and Water in South Australia. He has been involved in the investigation, monitoring and management of groundwater resources in SA for over 40 years and has contributed technical and policy input into ten groundwater management plans, which incorporate a variety of different aquifers and management issues. He is a past president of the Australian Chapter of the International Association of Hydrogeologists.

Cameron Holley is a professor at University of New South Wales Law and is a member of the Global Water Institute and Connected Waters Initiative, University of New South Wales Sydney. Cameron has worked closely with Australian and international government and non-government organisations on a range of water and natural resource management research projects. He currently holds ARC Discovery Grants (DP170100281 DP190101584) on Non-Urban Water Regulation and Integrating the Governance of Water and Coal Seam Gas that supported this chapter. He is an editorial board member on the *Environmental and Planning Law Journal* and in 2016 was the guest editor of a Special Issue (EPLJ Vol 33 Part 4), entitled Rethinking Water Law and Governance.

Marielle Montginoul is senior researcher in Economics at the National Research Institute for Agriculture, Food and Environment (INRAE – previously IRSTEA) and she is a member of the Joint Research Unit G-Eau. Her work focuses on understanding and modeling farmers and households' water consumption behaviors. She more specifically studies instruments that can be used to reveal these behaviors when information is incomplete. Her research also focuses on economic tools to manage water withdrawals, with a focus on water pricing. She mobilizes a wide range of methodologies including surveys, experimental economics, and scenarios workshops. Marielle is member of the scientific council of the Rhone Méditerranée and Corsica Water agency. She coordinates a Master in Social sciences applied to water management in Montpellier University.

Jean-Daniel Rinaudo is a researcher at Brgm, Montpellier University, where he coordinates the scientific programme on environmental and risk economics. Initially trained as an agricultural engineer (Montpellier SupAgro 1994), he specialised in agricultural and resource economics (PhD University of Auvergne, 2000). Prior to joining Brgm, he worked for the International Water Management Institute in Pakistan, where his research focused on the political economy of irrigation management reforms. His current research mainly focuses on the institutional economic dimension of groundwater management. Most of his research is conducted in France, but he also works in Morocco and Chile. He is currently developing new research activities in the field of natural disaster economics, focusing on the methods for assessing economic vulnerability and resilience. Dr. Jean Daniel Rinaudo is also a member of the Scientific Council of the Adour Garonne River basin agency.

Contributors

Charlotte Alcouffe INRAE – UMR G-EAU. Montpellier University, Montpellier, France

Luc Arnaud BRGM (French Geological Survey), Orléans, France

Pascal Billault Agence de l'Eau Loire Bretagne, Orléans, France

Elisabetta Carrara Bureau of Meteorology, Melbourne, VIC, Australia

Juan Castilla-Rho UTS (University of Technology Sydney), Sydney, NSW, Australia

Laurence Chery BRGM (French Geological Survey), Orléans, France

Pierre Le Cointe BRGM, Ramonville-Saint-Agne, France

Phil Commander Consultant, Perth, WA, Australia

Peter Cook Flinders University, Adelaide, SA, Australia

Roger Cranswick Department for Environment and Water, Adelaide, SA, Australia

Guillermo Donoso Harris Deparment of Agricultural Economics, Pontificia Universidad Catolica de Chile, Santiago, Chile
Water Law and Management Center, Pontificia Universidad Catolica de Chile, Santiago, Chile

Bruno de Grissac SMEGREG, Bordeaux, France

Olivier Douez BRGM (French Geological Survey), Bordeaux, France

Jean Eudes du Peuty EPTB Marais Poitevin, Luçon, France

Clive Hampton Water Corporation of Western Australia, Perth, WA, Australia

Megan Hancock Lane Department for Environment and Water, Adelaide, SA, Australia

Nikki Harrington Innovative Groundwater Solutions Pty Ltd (IGS), Victor Harbour, SA, Australia

Thomas Harter Department of Land, Air, and Water Resources, University of California, Davis, CA, USA

Amanda Kennedy Faculty of Law, Queensland University of Technology, Brisbane, QLD, Australia

Ann Kollmorgen Department of Environment, Land, Water and Planning, Melbourne, VIC, Australia

Frédéric Lapuyade SMEGREG, Bordeaux, France

Gaïa Lassaube Centre Emile Durkheim, Institut Français de Pondichery, Pondicherry, India

Daniel Lepercq CACG, Tarbes, France

Elisabeth Lictevout Universidad de Concepcion, Concepción, Chile Carpe Science, San Pedro de la Paz, Chile

Benjamin Lopez BRGM (French Geological Survey), Orléans, France

Julia De Luca Department of Environment Land Water and Planning (DELWP), Melbourne, VIC, Australia

Pierre Marchet Agence de l'Eau Adour Garonne, Toulouse, France

Jean-Christophe Maréchal BRGM, Montpellier University, Montpellier, France

Mal McGivern Water Corporation of Western Australia, Perth, WA, Australia

Tariro Mutongwizo University of New South Wales, Sydney, NSW, Australia

Rebecca Nelson Melbourne Law School, University of Melbourne, Melbourne, VIC, Australia

Vorlette Nuttinck Direction Départementale des Territoires de Tarn-et-Garonne, French Government, Montauban, France

Daniel Pierce Department for Environment and Water, Adelaide, SA, Australia

Jon Philippe Pigois Department of Water and Environmental Regulation, Perth, WA, Australia

Lindsay Preece Department of Water and Environmental Regulation, Perth, WA, Australia

Susan Pucci NSW Department of Planning, Industry and Environment – Water, Sydney, NSW, Australia

Gabriela Cuadrado Quesada Department of Integrated Water Systems and Governance, IHE-Delft Institute for Water Education, Delft, The Netherlands

Audrey Richard-Ferroudji French Institute of Pondicherry, Pondicherry, India UMR G-EAU, Montpellier, France

Stuart Richardson CDM Smith, Adelaide, SA, Australia

Josselin Rouillard BRGM, Montpellier University, Montpellier, France

Marc Saltel BRGM (French Heological Survey), Bordeaux, France

Stephanie Schulte NSW Irrigators Council, Sydney, Australia

Ken Schuster Farmer, Founding Member of the Murrumbidgee Groundwater Preservation Association, Darlington Point, NSW, Australia

John Sharples Bureau of Meteorology, Melbourne, VIC, Australia

Craig T. Simmons National Centre for Groundwater Research and Training, Flinders University, Adelaide, SA, Australia

Darren Sinclair Institute for Governance and Policy Analysis (IGPA), University of Canberra, Canberra, Australia

Frederic Verley DREAL, French Government, Orléans, France

Glen Walker Grounded in Water, Mount Osmond, SA, Australia

David Williamson Department for Environment and Water, Mount Gambier, SA, Australia

Chapter 1
Sustainable Groundwater Management in France and Australia: Setting Extraction Limits, Allocating Rights and Reallocation

Cameron Holley, Jean-Daniel Rinaudo, Steve Barnett, and Marielle Montginoul

Abstract This chapter briefly introduces the main policy developments from both France and Australia regarding groundwater management and their particular approach to setting caps, allocating rights and allowing reallocations. It then presents the objectives of the book and explores the book's contributions under four key themes, namely groundwater and policy approaches in France and Australia, capping water use and defining sustainable abstraction limits, reducing entitlements to sustainable limits, and comparisons between France, Australia and other international groundwater developments.

Keywords Groundwater · France · Australia · Capping resource use · Allocating use rights · Reallocation · Adaptation · Chile · USA

C. Holley (✉)
Faculty of Law, University of New South Wales Sydney, Sydney, NSW, Australia
e-mail: c.holley@unsw.edu.au

J.-D. Rinaudo
BRGM, Montpellier University, Montpellier, France

S. Barnett
Adelaide, SA, Australia

M. Montginoul
INRAE – UMR G-Eau, Montpellier University, Montpellier, France

J.-D. Rinaudo et al. (eds.), *Sustainable Groundwater Management*,
Global Issues in Water Policy 24, https://doi.org/10.1007/978-3-030-32766-8_1

1.1 Introduction

During the last three decades, economic development of both urban and rural areas has increasingly relied on groundwater resources, which supply water for around 40% of irrigated lands, half of all drinking water, and are impacted by the growth of unconventional oil and gas projects (WWAP, 2015; Holley and Kennedy, 2019). However, this development has often taken place in a context of "weak" governance (Faysse, Errahj, Imache, Kemmoun, & Labbaci, 2014), in which groundwater was often considered as an open access resource. In many regions around the world, individual water users acting independently according to their own self-interest, without considering the aggregate impact of their decisions on the sustainability of the resource, have depleted groundwater, illustrating the tragedy of the commons (Hardin, 1968). Due to excessive pumping, groundwater levels have been declining, affecting dependent ecosystems, in particular by reducing river base-flows, disconnecting rivers from aquifers and lowering water levels in wetlands (WWAP, 2015). Overdraft has led to land subsidence and increased cost of pumping, as well as irreversible deterioration of many aquifers through intrusion of saline or contaminated water from adjacent aquifers (FAO, 2016a; Fienen & Arshad, 2016; WWAP, 2015; Van der Gun, 2012). These trends have been documented in many semi -arid, but also temperate regions in Asia (China, India, Pakistan), America (Chile, the United States of America, Mexico), Europe (Spain), North-Africa and the Middle East (Morocco, Jordan) and to some extent, in both Australia and France.

While contributing to creating wealth and alleviating poverty in the short term, these problems arising from groundwater development could lead to the collapse of thriving agricultural economies which are strongly dependent on groundwater (Petit et al., 2017). These threats are a matter of increasing concern to many nation States that have supported agricultural development through subsidies and infrastructure development. Indeed, as many States and the global community now recognise (see e.g. Sustainable Development Goal 6), on-going groundwater overdraft could render these investments worthless and transform areas of former economic expansion into regions of poverty.

A critical issue for policy makers is ensuring that groundwater extraction is sustainable in the long term. Although there are large groundwater policy and governance gaps across the globe, where policies do not exist, attention is paid to models and success stories that could be replicated (FAO, 2016a; Molle & Closas, 2017). Many studies have been carried out into groundwater problems, and many technical solutions (e.g. recharge, water transfers, conjunctive use, water saving technologies) and institutional frameworks (e.g. collective and common pool resource approaches) (Giordano, 2009; Jakeman, Barreteau, Hunt, Rinaudo, & Ross, 2016; Ostrom, 1990; Van der Gun, 2012; Villholth, Lopez-Gunn, Conti, Garrido, & Van Der, 2017) have been proposed. Yet despite these institutional and technical tools, their actual implementation has remained a significant global challenge. As the FAO (2016b) has noted: "one thing is clear; it is not the formulation of laws and regulations that will make a difference, but their implementation and adoption …".

This edited collection accordingly provides insights by bringing together practitioners and academics to reflect on their experience with developing and implementing groundwater management policy. In this regard, the book focuses on a policy model and its implementation that a number of academics and international agencies have been promoting. This policy model consists of (i) capping total resource use, (ii) allocating use rights accordingly and (iii) defining rules to allow reallocation and adaptation to changing economic and climatic conditions. Capping consists of calculating and imposing a Sustainable Abstraction Limit (SAL), which when observed, guarantees the continuity of use for future generations and ensures the proper ecological functioning of groundwater dependent ecosystems such as streams and wetlands. The available resource defined by the SAL is then allocated to users via rights, which can either be individual or collective, defined in volume or pumping rate and taking the form of administrative permits, concessions or types of property rights. Those rights can be reallocated over time, based on either administrative procedures (e.g. waiting lists), market mechanisms (if rights are made tradable), or negotiated rules defined by users themselves (e.g. decentralized self-regulated management). This allows adaption of the initial allocation of rights in response to changing economic or demographic conditions, or to the exit or entry of users, with the primary objective of seeking maximum economic returns from use of the resource. Finally, rules are set-up to adjust water entitlements in the event of a reduction in the available resource.

This generic model underpins groundwater management policies implemented in a number of high or intermediate income countries such as Australia, Chile, the United States of America (particularly the Western United States), Spain, Mexico, and France. While this model is one that other countries, including less developed ones, could aspire to, it is important to highlight that it is not a rigid prescriptive model. It can be adapted to very diverse technical, social and political contexts and can accommodate different concepts of social justice, water rights, decentralisation and trade-offs between environment, economics and equity. It is equally important to note the difficulties likely to emerge during the implementation phase, whose duration is often measured in years, if not decades. This book highlights this diversity of implementation approaches, problems and successes, though a comparative analysis of several case studies in France and Australia, two countries which have a long history in groundwater management reforms and implementation.

In the early 1990's, both countries initiated a groundwater management policy reform which broadly followed the model presented above. As displayed in Figs. 1.1 and 1.2, both nations initially followed a broadly similar trajectory, that began with access regimes based around individual rights, before shifting in the twentieth century to the regulation and licensing of wells/bores, but with little consideration of sustainable extraction limits. It was during the late 1990's and early 2000's that both nations commenced major reforms based around the policy model of capping total resource use, allocating use rights and defining rules to allow reallocation and adaptation. Notwithstanding this commonality, as shown in Figs. 1.1 and 1.2 and throughout the book, both nations diverge in how this model was given effect in practice.

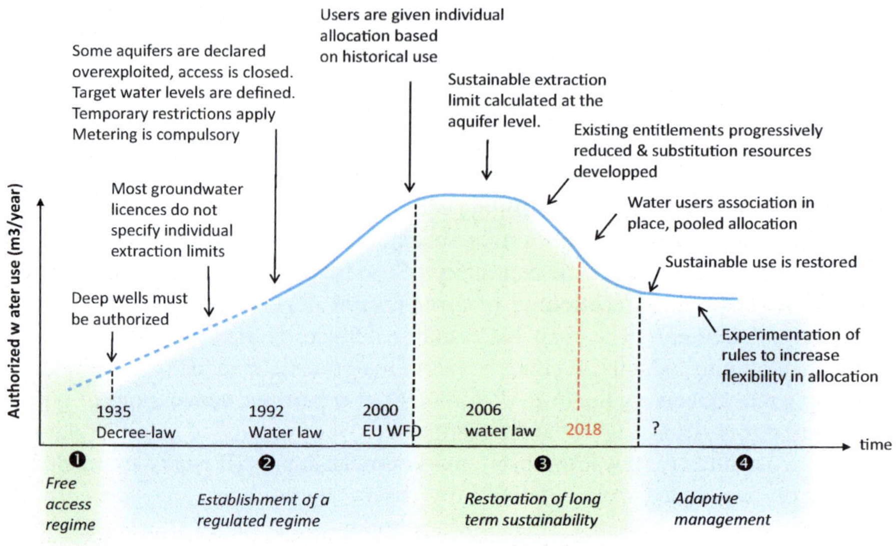

Fig. 1.1 The four policy phases for regulating groundwater abstraction in France

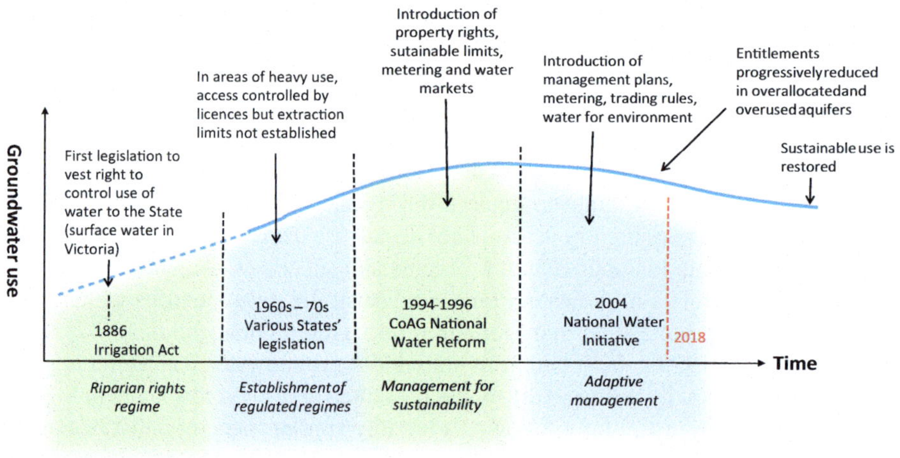

Fig. 1.2 The four policy phases for regulating groundwater abstraction in Australia

In the following discussion, we briefly introduce the main policy developments from both France and Australia regarding groundwater management and their particular approach to caps, rights and reallocations.

1.2 Groundwater Management Policies in France and Australia

1.2.1 Overview of the French Approach

In France, the historical evolution of groundwater development and management can be broken down into four major phases (see Fig. 1.1). The initial phase corresponds to a system of free access to the resource, in which landowners can freely appropriate the water located beneath their land. The proliferation of deep industrial boreholes and the rapid development of confined aquifers that occurred during the 1850's and early 1900's led to some occurrences of overexploitation. This threatened the resources regarded as being of strategic importance for supplying drinking water, which prompted the State to intervene.

The first groundwater legislation was subsequently passed in 1935. It involved setting up a permit system for wells and boreholes, which essentially aimed to control industrial use in order to protect the supply of drinking water. Between the end of the 1960s and the early 1990s, the increase in the number of agricultural boreholes, often tapping shallow aquifers, generated new cases of overexploitation and conflict over environmental protection issues. The 1992 Water Act provided a response to this crisis by strengthening the State provisions for controlling abstraction. In particular, it established the necessary conditions for volumetric management of water abstraction, including the obligation to record actual use (metering) and the allocation of individual abstraction quotas. Although the mechanisms were in place, overexploitation problems persisted due to over-allocation.

The third phase was initiated by European legislation, known as the Water Framework Directive. This Directive obliged member states to restore all their bodies of water to a satisfactory state in terms of quality and quantity. The French implementation strategy of that Directive was laid down in the 2006 Water Act which requires capping total abstraction and sharing the available resource among users. As the cap was lower than historical use in many groundwater and river basins, managers had to design rules to reduce entitlements. To do so, the 2006 Act encouraged the development of a collective approach to water allocation, notably through the creation of the Water Users' Associations (called OUGC). In the first step, this collective management was only implemented to manage agricultural users, which represent the highest number of users and frequently the highest share of resource use.

The final phase will involve developing new and flexible water management mechanisms capable of adapting to a rapidly changing economic and climatic environment.

1.2.2 Overview of the Australian Approach

Following thousands of years of Indigenous rules and concepts relating to water and the environment (Marshall, 2017), the transplantation of the Anglo common law riparian and capture rights granted landholders the ability to conditionally access and use water adjacent to and beneath their land. As demand for water by growing urban centres increased, the inadequacies of this approach became apparent. This prompted the first state legislation in 1886 which vested the right to the use, flow and the control of water in the state, marking the transition from rights to state legislative regimes (Gardner, Bartlett, Gray, & Nelson, 2017). Reflecting broadly similar developments in France, Australia's states progressively vested control over water in the Crown and abolished or displaced existing common law rights in response to increasing groundwater development in the 1960's and 1970's, creating a system of licencing (albeit one that did not pursue wide ranging caps on water use) (Holley & Sinclair, 2018).

Echoing comparable developments in France, Australia's modern water reform journey commenced in the early 1990's motivated by concerns about the efficiency and equity of water allocations and also with environmental sustainability. Under the Australian constitution, the states historically had primary responsibility for water management, but the initial reforms were founded on ideas of intergovernmental agreements and action through the Council of Australian Governments ('CoAG'). A national water framework was agreed to in 1994 (CoAG, 1994), closely followed by a similar 1996 Framework for Improved Groundwater Management.

These reforms created the emblematic aspect of Australia's approach, which is the creation of water rights (separated from land), within overarching sustainable limits set using scientific methods. Rules for the trading of water rights would support the intention that water would be used in the most efficient and productive way. The reforms also encouraged a system of regulatory enforcement. Perhaps the main contrast to the French approach is that the Australian policy model sets out aspirations for market-based reforms.

A subsequent 2004 Intergovernmental Agreement known as the National Water Initiative (NWI), consolidated the 1994 reforms and aimed to embed a nationally-compatible water market, progressively remove barriers to water-trading, facilitate efficient water use and address adjustment issues (Cwth of Aus., 2004). This next wave of reforms also aspired to return over-allocated or overused systems to environmentally-sustainable levels of extraction by encouraging the development and finalisation of aquifer and catchment based statutory water allocation plans, and making statutory provision for environmental and other public benefit outcomes. Community engagement, partnerships and consultation throughout plan development and review was deemed essential to this adjustment process.

1.3 Objectives and Scope of the Book

The main objective of this book is to describe and analyse a variety of possible approaches and policy pathways to implement sustainable groundwater management, based on a comparative analysis of selected case studies in France and Australia. The book strictly focuses on quantitative management and does not cover in detail water quality or pollution management issues.

One of the specific features of the book is that a majority of the contributors are water professionals who have been involved for several decades in groundwater policy making, planning and implementation of management plans. Most of the contributors to this book participated in a French – Australian workshop organised in Montpellier (France) in October 2016 where they presented and discussed case studies that are covered in more detail in the following chapters and represent a significant contribution to the empirical water management literature that has not been published elsewhere, even in grey literature.

Recognising that groundwater has become an interdisciplinary subject (Van der Gun, 2012, p i) the originality of the book also lies in the different disciplinary perspectives covered in many chapters (hydrogeology, economics, planning, law and social sciences in particular).

In addition to the case studies, the book also presents the results of a comparative analysis conducted by these French and Australian water professionals, supported by a group of academics. This dialogue, initiated during the Montpellier workshop, led to the identification of similarities but also fundamental differences which are analysed and presented as alternative policy options in the conclusion of the book – these differences being mainly related to the role of the State, the community and market mechanisms in groundwater management. Given the importance of linking the experiences of Australia and France to other global developments, we also invited leading water academics to reflect on groundwater management experiences in other countries, in particular in Chile and the USA (particularly California).

1.4 Structure of the Book

The book's contributions can be divided into four main themes across a total of 27 chapters. Below is a brief overview of the themes and chapters.

1.4.1 Theme 1: Groundwater and Policy Approaches in France and Australia

The first selection of chapters provides background information on the French and Australian groundwater policy context at Federal/national levels as well as at river basin and catchment levels, where long term planning and implementation of groundwater policy actually takes place. The contributors provide a general assessment of the situation of groundwater depletion in both countries, with a focus on drought years, including the Millennium Drought in Australia and its impact on groundwater resource in the Murray Darling Basin. Groundwater professionals also describe how policies have progressively developed over the last 25 years, using primary information accumulated from their experience in practice, with the support of academic authors providing conceptual models for policy analysis.

Chapters 2, 3, 4 and 5 outline groundwater and management contexts in France. Maréchal and Rouillard (Chap. 2) describe the status of groundwater resources in France. The chapter highlights the geology and types of aquifers, as well as use of groundwater resources across domestic use, industry and agriculture. It notes that although France has not yet faced extreme cases of aquifer depletion, the long-term management challenges relate to the decrease of recharge due to climate change, sea level rise along the coast, and future change in groundwater use. It concludes by suggesting three core adaptation strategies.

In Chap. 3, Rinaudo examines the development of groundwater policy in France. The chapter maps a shift from private property to increasing State regulation of its use, broadly akin to similar developments in Australia discussed in Chap. 7. The chapter characterizes the development of the 2006 water law as constituting a clear break in French water policy, and examines the changes it introduced and the subsequent shift from a private to a common property regime.

The groundwater planning process in France resulting from the 2006 water law is analysed in Chap. 4. Rinaudo et al. explore the framework of local plans (SAGE) and strategic master plans for managing river basins (SDAGE). This chapter describes how strategic blueprints are formulated and implemented, including a historical analysis of 20 years of groundwater planning in the Adour-Garonne and Loire-Bretagne river basin districts.

Transitioning from the basin to the local aquifer level, Chap. 5 highlights lessons from 20 years of local volumetric groundwater management in the Beauce aquifer. In this chapter, Verley draws on personal experience to describe the evolution of management mechanisms for water abstraction, the characteristics of the water resource, its various uses, the problem of overexploitation and how the management plan evolved. The chapter also reflects on prospects for change.

Chapters 6, 7 and 8 shift the focus from the northern to the southern hemisphere, with Barnett et al. introducing groundwater in Australia (Chap. 6). The chapter charts the social, economic and environmental features of groundwater resources, while discussing the various types of aquifers, their development and future management issues, including impacts of climate change, impacts of mining and declining government funding.

Building on the overview of Australia's groundwater resources, Nelson et al. (Chap. 7) chart the development of groundwater management in Australia, and how the experiences of other countries were taken into account. Recognising that the states and territories continue to be the primary managers of groundwater and are responsible for licensing processes and adopting legally enforceable plans to manage extraction, the chapter provides some case studies of differing approaches to groundwater management from different Australian states.

In Chap. 8, Walker et al. turn their attention to perhaps the most well known water management context in Australia, the Murray Darling Basin. The chapter describes the nature of groundwater systems in the Basin, noting that management of groundwater on a basin-scale had a lower priority compared to the more controversial surface water resources. It explains how a coordinated joint management plan for the increasingly important groundwater resources in the Basin was developed using a methodology to determine sustainable extraction limits across five states and territories. The chapter concludes its analysis by considering some of the challenges arising from this joint management approach.

Concluding this assessment of groundwater and policy approaches, Chaps. 9 and 10 focus on the dissemination and communication of groundwater information in both France and Australia. Sharples et al. (Chap. 9) use examples from Australia and France to discuss similarities and differences in the two nations' approaches to groundwater information systems, their history, and how these systems have been used to inform and improve groundwater management. A range of examples are explored including local management, national data standardization, online data sharing, and environmental impact assessment before summing up the future directions in this field.

Finally, in Chap. 10, Richard-Ferroudji and Lassaube draw on 11 case studies from France to report on a number of communication approaches and activities and how they were used to make groundwater "visible" for various stakeholders, including the general public, farmers and elected representatives. The chapter introduces a framework to analyse communication approaches and tools, before assessing the use of the tools, their benefits and limits, and concluding with recommendations.

1.4.2 Theme 2: Capping Water Use and Defining Sustainable Abstraction Limits

Building on the above overview, the second grouping of chapters examines the first part of the policy model, specifically looking at how water managers cap total water use by defining sustainable abstraction limits. These chapters investigate how this process is conceptually defined in the two countries, revealing the diversity of trade-offs made between environment and economic activities. They also provide a good overview of the tools and groundwater models used to estimate extraction limits at different geographic and time scales, considering climate variability and uncertainties about future changes.

Chapter 11 commences with a review of conceptual approaches, methods and models used to assess abstraction limits for unconfined aquifers in France. Based on the analysis of over 30 studies, Arnaud shows that the estimation of this limit, called Maximum Permissible Volume (MPV) in France, is complicated by numerous uncertainties, data availability constraints and simplified assumptions made by hydrogeologists. These technical limitations of hydrogeological studies allow users to contest the MPV, which are often renegotiated.

Chapter 12 then focuses on the challenges of setting abstraction limits in confined aquifers, based on experiences from the deep confined aquifers in the Bordeaux region in France. In this chapter, Lapuyade et al. explore the historical development of cap setting, noting that risks of overexploitation of these resources was a driver for the implementation of specific regulations. Implementation of management policies and investigations to improve knowledge and develop groundwater flow models are also examined, and as the chapter explains, the local stakeholders involved in aquifer management employed these modelling tools to create the principles and policies for controlling groundwater-abstraction.

Chapter 13 (Le Cointe et al.) continues the focus on France with an analysis of the process and tools for determining sustainable annual allocations in the Tarn-et-Garonne alluvial aquifer. Using the previous history of events, the authors demonstrate the complexity and lengthy period of time required to develop a groundwater flow model that can be used by a government agency to support water allocation decisions. This chapter depicts a unique French water management approach where groundwater allocations for water users are updated every year, based on observed resource conditions. The chapter concludes with some unique insights on a shift in responsibility for the allocation process from the State to collective water user associations.

The evolution of the concept of sustainable development for groundwater resources in Australia is discussed in Chap. 14 by Pierce and Cook. Originally, the "safe yield" approach was employed whereby the upper limit for extraction was determined by the estimation of recharge. However, due to the difficulties and uncertainties in estimating recharge, and the fact that this approach does not allow for environmental uses of groundwater, management plans are increasingly moving toward the notion of acceptable impacts based on specified resource condition limits. They discuss in depth the methods used to evaluate four main areas of risk namely: storage capacity, groundwater dependent ecosystems, groundwater quality and aquifer integrity.

In Chap. 15, McGivern and Hampton provide a useful case study of a Western Australian approach to establish sustainable pumping limits. The chapter draws on insights from the management of an aquifer in Perth's North West Urban Growth Corridor, where declining winter rainfalls, and an increase in average temperatures has complicated access to sustainable water resources for a fast growing population. McGivern examines how the sustainable yield of the aquifer was determined, and argues that both groundwater flow models and simple spread sheet analytical models using representative hydraulic parameters can play important roles. The chapter also highlights how co-operation between water providers and regulators, and flexibility in the management approach, are important ingredients for successful outcomes.

The Barossa Valley wine region is the subject of Chap. 16 where Pierce et al. describe a new responsive and participatory management approach using resource condition limits. Consultations were held with a representative community group to determine the level of risk of adverse impacts occurring as a result of groundwater extraction. The impacts considered included changes in water levels, groundwater discharge to streams and the ingress of higher salinity groundwater. A groundwater flow model was then used to determine what extraction rates would result in acceptable levels of risk.

1.4.3 Theme 3: Reducing Entitlements to the Sustainable Limit

Despite efforts to allocate entitlements and set sustainable limits for extraction, a common challenge in many nations, including France and Australia, is overallocation where the volume of entitlements exceeds the sustainable limit. The third theme of the book provides insights on how to reduce entitlements down to sustainable limits in over-allocated resources. A central theme across all these chapters is how water use rights are defined and allocated to users. The Australian chapters assess the results attained since management plans and water markets were introduced to reduce depletion and achieve sustainable abstractions limits. A comparison of the Australian and the French approaches reveals fundamental differences in the political and philosophical values in relation to water rights and to the role that user communities should play in reallocation.

In Chap. 17, Schulte and Cuadrado Quesada discuss Australia's policy pathways for reducing entitlements when groundwater resources are over-allocated. The chapter highlights definitional challenges that initially hampered progress within Australia's federated structure, before examining attempts to reduce over-allocation and over-use in Australia's numerous groundwater management plans. The chapter highlights the challenges that led to slower than expected progress in addressing over-allocation and over-use, as well as exploring the use of various mechanisms and tools, including phasing in allocation reductions and carry-over provisions, compulsory reductions of allocations with compensation, moratoriums, conjunctive forms of management through collective action, including donations of groundwater rights in return for surface-water rights, and water licence/entitlement purchases by governments in the water market.

Douez et al. (Chap. 18) turn their attention to approaches for developing alternative water resources as compensation for reduced groundwater entitlements.

In the case of the groundwater dependent Poitou Marshes in France, Douez et al. describe the relevant groundwater management policy and its response to the growth of irrigated agricultural as in other basins in central and western France (see Chaps. 5 and 13). The chapter examines the significant reduction in historical water entitlements and pinpoints the difficulties encountered in implementing this reduction in a context of extreme competition between economic uses (agriculture, urban uses, and tourism) and environmental objectives. The chapter also reports on the complexities in developing integrated water management plans for basins, providing insights on the requirements for success and exploring issues of coordination between the State, the local water management board and users associations where groundwater, rivers, wetlands, and canals are highly interdependent.

In Chap. 19, Barnett and Williamson examine approaches for allocation reductions and groundwater salinity management in South Australia. The chapter presents a case study of an exercise to reduce irrigation entitlements in an overallocated groundwater management area, driven by a longer-term risk to effective management of the resource. The chapter identifies a range of conditions that contributed to success, including establishing a good relationship and trust with the irrigators and staged reductions so that irrigators had time to adjust their operations.

Schuster et al. (Chap. 20) provide an additional example from Australia of approaches to reducing groundwater entitlements. Drawing on the history of events and the personal experience of Ken Schuster in the process of groundwater reductions in the Lower Murrumbidgee Groundwater Management Area, the case study provides lessons on water planning and policy approaches for reducing groundwater entitlements and the ensuing litigation by irrigators. The chapter points out the need to take local knowledge and concerns into account during the planning process, as well as providing adjustment mechanisms (e.g. economic compensation via Australia's Achieving Sustainable Groundwater Entitlements program) to ensure the long term sustainable management of groundwater.

In Chap. 21, De Luca and Sinclair offer some significant insights on Australia's innovative approach to managing entitlements, namely water markets. The chapter explores the challenges of using water markets to achieve sustainable water use, including physical and policy constraints that may determine where such markets operate. It examines how legal rights and water markets are used to manage groundwater in Victoria and other states throughout Australia, the success or otherwise of this policy approach, and its capacity to adapt to future pressures on water availability as a consequence of climate change.

The next two chapters address the issues of compliance and enforcement, an important component in ensuring any reduction in allocation is achieved in practice, and not undermined by groundwater theft or other illegal practices. In Chap. 22, Holley et al. draw on an empirical survey, regulator experiences and agent based modelling, to explore Australia's significant reform journey of compliance and enforcement policy over recent decades. They offer an analytical framework for studying groundwater compliance and enforcement and apply this frame to examine the experiences of a government regulator and water users. It concludes with a summary of challenges and policy implications for groundwater compliance and enforcement regimes.

A similar set of compliance challenges emerge in Montginoul et al.'s analysis of groundwater regulation, compliance and enforcement in France (Chap. 23). They characterise compliance and enforcement as the "Achilles heel" of French groundwater policy. Drawing on a review of existing grey and scientific literature and a series of interviews conducted with enforcement officers in 16 French counties, the chapter examines the regulations governing groundwater abstraction, followed by a description of how the law enforcement agencies are organised and how they operate. Montginoul et al. analyse the infractions observed by regulators and the factors that may explain compliance and non-compliance, before highlighting the issues that limit the effectiveness of groundwater policy enforcement.

This grouping of chapters concludes with a discussion by Rouillard of the role of sectoral policies to restore groundwater balance (Chap. 24). Based on an analysis of European and French agricultural policies, Rouillard shows that sustainable groundwater quantitative management does not only depend on implementing the right water policy instruments. It also relies on enabling sectoral policies that work in synergy with water policy objectives.

1.4.4 Theme 4: France, Australia and International Comparisons

The last selection of chapters broadens the perspective by examining the groundwater management approaches in Chile and California. Based on two contrasting case studies, Donoso et al. (Chap. 25) describes the implementation of a relatively sophisticated groundwater management framework in Chile which relies on a unique combination of State intervention, market mechanism and collective management. The two case studies presented by the authors also highlight the existence of problems common with France and Australia, in particular the occurrence of over-allocation, the lack of State resources to enforce existing regulation and difficulties to obtain support from users to reduce abstraction when aquifers are overexploited. Their chapter also sheds light on the political dimension of groundwater management, unveiling how strategic behaviours may impact management decisions. In Chap. 26, Harter presents the ongoing groundwater policy reform in California, which promotes the development of sustainable groundwater management plans at the local level, with the State having substantial oversight over the planning process. Harter shows that many issues currently under discussion in California are similar to those which are still debated in France, Australia and Chile. In conclusion, Chap. 27 draws together the lessons from the above chapters to offer a "big picture" and comparative assessment of the Australian and French approach to the problem of groundwater depletion, and discusses which methods have been successful and which have not.

References

Commonwealth of Australia (Cwth of Aus). (2004). *Intergovernmental agreement on a National Water Initiative*. Canberra, Australia: Commonwealth of Australia.

Council of Australian Governments (COAG). (1994). *Council of Australian Governments' Communiqué, 25 February 1994*. Hobart, Australia: COAG, Australian Government.

Council of Australian Governments (COAG). (1996). *National framework for improved groundwater management*. Australian Government.

FAO. (2016a). *Global diagnosis on groundwater governance* (p. 194). Rome: FAO.

FAO. (2016b). *Shared global vision for Groundwater Governance 2030 and a call-for-action*. Revised edition, FAO 12p.

Faysse, N., Errahj, M., Imache, A., Kemmoun, H., & Labbaci, T. (2014). Paving the way for social learning when governance is weak: Supporting dialogue between stakeholders to face a groundwater crisis in Morocco. *Society & Natural Resources, 27*(3), 249–264.

Fienen, M. N., & Arshad, M. (2016). The international scale of the groundwater issue. In Jakeman et al. (Eds.), *Integrated groundwater management* (pp. 21–48). Cham, Switzerland: Springer.

Gardner, A., Bartlett, R., Gray, J., & Nelson, R. (2017). *Water resources law*. Chatswood, Australia: LexisNexis.

Giordano, M. (2009). Global groundwater? Issues and solutions. *Annual Review of Environment and Resources, 34*, 153–178.

Hardin, G. (1968). The tragedy of the commons. *Science, 162*(3859), 1243–1248.

Holley, C., & Kennedy, A. (2019). Governing the energy-water-food nexus: Regulating unconventional gas development in Queensland, Australia. *Jurimetrics, 59*(2), 233–266.

Holley, C., & Sinclair, D. (Eds.). (2018). *Reforming water law and governance. From stagnation to Innovation in Australia*. Singapore: Springer.

Jakeman, A., Barreteau, O., Hunt, R. J., Rinaudo, J. D., & Ross, A. (2016). *Integrated groundwater management: Concepts, approaches and challenges*. Cham, Switzerland: Springer.

Marshall, V. (2017). *Overturning aqua nullius: Securing aboriginal water rights*. Canberra, Australia: Aboriginal Studies Press.

Molle, F., & Closas, A. (2017). *Groundwater governance: A synthesis. Project Report – IWMI 6* (p. 187). Colombo, Sri Lanka: International Water Management Institute.

Ostrom, E. (1990). *Governing the commons*. Cambridge, UK: Cambridge University Press.

Petit, O., Kuper, M., lopez-gunn, E., Rinaudo, J.-D., Daoudi, A., & LEJARS, C. (2017). Can agricultural groundwater economies collapse? An inquiry into the pathways of four groundwater economies under threat. *Hydrogeology Journal. 25*. 1549–1564.

Van der Gun, J. (2012). *Groundwater and global change: Trends, opportunities and challenges. UNWWAP*. Paris: UNESCO.

Villholth, K. G., Lopez-Gunn, E., Conti, K., Garrido, A., & Van Der Gun, J. (2017). *Advances in groundwater governance*. Leiden, Netherlands: CRC Press.

WWAP (United Nations World Water Assessment Programme). (2015). *The United Nations world water development report 2015: Water for a sustainable world*. Paris: UNESCO.

Dr. Cameron Holley is a Professor at University of New South Wales Law and is a member of the Global Water Institute and Connected Waters Initiative, University of New South Wales Sydney. Cameron has worked closely with Australian and international government and non-government organizations on a range of water and natural resource management research projects. He currently holds ARC Discovery Grants (DP170100281 DP190101584) on Non-Urban Water Regulation and Integrating the Governance of Water and Coal Seam Gas that supported this chapter. He is an Editorial Board member on the Environmental and Planning Law Journal, and in 2016 was the guest editor of a Special Issue (EPLJ Vol 33 Part 4), entitled Rethinking Water Law and Governance.

Dr. Jean-Daniel Rinaudo is researcher at Brgm, Montpellier University, where he coordinates the scientific program on environmental and risk economics. Initially trained as an agricultural engineer (Montpellier SupAgro 1994), he specialized in agricultural and resource economics (PhD University of Auvergne, 2000). Prior to joining Brgm, he worked for the International Water Management Institute in Pakistan where his research focused on the political economy of irrigation management reforms. His current research mainly focuses on the institutional economic dimension of groundwater management. Most of his research is conducted in France but he also works in Morocco and Chile. He is currently developing new research activities in the field of natural disasters economics, focusing on the methods to assess economic vulnerability and resilience. Dr. Jean Daniel Rinaudo is also member of the Scientific Council of the Adour Garonne River basin agency.

Steve Barnett is Principal Hydrogeologist at the Water Science and Monitoring Branch of the Department for Environment and Water in South Australia. He has been involved in the investigation, monitoring and management of groundwater resources in SA for over 40 years, and has contributed technical and policy input into ten groundwater management plans which incorporate a variety of different aquifers and management issues. He is a past-president of the Australian Chapter of the International Association of Hydrogeologists.

Dr. Marielle Montginoul is senior researcher in Economics at the National Research Institute for Agriculture, Food and Environment (INRAE – previously IRSTEA) and she is a member of the Joint Research Unit G-Eau. Her work focuses on understanding and modeling farmers and households' water consumption behaviors. She more specifically studies instruments that can be used to reveal these behaviors when information is incomplete. Her research also focuses on economic tools to manage water withdrawals, with a focus on water pricing. She mobilizes a wide range of methodologies including surveys, experimental economics, and scenarios workshops. Marielle is member of the scientific council of the Rhone Méditerranée and Corsica Water agency. She coordinates a Master in Social sciences applied to water management in Montpellier University.

Chapter 2
Groundwater in France: Resources, Use and Management Issues

Jean-Christophe Maréchal and Josselin Rouillard

Abstract This chapter describes the status of groundwater resources in France. French geology consists of a large variety of rock types, resulting in very different types of aquifers ranging from sedimentary basins, alluvial plains, limestone rocks, and crystalline rocks. Today, groundwater resources represent about 66% of Frances's domestic water supply, 31% of industrial water supply and 37% of total water use in agriculture. According to the European Water Framework Directive, about 33% of groundwater bodies were considered in good chemical status, and 10% were considered in a bad quantitative status in 2013. The main quality issues for groundwater are related to diffuse contamination by agricultural practices (i.e. fertilizers and pesticides). France has not yet faced the extreme cases of aquifer depletion experienced in many other countries. However, associated groundwater dependant ecosystems can be affected by groundwater abstraction. The long term challenges for groundwater management in France are related to the decrease of recharge due to climate change, sea level rise along the coast, and future change in groundwater use. The identified adaptation strategies are (i) new groundwater management policies, (ii) the development of managed aquifer recharge, and (iii) active groundwater management.

Keywords Aquifer · Karst · Water quality · Climate change · Abstraction · Agriculture · Adaptation strategy

---

J.-C. Maréchal (✉) · J. Rouillard
BRGM, Montpelier University, Montpellier, France
e-mail: Jc.marechal@brgm.fr; j.rouillard@brgm.fr

J.-D. Rinaudo et al. (eds.), *Sustainable Groundwater Management*,
Global Issues in Water Policy 24, https://doi.org/10.1007/978-3-030-32766-8_2

2.1 Introduction

On a national scale, France has abundant water resources, thanks to abundant precipitation (900 mm/year), extensive river systems flowing from numerous mountain ranges and large volumes of groundwater stored in aquifers. Every year, France receives 480 billion m^3 (480,000 GL) of precipitation plus 11 billion m^3 of surface water flowing in from neighbouring countries (including the River Rhine). From this quantity of water, about 75% is lost by evaporation and transpiration through vegetation. Consequently, about 170 billion m^3 is available for consumptive use which corresponds to about 2800 m^3/inhabitant/year[1] (AQUASTAT data from FAO).

An estimated 2000 billion m^3 is contained within aquifers with about 108 billion m^3 stored as surface water in lakes, dams and other reservoirs. However, water resources are not equally distributed throughout the country and availability can vary greatly according to the seasons. Mediterranean regions in the south have a dry and changing climate, while the southwest region is often affected by droughts. In this context, groundwater plays a crucial role in water supply especially for drinking water.

This chapter describes the diverse types of aquifers located on the French territory, how groundwater is used, groundwater management issues, and long-term challenges in a changing world.

2.2 Overview of the Groundwater Resources in France

French geology consists of a large variety of rock types (Fig. 2.1), resulting in very different types of aquifers ranging from sedimentary basins (depicted in orange to yellow), alluvial plains (light yellow), limestone rocks (blue and dark green) and crystalline rocks (red and brown). Three categories of aquifers are distinguished: (i) porous sedimentary aquifers located in alluvial valleys and large sedimentary basins where flow velocities are generally low, (ii) the heterogeneous aquifers with a fissure permeability constituted by limestone where flow velocities are generally high, and (iii) volcanic and crystalline rocks.

2.2.1 Alluvial Aquifers

These aquifers provide about 45% of France's groundwater use. They have a very important role in supplying the human needs of the country because they are located in alluvial plains where the most fertile agricultural lands and many cities are located (Fig. 2.2a). Because of their shallow depth, the aquifers provide high yields at low cost and play an important ecological role in supporting river baseflows and wetlands. In addition to the diffuse recharge from rainfall, the water balance of

[1] The term 'water stress' is applied when annual water resources are below 1700 m^3 per capita; the term 'water shortage' applies when the annual water resources drop below 1000 m^3 per capita.

Fig. 2.1 Geological map of France

alluvial aquifers is highly dependent on groundwater flow from neighbouring aquifers and interaction with surface water (Fig. 2.2b). The drawdowns induced by pumping from the alluvial aquifer often increase these inflows. This contributes to maintaining well yields but threatens groundwater quality due to the intrusion of poor quality surface water.

The largest abstractions are pumped from the Alsace alluvial aquifer (~500 Mm^3/year), the Lyon plain (~300 Mm^3/year) and Isere river valley (~180 Mm^3/year).

Example The Alsace aquifer underlies the alluvial plain of the Rhine Graben (Fig. 2.3a). The French portion extends from the Vosges mountain range and Sundgau area to the Rhine River at Lauterbourg. The aquifer thickness ranges from

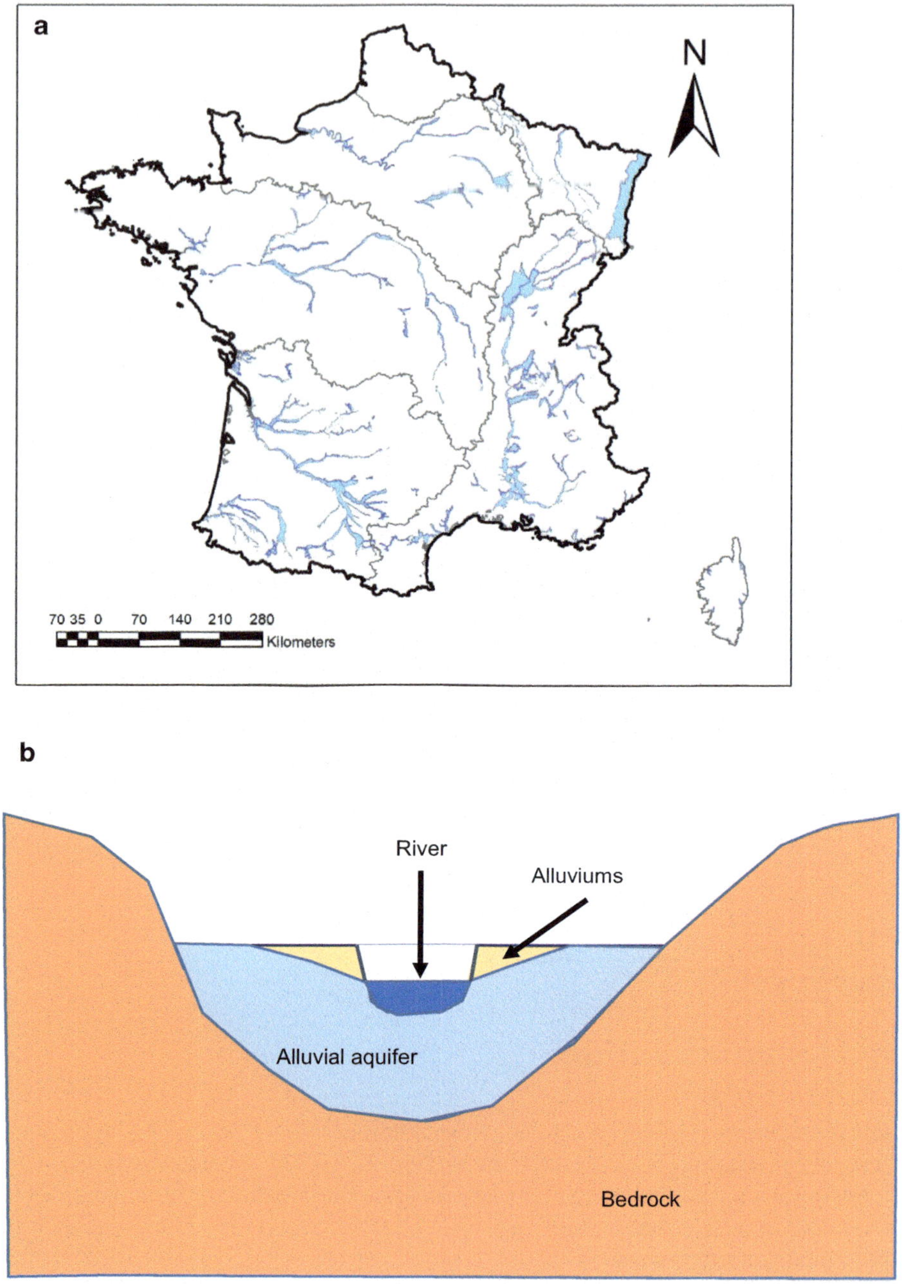

Fig. 2.2 (**a**) Alluvial aquifers in France (**b**) Geological section of an alluvial plain

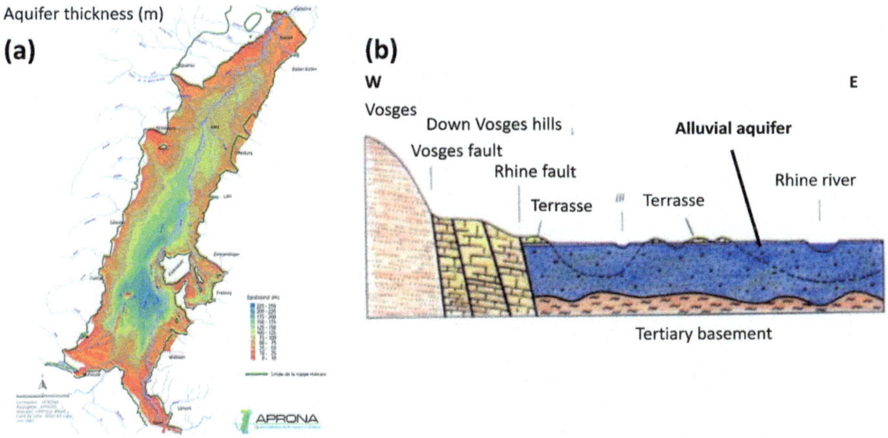

Fig. 2.3 (**a**) Map of Alsace aquifer thickness. (Modified after APRONA, 2008) (**b**) Alsace aquifer cross-section

a few meters to 200–250 m in the centre of the plain (Fig. 2.3a). Marls constitute the bottom of the aquifer (Fig. 2.3b). The sandy gravel alluvium is highly permeable, especially in the vicinity of the Rhine River. The aquifer is recharged by (i) indirect recharge from rivers and canals partly fed by water diverted from Rhine itself, (ii) infiltration from rivers flowing out of the Vosgian range onto the plain and (iii) rainfall on the plain.

The Alsace aquifer has a surface area of about 3000 km^2, with the average stored volume of groundwater ranging from 30 to 50 billion m^3 (AERM, 2009). The average yearly flow[2] is about 1.5 Mm^3/year (1.5 GL/year), which corresponds to a renewal rate of 3% per year (AERM, 2009). The total abstraction rate is about 500 Mm^3/year. Most of the bores tap the aquifer at shallow depths (20–100 m) while some of them reach 150 m depth. The well yields range from 20–30 m^3/h (5–10 L/s) on the margins, to 200–400 m^3/h on the plain. Large diameter wells can supply up to 3000 m^3/h.

2.2.2 Sedimentary Basin Aquifers

They are three main large sedimentary basins in France: the Paris, Aquitain and South-East basins (Fig. 2.4a). Aquifers in these basins can be classified into three types according to their structure and flow regimes:

- Large single-layer unconfined aquifers, mainly constituted by chalk and limestone rocks.

[2] Equal to natural recharge rate

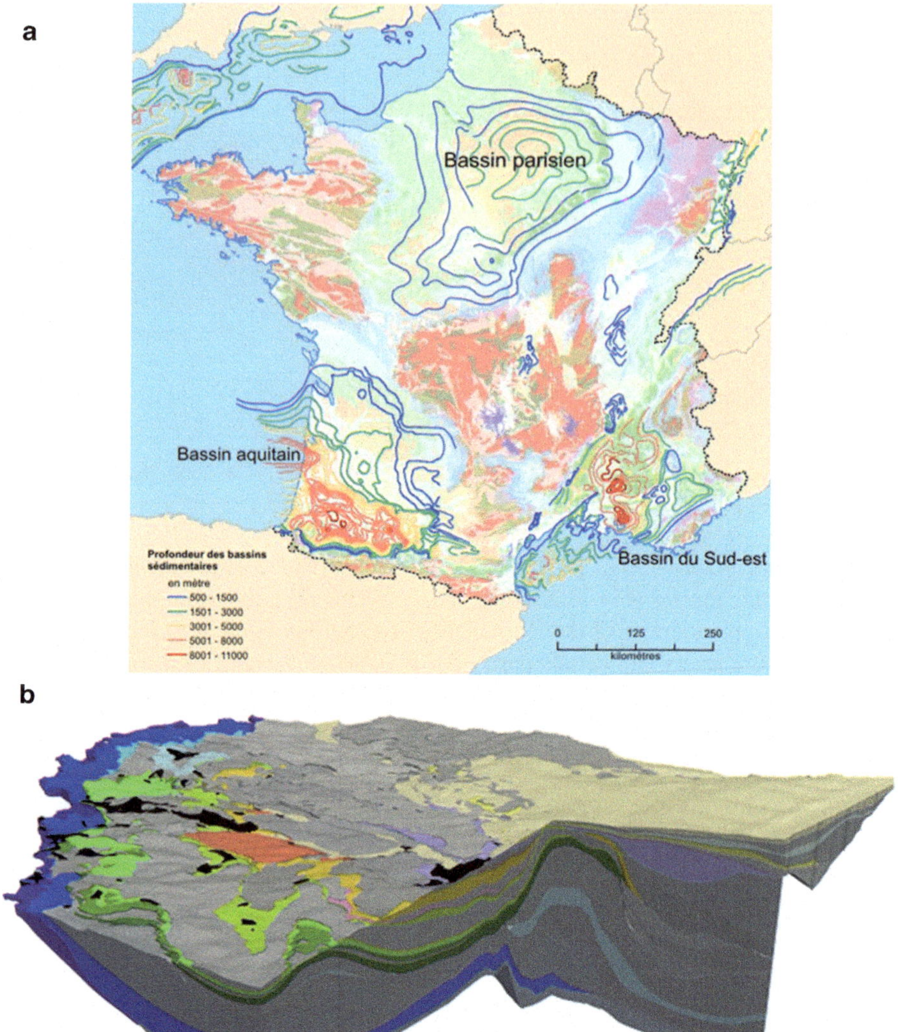

Fig. 2.4 (**a**) Location of the three main sedimentary basins (Paris, Aquitain and Southeast) (**b**) Geological cross-section of Aquitain sedimentary basin. (© brgm, SIGES http://sigesaqi.brgm.fr/Qu-est-ce-que-le-MONA.html)

- Multi-layered aquifers comprising heterogeneous Tertiary sediments located in the centre of Aquitaine (Fig. 2.4b) and Paris Basins that form a shallow unconfined aquifer and several deeper confined aquifers.
- Large deep confined aquifers mainly constituted by sands, sandstones (Albien aquifer in the Paris basin) or limestone (Carboniferous rocks in the North).

Initially artesian, these aquifers (Inferior Trias sandstones, inframollasic aquifer in Aquitaine basin) are now highly developed and artesian conditions have mostly disappeared.

The Paris and Aquitaine Basins contain the most productive sedimentary aquifers which provide high yields from permeable layers. Chalk aquifers within the Paris Basin in northern France provide ~360 Mm³/year, while multi-layered aquifers in the Aquitaine Basin supply ~350–450 Mm³/year.

The Paris Basin is the largest sedimentary basin in France. The sequence extends from Permian and Triassic sediments at the base to Tertiary deposits at the surface and contains at least seven major aquifers (Fig. 2.5b), the deeper of which are brackish. The main aquifers are: the chalk aquifer from Upper Cretaceous (light green on Fig. 2.5a), the Albien green sands (~18 Mm³/year, dark green), the Lower Jurassic limestone (light and medium blue) and the Vosges Lower Trias sandstones

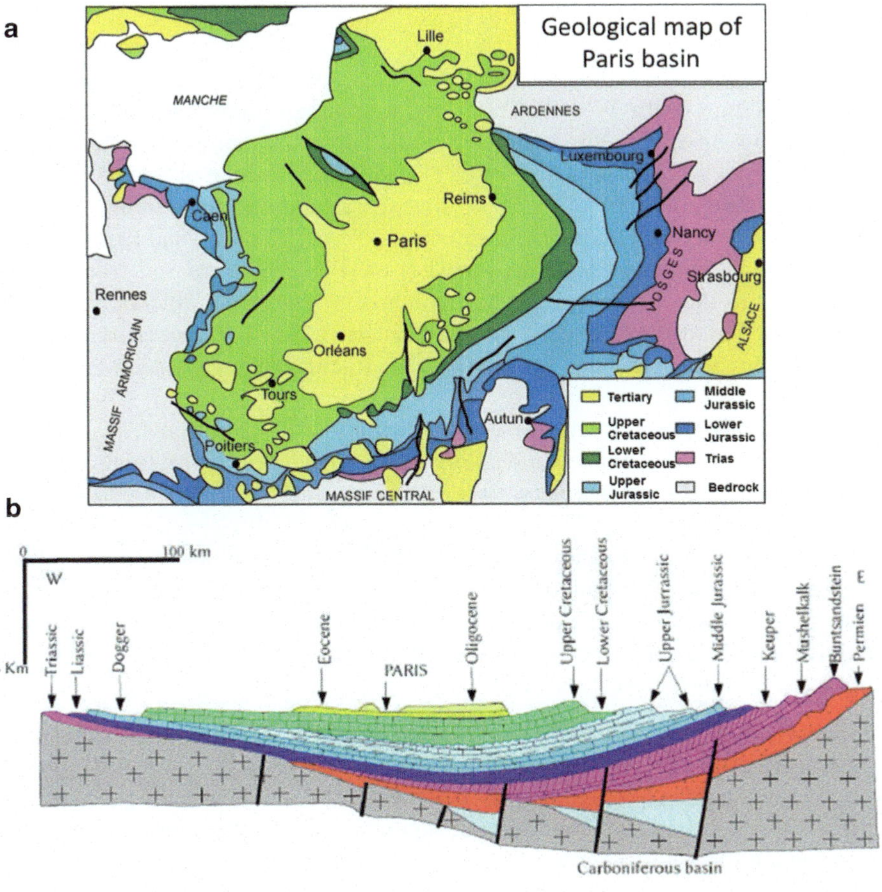

Fig. 2.5 (**a**) Geological map of the sedimentary Paris basin. (Modified after BRGM, 1996) (**b**) Cross- section of the Paris basin. (Modified after BRGM, 1996)

(~160 Mm^3/year, magenta on Fig. 2.5a). Large Tertiary aquifers (Beauce, Brie, yellow) lie at the surface of the basin.

The calcareous Beauce aquifer, located north of Orleans in the centre of France, is one of the largest aquifers in the country (see Chap. 5). The aquifer extends over an area of 9000 km^2 and contains an average storage volume of 20 billion m^3. This area is one of the largest producers of cereals in Europe, with agricultural land covering more than 70% of the total area. About 3000 km^2 is irrigated, which represents an increase of 50% since 1988, mainly driven by the production of cash-crops in the summer (Lejars et al., 2012). Not surprisingly, groundwater abstraction has also increased. The sustainability of the Beauce aquifer has been achieved thanks to the implementation of a sophisticated volumetric management approach described by Verley in Chap. 5 of this book.

2.2.3 Crystalline and Volcanic Rock Aquifers

Crystalline rocks are mainly located in two large mountain ranges: the Armoricain mountain range in the West and the Central mountain range in the centre of the country (Fig. 2.6a). The Vosges, Pyrenees and Alps mountains constitute other significant outcrops. The island of Corsica is also mainly formed of fractured rocks.

The typical geological profile in weathered crystalline aquifers follows the lithological description by Dewandel, Lachassagne, Wyns, Maréchal and Krishnamurthy (2006) which from top to bottom, consists of (Fig. 2.6b): red soil from the first decimeters to the first meter, sandy regolith of a few meters thickness, saprolite from about 3 m to 13–24 m deep, granite or gneiss rocks. The upper part of the hard rock is highly weathered and fractured but the fracture frequency decreases rapidly with depth.

In low lying areas (Brittany), these aquifers are exploited through shallow boreholes 50–100 m deep, while in mountainous areas (Pyrenees, Alps, Central Massif), water is obtained from natural springs. Water abstraction rates are generally low, only a few m^3/hour.

Volcanic rocks are mainly located in the Massif Central and overseas islands, such as La Réunion, Martinique, and Guadeloupe. The total amount of groundwater supplied by Massif Central volcanic rock aquifers is ~40 Mm^3/year. These aquifers provide low yields but they often represent the sole source of supply for small villages or agricultural farms.

Example In the Armorican Massif region (Fig. 2.7a), groundwater accounts for 25% of the drinking water supply from about 400 bores. In all, 40% of the groundwater supply comes from crystalline rock aquifers, while the remainder is obtained from alluvial, sedimentary and volcanic aquifers. Historically, groundwater extraction from the crystalline basement came from shallow wells (20–30 m deep) drilled into the upper part of the basement where regolith and shallow fractured rocks were sufficiently permeable (Roques, Bour, Aquilina, & Dewandel, 2016). During the

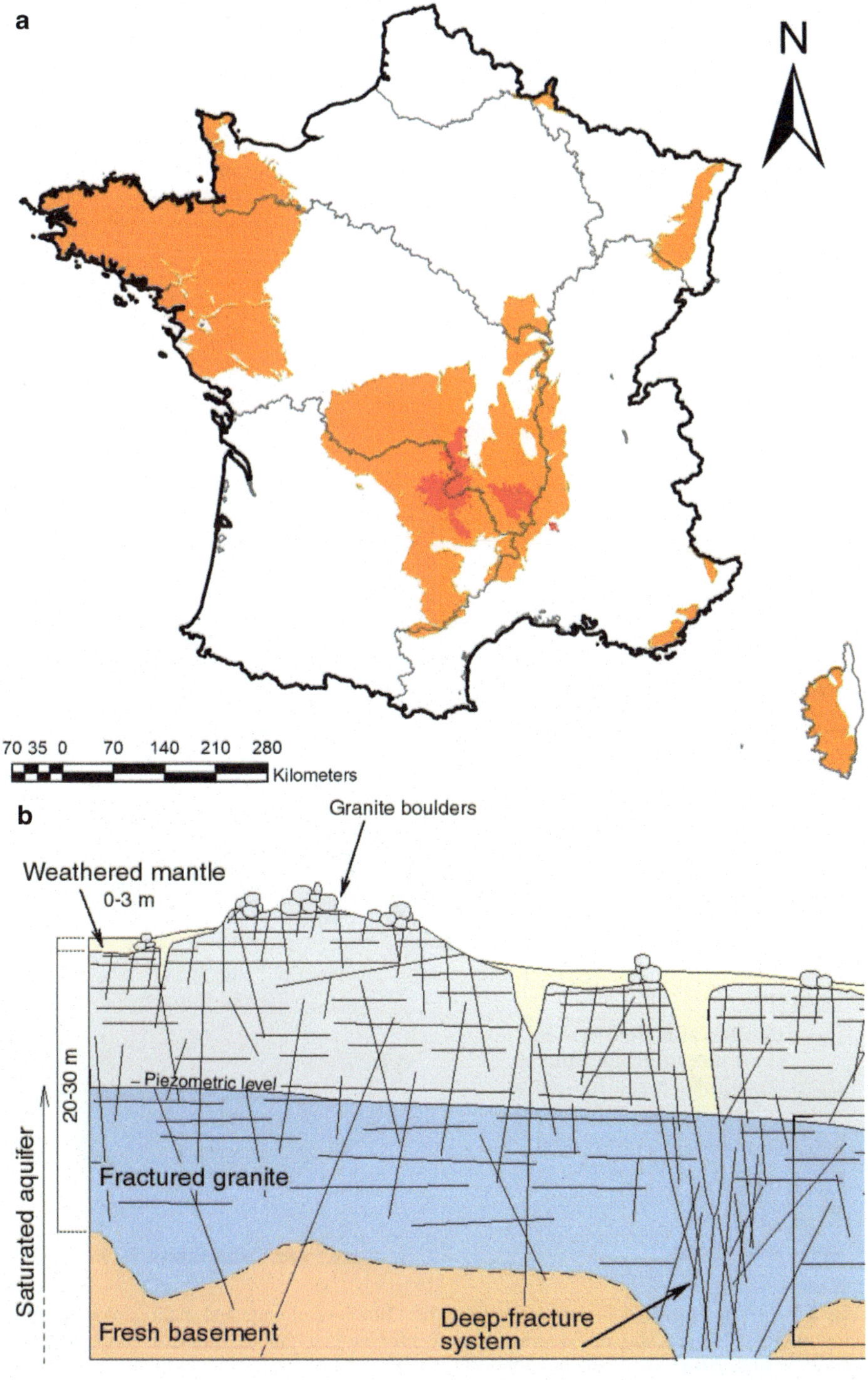

Fig. 2.6 (**a**) Map of crystalline rocks (red: granite, pink: gneiss, brown: schist, green: ophiolite, blue: basalts); (**b**) Hydrogeological cross section of a weathered crystalline aquifer. (Modified from Maréchal, Dewandel, & Subrahmanyam, 2004)

last few decades, deeper wells in the basement have been drilled to over 50 m deep to meet the increasing water demand, as well to avoid recurrent problems of surface water contamination. At the regional scale, the average borehole flow rate in the crystalline rocks is estimated to be around 9 m^3/h (Mougin et al., 2008). In this region, high yields can be obtained from local fault zones (Fig. 2.7b).

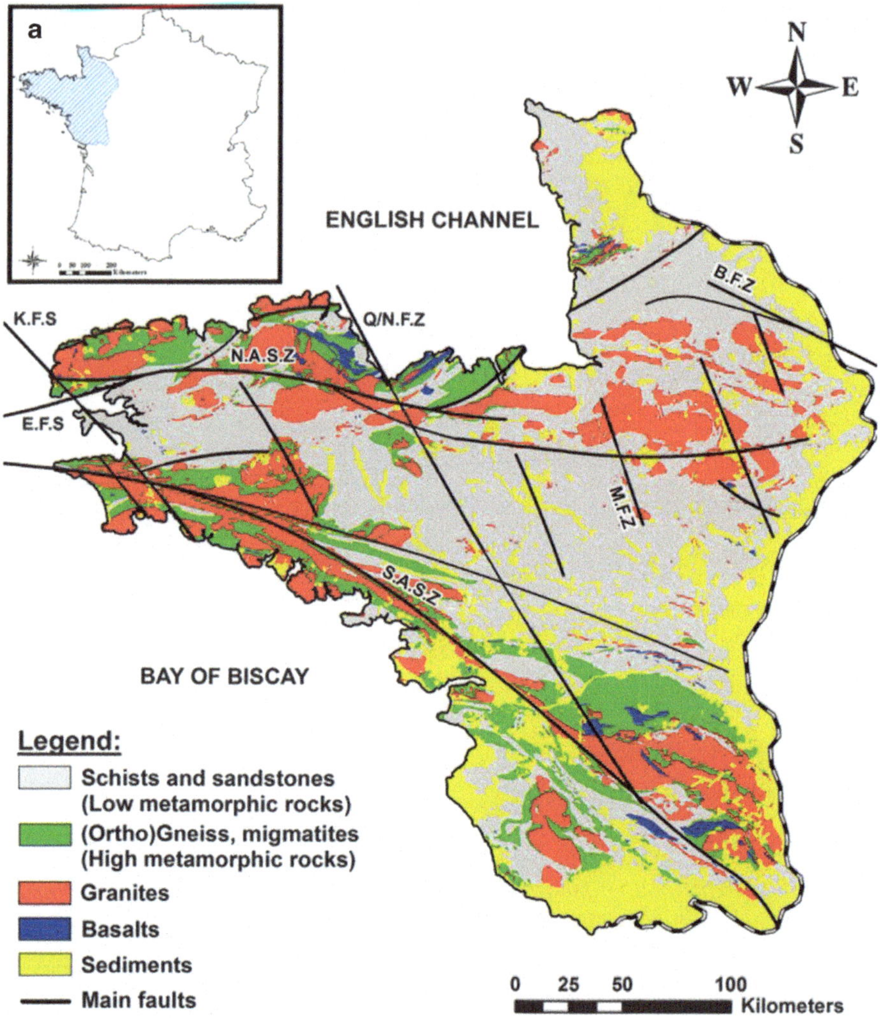

Fig. 2.7 (**a**) Simplified geological map of the main lithological units and main geological structures of the Armorican Massif. (Modified from Armandine-Les Landes, Aquilina, Davy, Vergnaud, and de Veslud, 2014). (**b**) Conceptual models of high-yield borewells due to fault zones Average specific capacity (SC; Q/s) and the range of exploitation flow rates (Qe) are displayed for each model. (Modified from Roques et al., 2016)

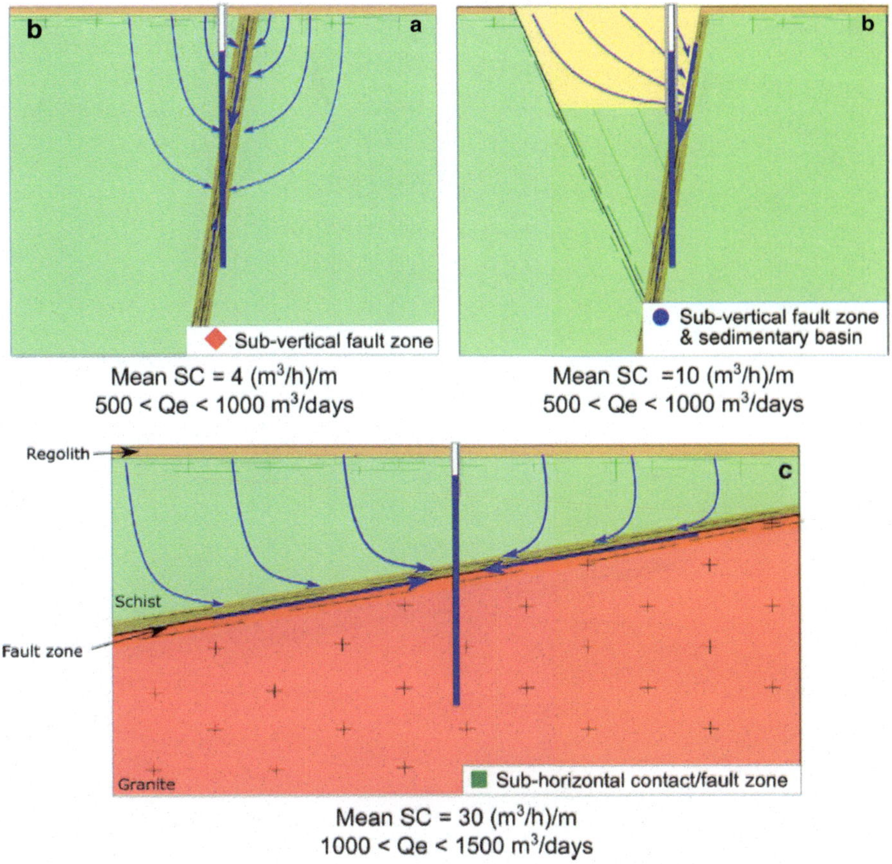

Fig. 2.7 (continued)

2.2.4 *Karst Aquifers*

Karst aquifers are widespread in France and supply 40% of the nation's drinking water supply. Most of these aquifers occur in the southern part of the country (Fig. 2.8a). Their main advantage is the high permeability of the karst drainage network that can supply very large volumes of water (Fig. 2.8b). They are replenished very quickly through diffuse and localized recharge.

Close to the Mediterranean coast, the limestone massifs have been affected by the Messinian salinity crisis which occurred 6 million years ago when the closing of the Strait of Gibraltar cause a lowering of the Mediterranean Sea level. This eustatic and tectonic phenomenon and the associated lowering of the regional water levels, has increased the erosion and karstification potential of rivers and groundwater in the associated region, creating deep karst cavities and karst drainage networks. An example is the well-known Fontaine de Vaucluse karst spring which has been

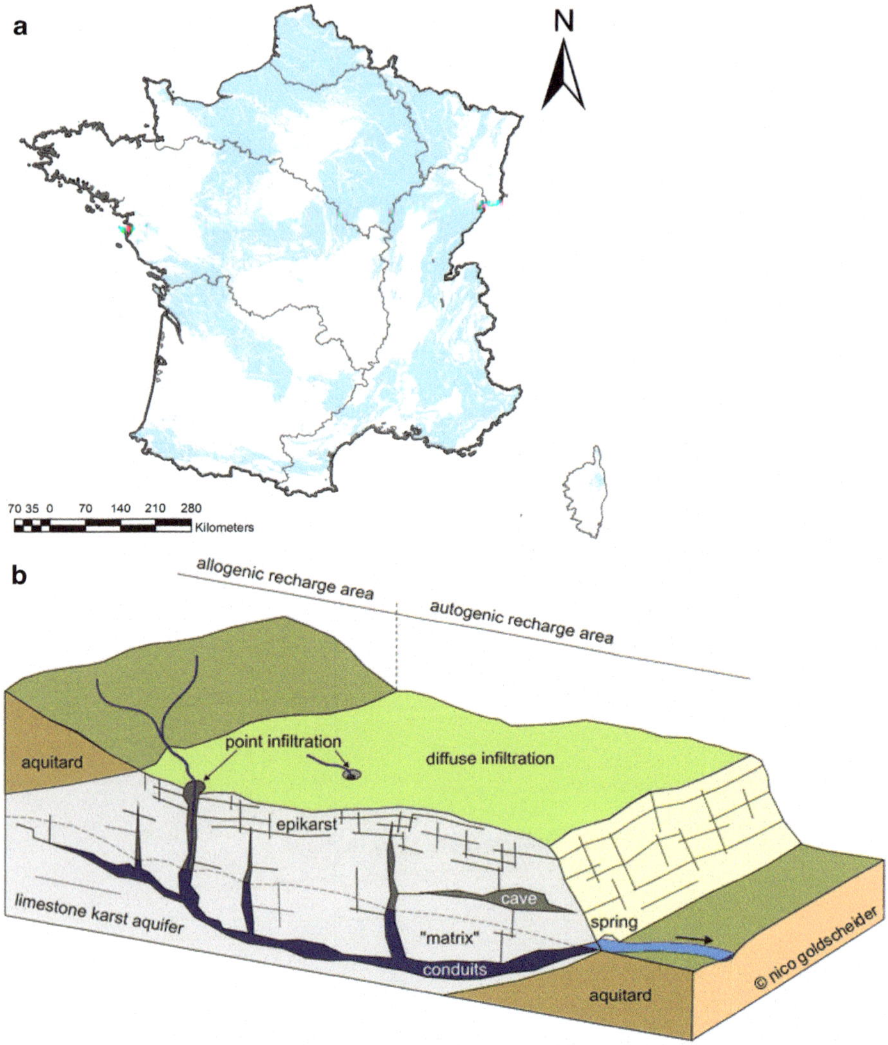

Fig. 2.8 (**a**) Map of karstic carbonate rocks (in blue), main karst springs and caves, and other aquifers. (Modified from Chen et al., 2017) (**b**) Karst aquifer simplified sketch. (Modified from Goldscheider & Drew, 2007)

explored to a depth of 315 m. This deep development of karstification leads to high volumes of stored groundwater which can be pumped at high rates from a single pumping station under active management schemes like the Lez aquifer. As a result of this Messinian crisis, the deeper karst systems are now located under low permeability rock cover (for example, the Arc karst aquifer close to Marseille).

Apart from the Mediterranean coast, other examples of highly productive karst aquifers include: La Rochefoucauld aquifer and la Touvre spring supplying Angoulème

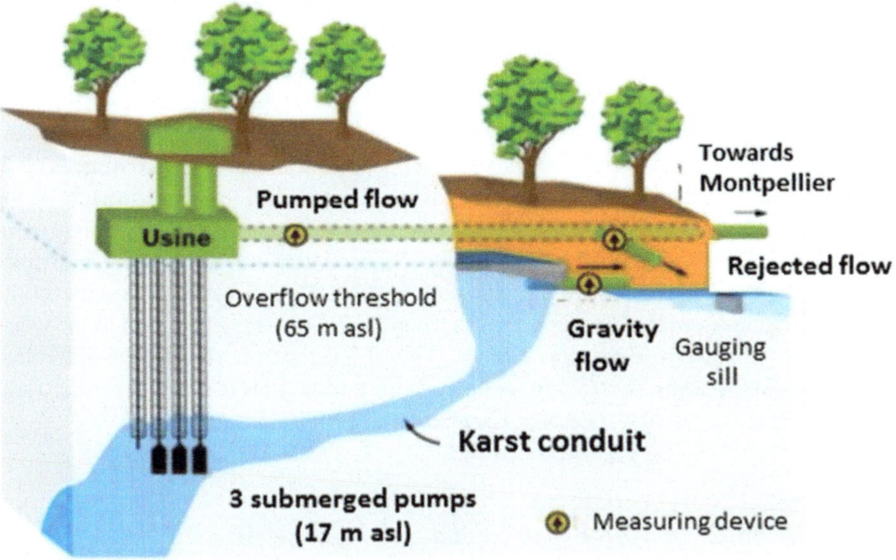

Fig. 2.9 The pumping system at the Lez karst spring © KFH Montpellier

city (~13 Mm^3/year), la Chartreux spring supplying Cahors city (~3.5 Mm^3/year), and the Arcier spring in Jura mountains supplying Besançon city (~5 Mm^3/year).

Due to rapid groundwater flows in the karst conduits and direct infiltration of water in sinkholes, karst aquifers are highly vulnerable to surface pollution.

Example The Lez spring system is one of the largest groundwater abstraction systems in the world, similar to the Figeh karst system which supplies water for the city of Damascus. It is currently tapped by four pumping units located in four vertical boreholes that intercept the main karst conduit (Fig. 2.9). The mean pumping abstraction rate of 34 Mm^3/year is sufficient to supply drinking water for a permanent population of around 340,000 inhabitants in the city of Montpellier. Part of the pumped water is diverted into the Lez river in order to assure a minimum discharge rate of 200 L/s for environmental purposes.

2.3 Groundwater Usage

2.3.1 Historical Development of Groundwater Use

France has a long history of groundwater use, with the first wells drilled in the tenth century in the Chalk confined aquifers of the Artois region in Northern France where the name "artesian well" originated (Barraqué, Chery, Margat, de Marsily, & Rieu, 2010). The increase in groundwater use in France was first associated with the development of drinking water supply systems and the industrial revolution in the nineteenth century, and was mainly localised in urban areas (e.g. Paris, Bordeaux). Today, groundwater represents about 66% of Frances's domestic water supply, 31% of industrial water supply and 37% of total water use in agriculture (CGDD, 2017).

Irrigation in France was first developed in the Mediterranean area from the diversion of river water from the Alpes, Massif Central and the Pyrenees into collective canal irrigation schemes. The large post-war development projects of the 1950s and 1960s increased the scale of these diversions and popularised the use of surface water pumps and collective distribution infrastructure. Groundwater abstraction through collective or individual boreholes first occurred at a large scale in the Beauce region for maize production (see Chap. 5) but quickly expanded in the western and northeastern regions of France. This trend towards the use of individual boreholes and pumps is ongoing, coupled with a reduction in the use of the traditional collective systems (Loubier, Campardon, & Morardet, 2013).

2.3.2 Trends in Water Use by Sector

In 2013, the total water abstraction in France was about 38.5 billion m^3, with the vast majority (70%) sourced from surface water to serve as cooling water for electricity production (21.6 billion m^3) and to supply navigation canals (5.5 billion m^3) (AFB, 2017). Other uses (drinking water, agriculture, industry) comprise a total of almost 12 billion m^3, of which about 50% is supplied by groundwater. Table 2.1 presents the contribution from surface water and groundwater for water use by the various sectors in 2013.

About 66% of abstracted water for drinking water is from groundwater which is a strategic resource given its higher quality compared to surface water, and consequently has lower treatment costs. Groundwater represents about 36% of water abstracted for agriculture and 31% of abstraction for industrial use, which includes factories, commercial firms and various public buildings.

Overall, total water abstraction for drinking water has reduced by 15% between 2003 and 2013 as shown in Fig. 2.10. Industrial water abstraction has similarly reduced by 27% between 1998 and 2013. No significant evolution in overall water abstraction in agriculture can be seen since 2008 when monitoring and reporting became more consistent nation-wide.

Unlike domestic and industrial use which eventually recycles most of abstracted volume back to surface waters as wastewater, irrigation water applied by sprinkler

Table 2.1 Water use in 2013 and trends over time (Mm^3)

	Drinking water	Agriculture	Industry	Total
Groundwater	3700	1035	930	5665
Surface water	1866	1766	2700	6332
Total	5566	2801	3630	11,997

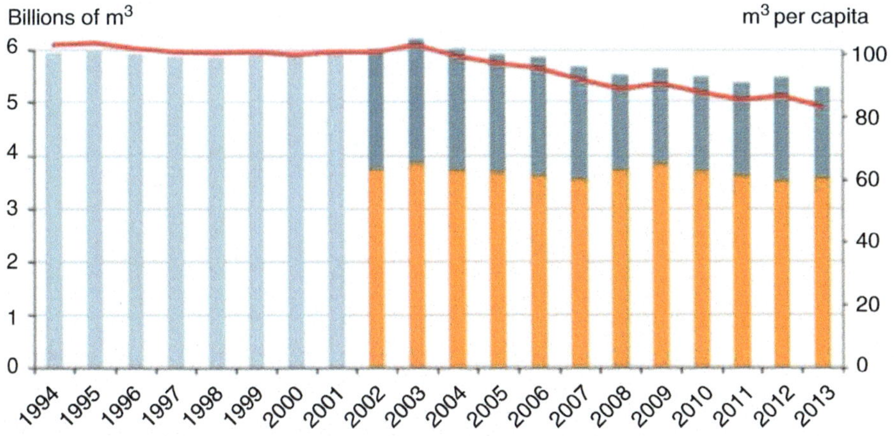

Fig. 2.10 Trend in water abstraction for drinking water supply. In light blue: total abstraction. In dark blue: surface water abstraction. In orange: groundwater abstraction. Red line: abstraction water per capita. (Source: modified from Banque nationale des prélèvements quantitatifs en eau, ONEMA-SOeS on 2013 data (CGDD, 2017))

or micro-irrigation is mostly consumed by evapotranspiration and as a result, agriculture is the largest net consumer of water e.g. 58% of water consumption in the Adour-Garonne basin (Ayphassorho, Caude, Mathieu, Groslaude, & Renoult, 2015).

2.3.3 *Groundwater Use in Agriculture*

Generally, irrigation in France is not essential for agricultural production as is the case in arid countries such as Australia; rather, it is used to (1) secure yields against climate risks such as drought, (2) increase average yields, and (3) improve product quality. The irrigation of crops consumes 80% of water used in agriculture in France, while the remaining 20% is used for livestock water supply and cleaning. Accurate figures on irrigation use are difficult to obtain because water meters are not installed on all individual water pumps yet, and the reporting of abstracted volumes is not systematic.

Most agricultural land equipped for irrigation is situated in the southwest, west, and northeast of France (Fig. 2.11). The main irrigated crops are maize and cereals, as well as potatoes, vegetable cropping and fruit production. Maize represented 41% of all irrigated land in 2010, down from 50% in 2000. This trend is partly related to the reduction in European subsidies for this crop (see Chap. 24), as well as stricter restrictions on water use (see Chap. 3) and higher prices for other cereal crops, in particular wheat.

The area of irrigated land has steadily increased from 500,000 ha in 1970, to a maximum of 1.57 million ha reached in 2000 (representing 6% of the total agricul-

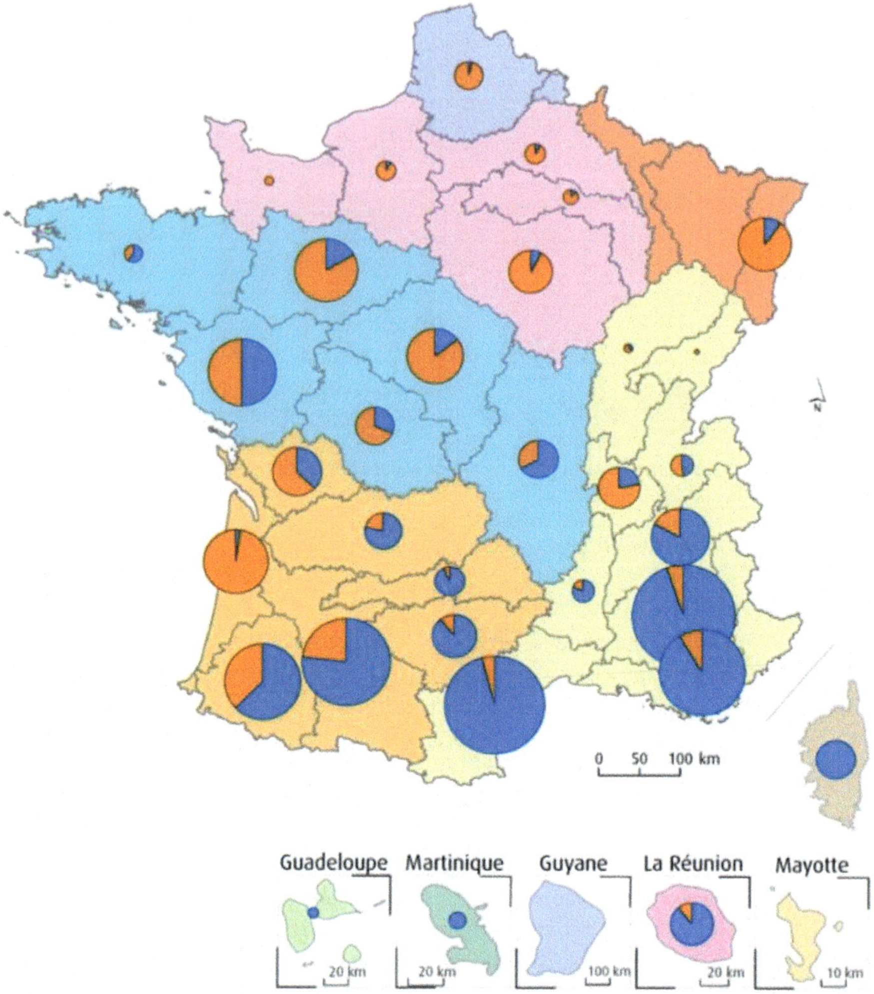

Fig. 2.11 Agricultural water abstraction by administrative district. In orange: abstraction from groundwater. In blue: Abstraction from surface water. The size of the circle represents the relative abstracted volume. (Source: modified from Banque nationale des prélèvements quantitatifs en eau, ONEMA-SOeS on 2013 data (CGDD, 2017))

tural land). In 2010, the area of irrigated land had not changed significantly while the total area of used agricultural land had reduced by 900,000 ha (or 3.5% of the total area). Overall, irrigation appears to be maintained where it is regularly used, and may help to keep small agricultural holdings economically viable in a context of general consolidation of holdings and abandonment of agricultural land (Loubier et al., 2013).

The internal irrigation rate of agricultural holdings practicing irrigation in 2010 was 32%, a number that has slightly increased since 2000 and indicating that irrigation is becoming a more important part of some agricultural units. On average, irrigation is responsible for about 2000 m^3 of water abstracted per ha.

In 2010, a sharp reduction in area equipped with collective irrigation systems was observed, while areas equipped with individual irrigation systems has continued to increase (Loubier et al., 2013). This trend is also occurring in the Mediterranean region where most irrigation is traditionally carried out through collective canal systems. At the time, a reduction of 50% in surface water irrigation is observed in these regions. These trends indicate a move towards more water efficient systems, although it also poses local challenges due to a reduction in groundwater recharge via the reduced seapage from distribution canals and surface irrigation practices.

The development of irrigation has led to increasing societal conflicts in the agriculturally productive regions in the west and south west of France which underwent a significant increase in irrigation for maize and cereal production in the 1980s and 1990s. Assuring minimum ecological flows is a significant challenge resulting from the cumulative pumping from rivers and extractions from individual boreholes in alluvial and sedimentary aquifers (Ayphassorho et al., 2015).

2.3.4 Groundwater and Drinking Water Supplies

The drinking water supply network provides water to domestic users, public services (e.g. schools, hospitals, hotels, sports, etc.), and small businesses and industries. In 2013, the water abstraction per capita in France was 85 m^3, a reduction of 20% compared to 2003.

The volume of water use is mainly dependent on the size of the residential population; however some groundwater basins experience large seasonal variations in population due to tourism. This can pose supply challenges in Mediterranean basins during the low flow season similar to those faced by irrigation. Drinking water supply is given the highest priority use during crisis. Water shortages have not yet caused restrictions on drinking water use in France, however restrictions on garden watering are regular.

The vast majority of the population (98%) have water delivered to their homes by public water suppliers, however since the 1990s, an increasing number of households in detached or semidetached housing units have drilled private supply bores due to various economic, political and ethical reasons (Rinaudo, Montginoul, & Desprats, 2015). Typically, households use alternative water supplies for gardening and other non-consumptive uses (e.g. toilet flushes) in order to reduce their water bill. According to Montginoul and Rinaudo (2011), the presence of domestic bores and shallow wells is reported in a majority of French counties in both southern and northern France, and is expected to significantly increase in the coming decade as a result of increased water scarcity, higher prices from public water suppliers, and the decreasing cost of alternative supply technologies. Furthermore, there is currently little regulation controlling drilling to provide private domestic supplies.

Fig. 2.12 Examples of groundwater resources reserved for their insurance role. (Hérivaux & Rinaudo, 2016)

Overall the main challenge regarding groundwater abstraction for drinking water supply purposes relates to water quality and the risks posed by pollution, mainly from agricultural activities.

2.3.5 Strategic Groundwater Resources

Recent work has investigated the role that groundwater may play as insurance for drinking water supplies in case of extreme events or potential failures of supply systems in the future due large storms, flooding, earthquakes or large technological disasters such as nuclear war or accidents (Hérivaux & Rinaudo, 2016). In the long term, drinking water supplies may be exposed to the impacts of climate change or from other progressive changes such as large scale pollution and increased operational costs. In the same way as typical insurance, the preservation of specific groundwater resources may result in increased costs in the short term but will provide long term guarantees which are of higher value than other types of resources.

According to Hérivaux and Rinaudo (2016), three approaches have been implemented in France to secure drinking water supplies for such situations (see Fig. 2.12):

- Creation and maintenance of emergency boreholes;
- Increasing the share of abstraction from confined aquifers that are naturally protected from human pressures;
- Preservation of groundwater resources that are currently not exploited (or are developed to a limited extent) for future use.

One such example is the Albien-Néocomien aquifer beneath Paris which has been identified in the river basin management plan as a strategic resource for an emergency drinking water supply. This large confined aquifer is up to 1000 m deep and can provide good quality water to supply the population of Paris for several months in case of an major emergency that would disrupt the existing supply system. In order to maintain its "insurance value", public authorities have established regulatory controls on current abstraction to preserve sufficient volumes for cases of emergency. In addition, deeper pumping infrastructure was installed in order to secure supplies over a long period if necessary.

2.4 Groundwater Management Issues

2.4.1 Quality Issues

According to the European water framework directive, 67% of groundwater bodies in France are considered to be in a good chemical state in 2013 (Fig. 2.13a). Bad status is often linked to polution resulting from human activity. Generally, the quality of groundwater is not adversely affected by water-rock interaction. Despite the existence of evaporites deposited during the Trias period, salinity contamination of aquifers has not occurred because these rocks are mostly isolated from aquifers by low permeability sediments.

The main quality issues for groundwater are related to diffuse contamination by agricultural practices (i.e. fertilizers and pesticides). High nitrate levels are observed in the crystalline aquifers of Brittany, due to effluent from intensive pork breeding and agriculture.

In karst aquifers, the issue is related to rapid flows during floods bringing turbidity and bacteria through the karst network to springs. It is necessary to properly map the vulnerability to pollution of such aquifers and accordingly define appropriate land-use rules. Wastewater treatment plants and sewage networks must be well designed and maintained especially for such flood events.

In coastal areas, especially in overseas islands, sea water intrusion can be induced by high levels of pumping for drinking and irrigation purposes in densely populated regions. Chapters 12 and 19 describe two case studies where seawater intrusion was successfully managed. Monitoring of the geometry and fluctuations of the salt-water interface is necessary in order to properly manage water abstraction. In sev-

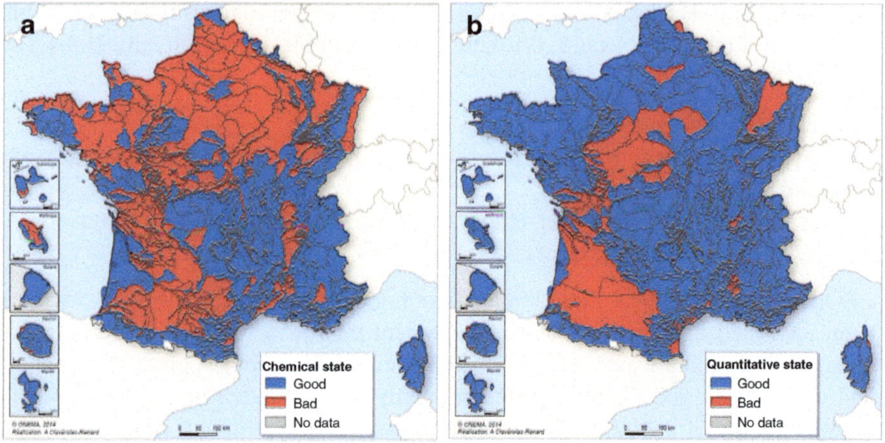

Fig. 2.13 Chemical (**a**) and quantitative (**b**) states of groundwater bodies in France in 2013, from Petit et Michon (2015)

eral cases, managed aquifer recharge is used to prevent saline intrusion. Compared to other countries (e.g. Spain), seawater intrusion is well managed in France.

In alluvial aquifers, increasing pollution of rivers lead to the potential contamination of groundwater especially when pumping wells are located close to rivers. The potential filtration impact of riverbanks and aquifer is a rising scientific issue.

2.4.2 *Quantity Issues*

Under the European water framework directive, water agencies analyse the state of groundwater aquifers. In 2013, 10% of groundwater bodies was considered in a bad quantitative state (Fig. 2.13b).

There is a wide range in present exploitation rates[3] of aquifers (Margat, 2006):

- The most exploited aquifers are not necessarily those with the highest bore yields. For example, large unconfined aquifers with high recharge rates and large storage volumes (e.g. karst aquifers), are often underutilised;
- Exploitation rates reaching 100% are rare: in most cases, the extraction has increased the infiltration from rivers which counter balances the losses due to abstraction.
- In most of the aquifers, the abstraction volume is less than a tenth of the natural groundwater flow.

Watertable lowering is observed in a few aquifers where water flows are unbalanced (i.e. the pumping rate being higher than the replenishment of the aquifer

[3] Exploitation rate is the ratio between abstraction flow and natural groundwater flow

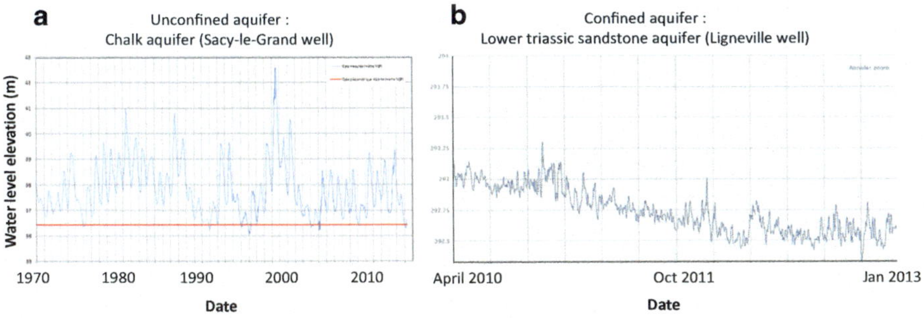

Fig. 2.14 (**a**) Watertable fluctuation in an unconfined aquifer (**b**) Water pressure fluctuations in a confined aquifer

through natural recharge). In these cases, studies have been launched by water agencies in order to define a sustainable extraction (volumetric) limit ("volumes prélevables" in French, Chap. 11). Declines in groundwater levels can induce several adverse impacts.

In alluvial aquifers (or other shallow aquifers connected to surface water), the watertable decline induced by pumping can temporarily reduce groundwater contribution to river base flows, which in turn threatens aquatic life and decreases water quality. This occurs for example in several rivers in the Beauce region, and in north of France (see Chap. 5). In regulated rivers, a minimum discharge rate in the river can be maintained by water released from reservoirs managed by electricity companies. In some cases, watertable decline could reverse the natural exchange flow between the aquifer and surface water leading to a deterioration of groundwater quality. Research is on-going in karst aquifers where flow exchanges between surface and ground water are very complex. Most of the time, in unconfined aquifers, the induced increased recharge counterbalances the pumping and water table remains stable (Fig. 2.14a).

The lowering of the watertable can also threaten wetlands. Maintaining a watertable that stands at or near the land surface for a long enough period each year to support aquatic plants is a challenging issue for some wetlands. For example, the historically well-known Marais Poitevin marsh has a high risks of adverse impacts during droughts due to groundwater pumping for maize irrigation (see Chap. 18).

In a few confined aquifers, the absence of direct or indirect recharge induces long-term water table depletion (Fig. 2.14b). In such multi-layered aquifers, the vertical hydraulic exchanges are dependent on the hydraulic gradient between the aquifers. The natural upward flow from a deep confined aquifer to a shallow unconfined aquifer can be reversed by water level declines induced in the deep aquifer by high pumping rates (Fig. 2.14). In that case, possibly poorer quality water from the shallow aquifer can contaminate water in the deep aquifer. It is therefore important to properly manage pumping in both aquifers in order to avoid such contamination (see Chap. 12).

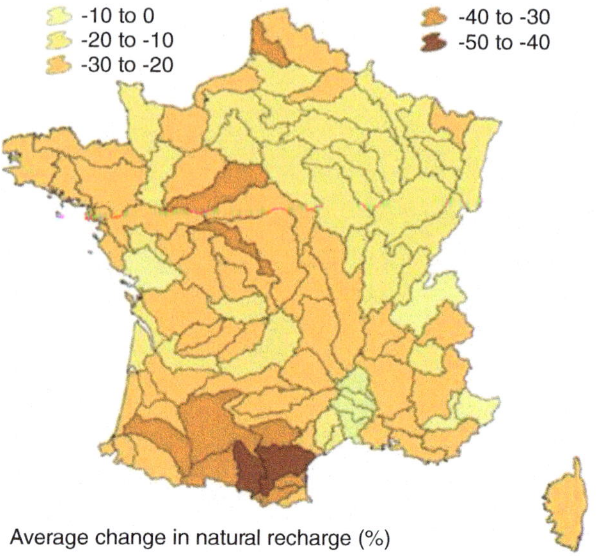

Fig. 2.15 Impact of climate change on natural recharge rates of aquifers from now to 2045–2065 period. (From MEDD, 2015)

In summary, France does not face the problems of aquifer depletion which are experienced in many other countries. However, the associated groundwater dependant ecosystems are more likely to be affected by the abstraction of groundwater.

2.4.3 Long Term Challenges

2.4.3.1 Climate Change and Recharge

The Explore 2070 project[4] has developed and assessed strategies to adapt to climate change impacts on hydrological systems and coastal environments in mainland and overseas France up to 2070, based on different climatic, demographic and socio-economic scenarios.

Rises in temperature (and consequently evapotranspiration) combined with decreasing rainfall, will lead to a decrease of effective precipitation in the future. The application of seven climate models using the median GHG emission scenario (A1B, fourth GIEC report) enabled an estimate of the change in natural recharge rates (Fig. 2.15). With predicted recharge variations of +10 to −30% in the optimistic scenarios, and −20 to −55% in the pessimistic scenarios, a decline of similar proportions in groundwater levels would be expected, and therefore groundwater resources are likely to decline significantly overall by 2070. Two areas which are likely to be more severely affected are:

[4] http://www.brgm.eu/project/explore-2070-rising-to-climate-change-challenge

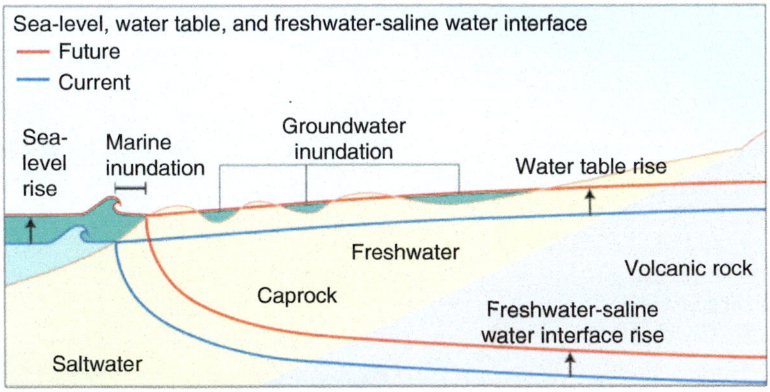

Fig. 2.16 Impacts of sea level rise, from Rotzoll and Fletcher (2013)

- the Loire basin with a 25–30% recharge decline across half of the catchment area,
- the south-west of France with a 30–50% decline in recharge.

All of the scenarios also show a decline in average river flow by 2065, which varies from a 10 to 40% reduction in the northern half of the country, and a 30–50% reduction in the southern half, with local extremes of up to 70%. Despite this relative decline in river flow, some models show that very high surface water levels are nevertheless possible during the winter in some catchments (e.g. the Somme and Rhine Rivers), confirming the likelihood of lengthy periods of flooding.

2.4.3.2 Climate Change and Sea Level Rise

Sea level rise is expected to have several impacts (Fig. 2.16):

- An increased risk of sea water intrusion and variable inland migration of salt water, especially in karst aquifers
- A rise of the salt/fresh water interface in estuaries and the infiltration of salt water into unconfined aquifers: this phenomenon could be accentuated by the decline of river flow rates due to climate change or increases in groundwater abstraction
- The submersion/inundation of lowlands by seawater and infiltration of salt water into unconfined aquifers.

The risk of inundation is especially high along the Mediterranean coastline where significant areas are likely to be flooded due to the very low elevation.

Table 2.2 Predicted changes in groundwater abstraction (Mm^3)

Year	Drinking water	Irrigation	Industry	Total
2006	3631	1276	1436	6344
2070	3100	1271	679	5050
Change (%)	−14.6	−0.4	−52.7	**−20.3**

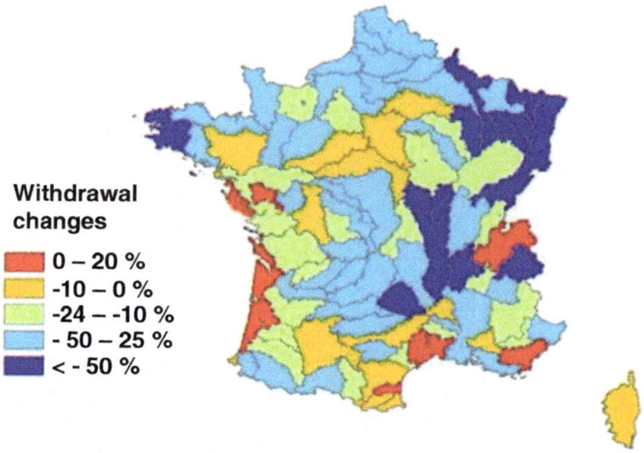

Fig. 2.17 Expected changes in groundwater abstraction (from 2006 to 2070)

2.4.4 *Future Changes in Groundwater Use*

The main scenario for change is an overall decrease in groundwater abstraction of 20% by 2070 (Table 2.2). Drinking water abstraction is expected to decrease by 15% due to decreases in individual consumption rates and network losses, and industrial consumption will decrease by 53% due to recycling efforts and production changes. Agricultural demand is expected to remain stable. The regional distribution of groundwater abstraction changes is illustrated in Fig. 2.17.

The climate change induced changes in groundwater abstraction are expected to lead to a small decrease of rates of water level declines generally but in coastal areas, an increased decline in water levels and increased risk of saline intrusion is predicted due to population increases and increased water demand (Fig. 2.17).

2.4.5 *Emerging Adaptation Strategies*

2.4.5.1 New Groundwater Management Strategies and Policies

An increasing number of basins are expected to be confronted with a mounting imbalance between available resources and demand for those resources. The management model using volumetric limits which is currently deployed in a limited number of groundwater basins will need to be implemented in many others. Allocation policies will also need to be more flexible (see Chaps. 22 and 23).

2.4.5.2 Managed Aquifer Recharge

"Good quantitative status" of groundwater is achieved when the volume of water withdrawn is less than the renewal capacity of the aquifer, and when the connected surface ecosystem health (e.g. wetland, river) is maintained. With continual population growth combined with climate change, the management of groundwater resources in France is mostly focused on more efficient water use. But it is likely that these actions shall not be sufficient to cope with water shortages in some areas, and Managed Aquifer Recharge (MAR) could be a novel and efficient way to maintain and improve groundwater quality and quantity.

In Europe, France has the third highest number of MAR sites after Germany and Netherlands. Surface water spreading via infiltration basins and induced bank filtration are the most numerous MAR systems (Casanova et al., 2015). Most of these are located in alluvial plains or sedimentary aquifers, but they contribute only 3% of drinking water supplies. Currently, the MAR technique is used to maintain water levels in pumped aquifers but also for improving groundwater quality: i.e. to stop saline intrusion in coastal aquifers (example of Hyeres Aquarenova project), or to improve the quality of water pumped in alluvial plains close to a polluted river using a double pumping/injection system.

In the future, the MAR technique could constitute an alternative solution to deal with decreases in natural recharge rates under climate change. The cost of recharge (including the maintenance of infiltration systems) and the recovery efficiency (the volume of infiltrated water that can be recovered for any use) are still the main issues for implementing such management solutions.

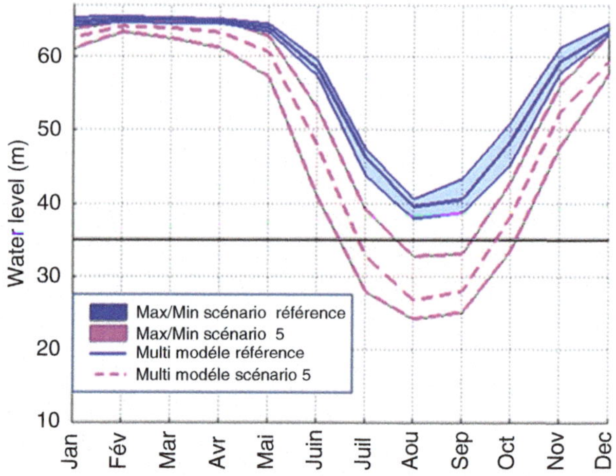

Fig. 2.18 Modelling results at the Lez karst system spring – Comparison of present water levels under present abstraction rate (33 Mm^3/year) with future water levels under climate change and increased abstraction rate (44 Mm^3/year) – horizontal black line corresponds to the authorized minimum water level. (Modified from Maréchal et al. 2014)

2.4.5.3 Active Groundwater Management

Karst aquifers serve as major underground water reservoirs that supply the ever-increasing water demand arising from the population boom in the coastal region and main cities especially in the Mediterranean Basin. Through "active groundwater management" (Collin, 1994), water resources in these systems can be optimally utilized by overcoming the negative effects of high variations in spring flow variations. In the dry season, this requires pumping at a rate that is higher than the replenishable flow so as to tap aquifer water reserves, which in turn will be replenished during the following rainy season. This also reduces the intensity of floods at the onset of the rainy season. Karst aquifer systems in France are often located upstream of a coastal stream catchment which has been urbanized and is subject to Mediterranean-type flash floods. The latter cause flooding and property damage, which in turn may have considerable economic impacts, or lead to loss of human life, as occurred in Nîmes in 1988.

Conservation of coastal river ecosystems is also a key issue, especially during low flow periods when some of the pumped groundwater is diverted to the river to help maintain the streamflow and consequently the stream's ecological balance. Multipurpose management of such aquifers is an integrated way to address these seemingly conflicting issues: supplementing drinking water supply demand, reduction of flooding hazards, and conservation of aquatic environments.

In the Lez karst aquifer under active management, described above (Fig. 2.9), a model has been used to simulate several scenarios of pumping under present and future (2045–2065) climate (Ladouche, Caballero, & Maréchal, 2013). It is useful to determine the sustainable pumping rate allowing to respect the authorized drawdown and pumps elevation. However, even when considering the present water pumping rate being continued into the future, the aquifer would be totally recharged and the groundwater level recovered to the overflow level each year (Fig. 2.18).

2.5 Conclusions

Although France is blessed with abundant surface water and groundwater resources and appear favourably in international terms, the high water demand and the widespread emission of pollutants, together with the predicted impacts of climate change, put significant pressure on the long term sustainability of groundwater reserves. The existing regulatory and policy instruments provides a comprehensive framework for managing groundwater. However, improved implementation and further political awareness of the long term value of protecting groundwater resources is necessary. Given the high spatial heterogeneity of situations across the country, future implementation will also need to be flexible to enable delivery of national strategies in a targeted manner appropriate to diverse situations.

Acknowledgement The contribution of Josselin Rouillard to this work is part of the RURECO project ("Institutions for groundwater dependent rural economies"), funded by the Marie Skłodowska-Curie Action Individual Fellowship of the EU's Horizon 2020 Programme for Research, Technological Development and Demonstration under Grant Agreement no. 750553.

References

AERM (Agence de l'Eau Rhin-Meuse). (2009). 2009: SDAGE Rhin.

AFB (Agence Francaise pour la Biodiversité). (2017). Bulletin n°2: prélèvements quantitatifs sur la resource en eau. Edition Mars 2017, donnéees 2013. Available at: http://www.eaufrance.fr/IMG/pdf/bnpe_2013_201703.pdf

APRONA (Association de Protection de la Nappe phréatique d'Alsace). (2008). Epaisseur de la nappe aquifère d'Alsace. http://www.aprona.net/

AQUARENOVA Project. http://www.aquarenova.fr/le-projet-aquarenova/

Armandine-Les Landes, A., Aquilina, L., Davy, P., Vergnaud, V., & de Veslud, C. L. C. (2014). Time scales of regional circulation of saline fluids in continental aquifers (Armorican massif, Western France). *Hydrology and Earth System Sciences Discussions, 11*, 6599–6635.

Ayphassorho, H., Caude, G., Mathieu, N., Groslaude, J. Y., & Renoult, R. (2015). *Evaluation de la mise en œuvre des protocoles Etat-profession agricole conclus en 2011 dans le bassin Adour-Garonne pour la gestion quantitative de l'eau*. CGEDD-CGAAER, Paris, 171 pp.

Barraqué, B., Chery, L., Margat, J., de Marsily, G., & Rieu, T. (2010). *Groundwater in the Southern Member States of the European Union: An assessment of current knowledge and future prospects*. Country report for France. European Academies Science Advisory Council. Available at: https://easac.eu/fileadmin/PDF_s/reports_statements/France_Groundwater_country_report.pdf

BRGM. (1996). Carte géologique de la France à 1 000 000. Éd. BRGM, *Bureau de recherches géologiques et minières*. Orléans – 6e édition.

Casanova, J., et al. (2015). Managed Aquifer Recharge: An Over-view of Issues and Options. In Jakeman et al. Available at: http://www.springer.com/us/book/9783319235752

CGDD (Commissariat Général au Développement Durable). (2017). Les prélèvements d'eau douce en France: les grands usages en 2013 et leur évolution depuis 20 ans. Service de l'observation et des statistiques, MEEM, France, 26pp.

Chen, Z., Auler, A. S., Bakalowicz, M., Drew, D., Griger, F., Hartmann, J., et al. (2017). The World Karst Aquifer Mapping project: Concept, mapping procedure and map of Europe. *Hydrogeology Journal, 25*(3), 771–785. https://doi.org/10.1007/s10040-016-1519-3

Collin, J.-J. (1994). La gestion active des aquifères. *Courants, 25*, 36–42.

Dewandel, B., Lachassagne, P., Wyns, R., Maréchal, J. C., & Krishnamurthy, N. S. (2006). A generalized 3-D geological and hydrogeological conceptual model of granite aquifers controlled by single or multiphase weathering. *Journal of Hydrology, 330*(1–2), 260–284. https://doi.org/10.1016/j.jhydrol.2006.03.026

Goldscheider, N., & Drew, D. P. (2007). *Methods in Karst hydrogeology*. Hoboken: Taylor & Francis Group.

Hérivaux, C., & Rinaudo, J. D. (2016). – Pourquoi et comment préserver les eaux souterraines pour leur rôle d'assurance ? Tour d'horizon de l'expérience française. Rapport final. BRGM/RP-65631-FR, 60p.

Ladouche, B., Caballero, Y., & Maréchal, J.-C. (2013). Evaluation de l'impact du changement climatique sur la ressource en eau souterraine du système karstique du Lez. *Karstologia, 62*, 57–64.

Lejars, C., Fusilier, J. L., Bouarfa, S., Coutant, C., Brunel, L., & Rucheton, G. (2012). Limitation of agricultural groundwater uses in Beauce (France): What are the impacts on farms and on the food-processing sector. *Irrigation and Drainage, 61*(S1), 54–64.

Loubier, S., Campardon, M., & Morardet, S. (2013). L'irrigation diminue-t-elle en France? Premiers enseignements du recensement agricole de 2010. *Sciences Eaux & Territoires, 2*, 12–19.

Maréchal, J. C., Dewandel, B., & Subrahmanyam, K. (2004). Use of hydraulic tests at different scales to characterize fracture network properties in the weathered-fractured layer of a hard rock aquifer. *Water Resources Research, 40*(11), W11508. https://doi.org/10.1029/2004WR003137

Maréchal, J. C., Ladouche, B., Batiot-Guilhe, C., Borrell-Estupina, V., Caballero, Y., Cernesson, F., et al. (2014). Projet gestion multi-usages de l'hydrosystème karstique du Lez –synthèse des résultats et recommandations – Rapport BRGM/RP-61051-FR, 126 p.

Margat, J. (2006). Exploitation des eaux souterraines. IN: *Aquifères et eaux souterraines de France écrit par Sous la direction de J.C.ROUX*, éditeur BRGM/AIH, 2006, isbn 9782715909809.

MEDD (Ministère de l'Ecologie, du Développement Durable et de l'Energie). (2015). *Synthèse du projet Explore 2070: hydrologie souterraine*. Ed. Ministère de l'Ecologie, du Développement Durable et de l'Energie http://www.gesteau.fr/sites/default/files/explore2070-hydrologie-souterraine.pdf

Montginoul, M., & Rinaudo, J. D. (2011). Controlling households' drilling fever in France: An economic modeling approach. *Ecological Economics, 71*, 140–150.

Mougin, B., Allier, D., Blanchin, R., Carn, A., Courtois, N., Gateau, C., & Putot, E. (2008). SILURES Bretagne (Système d'Information pour la Localisation et l'Utilisation des Ressources en Eaux Souterraines) [SILURES Bretagne (Information system for the location and use of groundwater resources in the Armorican Massif)]. Final report, BRGM/RP-56457. www.brgm.fr. Accessed July 2016.

Petit, K., & Michon, J. (2015). *L'état des eaux de surface et des eaux souterraines*, Ed. Office national de l'eau et des milieux aquatiques (Onema), 12 p. http://www.eaufrance.fr/IMG/pdf/evaluation_2010-2013_201506.pdf

Rinaudo, J. D., Montginoul, M., & Desprats, J. F. (2015). The development of private borewells as independent water supplies: Challenges for water utilities in France and Australia. In *Understanding and managing urban water in transition* (pp. 155–174). Dordrecht, Netherlands: Springer.

Roques, C., Bour, O., Aquilina, L., & Dewandel, B. (2016). High-yielding aquifers in crystalline basement: Insights about the role of fault zones, exemplified by Armorican Massif, France. *Hydrogeology Journal, 24*(8), 2157–2170. https://doi.org/10.1007/s10040-016-1451-6

Rotzoll, K., & Fletcher, C. H. (2013). Assessment of groundwater inundation as a consequence of sea-level rise. *Nature Climate Change, 3*, 477–481. https://doi.org/10.1038/nclimate1725

Jean-Christophe Maréchal is an experienced hydrogeologist specialized in crystalline and karst aquifers, with a strong knowledge in hydro-dynamics and modeling applied to such complex aquifers. Since more than 15 years, he has been participating to the development of methodologies and tools specifically devoted to the survey, management, protection of these specific aquifers, through several research or operational projects, in France (Massif central, Mediterranean karsts), French overseas departments (French West Indies) and abroad (India, Africa). These methodologies are mainly based on a multidisciplinary approach (geology, geophysics, hydrodynamics, hydrochemistry, database, modeling, etc.), with an emphasis on a strong geo-logical component (lithology, weathering).

Josselin Rouillard is currently a recipient of a Marie-Skłodowska Individual Fellowship working at the French geological survey (Brgm) on quantitative groundwater management and agriculture. His work in the past 12 years has focused on the assessments of governance arrangements and economic instruments in water management, focusing on the EU with some experience in Asia, North Africa and South America. His research work draws on institutional economics, collective action theories and the policy sciences. He holds a PhD (Dundee/Edinburgh) in the management of environmental systems, a MSc. (Oxford) in environmental change and management, and BSc. (Reading) in environmental sciences of the earth and atmosphere.

Chapter 3
Groundwater Policy in France: From Private to Collective Management

Jean-Daniel Rinaudo

Abstract According to 1804 French Civil Code, groundwater is considered as a private property. However, after this resource started to be intensively exploited by industries in the 1850's, the State increasingly regulated its use. In 1935, a system of individual access and withdrawal rights, managed by the State, was established to protect deep confined aquifers which were showing signs of overexploitation. This system of use right was later on extended to unconfined shallow aquifers with the 1992 water law, mainly to protect the environment. A new management approach, based on individual volumetric entitlements, was then developed and tested in several French groundwater basins, subsequently obtaining a legal basis in the early 2000's. The 2006 water law constitutes a clear break in French water policy. The system of individual volumetric entitlements managed by the State was cancelled and users asked to form Water Users' Associations at the catchment level. Associations became the recipients of pooled water use entitlements, which they must share among their members using rules agreed collectively. Although this reform only applies to the agricultural sector, its represents a clear shift from a private to a common property regime.

Keywords Allocation policy · Private property · Common property · Users' associations · Water trading

3.1 Introduction

In France, as in other Latin countries, groundwater has long been regarded as *Res Nullius*, i.e. it has no master and is subject to private appropriation by those who own the land that overlies the resource. This principle was enshrined in law as of

J.-D. Rinaudo (✉)
BRGM, Montpellier University, Montpellier, France
e-mail: jd.rinaudo@brgm.fr

J.-D. Rinaudo et al. (eds.), *Sustainable Groundwater Management*,
Global Issues in Water Policy 24, https://doi.org/10.1007/978-3-030-32766-8_3

1804 (articles 552 and 641 of the civil code) and remains unchanged to this day. Groundwater use has developed in the context of this unrestrictive institutional framework (Gazzaniga & Larrouy-Castéra, 2010; Guttinger, 1992).

Until the mid nineteenth century, groundwater use remained limited to spring-water catchments or abstraction through drainage tunnels to supply large cities (e.g. Paris, Nancy). With advances in hydrogeology and drilling technologies, confined aquifers were developed progressively, particularly for industrial activities in the north of France and the Paris Basin. However, the proliferation of boreholes caused a rapid decrease in potentiometric levels. As a result, the State regulated access to deep aquifers with the publication of the first legislation, the 1935 Water Act.

From the 1950s onwards, the development of drinking-water supply systems led to a new wave of borehole construction that provided water of a better overall quality and reliability than streams. This development was facilitated by the rural code, with the aim of developing and improving living standards in the countryside.

Following the severe drought of 1976 and subsequent droughts in the late 1980s, agriculture joined the race to exploit groundwater resources. Thousands of boreholes were drilled on farms for the development of irrigation, primarily from easily accessible unconfined aquifers. The increase in agricultural abstraction sometimes had a significant impact on aquatic ecosystems, causing springs to dry up, the draining of wetlands, and the reduction of water levels in rivers and streams during low-flow periods. This urged the State to intervene once more with the 1992 Water Act, which laid the foundations for a quantitative management policy. In 2000, the European Water Framework Directive strengthened the protection requirements for aquatic systems, which led to the new 2006 Water Act.

This chapter provides a detailed chronological description of the changes. Above all, it describes the type of regulatory tools implemented, their limitations and their gradual improvement. We also analyse how the State has progressively restricted owners from exercising their rights in order to protect public interest, which was continually redefined over time. In addition, it shows how the State has gradually involved users in the process of groundwater management.

3.2 Protecting Deep Aquifers for the Public Good

3.2.1 Science Discovers How Groundwater Flows

The art of capturing springs and channelling water by gravity to large cities goes back to ancient times, as shown by the numerous aqueducts that date from the Roman era in France. However, a scientific approach to groundwater and how it flows through aquifers did not actually emerge until the nineteenth century, along with the development of drilling techniques to capture groundwater.

The first hydrogeological theoreticians and practitioners in France were engineers, including Arago, Belgrand, Dausse and Darcy, who were interested in the establishment of urban drinking water supply systems. These pioneers of hydrogeology observed and attempted to explain the variations in groundwater levels. The foundation of quantitative hydrogeology was the derivation of Darcy Law (1856), which Dupuit applied to the flow of water in aquifers 10 years later.

In parallel, the techniques for drilling at greater depths also progressed, partly due to advances in the understanding of geology and the structure of sedimentary basins. The first deep boreholes were drilled in Tours, Lille and then Paris where the Grenelle artesian borehole was sunk between 1833 and 1841. It was 548 m deep, with an artesian flow rate of 160 m^3 per hour. The number of deep boreholes in the major sedimentary basins soon increased, primarily for industrial supplies.

It was not until the beginning of the twentieth century that scientists began working on a national survey of groundwater resources (Margat, Pennequin, & Roux, 2013). In 1909, E. Imbeaux presented the first description of the "aquifers of France" to the French Geological Society, followed by a series of regional hydrogeological monographs in 1930. The first potentiometric maps were compiled for the Somme area in 1933. Lemoine, Humery and Soyer (1939) published a comprehensive inventory of the deep boreholes in the Paris Basin in 1939, which was a prelude to the databases that exist today. The first map of the groundwater resources of France only became available in 1964, with a detailed atlas following in 1970.

3.2.2 The First Regulation: The 1935 Water Act

Parallel to the advances in knowledge, the use of deep groundwater spread rapidly throughout France with abstraction from deep aquifers rising from 60 to over 300 million m^3/y in the first half of the twentieth century. The same phenomenon occurred in the French colonies of North Africa (Margat et al., 2013).

This resulted in a rapid drop in pressure in several deep aquifers, which triggered state intervention to regulate groundwater use. With the Water Act of 8 August 1935, the State introduced a procedure prohibiting the construction of wells and boreholes exceeding a depth of 80 m without prior authorization. This allowed the State to prevent further abstraction from overexploited aquifers. The permits granted to existing users were not subject to alteration, which amounts to recognising historical rights. In contrast, the permits were not transferable.

The aim of the regulation was obviously not to protect water-dependent ecosystems, but, instead, it set out to avoid depleting a resource that was clearly perceived as a strategic asset for the Paris region and to ensure that industrial use remained compatible with the necessity of supplying drinking water to a growing population in the future.

The 1935 Water Act therefore primarily focused on controlling abstraction in the industrial sector. Agricultural use was not affected, given that irrigation relying on groundwater was virtually non-existent until the early 1960s. The law also did not apply to boreholes used for supplying drinking water, which were authorised by a different procedure, namely the declaration of a public utility in accordance with article 113 of the former rural code. The supply of drinking water was already regarded as a high-priority use, which served the public good.

3.2.3 *Extending the Scope of the Water Act: 1935–1985*

For 50 years, the 1935 Water Act was the only regulatory tool available for managing groundwater abstraction. It initially targeted three counties in the Paris region, but was subsequently applied to seven other counties over a 50-year period, with slightly different procedures Fig. 3.1).

Gradual application of the law to other counties coincided with an endeavour to improve knowledge of the groundwater resources. In the late 1950s, the State developed a network for monitoring pressure levels in the deep aquifers of northern France, the Paris Basin, and the Aquitaine Basin. It also encouraged the development of mathematical models designed to assess the impact of pumping scenarios. In the late 1960s, the first mathematical models were developed for the most significant confined and alluvial aquifers. (Margat et al., 2013).

The effectiveness of policies implemented under the 1935 Water Act varies from region to region. In the Paris region, the measures applied made it possible to stabilize and later partially recover previously declining water levels in the Albian greensands aquifer (Fig. 3.2). In some counties, the legal provisions were not strictly applied because of insufficient resources allocated to law-enforcement activities. In other counties, abstraction continued to increase even though the number of boreholes was controlled (e.g. the Gironde aquifers). In the Nord Pas de Calais region, the legislation only had a limited impact on the Carboniferous limestone aquifer because no restrictive measures covered the Belgian section of this cross-border aquifer. In the Roussillon region, many deep boreholes were drilled for agricultural purposes without authorisation.

3.3 The Emergence of an Integrated Approach to Surface and Groundwater Management

3.3.1 *The Development of Groundwater Use in Agriculture*

The 1935 Water Act deliberately excluded abstraction for agricultural purposes from its provisions because, at the time, the volumes were very limited compared to other uses. However, since the 1960s there has been major growth in irrigation using shallow groundwater and, to a lesser extent, deep aquifers. Many boreholes were drilled in the cereal-growing plains in central and western France, where access to groundwater made it possible to increase yields and diversify production (Martin, 1972). Following the droughts of 1976, 1985–86 (Ollivier, 1989) and 1988–89

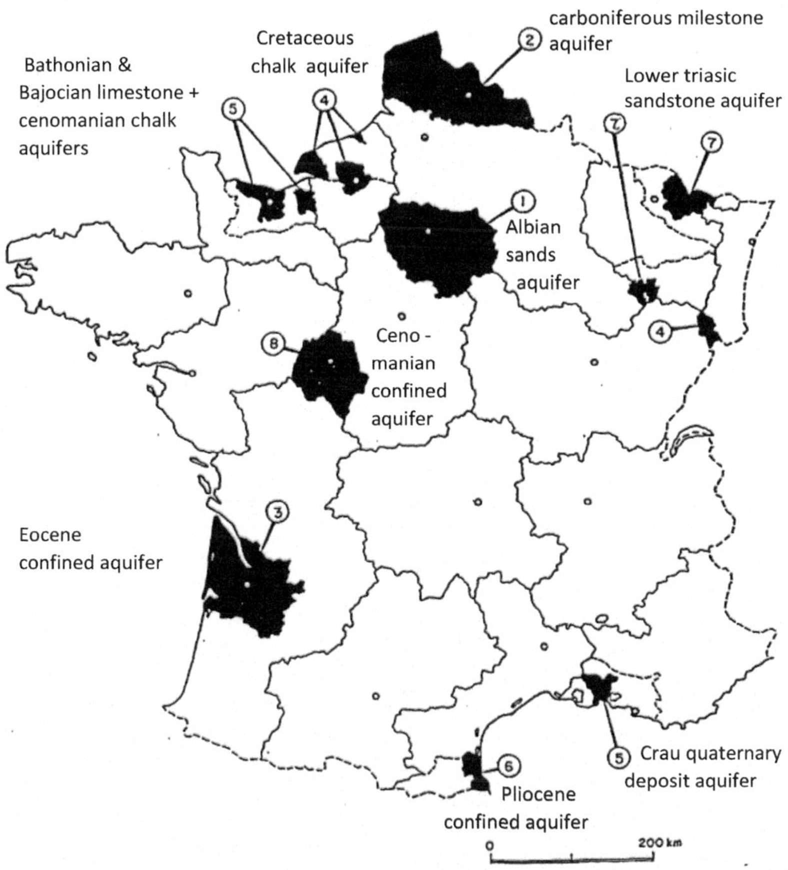

	Region or counties	Date	Authorisation required:
(1)	*Paris Region: 8 counties*	1935	If depth > 80m
(2)	*Nord Pas de Calais*	1958	If depth > 80m or if pumping rate < 250 m3/h and depth > 5m
(3)	*Gironde county*	1959	If depth > 60 m
(4)	*Seine Maritime & Territoire de Belfort*	1973	If depth > 80m or if pumping rate > 8m3/h and depth > 10m
(5)	*Bouches du Rhône & Calvados*	1973	If depth > 80m or if pumping rate > 8m3/h and depth > 2m
(6)	*Pyrénées Orientales*	1973	If depth > 80m or if pumping rate > 8m3/h and depth > 30m
(7)	*Moselle & Vosges*	1981	If depth > 80m or if pumping rate > 8m3/h and depth > 40m
(8)	*Indreet - Loire*	1985	If depth > 40 m

Fig. 3.1 French counties where the 1935 Water Act restricting groundwater use was applied between 1935 and 1985

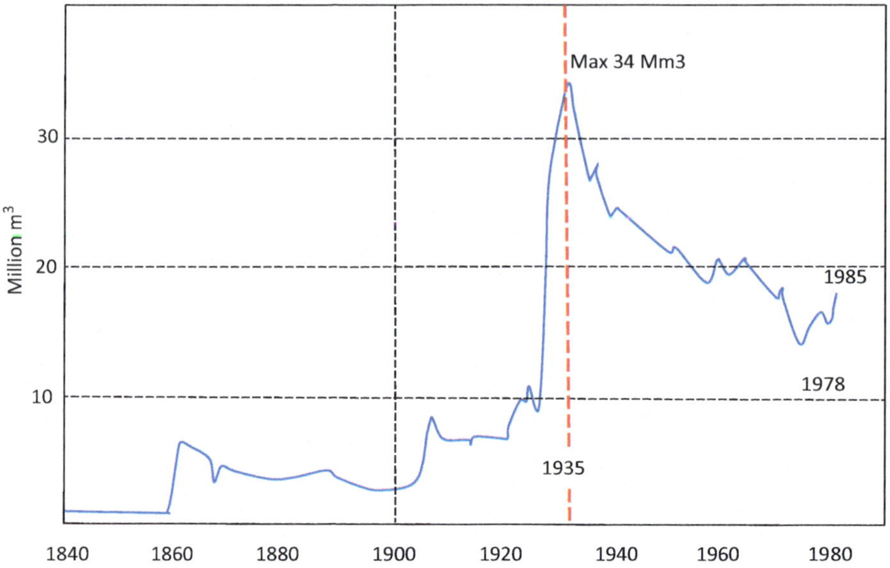

Fig. 3.2 Groundwater abstraction (m³/year) in the Albian greensand aquifer in the Paris region. (Adapted from Risler and Roux, 1993)

(Merillon & Chaperon, 1990), numerous farmers installed boreholes as a safeguard against the risk of drought. This development was encouraged by public funding through Common Agricultural Policy incentives until 1992 (see Chap. 24), as well as by food-processing stakeholders in some sectors (especially maize production). The phenomenon was particularly common in western and central France (Loubier et al., 2013).

As most agricultural boreholes were shallower than the limit imposed by the 1935 Water Act, the regulatory provision for them merely stipulated that boreholes must be declared under the Mining Code if their depth exceeds 10 m.[1] The only main constraint likely to limit the volume abstracted by each user was yield characteristics of the various aquifers.

In 1964, a new water law was passed, establishing water agencies (Barraqué et al., 2018) and several provisions related to water use. It required that water abstraction be declared if the borehole flow rate exceeds 8 m^3 per hour but this provision did not alter this virtually unlimited access to shallow aquifers. It also introduced a water abstraction tax but its level was too low to represent an economic constraint likely to limit groundwater use, which continued rising.

In some regions, the development of agricultural wells and boreholes was such that it lowered groundwater levels, which in turn caused some springs to dry up and drained watercourses and wetlands. The environmental impacts created tension with people who used watercourses for recreational purposes (e.g. fishermen) and

[1] Apart from several counties, where a preliminary authorisation is required beyond a certain depth (50–80 m) pursuant to the Water Act of 8 August 1935.

environmental-protection groups. This prompted the State to gradually strengthen the legal and regulatory framework.

The law of 1984[2] provided a preliminary response to the social demand, claiming that "*the preservation of aquatic environments and fish habitats was in the public interest*". In practice, this principle led to the establishment of environmental flow for watercourses. The administration was given the power to restrict the abstraction of surface water when environmental flow was not being maintained. The restrictions were in the form of water rotations, whereby the irrigators had to irrigate successively (by geographic sector) with a weekly time period that became shorter and shorter. The law only applied to surface-water abstraction. Thus it failed to resolve the conflicts in catchment areas where low river flows were declining due to the overexploitation of connected alluvial aquifers. The 1989 drought clearly revealed the weaknesses in the regulations (Merillon & Chaperon, 1990).

3.3.2 Water Becomes the Heritage of the Nation (1992)

The 1992 Water Act marked a major shift in the policy for water-resource management. While it did not fundamentally negate the right of landowners to abstract groundwater, it included several provisions that severely restricted the ability of landowners to exercise this right.

The first provision consisted in establishing a declaration/authorisation system[3] applicable to all water use (for volumes exceeding 8/80 m^3/h, respectively). The installation of a water-metering device became obligatory at each abstraction point and users were also required to keep a quarterly record of the volumes abstracted. In addition, the law introduced the possibility of lowering the authorisation thresholds to 8 m^3/h in areas considered to have extraction exceeding recharge. These were known as Water Restriction Areas (Zones de Répartitions des Eau or ZREs). In the ZREs, the State could also prohibit the construction of new boreholes. Extraction for the agricultural sector was now directly brought under State control.

The second provision of the water act underlined the uniqueness of the water resource, i.e. that groundwater is an integral part of the water cycle. The legal distinction between surface and groundwater resources, inherited from Roman law and formalised in the Civil Code, became less apparent with the emergence of a more holistic vision of how aquatic environments and water resources interact. Henceforth, the State could legitimately impose restrictions on groundwater use, which earlier had not been possible, and thus substantially weaken groundwater ownership rights. It is important to note that the legislative changes in France reflected the reforms that were being introduced in other European countries at the same time, especially in Spain with the 1985 legislation and Portugal with the 1994 legislation (Barraque, 2004).

---

[2] Law n° 84-512 of 29th June 1984 relating to freshwater fishing and the management of fish resources. Journal Officiel de la République Française, n°30 June 1984, France, p. 2039–2045.

[3] The 1935 Water Act was repealed and replaced by the 1992 Water Act.

The third provision imposed the development of water-management plans on a catchment scale, requiring the creation of Local Water Management Plans (Schémas d'Aménagement et de Gestion des Eaux or SAGEs). The plans are established in consultation with the stakeholders in order to reconcile the demands of different water users and to maintain the quality of the aquatic environments (see Chap. 4). For quantitative water-resource management, particularly groundwater, the plans had to restore the balance between use and the available resources by modifying or withdrawing certain abstraction permits where necessary. The law specifies that, when required, permits can be withdrawn or modified without State compensation, for guaranteeing drinking-water supplies or for protecting aquatic ecosystems.

In this way, quantitative management planning was introduced in several catchment areas where groundwater was a major component. The approach involved defining the groundwater-level warning indicators, which, when exceeded, triggered measures to restrict water use (to the possible extent of a total ban). Therefore, it extended the scope of the surface-water provisions of the 1984 act to include groundwater. Indicators of this type were established in many regions to manage both confined and unconfined aquifers. This is illustrated by the case of the Beauce aquifer (Chap. 5), the aquifers in the south of the Vendée with respect to the Poitevin marshes (see Chap. 18), the Rochefoucauld karst in Poitou Charentes and the Albian aquifer (Seguin et al., 2009). Figure 3.3 shows the location of ZREs in 2017.

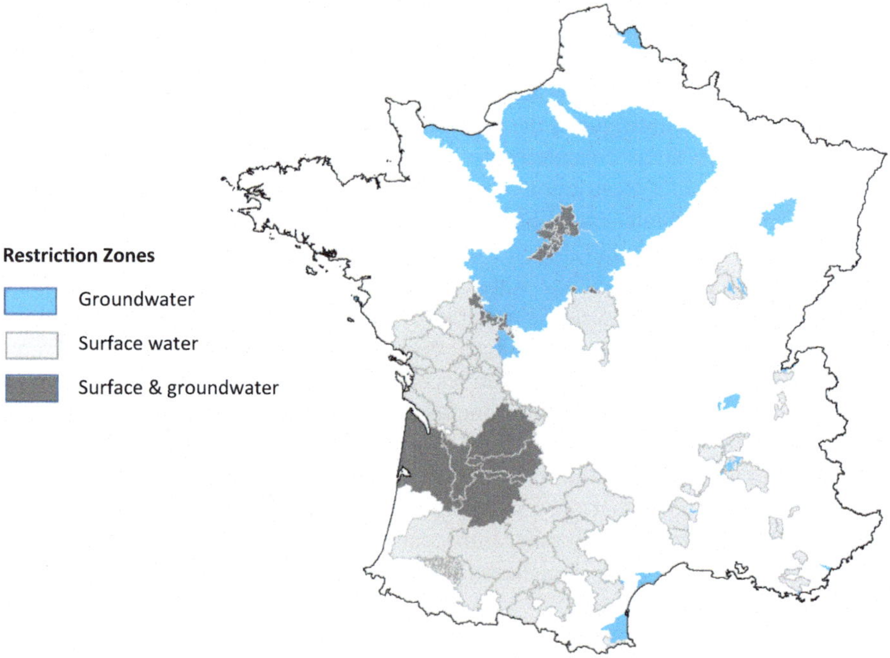

Fig. 3.3 Surface- and groundwater use Restriction Zones (ZRE) in 2017

3.3.3 The Problems of Implementation

Applying the provisions set out in the 1992 Act to groundwater was not an easy task (Compte et al., 1995). The State lacked the means to enforce irrigation restriction measures, to check the installation and accuracy of meters, to monitor the abstraction registers as laid down by law, or even to control the pumping capacities (see Chap. 23). A 1996 Parliamentary report stated: "*It is clear that the texts relating to groundwater policing are not being applied, far from it, and implementing them would require staff that the administration does not seem to have*" (Martin, 1996). The requirement for meters was not well received by the agricultural community and consequently they were installed much later than the deadlines set out by the Act, particularly in the Adour Garonne Basin. A large number of abstraction boreholes was not declared. This situation persisted locally until the mid-2000s (Brun, 2003; Montginoul & Rinaudo, 2010). The administration gained additional leverage following the 1999 Common Agricultural Policy reform that introduced the principle of cross-compliance, which made direct payments to farmers dependent on their compliance to install water meters (Chap. 24).

3.4 Towards Volumetric Management

3.4.1 The Emergence of Volumetric Management

The limitations of management based on water-level indicators soon became apparent when two adverse impacts were observed. First, when groundwater level falls and approaches the warning threshold, farmers increase their irrigation to build up soil moisture reserves, resulting in further reductions in groundwater levels and a waste of the resource. Second, to overcome the effect of restrictions on the time period for irrigation, farmers invest to improve their irrigation capacity so they could irrigate all their land in a shorter time, resulting in no reduction in the quantity of water abstracted. Both responses accelerate the rate of abstraction, hasten the onset of a crisis during the irrigation season and increase the frequency of such crises. Therefore, the risk of a water shortage (and the associated agronomic and economic implications) remains high and difficult for farmers to predict and plan for.

In the late 1990s, this failure led to several managers adopting a new approach: volumetric management. This involves capping the total volume that can be abstracted from groundwater resources that are deemed to be overexploited, and dividing it between the users in the form of individual abstraction quotas. As each farmer has access to a set volume for the whole irrigation season, they were encouraged to manage their water use efficiently to maximise their economic return.

In most cases, the total abstraction limit was initially established on the basis of resource use at the time of the reform ("grandfathering"). The objective was only to prevent any new increases in abstraction that might have adverse impacts on the resource and not to align use with recharge. In the case of the Beauce aquifer, the quotas negotiated by the agricultural water users in 1998 corresponded to the maximum volume used over the last decade (Chap. 5; Bouarfa et al., 2011).

The advantage of this volumetric approach is that it reduces uncertainty for the irrigators, while giving them greater responsibility for managing their annual volume. In addition, the system may allow the irrigator to carry forward the unused volume from one year to the next. This encourages farmers to improve the technical and economic efficiency of their water use. This inter-annual transfer is only possible if the aquifer is robust (if the ratio of available water storage over annual recharge is high). Such volumetric management was introduced towards the end of the 2000s in several catchment areas and counties.[4]

3.4.2 Individual Appropriation of the Resource

The introduction of volumetric management had a paradoxical effect on groundwater ownership. Initially, it weakened the landowners' right to freely use the groundwater located beneath their property because it capped the volume used. However, by limiting the total usable volume, the new management system created a situation of shortage, which increased the value of the water quota.

Farmers did everything within their power to ensure that the individual volume allocated to them was tied to their farm as a way of enhancing the value of their landholding. They managed to persuade the State agencies to transfer the quotas from the previous to the new owner when farms were sold, instead of reallocating the water to the newcomers registered on a waiting list. This transfer occurred not only in the case of family successions, but also in the case of a sale of the whole farm (land, machinery, buildings) to a third party. From then on, the income associated with quotas was built into the value of the land. Although a farmer who transfers his volume (or quota) cannot officially receive financial compensation from the new beneficiary for transferring the authorised volume, he actually recovers its value by selling his farm at a high price. By tolerating this practice, the administration contributed to a situation where the first groundwater users who received free permits, ended up appropriating the economic rent associated with groundwater.

Various discussions in the 1990s show that the idea of private appropriation of water resources was already very clear in people's minds. Several French economists debated the possibility of formalizing the market for exchanging individual water quotas through a proposal that involved decoupling the rights to use a resource from land ownership, thus creating a market where they could be traded freely (Kosciusko-Morizet et al., 1998; Strosser & Montginoul, 2001). At about the same

[4]The main examples are the Beauce aquifer, the Yèvre-Auron Basin (Cher county), the alluvial aquifers in the plains in Garonne-Tarn-Aveyron and Ariège, the Sud-Vendée aquifers, Vienne county, Charente county, and in the catchments located in the counties of Aisne, Aube, Somme and Sarthe.

time, experiments were being conducted on "water banks" in western U.S. states following the 1989–91 drought, and Australia adopted the proposal in its draft water reform policy of 1994 (see Chap. 21). This proposal continued to spark debate in France for a decade (Petit, 2004; Rinaudo, 2014), but was never implemented, the State considering that "*the transfer of this type of system [from Anglo-Saxon countries to France] is not desirable in France because it is contrary to our concept of water, which goes beyond that of a purely economic good*" (Conseil Economique et Social, 1991).

3.5 Towards Collective Management

In March 2000, the European Water Framework Directive (WFD) came into force. It obliged member states to take all the necessary action to ensure that water resources and associated aquatic ecosystems were restored to a satisfactory qualitative and quantitative state within 15 years. This imperative to achieve results put pressure on the French State to review its legislation. First of all, the directive was converted into French law (the law of 21 April 2004), and then in 2006, a new law for water and aquatic environments was passed that aimed at providing France with the necessary tools for achieving the satisfactory state in 2015, as set out by the WFD.

3.5.1 The 2006 Water Act

The 2006 Water Act[5] introduced a major change with regard to water distribution among users and enhanced the rationale of volumetric management.

One of the law's main provisions concerning quantitative management was the obligation to restore a balance between abstraction and the available resources for all catchment areas (surface- or groundwater) considered to be overdeveloped. The administration was responsible for assessing the maximum abstraction volume, which ensured that the environmental goals were achieved in at least 4 years out of 5.[6] The authorised abstraction should not exceed this maximum volume. The studies on the maximum volumes to be abstracted generally concluded that it was necessary to reduce abstraction by 10–20% in most catchments and by over 50% in some cases (Chap. 11). The reductions were to be achieved with no financial compensation.

The law also redefined how the maximum volumes to be abstracted should be shared amongst the existing users. Drinking-water supplies were considered a prior-

[5] Law n° 2006-1772 of 30th December 2006 on water and aquatic environments, Journal Officiel de la République Française, n° 303 of 31 December 2006, France, Text n° 3/175.

[6] The volume to be abstracted is defined in the circular of 03/08/2010 and available online at: http://circulaire.legifrance.gouv.fr/pdf/2010/08/cir_31709.pdf

ity and operators received a volume corresponding to a technically efficient supply for the demand. The remaining volume was shared between the industrial sector (which often had few users) and the agricultural sector. Given the large number of farmers, the State allocated the responsibility of sharing the volume designated for farming to an intermediary institution named Organisme Unique de Gestion Collective or OIUGC (Unique Collective Management Organisation). This OUGC is an agricultural water user association that was made obligatory in the 2006 Water Act (Figureau et al., 2012; Lafitte et al., 2008). From a regulatory point of view, the State cancelled all individual water use permits previously granted to each farmer and replaced them with a single aggregate permit that was attributed to the OUGC.

The OUGC was made responsible for preparing a plan to allocate the total volume between the users. This involved devising its own set of management rules, which had to be validated by the administration. For example, the allocation rule had to include admission procedures for newcomers, the priority rules in the event of drought, etc. This provided the basis for the annual allocation scheme, which is approved by the State each year. The rationale underlying this transfer of responsibility is that the users, brought together in the OUGC, are the best placed when it comes to adjusting water sharing among farmers and accounting for the local technical and economic circumstances.

3.5.2 Gradual and Differentiated Implementation

In practice, the "quantitative management" component of the 2006 water law met with considerable opposition on the part of the farming profession, particularly in south-western France (Hébert et al., 2012). Some farmers perceived the pooled allocation as a form of expropriation and were quick to make the parallel with agricultural collectivization in Russia in the early 1920s. The feeling was heightened by the fact that, almost invariably, they had invested individually to access the resource.

The sustainable extraction limits (specifying the maximum volumes to be abstracted) were also challenged because of doubts surrounding how they were estimated. The values that were ultimately agreed upon were negotiated on economic and political, rather than scientific, grounds (see Chap. 11). In some basins, the negotiation also allowed the agricultural users to obtain public subsidies to finance the construction of reservoirs for storing surface water or groundwater abstracted in winter when it is more abundant, for later use in summer. The allocation of public subsidies represented an implicit compensation for the reduction in the volumes to be abstracted.

At the end of 2016, about 30 OUGCs had been established and they started applying their allocation rules for the first time in 2017. A preliminary analysis of the new rules highlights the diversity of the different approaches adopted by the OUGCs. The differences concern the identification of beneficiaries—the farmers who can legitimately benefit from access to the water resource—and the criteria used to determine the share of the volume that each beneficiary can claim.

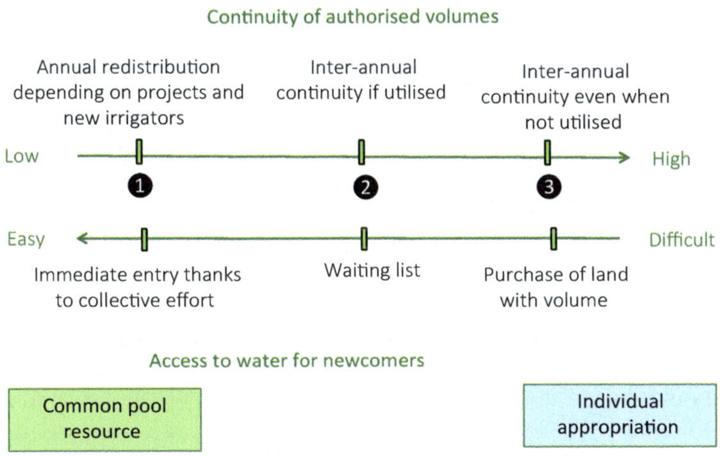

Fig. 3.4 Different approaches for choosing the beneficiaries that can use the volume allocated

The following three examples demonstrate how beneficiaries are identified (Fig. 3.4). In some OUGCs, represented by case ❶ in the figure, stakeholders consider that the water resource is a common good to which every farmer should have access to if he needs it. Therefore, it cannot be subject to individual appropriation or even tacit entitlements. The number of farmers that share the authorised volume is likely to vary each year. Consequently, the volume allocated to each farmer varies because everyone has to limit their share in order to meet the demands of new users. The advantage of this system is that it allows young farmers to establish a new farming enterprise and enables farm strategies to be adjusted to market fluctuations. The disadvantage is that it generates uncertainty in terms of the volume that each farmer can expect to obtain, which is incompatible with perennial crop production or new investment in irrigated crop production. This flexible approach is currently applied in the Aisne county where farmers are very reactive to market changes and water still relatively abundant.[7]

Case ❸ represents a situation that is diametrically opposite, and in which the volumes attributed to each individual are automatically renewed each year, regardless of whether or not the volume has actually been used. The underlying logic is that rights are acquired on the basis of previous use. The volume attributed to the farmer actually becomes tied to the land, thereby increasing its value. A newcomer who wants access to a volume of water must buy land with an individual volume attached, at a price that is two or three times higher than for the same land without a water entitlement. In this type of situation, water is implicitly privatised. This management method provides considerable security to the holders of a historic water right, which means they can optimise their investments (equipment, planting). Its main disadvantage is that it makes it harder for young farmers to establish new

[7] A reader not familiar with the French context should keep in mind that rainfed agriculture is possible over the entire French territory. An interannual variation of water allocation therefore does not systematically threaten the economic viability of farms.

irrigation (due to the cost of accessing water), and does not necessarily encourage efficient water use. In fact, the volume allocated to a farm based on historical use may be completely different to irrigation practices 10 years later. In an extreme case, significant unused volumes of water are needlessly frozen and cannot be reallocated to and used by other farmers. This approach to groundwater management currently applies in the county of Tarn-et-Garonne.

Example ❷ represents an intermediate approach. The volumes attributed are automatically renewed within the limits of the volume abstracted the previous year. Thus a farmer whose usage does not reach the full allowance that was requested in 1 year, will see his allowance decrease to the same volume of usage in the following year.[8] The advantage of this management method is that it allows dormant volumes to be reallocated to newcomers on the waiting list, while guaranteeing a degree of continuity in terms of the volume allocated to users. However, it fails to encourage farmers to improve the efficiency of their irrigation practices because, if they save water, their volume is likely to be reduced. This type of situation can be found in the Clain Basin in the county of Vienne.

The criteria chosen to determine the volume attributed to each beneficiary also vary from one OUGC to another. Diagrammatically, the existing approaches can be spread along an axis with two opposite poles: distribution based on an analysis of the agronomic requirements on one pole and distribution based on the farm's historic activity on the other. The first approach (on the left of Fig. 3.5) involves distributing the volume in relation to the beneficiaries' real needs. Water is allocated proportionally to the theoretical requirements of the crops they plan to grow. The agronomic calculation of water requirements may account for the soil characteristics, which means that the volume per hectare can be increased for soils with a low water holding capacity, for example (case ❶). The allocation rule is seen as a tool that compensates for natural inequalities and provides a level playing field for beneficiaries. It should be noted that the calculation is based on the assumption that

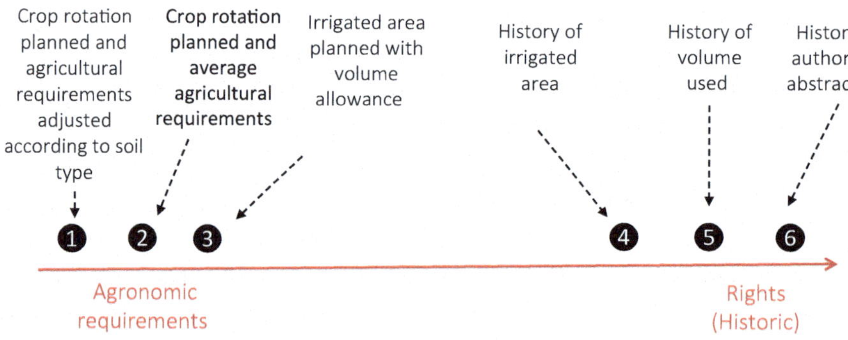

Fig. 3.5 Criteria for allocating the volume to beneficiaries

[8]This rationale is also applied in the western U.S. where the holder of a water right must use it ("beneficial use") to avoid losing it ("*use it or lose it*"). Hanak & Stryjewski, 2012).

irrigation practices are efficient, and therefore it penalises inefficient farmers and encourages water saving. A simplified version of this approach does not take account of soil differences and bases the calculation for distribution on the average agronomic water requirements per crop type (case ❷). A further simplification ignores crop differences and allocates water as a function of the number of hectares to be irrigated in the following season (case ❸).

In the second approach (on the right of Fig. 3.5), the volume is distributed on the basis of the irrigation history of each farm. This approach implicitly acknowledges that historical use creates a right. Here, the primary method used involves allocating volumes as a function of the area of land irrigated historically, irrespective of the types of crops (case ❹). The volume can also be allocated on the basis of former water abstraction (case ❺). Lastly, the volume allocated may be proportional to the volume or the authorised abstraction at the time when the well or borehole was installed, regardless of the actual volume of water used more recently (case ❻). These variations can be found in the various management policies implemented in different French OUGCs (Figs. 3.5).

The previous paragraphs and figures illustrate the range of the strategies adopted by French stakeholders to share the maximum volume that can be abstracted in overdeveloped catchments. The diverse range of approaches is in response to the large variation in the hydrogeological and agricultural situations, the level of pressure on the resource, and also the history of water management in each catchment. Thus when individual volumes have been allocated to users for several years with automatic annual renewal, the notion of rights becomes established in people's minds, leading to the individual (albeit implicit) appropriation of water. In these catchment areas, farmers are likely to perceive any reduction in their volume as an expropriation. The situation is very different in catchments where volumetric management is a fairly recent phenomenon and where demand on resource access is moderate. A wide range of scenarios can be considered in these areas, which gives the OUGCs a broad scope for developing a set of rules for distribution that is acceptable to the vast majority of users.

3.6 Discussion

3.6.1 Refusing Individual Appropriation

By describing the changes in the water management policy and its practical application, this chapter has revealed the permanent tension that exists between private property and the common heritage. This tension is reflected by several contradictions between the theoretical approach as set out by law, and its practical application.

The first contradiction is of a legal nature. According to the Civil Code, groundwater is still considered as a private resource linked to the land. Yet in practice, the regulations established between 1935 and the present day have systematically sought to reduce the exercise of this property right to protect the public interest. In the 1992 and 2006 Acts, water was attributed the status of a common heritage of the nation, without revoking the article in the Civil Code.

The second contradiction is linked to the status of the abstraction permits granted by the State. The permits have been temporary and annually renewable since 1992 and, theoretically, can be reduced without compensation. Yet when the permits had to be reduced in overdeveloped catchments, the State indirectly (partially) compensated for the losses incurred by the users by subsidising the provision of alternative water resources. This primarily involved the development of small dams capable of storing excess winter surface water, or groundwater pumped from the aquifer when it "overflows" in winter. The water agencies subsidised investments by up to 40–70%. This funding was granted subject to three conditions: (1) the irrigated area should not increase; (2) water savings were made by improving the technical efficiency of irrigation; and (3) a regional development project was drafted, proposing a shift in agriculture towards a less water-dependent model. The Poitevin marshes provide a perfect illustration of this situation, which is discussed in Chap. 18. The notion of regional project development is also discussed in Chap. 24.

The third contradiction is linked to the fact that the abstraction permits are non-transferable (for a more detailed discussion see Hérivaux et al., 2019. In the late 1990s, the allocation of individual quotas and the increasing scarcity of water created the necessary conditions for the emergence of a water market by the end of the 2000s. However, national legislators systematically refused to formalise it. This resulted in a land-based water market, in which water abstraction permits were transferred when land changed hands and the price of the land reflected the value of the associated abstraction permits. The desire to avoid using a market mechanism for groundwater users was clearly stated in the 2006 Water Act, which promoted collective management for the volumes to be abstracted, at least in the agricultural sector. Collective management represented a mechanism with the potential to introduce flexibility into resource allocation. Equity remained a goal, but efficiency was sidelined. Therefore, there was a clear move towards negotiated and subsidiary policies, based on the strengthening of the role of the user communities.

The tension between appropriation of water-use rights and defending the public good has now eased with the establishment of collective water resource management mechanisms, based on the creation of intermediary institutions and the notion of common property.

3.7 Future Challenges

For years, groundwater was invisible and regarded as abundant, cheap and freely available. Now, it is a resource that must be shared amongst competing users, including the environment. In France, this situation has sparked conflicts that are likely to worsen in the future, especially when the impacts of climate change become apparent. Rising temperatures and increased evapotranspiration will lead to an increase in water requirements, especially for agriculture. Aquifer storage volumes could also diminish as abstraction increases to replenish surface water. The frequency of droughts will increase in the north and south of the country, even though the northern zone may also have higher winter rainfall. Research projections suggest that a decrease of 10–25% in recharge is to be expected. Two zones, the Loire Basin and south-west France, will be more severely affected (see Chap. 2).

To deal with these future challenges, there is a consensus among French stakeholders on the need to adapt by striving to optimise consumption in all sectors, improving the management of the available resources and seizing all opportunities offered by technological progress. The simplest, most immediate and least expensive response is to avoid wastage of water, particularly by reducing leaks in the drinking water distribution system and raising consumer awareness about how to save water. The example of the Gironde aquifer (presented in Chap. 12), illustrates the enormous potential of this preliminary adaptation. Agriculture must also change to become less water dependent and therefore, less vulnerable to the risk of drought. In France, alternative strategies based on agroecology and agroforestry have attracted particular interest. Varietal selection of plants with drought resistance is one such approach. Reorienting some sectors is probably inevitable, for example, substituting sunflower crops with maize which has a lower water use. However, this strategy has limitations. Sorghum will not replace high value-added fruit and vegetable crops which have a higher water use but create greater employment in the production chain.

A second option for adaptation involves managing the state of the catchment surfaces. Drainage netwoks, the sealing of open spaces (e.g. car parks) and hedge removal have accelerated water runoff and reduced infiltration into the soil and aquifers. Regional development should be reviewed so that water is held and infiltrates where it falls. Aquifers can be used as natural reservoirs, which could be artificially recharged when surface water is abundant. When these aquifers are full, excess water can be stored in surface reservoirs. The State is actively encouraging this course of action, but the limitations are also being examined. To what extent can groundwater extractions in winter be used to fill surface reservoirs without impacting on the groundwater resources available in the following summer?

Finally, the pressure on groundwater resources could always be eased by the development of seawater desalination plants. As yet, there are none on French soil, which suggests that the crises experienced in Australia and western U.S. states are still only a remote prospect for France.

Acknowlegdments I warmly thank Steve Barnett, for his very detailed and useful comments on a previous version of this chapter. The contribution from Rebecca Nelson and Josselin Rouillard is also acknowledged. The research presented in this chapter has been supported by Brgm (OPTIQUANT project).

References

Barraque, B. (2004). Groundwater management in France: From private to common property? Managing common pool groundwater resources, an international perspective. M. Brentwood, & S. Robar, Praeger: 85–96.

Barraqué, B. O., Laigneau, P., & Formiga-Johnsson, R. M. (2018). The rise and fall of the french agences de l'eau: From german-type subsidiarität to state control. *Water Economics and Policy, 4*(03), 1850013.

Bouarfa, S., Brunel, L., Bouarfa, S., Brunel, L., Granier, J., Mailhol, J. C., et al. (2011). Evaluation en partenariat des stratégies d'irrigation en cas de restriction des prélèvements dans la nappe de Beauce (France). *Cahiers agricultures, 20*(1), 124–129.

Brun, A. (2003). Aménagement et gestion des eaux en France: l'échec de la politique de l'eau face aux intérêts du monde agricole. *VertigO-la revue électronique en sciences de l'environnement, 4*(3). Available at: https://journals.openedition.org/vertigo/3779

Compte, J., Retkowsky, Y., et al. (1995). *Les SDAGE, les SAGE et l'eau souterraine*. Paris, France: Conseil Général des Mines.

Conseil Economique et Social. (1991). L'eau, gestion des ressources et protection de la qualité. Rapport présenté par M. Philippe Mangin, séances des 12 et 13 mars 1991, Paris, Direction des Journaux Officiels, brochure n° 4166.

Figureau, A. G., Montginoul, M., et al. (2012). Gestion quantitative de l'eau d'irrigation en France: Bilan de l'application de la loi sur l'eau et les milieux aquatiques de 2006. Orléans, BRGM. Report BRGM/RP-61626-FR: 50.

Gazzaniga, J.-L., & Larrouy-Castéra, X. (2010). Droit de l'Eau en France entre Permanences et Mutations, Le. C. de D. 51: 899.

Guttinger, P. (1992). Le statut juridique des eaux souterraines. *Economie Rurale, 208–209*, 66–69.

Hanak, E., & Stryjewski, E. (2012). California's water market, by the numbers: Update 2012. P. P. I. o. California. San Francisco.

Hébert, N., Grandmougin, B., et al. (2012). Réforme des autorisations de prélèvement dans le bassin Adour-Garonne: impacts sur l'économie agricole. *Agronomie, Environnement et Sociétés, 2*(2), 113–126.

Hérivaux, C., Rinaudo, J. D., & Montginoul, M. (2019). Exploring the potential of groundwater Markets in Agriculture: Results of a participatory evaluation in five French case studies. *Water Economics and Policy*, 1950009.

Kosciusko-Morizet, N., & Lamotte, H. et al. (1998). Que peut on attendre de la mise en place de quotas individuels échangeables de prélèvement sur la ressource en eau en France? L'exemple de l'agriculture irriguée. Direction de la Prévision: 1 à 29.

Lafitte, J.-J., Devos, P., et al. (2008). Les organismes uniques d'irrigation. Paris, Ministère de l'Ecologie, de l'Energie, du Développement Durable et de l'Aménagement du Territoire, Ministère de l'agriculture et de la pêche.

Lemoine, P., Humery, R., & Soyer, R. (1939). *Les forages profonds du bassin de Paris*. Paris, France: La nappe artésienne des sables verts. Mémoires du Muséuml. 700 p.

Loubier, S., Campardon, M., et al. (2013). L'irrigation diminue-t-elle en France? Premiers enseignements du recensement agricole de 2010. *Sciences, Eaux et Territoires, 11*, 12–19.

Margat, J., Pennequin, D., et Roux, J.-C. (2013). *Histoire de l'hydrogéologie française* (Assoc. Internat. des Hydrogéologues (AIH) and Comité Français d'Hydrogéologie (CFH), Eds.). Paris.

Martin, J. (1972). L'évolution de l'irrigation en France. *Revue de géographie alpine, 60*(3), 419–443.

Martin, Y. (1996). *Rapport sur la gestion durable des eaux souterraines*. Paris, France: Conseil Général des Mines.

Merillon, Y., & Chaperon, P. (1990). La sécheresse de 1989. *La Houille Blanche*, n°*5*, 325– 340

Montginoul, M., & Rinaudo, J. (2010). Quels instruments pour gérer les prélèvements en eau souterraine? Le cas du Roussillon. *Economie Rurale, 310*, 40–56.

Ollivier, J. Y. (1989). Conséquence des sécheresses de 1985 et 1986 dans le sud-ouest de la France: élaboration d'un programme de lutte. *La Houille Blanche*, n°. *7–8*, 530–533
Petit, O. (2004). La nouvelle économie des ressources et les marchés de l'eau: une perspective idéologique? *VertigO-la revue électronique en sciences de l'environnement, 5*(2). Available at : https://journals.openedition.org/vertigo/3608
Rinaudo, J.-D. (coord). (2014). *Water market scenarios for southern Europe: New solutions for coping with increasing water scarcity and drought risk?* Final project report. Available at https://sites.google.com/site/capandtraderesearch/final-report, IWRM Net
Risler, J. J, & Roux, J. C. (1993). Enquête sur l'application du décret-loi du 8 août 1935 sur la protection des eaux souterraines. Rapport R 37914. BRGM: Orléans.
Seguin, J.-J., & Gomez, E., et al. (2009). Les indicateurs piézométriques: un outil dans la gestion des hydrosystèmes. Orléans, BRGM/RP-58139-FR.
Strosser, P., & Montginoul, M. (2001). Vers des marchés de l'eau en France? Quelques éléments de réflexion. *Annales des Mines, Responsabilité et Environnement, 23*, 13–31.

Dr. Jean-Daniel Rinaudo is researcher at Brgm, Montpellier University, where he coordinates the scientific program on environmental and risk economics Initially trained as an agricultural engineer (Montpellier SupAgro 1994), he specialized in agricultural and resource economics (PhD University of Auvergne, 2000). Prior to joining Brgm, he worked for the International Water Management Institute in Pakistan where his research focused on the political economy of irrigation management reforms. His current research mainly focuses on the institutional economic dimension of groundwater management. Most of his research is conducted in France but he also works in Morocco and Chile. He is currently developing new research activities in the field of natural disasters economics, focusing on the methods to assess economic vulnerability and resilience. Dr. Jean Daniel Rinaudo is also member of the Scientific Council of the Adour Garonne River basin agency.

Chapter 4
Groundwater Management Planning at the River Basin District Level: Comparative Analysis of the Adour-Garonne and Loire-Bretagne River Basins

Jean-Daniel Rinaudo, Pierre Marchet, and Pascal Billault

Abstract In France, water resource management issues are addressed in the framework of blueprints which set out the strategic master plans for managing river basins (SDAGE). Master plans are then refined at local level (SAGE) in Local Plans jointly developed by stakeholders. This chapter describes how strategic Master Plans are formulated and implemented. It focuses on quantitative groundwater management issues and the legal and regulatory framework which defines planning objectives and practices. In addition, the chapter provides an historical analysis of 20 years of groundwater planning in the Adour-Garonne and Loire-Bretagne river basin districts, based on two of the authors' personnal experience.

Keywords River basin · Plan · Water agency

4.1 Introduction

Since the beginning of the 1990's, French policy makers have increasingly recognized the need to better integrate the different aspects of groundwater and surface water management, and the protection of aquatic habitat and ecosystems (Piegeay, Dupont, & Faby, 2002). This integration has progressively been achieved through

J.-D. Rinaudo (✉)
BRGM, Montpellier University, Montpellier, France
e-mail: jd.rinaudo@brgm.fr

P. Marchet
Agence de l'Eau Adour Garonne, Toulouse, France
e-mail: pierre.marchet@eau-adour-garonne.fr

P. Billault
Agence de l'Eau Loire Bretagne, Orléans, France
e-mail: pascal.billault@eau-loire-bretagne.fr

J.-D. Rinaudo et al. (eds.), *Sustainable Groundwater Management*,
Global Issues in Water Policy 24, https://doi.org/10.1007/978-3-030-32766-8_4

Fig. 4.1 Regions covered by the six water agencies in metropolitan France

the development of blueprints, called *Schéma Directeurs de Gestion et d'Aménagement des Eaux or* SDAGE (Water Development and Management Master Plans). Established for each of the six major river basin districts (see Fig. 4.1), these blueprints outline how to implement the national legal and regulatory framework in operational terms. Each SDAGE takes into account all surface water environments (watercourses, canals, bodies of water and so-called transitional coastal and brackish waters) and groundwater (confined and unconfined aquifers). It deals with the problems relating to quantitative management, pollution, the ecological quality of aquatic habitats, as well as issues of flooding. It also tackles broader issues, such as the governance, organisation and dissemination of data relating to water.

This chapter describes how the SDAGEs have taken groundwater into account, by examining the case study of the Adour-Garonne and Loire-Bretagne river basin districts. The first section examines water management planning in France in a historical context, while the second section presents the main characteristics of the Adour-Garonne and Loire-Bretagne river basins and focuses on the mechanisms involved in quantitative groundwater management. The water management plans are studied in Sects. 4.3 and 4.4. The conclusion compares the two examples and proposes several recommendations drawn from over 20 years of experience of two the authors in the two river basins studied.

4.2 The French Approach to Water Management Planning

4.2.1 The Creation of River Basin Agencies (1964)

Water management planning on the scale of the major river basins was introduced in France by the 1964 Water Act, which was reinforced by the 1992 law. Inspired by the model implemented in the Ruhr in Germany (Barraqué, Laigneau, & Formiga-Johnsson, 2018), the 1964 law created six large river basin agencies (Fig. 4.1). Their mission was to promote the joint and cohesive management of water resources, based on three main principles: (i) an integrated approach to all issues of water management (irrigation, sanitation, drinking water, aquatic environments and flooding); (ii) a mutualisation of financial resources at the river basin level to protect or restore water resources; and (iii) the participation of representatives of water users in management decisions. Unlike similar institutions established in Germany, Holland or Spain, the water agencies are not directly involved in the creation or operational management of infrastructure.

The 1964 water law authorised the river basin agencies to collect a fee from water users. The fee is proportional to the quantity of water abstracted and/or the quantity of pollution discharged and all the revenue raised is used to support projects (in the form of a grant or an interest-free loan), which aim to improve the state of water resources. The projects may be run by public or private stakeholders. Once established, the agencies operate as "mutual savings banks" for water users (Barraqué et al., 2018). The budget allocation for each agency is based on a five year plan drawn up by the river basin committee which defines the priorities for action and the corresponding budget allocations. The committee is composed of equal numbers from three groups; water user representatives, mayors and local councillors and officials from the state agencies. The fees are set by the agencies' board of directors where the state has the majority, and then approved by the river basin committee[1]. Today, 85% of the revenue comes from fees paid by domestic users. The overall budget managed by the six agencies amounts to 1.8 billion euros per year (Roche, Guerber, Nicol, & Simoni, 2016). The amount allocated to the quantitative management of surface and groundwater resources only represents 6.5% of the budget (Table 4.1).

4.2.2 The Introduction of Management Plans (1992)

The 1992 Water Act consolidated the role of the river basin agencies by significantly strengthening their powers with regard to water management planning and making them responsible for preparing the Water Development and Management Master

[1]As the fees were considered to be tax levies, parliament has directed the agencies' budget and capped their expenditure since 2006.

Table 4.1 Breakdown of subsidies granted by the six water agencies

Water management issue	Budget share %
Urban domestic wastewater treatment	55.8
Drinking water supply	8.4
Policy implementation	6.8
Industrial pollution control	5.7
Restoring the quality of aquatic environments	10.3
Agricultural pollution control	6.5
Quantitative water resource management	6.5

Source: http://www.lesagencesdeleau.fr/en/les-agences-de-leau/les-leviers-daction-des-agences-de-leau/ (consulted 10/10/2018)

Plans. The aim of these blueprints was to build a joint action framework for all the water stakeholders in the river basin in order to restore and maintain the good status of the water resources and aquatic habitats (which are considered to be a common heritage), with a view to providing a long term guarantee for all uses. In addition to the financial planning mission assigned to the agencies by the 1964 law, they were also given a technical planning mission. The SDAGE converts the sectorial public policy guidelines and the European directives into coherent, operational and localised action, by concentrating resources on priority objectives. The first SDAGEs were approved in 1996 (Buller, 1996).

The SDAGEs have a legal standing similar to urban planning documents. They are legally binding for the administration, which means that all decisions taken by public administrations must comply with the provisions set out in the SDAGE. Thus, they direct the action of numerous other administrations involved in local water management. They also define coherent units for the water basins, which provide the basis for the future Local Water Management Plans (Schémas d'Aménagement et de Gestion des Eaux or SAGE in French).

The SDAGE sets out provisions, recommendations and reminders. The provisions correspond to the river basin committee's major goals or main priorities, while the recommendations concern the partners and act as incentives. Lastly, the reminders highlight the legislation or regulations related to water management issues addressed in the plan.

The SDAGE's broad guidelines are deployed on a local level within the framework of the local water management plans, the SAGEs. Each SAGE is developed at the scale of sub-basins, aquifers, lakes, etc. and its boundaries usually does not correspond to administrative boundaries (Fig. 4.2). Some SAGEs, such as the Beauce Aquifer SAGE (see Chap. 5), straddle two agencies, several regions and several counties. The SAGE carries out an initial assessment of water and environmental conditions, sets goals for the use and development of the resources and the aquatic environments, identifies the priority actions to achieve the targets and assesses the technical and financial resources needed. The SAGE establishes operational management rules that are legally binding for the administration following the 1992

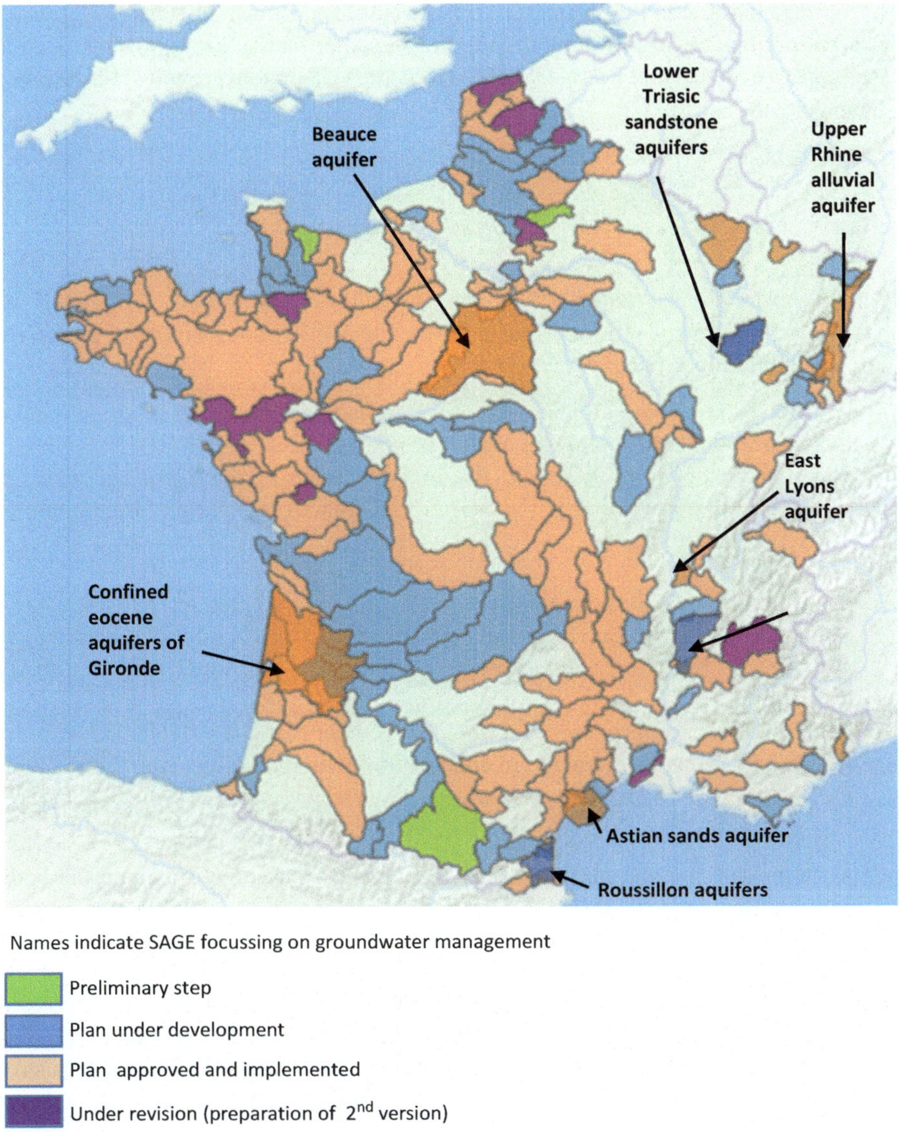

Fig. 4.2 Map of local water management plans in France

Water Act. In 2006, the rules defined by the SAGE also became legally binding for third parties. The SAGE is established by a local water commission composed of local councillors, user representatives and state service officials, and is approved by the state after verifying that it is consistent with the provisions of the SDAGE.

With regard quantitative groundwater management, all the groundwater pumping authorisations granted to users must be compatible with the provisions set out in the SDAGE and the SAGE. The SDAGE and/or the SAGE can restrict total use by setting abstraction limits and set groundwater level thresholds that trigger water use restrictions or prohibition if they are breached. Chapter 5 presents a good example of a SAGE that was established to ensure sustainable groundwater resource management.

4.2.3 Planning, a New European Obligation

The European Water Framework Directive (WFD) adopted in 2000 strengthens the planning objectives at the river basin level. The WFD imposes several principles already applied in France:

1. it promotes an integrated approach of water management issues relating to quality, quantity and the ecological quality of natural habitats;
2. planning must be formalised with the joint development of a river basin management plan (corresponding to the SDAGE in France) including a programme of measures describing the practical actions to be implemented over a six year period;
3. the programme of measures must be scaled in order to guarantee that the environmental goals set out in the management plan are achieved (performance requirement) by 2015 (with the option of a 6–12 year additional delay if properly justified);
4. the programme of measures and the management plan are established with the stakeholders' involvement and in consultation with the public; and
5. the programme of measures is subject to an economic assessment (Laurans, Bouni, Courtecuisse, Dubien, & Johannes, 2001).

After the WFD came into force, the river basin committee remained the body responsible for preparing the SDAGE. It relied on thematic or geographic commissions, made up of members of the river basin committee and invited outside experts. The technical secretariat was generally managed by the water agency and the regional administration (DREAL) which represents the Ministry of Ecology. Technical and scientific studies are either conducted by the Water Agency's teams or assigned to consultancies or research organisations. The programme of measures is established by the state services, under the authority of the prefect (government official), who coordinates the basin.

On a technical level, the WFD also modified the SDAGE's content. It imposed performing an in-depth appraisal of the initial situation, which went further than the appraisal conducted by the SDAGE in 1996. This initial appraisal was conducted systematically using a new hydrogeological mapping that identifies major groundwater bodies, which consist of groups of aquifers that can be managed jointly. The multi-layered confined aquifers were divided vertically (Brugeron, Schomburgk, & Chery, 2013; EC, 2004). For each water body, the assessment summarised its characteristics, hydrogeological function, as well as its qualitative and quantitative status. The appraisal identified bodies of water with a poor status (quantitative or qualitative), which needed to be remedied by the provisions in the programme of measures. According to the WFD, a water body has a good quantitative status if "*the level of groundwater in the body of water is such that the average annual abstraction rate in the long term does not exceed the available resource in the groundwater body*".

When it comes to monitoring the status of groundwater bodies, the WFD requires member states to set up several monitoring networks, which include two main networks. The *surveillance* network is designed to provide an overview of the general state of water at a European level. The *operational control* network is geared to monitoring the bodies of water unlikely to meet the good status target (see Chap. 9), and has additional stations in problematic zones. The operational controls cease when the body of water achieves a satisfactory status. Quantitative groundwater monitoring is based on stations that monitor groundwater levels (piezometers), spring flow and the flow of watercourses which depend on baseflow from aquifers.

The SDAGE is a document that results from extensive consultation with local councillors, state officials and user representatives. It is the outcome of several years of consultation involving the river basin committee's thematic and geographic commissions, discussions with water stakeholders, consultations with the regional and county assemblies, as well as a consultation with local residents in the river basin. Priority actions identified in the SDAGE reflect a compromise that was reached by the stakeholders involved in preparing the document. The approval of the SDAGE is subject to a vote by the river basin committee (absolute majority required) before being jointly signed by the prefect of the river basin representing the state, and the president of the river basin committee.

The following two sections describe how the planning process described above was implemented in the Adour-Garonne and Loire-Bretagne river basins. They present the content of the SDAGEs and how they evolved over the period from 1996 to 2018. The two river basins are characterised by intense groundwater use and problems of quantitative management which are more serious than elsewhere in France. In both cases, the main resources developed are deep confined aquifers in the Aquitaine and Parisian sedimentary basins and the alluvial aquifers associated with major watercourses (see Chap. 2). Table 4.2 shows some data relating to the water uses in these river basins.

Table 4.2 Main characteristics of the Adour-Garonne and Loire-Bretagne river basins

	Adour-Garonne	Loire-Bretagne
Population	7 million	13 million
Agriculture (cultivated area)	116,000 ha	155,000 ha
Volume abstracted		
for irrigation	1000 M m^3	500 M m^3
for drinking water	2000 M m^3	1000 M m^3

4.3 Groundwater Management Planning in the Adour-Garonne Basin

4.3.1 The Emergence of the "Groundwater" Problem

At the end of the 1980s, the Adour-Garonne Water Agency, like other French Water Agencies, acted primarily as a funding body for the water sector. It was organised into two main departments; one that managed issues related to water resources and the drinking water supply and the other that was involved with industrial and urban wastewater treatment. Each department independently planned their programs which were largely based on financial considerations. In the early 1990s, an agency-wide planning group was created which included a few specialists and was directly supervised by the agency's directors. By the end of the 1990s, this group had expanded to a large team.

At the time, diffuse groundwater pollution was not yet considered to be a serious problem. The question of nitrates was tackled by a national committee (named CORPEN) and pesticides were not recognised as an issue. Therefore the agency focused on quantitative management issues and above all, the construction of water storage facilities which it subsidised heavily. This program was implemented under the framework of the 10-year plan for water resource management (the Plan Décennal de Gestion des Ressources en Eau or PDRE in French). The plan's overall aim was to build reservoirs with a total capacity of 1 billion m^3, through a combination of large dams (like the Charlas project with a capacity of 110 million m^3), collective hill reservoirs and small individual reservoirs. The PDRE's goal was to sustain river flow during dry weather periods to compensate for the impacts of abstraction for irrigation purposes, which increased in the late 1980s. The 10-year plan did not include any action linked to groundwater. At the time, the groundwater resource did not seem capable of meeting the challenge of providing 1 billion m^3. Alternative solutions such as artificial groundwater recharge or water saving programs, had not yet been considered.

At the time, the Adour-Garonne Water Agency did not have an overall vision of the issues of groundwater management in the river basin. However there were some local initiatives, for example in the Gironde region, where stakeholders were involved in developing a management system for the deep aquifers (see Chap. 12). These initiatives made the agency aware of the need to examine the issues of groundwater management in the river basin before developing a strategy. This mission was assigned to one of the agency's hydrogeologists, Michel Plaud (Plaud, 1996), who spent 2 years working with the French Geological Survey (Bureau de Recherches Géologiques et Minières). The resulting inventory was made up of five regional assessments and provided a scientific basis for the discussion between stakeholders at the river basin level, as well as nationally (Martin, 1996).

When the preparation of the first SDAGE began, the river basin committee set up seven regional commissions responsible for identifying major water management issues and formulating strategic proposals to deal with them. The river basin committee was aware that tackling the problems relating to the confined aquifers was no easy matter, especially since they are extensive in area and often located beneath regions managed by several territorial commissions. As a result, it created an eighth commission responsible for managing the confined aquifers in the Aquitaine Basin. The commission produced a report in the form of a "*geographic notebook*", which focuses on the deep confined aquifers. The associated problems were already apparent because they concerned the drinking water supply for Bordeaux. The analysis conducted by the seven other territorial commissions barely mentioned groundwater. There were particularly large gaps in the knowledge about agricultural usage volumes, which meant that the analysis was not very relevant.

4.3.2 Groundwater in the First SDAGE (1996)

The 1996 SDAGE was organised around six water management issues (Comité de bassin Adour Garonne, 1996): (A) Management and protection of coastal and aquatic environments; (B) Qualitative resource management; (C) Quantitative resource management; (D) Flood control; (E) Organising and managing information relating to water; and (F) Organising integrated management. The measures relating to groundwater management are essentially found in C, E and F.

In terms of quantitative management, which is the subject covered in this chapter[2], the SDAGE mainly highlighted the issues linked to the deep aquifers. In particular, it identified the threat of salt water intrusion in the confined aquifers near the coast or estuary (the Eocene aquifer in the Gironde), which was caused by the intensity and concentration of groundwater pumping. It also highlighted the limited understanding of groundwater due to the inadequate groundwater monitoring network.

The SDAGE's three main provisions regarding groundwater were the following (Fig. 4.3). Firstly, it defined the outlines of a technical and institutional management scheme for ensuring sustainable groundwater extraction; this model was intended to guide the development of local water management plans (SAGE). Secondly, it encouraged the production of consistent scientific information about the groundwater resources and its dissemination. Finally, it established a Commission responsible for coordinating groundwater management actions at the basin level. These three main provision were translated into 10 operational proposals for action (measures) that provided the foundations for a groundwater management strategy (see Table 4.3). These measures were essential because they determined what programs the agency could help finance, and also provided a regulatory framework which directed the actions of the public agencies, as well as the stakeholders responsible for developing local water management plans.

The groundwater management scheme proposed in the SDAGE relies on three pillars: the identification of monitoring points that will be used to assess the state of each

[2]The SDAGE identified nitrate pollution in the aquifers as a major concern (p22).

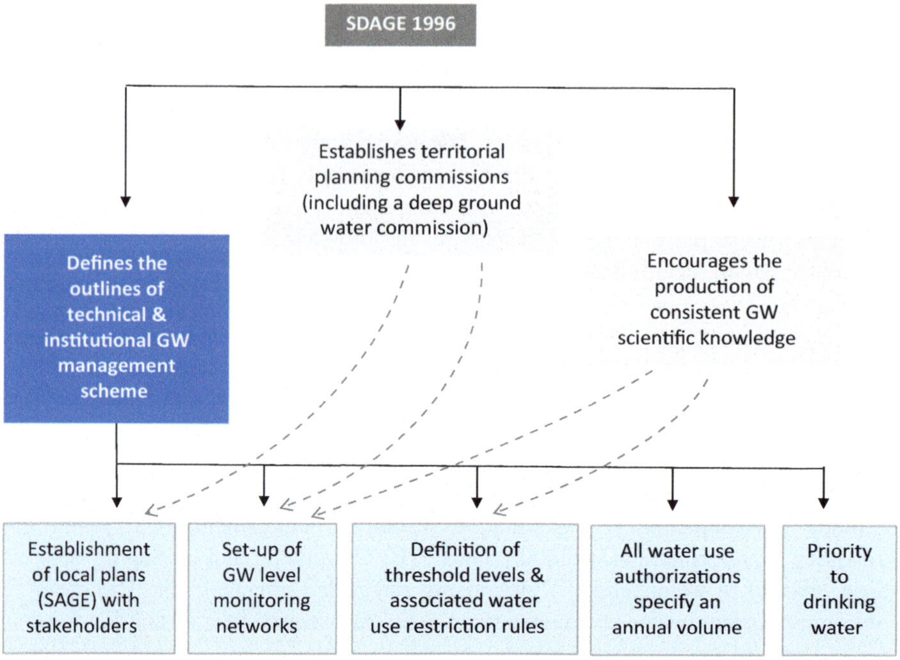

Fig. 4.3 Main provisions in the first Adour-Garonne SDAGE (1996)

resource and to monitor any changes over time; the definition of sustainable groundwater threshold levels that should be maintained over time; and the design of rules of use ensuring that abstraction remains compatible with the water level objectives.

For each monitoring site, the SDAGE (measures C18 and C20) required defining two types of threshold associated with these sites: a *target threshold level* (TTL) and a *crisis threshold level* (CTL). The target level is the groundwater level above which all uses can coexist normally while maintaining a satisfactory state of the aquifers and dependent ecosystems. The State should subsequently manage the allocation of abstraction licences in such a way that this threshold is not breached. The crisis threshold is the groundwater level that should never be exceeded. Its aim is to prevent the occurrence of major adverse impacts on the groundwater resource (e.g. the risk of salt water intrusion and contamination of shallower aquifers), the drinking water supply, or aquatic environments dependent on groundwater. If the crisis threshold is exceeded, the State gradually implements a series of measures ranging from restricting the timing of extractions, to a possible total pumping ban.

The TTL and CTL values can be set in the SDAGE for the main aquifer systems. The TTLs in confined aquifers are set to ensure that the aquifers remain confined and do not under any circumstances, allow inflow of lower quality water from other sources. The TTLs for coastal aquifers are set to prevent saline water intrusion and are based on the level that corresponds to the highest tides. Lastly, TTLs can be fixed for unconfined shallow aquifers where they significantly contribute to the recharge of deeper aquifers. To monitor groundwater levels, the SDAGE recom-

Table 4.3 Measures for groundwater quantitative management in the 1996 Adour-Garonne SDAGE (Comité de bassin Adour Garonne, 1996)

Code	Measure description
C17	Groundwater, particularly when abstracted from confined aquifer, should be used as a priority for drinking water supply, then thermal activities and finally agriculture and industry. This priority ranking should be observed by State agencies when delivering abstraction permits. They should also be reasserted in all SAGEs
C18	For each major aquifer, water level monitoring points should be defined. Target and Crisis threshold water levels should be defined for each of these points. A definition of TTL and TCL is given in the text
C19	The state, the relevant regional authorities and the water agency should jointly develop a common network of monitoring stations for the main aquifer systems. Areas where monitoring network should be deployed in priority are listed
C20	Water extraction regulation rules should be specified and implemented in a number of aquifers (listed) in the 2 years that follow the publication of the SDAGE. These rules aim at ensuring that TTLs are respected. Water restrictions rules are implemented if necessary
C22	The SAGE can further elaborate GW management rules. They can in particular, define TTL/TCLs as well as water use restriction rules. They should as much as possible account for interactions between surface and groundwater resources
C23	County and regional councils are encouraged to contribute to the collection and dissemination of groundwater data and information
C27	All water abstraction licences granted by government administration should specify maximum authorized flow rate and volume
B25	To undertake an inventory of aquifers representing a strategic interest for current and future water supplies and to define of programme of actions to protect those aquifers
E11	Research programs likely to contribute to the objectives of the SDAGE should be financially supported. A list of research themes is provided in the text
F4	The Territorial Commission in charge of deep groundwater management issues should establish a stakeholder platform for developing a global strategy and common governance for all the confined aquifers. This platform prepares decisions of the River Basin Committee concerning deep aquifers. It identifies studies that should be conducted

mends that the state, the relevant regional authorities and the water agency jointly develop a common network of monitoring stations for the main aquifer systems (measure C19). The SDAGE identifies priority zones for developing the networks.

With regard to water abstraction regulation, the SDAGE recommended that rules be established for the all priority confined aquifers in the 2 years following its approval (measure C20). These rules could be more precisely defined at a local level during the preparation of the local plans (SAGEs) with input from stakeholders. The SDAGE also required that the abstraction permits granted by the state specify a maximum volume for abstraction for each well or borehole (measure C27). Thus, it laid the foundations for volumetric management which was not imposed explicitly by the water law. The SDAGE also confirmed that the use of groundwater for drinking water was a priority (measure C17). This priority should be taken into account when the state services deliver administrative authorisations for abstraction. A complementary measure recommends conducting an inventory of the aquifer systems of strategic interest at the river basin level, for the current and future drinking water supply (measure B25).

A key element in the 1996 SDAGE groundwater strategy was to encourage co-operation between the participants. The county councils and the regional councils were invited to take part in the collection and dissemination of the data required for collective groundwater management. They were also invited to participate in information and awareness-raising programs targeting different audiences (including the general public), with the aim of improving the understanding of how the groundwater systems function and the related issues requiring management (measure C23). This resulted in the development of several regional web based information systems for groundwater management (SIGES - Systèmes d'Information pour la Gestion des Eaux Souterraines in French, see Chap. 9). The SDAGE also identified strategic aquifers for current and future drinking water supply, which should be studied as a priority. Lastly, the SDAGE required the deep aquifer commission to coordinate and develop a global strategy for all the confined aquifers within the basin, with the implicit aim of developing a global approach to governance for the deep aquifers.

Overall, the 1996 SDAGE provided the impetus for a developing a coherent groundwater policy for the Adour-Garonne Basin, in close consultation with representatives from the state and user groups. Its main limitation was that it gave almost total priority to the deep groundwater in the Gironde, with no mention of the real problems that affected the confined and unconfined aquifers in the sedimentary region of Poitou-Charentes-Dordogne or the major unconfined alluvial aquifers (Conseil Scientifique, 1999). This approach can be explained by the economic importance of the Gironde deep aquifers, but also by the fact that they had been the focus of studies and management endeavours for 20 years. To compensate for this shortfall, the state was forced to adopt unilateral measures prohibiting the construction of any new boreholes in the aquifers not covered by the SDAGE (except boreholes used for supplying drinking water).

4.3.3 The Revisions of the SDAGE from 2010 to 2016

The 1996 SDAGE for the Adour-Garonne Basin was updated in 2009 (approved 2010, see Comité de bassin Adour Garonne, 2009) and again in 2015 (approved in 2016, Comité de bassin Adour Garonne, 2015). Overall, the management strategy defined in the 1996 SDAGE was consolidated, with the recommended measures focussing on the issues which were not sufficiently covered in the first version. The SDAGE was also adapted to meet the requirements of the European Water Framework Directive.

The 2010 SDAGE stated that groundwater resources constitute an extremely important natural heritage for the river basin for two main reasons. Firstly, they significantly contribute to ensuring the good ecological conditions of rivers and dependent ecosystems (e.g. link between wetlands and the alluvial aquifers). Secondly, they make up a large proportion of the strategic resources for supplying drinking water today and in the future. Therefore, they should be protected and developed to supply drinking water as a priority. To achieve this, the 2010 SDAGE emphasised the need to improve our understanding of the interactions between surface and groundwater in particular (measure C1 and E10). It also recommended developing decision-making and modelling tools (measure C3) and conducting forecasts in order to anticipate the effects of climate change to facilitate adaptation (measure E12).

Unlike the 1996 SDAGE, which focused mainly on the confined aquifers, the 2010 SDAGE also dealt in depth with the problems of managing unconfined aquifers. A new assessment distinguished 105 groundwater bodies, including 20 confined aquifers. The initial appraisal and the programme of measures treat the two types of bodies of water separately. Overall, 30 of the 105 bodies of water are subject to over-abstraction and are classified as depleted, comprising 12 confined aquifers and 18 unconfined aquifers (see Fig. 4.4). Eleven of these bodies of water are classified as having a poor quantitative status because they were failing to supply sufficient water to the watercourses as a result of overexploitation. This approach illustrates how the SDAGE genuinely integrated surface and groundwater management. The analysis of the issues relating to confined aquifers was also more detailed

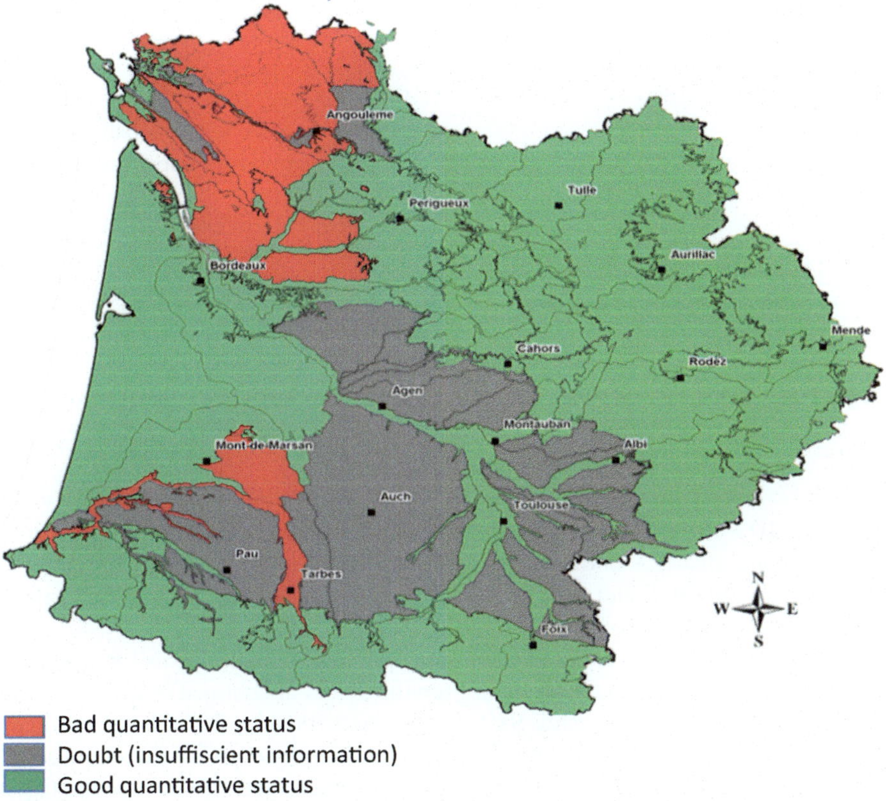

Fig. 4.4 Map showing the quantitative status of the unconfined groundwater bodies (SDAGE 2010–2015: Comité de bassin Adour Garonne, 2009)

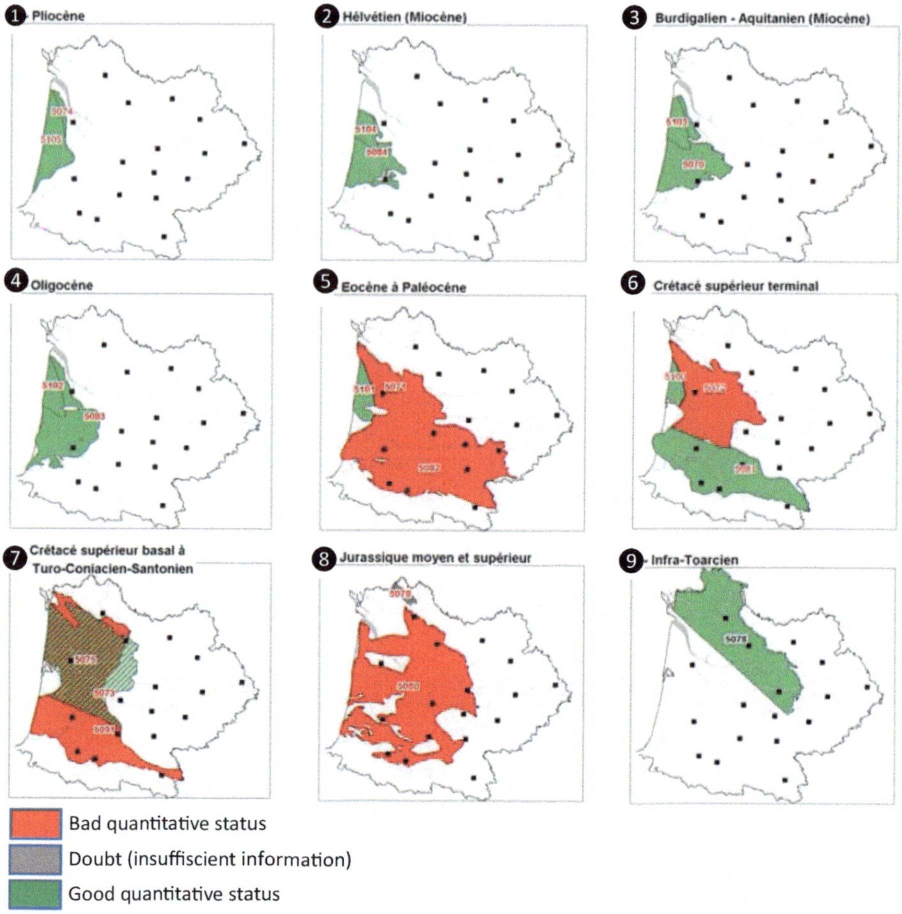

Fig. 4.5 Map of the confined groundwater bodies showing their quantitative status (SDAGE 2010–2015). Numbers indicate the position of each aquifer layer in the geological sequence (9 is the deepest)

in the 2010 version than previously. The state of the confined bodies of water (Fig. 4.5) are clearly identified on the basis of the better understanding obtained since the first SDAGE and with the use of large multi-layered groundwater models (Wuilleumier et al., 2016).

The 2010 SDAGE set extremely ambitious and fairly precise goals for groundwater: 95% of the 105 groundwater bodies (confined and unconfined) should achieve a good quantitative status by 2015, 98% by 2021 and 100% by 2027. It clearly identified five bodies of water (in the confined aquifer category) that would not be able to achieve a good status by 2015. However, it was rather vague about how the targets would be achieved. The actions listed in the programme of measures were very general, for example "*restore the balance between abstraction and recharge*" or "*limit the risk of saline intrusion*". The SDAGE stated (provisions C7 and E3) that the problems of quantitative management would be resolved by implementing the new regulations resulting from the 2006 water law. The law stipulated that the maximum volume to be abstracted (sustainable extraction limit) should be calculated for each body of water and that the state should adjust the abstraction permits accordingly (see Chap. 3). The provisional cost of the measures linked to quantitative groundwater management was relatively low (63 million euros or 1.5% of the total cost of the programme of measures, equivalent to 1.5 euros per inhabitant per year).

In early 2016, a new version of the SDAGE came into force after a second update involving stakeholders. Overall, it was an extension of the previous version but included more specific measures and practical details. Measure C1 relates to improving knowledge of the groundwater-river interactions and recommends that the SAGE should conduct studies to delineate the aquifers that interact with the watercourses and establish how the karstic aquifers function. Measure C5 specifies the methodology that can be used to identify the basins with a water deficit. One of the innovative elements of this SDAGE was its emphasis on taking climate change into account. In particular, it recommended conducting an assessment of the impact of climate change on the resources (measure A15) and the development of an adaptation plan (A16). In addition, it recommended conducting regional forecast exercises (A18 and A19) and taking into account the interactions between water and energy policies.

4.4 Groundwater Management Planning in the Loire-Bretagne Basin

4.4.1 The Context Leading Up to the Implementation of the SDAGE

In the early 1990s, groundwater resources in the Loire-Bretagne Basin were already heavily exploited during dry periods, particularly for irrigation. After the summer and winter droughts of 1989–1992, this intensive use reduced baseflow discharge to streams during low-flow periods, causing certain watercourses to dry up. The drop in groundwater levels observed in spring and summer also caused many springs to dry up completely, which severely affected certain wetlands, particularly, the Poitevin Marshes (see Chap. 18).

When work started on preparing the first SDAGE, the Loire-Bretagne Basin stakeholders were already aware of the specific challenges of groundwater management and the importance of the interactions between groundwater and surface water resources. Representatives from the agricultural sector were also convinced of the need to implement mechanisms to manage abstraction. The first Loire-Bretagne SDAGE was thus drafted in a more favourable political context than was the case for the Adour-Garonne SDAGE.

Moreover, pilot groundwater management experiments had already begun at local level in the basin, for example in Beauce, where an innovative quantitative management mechanism was introduced before the 1992 water law came into force (see Chap. 5). Piezometers had already been installed there to observe the change in the groundwater levels and us restrictions were imposed when the level dropped below the alert thresholds. The principles of this pilot experiment were incorporated in the SDAGE and subsequently inspired the quantitative management systems introduced in many other river basins (see Chap. 18).

The stakeholders in the Loire-Bretagne Basin were also pioneers in terms of the regulatory framework for monitoring resources. In 1990, the agency's board of directors had already agreed to finance the creation of a groundwater level monitoring network which began in 1993. It was a gradual process with a goal of creating 30 monitoring sites in each county.

4.4.2 Groundwater in the First SDAGE (1996)

As far as groundwater was concerned, the first SDAGE focused primarily on the need for preventive action to avoid the irreversible depletion of the groundwater resources (Comité de bassin Loire Bretagne, 1996). To achieve this, the SDAGE underlined the importance of improving the understanding of aquifers including their geometry and hydrodynamic characteristics, recharge rates and their interaction with watercourses. The SDAGE recommended conducting one-off campaigns to measure the groundwater levels until a denser groundwater monitoring network was in place. It also recommended conducting an inventory of the abstraction points, with detailed information on the aquifers that are exploited and most importantly, the volume abstracted for irrigation (provisions VII-2-11 and VII-3-1). The goal was to provide the necessary information to develop groundwater flow models that the managers could use as a management tool where necessary.

One of the main contributions that the 1996 SDAGE made to the quantitative management of groundwater resources was to identify six major aquifers or catchment basins where abstraction exceeded the renewable resource (provision VIII-3-1, see Fig. 4.6). These were referred to as the intensely exploited aquifers. The SDAGE underlined the fact that greater effort was needed to improve our understanding and management of these intensely exploited zones and recommended applying rules to manage abstraction. The idea was that local stakeholders would develop rules within the framework of the SAGE. These locally defined rules would then be incorporated into the SDAGE to consolidate their regulatory status. The water agency integrated the zoning of intensively exploited aquifers into its grant policy, by awarding greater subsidies to all the studies and actions designed to improve the resource management in these zones.

As far as the management rules were concerned, provision VII-2-11 of the SDAGE states "*the management objective is to reconcile the different uses in the best possible conditions and at the same time, ensure that the heritage is preserved* [by maintaining] *a minimum groundwater level*". As in the case of the Adour-Garonne Basin, this involves defining the target groundwater threshold levels (TTL) and the crisis threshold levels (CTL) that would trigger restrictions of use. The reference to this management principle in the SDAGE (provision VII-3-1) specifies

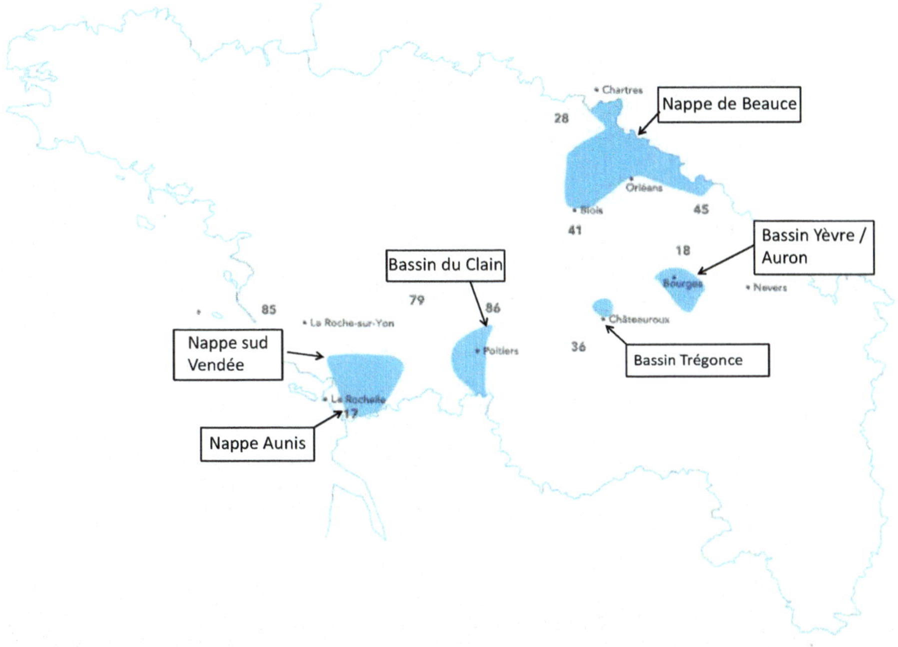

Fig. 4.6 Map of the intensively exploited aquifers (Source Comité de bassin Loire Bretagne, 1996)

how the 1992 water law is applied. In this sense, the SDAGE defines the strategy to implement the law which was then be implemented by State agencies in several river basins (the Clain, Sèvre Niortaise and Yèvre Auron basins). The introduction of this management scheme actually took several years because of the time involved in setting up the groundwater monitoring networks, collecting data and conducting the necessary studies to identify the TTLs and TCLs. In the case of the Beauce region, the existence of a groundwater monitoring network since 1974 meant that a TTL level could be defined directly in the 1996 SDAGE (provision VIII-3-3).

The droughts created favourable political condition for implementing a new groundwater management approach and the first water meters were installed at agricultural abstraction sites in 1994. The agency subsidised their installations at a rate of 80%, which facilitated their adoption especially in Beauce region. By 1999, almost 100% of the surface and groundwater abstraction points in the Loire-Bretagne Basin were equipped with a meter. The agricultural institutions supported the introduction of volumetric quotas for individual abstraction because they considered that the management model, although restrictive, would help reduce the frequency of crises and the associated bans on use. The volumetric management principle was made operational for the first time in Beauce in 1999. It is important to underline that in the remainder of the Adour-Garonne Basin, representatives from the agricultural sector are still reluctant to accept the principle in 2018.

The 1996 SDAGE also identified the aquifers that should be reserved for the drinking water supply. The provision VIII-3-2 lists these aquifers, most of which are confined (Fig. 4.7). Although existing agricultural boreholes can be maintained, this SDAGE provision prevents the state services from issuing new abstraction licences for agricultural use. The application of this provision did not pose major difficulties because there are generally unconfined aquifers that can be used for irrigation above the reserved confined aquifers. The main target here was to preserve the good

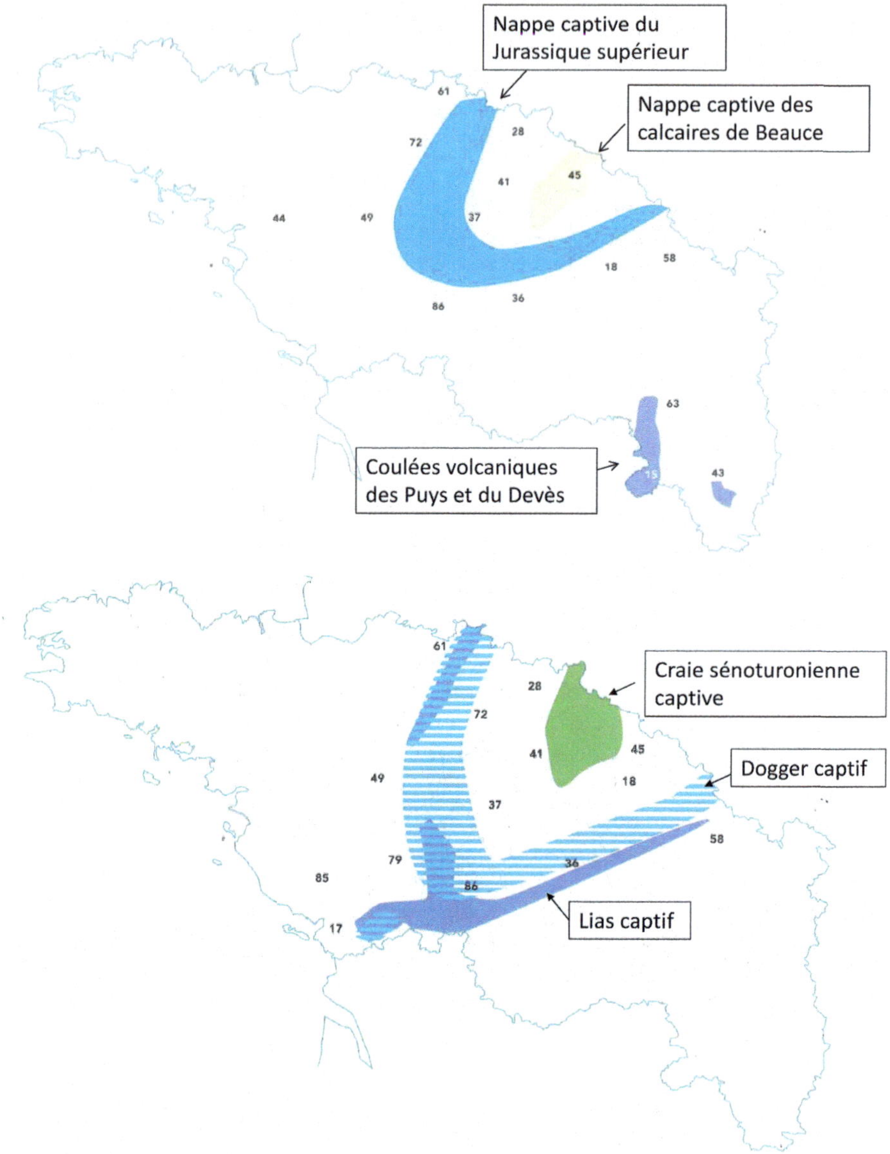

Fig. 4.7 Maps of the aquifers reserved for the drinking water supply (Source Comité de bassin Loire Bretagne, 1996)

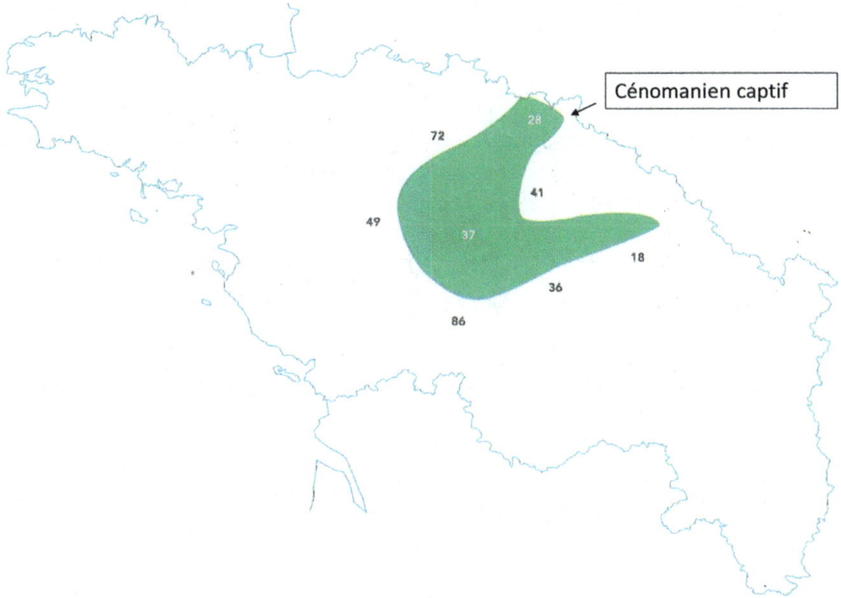

Fig. 4.7 (continued)

natural quality of these protected aquifers, by preventing the installation of works that could established undesired hydraulic connections between the polluted unconfined and good quality confined aquifers.

4.4.3 The Revisions of the SDAGE from 2010 to 2016

In 2003, all the aquifers that the SDAGE had identified as intensely exploited were decreed water restriction zones (Zones de Répartition des Eaux in French, Chap. 3) by the Council of State. Thus, the SDAGE facilitated the introduction of this regulatory tool which gave government agencies greater control over all abstraction. With the enforcement of the 2006 law on water and aquatic environments, a ministerial circular in 2008 requested that Sustainable Abstraction Limits (SAL) should be estimated (in annual volume) for all the aquifers under restriction. The SAL represents the maximum volume of water that can be abstracted annually for each body of water, without compromising its good ecological status (see Chap. 11). The SALs were defined on the basis of hydrogeological studies, which were supervised by the water agency if there was no SAGE in place. The SALs were included in the SAGE, which made them legally binding and enforceable. Henceforth, the state services that delivered abstraction licences had to ensure that the sum of individual allocations did not exceed the SAL. In most situations however, the current volumes actually abstracted considerably exceeded the SAL, which meant they had to be reduced significantly.

The Loire-Bretagne SDAGE was first updated in 2009 (approved in 2010) without significantly changing groundwater management policy priorities (Comité de bassin Loire Bretagne, 2009). As the first generation of problems had been tackled between 1996 and 2009, the SDAGE for the period 2010–2015 concentrated on new problem areas, where management rules had to be introduced to limit abstraction. Thus, the SDAGE identified the "*basins that require greater protection during low-flow periods*". This involved capping abstraction at 2009 levels and introducing water saving and conservation measures in the sectors of agriculture and urban water supply (measure 7A-1). The SDAGE also designated "*basins where prevention was required to prevent the appearance of quantitative deficits*", by capping annual abstraction at current levels (measure 7A-2). These provisions were to be confirmed in the 2016–2021 SDAGE (see Fig. 4.8), which includes additional zoning (7B-2) where a limited increase in abstraction is possible, which was specified in m^3 per square kilometre.

The SDAGE also sets out in detail how to implement its recommendations in three specific regions: the Beauce, the Poitevin Marshes and the Cenomanian Aquifer. In these basins, the state services and the water agency had already taken the initiative to introduce fairly elaborate local management rules. As the local stakeholders had not yet validated these rules in the framework of the SAGE, the SDAGE specified them to give them a legal status.

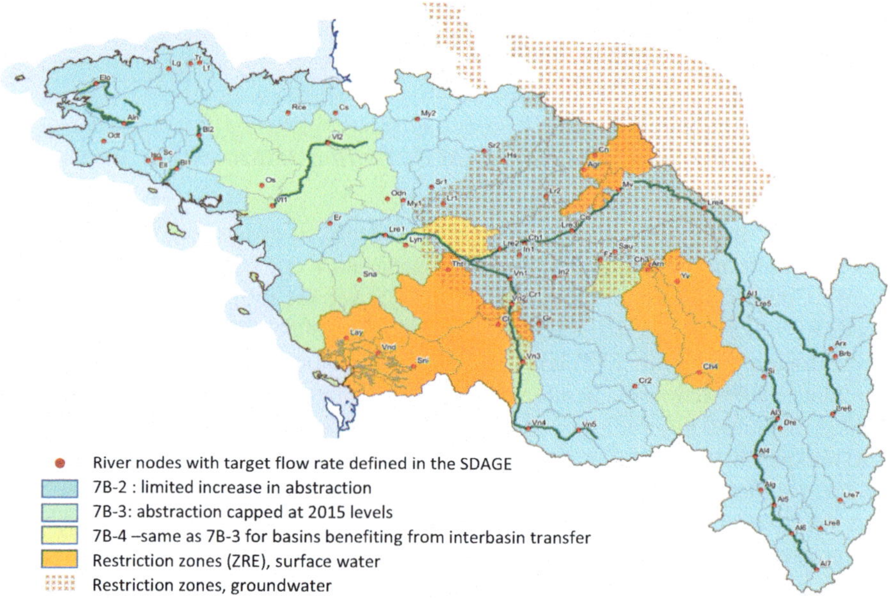

Fig. 4.8 Map of basins with a limited increase in abstraction at low water (7B-2) and with abstraction capped at current levels (7B-3, 7B-4). SDAGE 2016–2021 (Comité de bassin Loire Bretagne 2010)

- The management rules for the Beauce Basin were specified in provision 7C-3, which sets the target and crisis groundwater threshold levels (TTL and CTL) as well as an intermediary Alert Threshold Level (ATL) at which restrictive measures should gradually be implemented. The measure also sets the volumes to be abstracted per economic sector (drinking water, irrigation and industry) and the restrictions applied to groundwater abstraction in the event of a reduction in the flow of watercourses receiving discharge from the aquifer.
- The Poitevin Marshes are also subject to a specific measure (7C-4), which sets out the management rules, particularly the details of the alert and crisis thresholds levels for water in the marshes and the aquifers. These rules are established in the SDAGE provisionally, until such time as they are specified in the SAGE.
- Lastly, the management rules for the confined Cenomanian Aquifer are set out in provision 7C-5, which specifies the maximum volumes for abstraction per sector. These volumes were determined on the basis of the groundwater modelling study conducted by the water agency. Interestingly, local water users were reluctant to set up a SAGE because it was considered a too complex process for a region covering 25,000 km^2.

After its second revision (2010–2015), the SDAGE included a framework to regulate the development of new storage facilities that started to be constructed in the late 2000s. They consist in small to medium size reservoirs (with capacities up to 800,000 m^3) which are filled with groundwater during winter when aquifer levels are high and then used for irrigation in summer, when there is a risk of groundwater use restriction. The SDAGE encourages the construction of such reservoirs in the basins classified as water restriction zones, under the conditions that users would totally cease pumping groundwater in summer and that they would use 20% less water from the reservoirs than they took from groundwater previously (measure 7D-1). In addition, the SDAGE states that an abstraction licence must be obtained to fill the reservoirs, and this licence must specify the groundwater threshhold level below which all abstraction for filling the reservoir is prohibited (measure 7D-2).

The third version of the SDAGE for the period 2016–2021 is a continuation of the first two versions (Comité de bassin Loire Bretagne, 2015). It underlines the need to adapt to climate change by recommending that the definition of target groundwater levels (measure 7A-1) should take climate change into account and if necessary, abstraction licences should be reviewed and reduced accordingly (7A-6). The measures 7A-3 to 7A-5 also highlight the importance of saving as much water as possible for the different uses. The introduction of water saving measures was clearly set as a prerequisite for the construction of new water reservoirs.

4.5 Discussion

In the two river basins studied in this chapter, the planning approach outlined by the SDAGE significantly contributed to the development of quantitative groundwater management. The SDAGE made it possible to specify operational approaches to implement the legal framework, taking into considerations specific characteristics of each basins and the aspirations of stakeholders. More precisely:

- The SDAGE helped identify aquifers where urgent action was required, such as the protection of confined aquifers or the implementation of rules controlling abstraction where sustainability was threatened. By doing so, it often triggered action when local water users did not have the capacity to initiate control measures.
- It contributed to an increase in the technical and scientific knowledge relating to aquifers and the groundwater flow systems. The little known zones were investigated and groundwater flow models were developed to understand and subsequently, manage the aquifers that extend over tens of thousands of km^2.
- It led to the establishment of targets for the flow rate in the major watercourses at low water levels, as well as the corresponding groundwater level thresholds for several aquifers in the river basins that have high connectivity with surface water.
- In regions considered a priority, it encouraged and imposed the development of local management plans (SAGE) in which practical groundwater management rules were formulated.
- It also boosted efforts to ensure that groundwater monitoring was consistent throughout the region.
- Lastly, the SDAGE accelerated the process of making information on groundwater available to the public, notably through the groundwater management information systems (see Chap. 9).

The initial SDAGEs were not perfect and were progressively improved in the subsequent revisions:

- In the first SDAGE for the Adour-Garonne Basin, the implementation of the measures was limited because they were not specified with sufficient details, and time scales were not indicated. Rules and recommendations were vague and the river basins where specific measures should have been applied, were not clearly identified. In addition, no enforcement regime was planned. These shortcomings were corrected in subsequent versions.
- One of the SDAGE's goals was to provide a framework to coordinate and create collaboration between the different public organisations involved in the water sector (regions, counties, water agency, and towns). Observers who experienced this period have described how difficult it was for the participants to change their own planning strategies and integrate the SDAGE's goals into their action plans. The different branches within the water agency struggled to coordinate their actions because they had previously planned their programs independently. The collaboration which the SDAGE had hoped to develop, took a considerable amount of time to achieve.
- In the initial SDAGEs, very few TTLs and CTLs, and even fewer SALs were set, particularly in the Adour-Garonne Basin. This are several reasons for this omission. The concepts of TTL and CTL are difficult to apply to the confined aquifers

(see Chap. 1), and the lack of hydrogeological expertise in the state services and water agencies resulted in difficulties in understanding how confined aquifers functioned. This lack of understanding made it difficult to convince the local stakeholders of the need for management intervention. The SDAGE acted as a catalyst for local action by giving local stakeholders the means to act if the political will exists. However, it had little impact if the political will was lacking.

- The management of unconfined aquifers is still in its embryonic stage because the participants that develop and approve the SAGE (the local water commission and the state) lack expertise. Although measures are created, they are not necessarily enforced because the water police consider that surface water issues have a greater priority than groundwater issues (see Chap. 23). This situation is also due to the agricultural sector's hostility towards the introduction of groundwater management rules as farmers that are subject to severe restrictions for surface water use are keen to maintain the relative freedom they have to use unconfined aquifers, even though they are connected to the rivers.
- It is also highly likely that the imposed SALs are still too high in some cases. Under these circumstances, it is not possible to achieve the targets for groundwater levels and watercourse flow rates and consequently, it is not possible to attain the good status of the natural environment in both the river basins studied in this chapter.

4.6 Conclusion

In France, the development of blueprints for the major river basins played an important role in the implementation of the water policy. First of all, the planning process allowed local government and user representatives to be involved in the implementation of the water policy. This participatory approach made it possible to take into account the specific hydrogeological, economic and socio-political characteristics of the regions concerned. It enhanced the legitimacy and acceptability of the measures introduced, as well as ensuring the participants' commitment when it came to implementing the decisions that they had helped to negotiate. Lastly, one of the SDAGE's strengths is that it is a regulatory document which is endorsed by the state and specifies the legal obligations for water users in a major river basin.

By identifying priority knowledge gaps, the SDAGEs have also helped ensure that the water agencies concentrate their funding in problematic regions or on appropriate issues. Without this process, it is likely that the SDAGEs would not have been effective. The SDAGE also led to the development of cooperation between public funding bodies (agency, state and local governments), who had not previously coordinated their policies.

The SDAGEs also facilitated the creation of local water management plans (SAGEs). By setting targets, the SDAGEs gave local stakeholders considerable freedom to draft the measures that were needed to achieve the targets. They also facilitated local actions by producing consistent scientific knowledge for the different regions and by helping to establish groundwater monitoring networks and create stakeholder consultative networks.

Acknowlegdments The authors warmly thank Steve Barnett, for his very detailed and useful comments on a previous version of this chapter.

References

Barraqué, B., Laigneau, P., & Formiga-Johnsson, R.-M. (2018). The rise and fall of the French Agences de l'Eau: From German-type Subsidiarität to state control. *Water Alternatives, 4*(3), 1850013.

Brugeron, A., Schomburgk, S., & Chery, L. (2013, August). *WFD Groundwater bodies: Consequences of the new national aquifer reference system BDLISA for their delineation in France WFD groundwater bodies*. International workshop on groundwater systems in Europe, Berlin, Germany. https://hal-brgm.archives-ouvertes.fr/hal-01163584

Buller, H. (1996). Towards sustainable management: Catchment planning in France and Britain. *Land Use Policy, 13*(4), 289–302.

Comité de bassin Adour Garonne. (1996). Schéma directeur d'aménagement et de gestion des eaux du bassin Adour-Garonne. Agence de l'Eau, Toulouse, France. 113 p. + ann.

Comité de bassin Adour Garonne. (2009). Schéma directeur d'aménagement et de gestion des eaux du bassin Adour-Garonne 2010–2015. Agence de l'Eau, Toulouse, France. 144 p + ann.

Comité de bassin Adour Garonne. (2015). Schéma directeur d'aménagement et de gestion des eaux du bassin Adour-Garonne 2016–2021. Agence de l'Eau, Toulouse, France. 281 p. + ann.

Comité de bassin Loire Bretagne. (1996). Schéma directeur d'aménagement et de gestion des eaux du bassin Loire Bretagne. Agence de l'Eau, Orléans, France.

Comité de bassin Loire Bretagne. (2009). Schéma directeur d'aménagement et de gestion des eaux du bassin Loire Bretagne 2010–2015. Agence de l'Eau, Orléans, France. 248 p.

Comité de bassin Loire Bretagne. (2015). Schéma directeur d'aménagement et de gestion des eaux du bassin Loire Bretagne 2016–2021. Agence de l'Eau, Orléans, France. 355 p. Accessible at: http://www.eau-adour-garonne.fr/fr/programme-d-intervention-de-l-agence/un-cadre-le-sdage/sdage-pdm-2016-2021.html

Conseil Scientifique du comité de bassin Adour Garonne. (1999). Avis sur le programme eaux souterraines de l'agence de l'eau Adour Garonne. Mai 1999, Toulouse, Agence de l'eau Adour Garonne. 10 p.

EauFrance - The status of surface water and groundwater. (2015). http://www.eaufrance.fr/IMG/pdf/evaluation_2010-2013_201506_EN.pdf

European Commission. (2004, April 11). Common Implementation strategy for the water framework directive (2000/60/EC) - Groundwater body characterisation - Technical report on groundwater body characterisation issues (as discussed at the workshop of 13th October 2003), Final draft. https://circabc.europa.eu/sd/a/157c2240-b988-417b-9137-a14e89db41d8/Groundwater%20characterisation%20report.pdf

Laurans, Y., Bouni, C., Courtecuisse, A., Dubien, I., & Johannes, B. (2001). L'évaluation économique de la théorie à la pratique: l'expérience des SDAGE en France. *Natures Sciences Sociétés, 9*(2), 17–28.

Martin, Y. (1996). Rapport sur la gestion durable des eaux souterraines. Conseil Général des Mines. Ministry of Industry, Paris.

Piégay, H., Dupont, P., & Faby, J. A. (2002). Questions of water resources management. Feedback on the implementation of the French SAGE and SDAGE plans (1992–2001). *Water Policy, 4*(3), 239–262.

Plaud, M. (1996). SDAGE et SAGE pour les eaux souterraines en Adour Garonne. *La Houille Blanche, 3*(1996), 57–60.

Roche, P.-A., Guerber, F., Nicol, J.-P., & Simoni, M.-L. (2016). Eau potable et assainissement: à quel prix ? (No. 10151–1) 561. Paris: CGEDD-IGA. Ministère de l'Environnement, de l'Energie et de la Mer, Ministère de l'Intérieur.

Wuilleumier, A., Saltel, M., Douez, O., Cabaret, O., Abasq, L., & Pedron, N. (2016). *A platform to harmonize the regional hydrodynamic models in the south-western part of France*. IAH congress, Montpellier. https://www.cfh-aih.fr/images/DOCS/

Dr. Jean-Daniel Rinaudo is researcher at Brgm, Montpellier University, where he coordinates the scientific program on environmental and risk economics Initially trained as an agricultural engineer (Montpellier SupAgro 1994), he specialized in agricultural and resource economics (PhD University of Auvergne, 2000). Prior to joining Brgm, he worked for the International Water Management Institute in Pakistan where his research focused on the political economy of irrigation management reforms. His current research mainly focuses on the institutional economic dimension of groundwater management. Most of his research is conducted in France but he also works in Morocco and Chile. He is currently developing new research activities in the field of natural disasters economics, focusing on the methods to assess economic vulnerability and resilience. Dr. Jean Daniel Rinaudo is also member of the Scientific Council of the Adour Garonne River basin agency.

Pierre Marchet holds a PhD in hydrogeology (1991). He joined the Adour-Garonne Water Agency in 1987 as a "junior" hydrogeologist. He has spent his entire career there, and currently holds the position of groundwater expert (senior hydrogeologist). He is a member of the board of directors of the Comité Français d'Hydrogéologie (International Association of French Hydrogeologists chapter), after having held various positions in the office

Pascal Billault is hydrogeolist at Agence de l'Eau Loire Bretagne since 1982. He is in charge of groundwater policy for the Loire Bretagne river basin district. He designed the groundwater monitoring network of the basin, contributed to the develoment of the groundwater chapter of the River basin Master Plans (SDAGE) of 1996, 2009 and 2015).

Chapter 5
Lessons from Twenty Years of Local Volumetric Groundwater Management: The Case of the Beauce Aquifer, Central France

Frederic Verley

Abstract With an area of nearly 10,000 km^2, the Beauce aquifer is one of France's major groundwater reservoirs. This unconfined aquifer is used for drinking water, industry and for irrigation. Following a succession of dry years in the 1990s, the water table significantly dropped achieving the lowest groundwater levels ever observed and drying up several groundwater dependent watercourses. This triggered the development of an innovative groundwater management scheme. Volumetric meters were installed in 1994 and volumetric quotas were allocated in 1998. Quotas are now adjusted depending on resource condition at the beginning of the year. Allocation rules have been validated by the Local Water Management Commission which represents all users and the State.

Keywords Beauce groundwater · Irrigation · Volumetric management · Collective management · Planning

5.1 Introduction

The Beauce aquifer is one of the major groundwater reservoirs in France. Located in a zone of relatively low rainfall, the aquifer supplies large volumes of water for agriculture and, to a lesser extent, for industrial and domestic use (drinking water). The sharp decline in the groundwater level recorded in the 1990s and the damage caused to dependent aquatic ecosystems led stakeholders to develop a volumetric system for managing agricultural abstraction for irrigation. The management approach implemented for the Beauce aquifer was the first practical application of

F. Verley (✉)
DREAL, French Government, Orléans, France
e-mail: frederic.verley@developpement-durable.gouv.fr

J.-D. Rinaudo et al. (eds.), *Sustainable Groundwater Management*,
Global Issues in Water Policy 24, https://doi.org/10.1007/978-3-030-32766-8_5

the new tools established by the 1992 Water Act. Consequently, it has been a source of inspiration on a national scale. The system has evolved and is now integrated within the framework of a Local Water Management Plan (SAGE in French) which gives it legal authority (see Chap. 4).

This chapter describes the mechanism for the management of water abstraction which was gradually implemented in the early 1990s, and its subsequent evolution. The analysis presented in this chapter is drawn from the author's personal experience after almost 20 years of working with the state services to develop the SAGE (Verley, 2014). The chapter is organised as follows: Section 5.2 presents the characteristics of the water resource, its uses and the problem of overexploitation which emerged after the drought that occurred between 1989 and 1993. This is followed by a chronological account, divided into three phases, of how the management plan evolved. The prospects for change are presented in the conclusion.

5.2 The Beauce Aquifer

5.2.1 The Resource

The Beauce aquifer stretches across an area of almost 10,000 km^2 southwest of Paris, on a plateau located between several rivers: the Loire, the Loing, the Seine, the Eure and the Loir (Fig. 5.1). It covers six counties, two regions (Centre-Val de Loire and Ile-de-France) and straddle the border between two River Basin District Agencies (Loire Bretagne and Seine Normandie). Thus implementing a single coherent system for managing this resource is no simple matter.

The term "Beauce aquifer" refers to an aquifer system that is complex from both a geological and hydrogeological point of view. The system includes 14 geological units from the Tertiary period and the end of Secondary period, which are up to 200 m thick in the Pithiviers sector (Fig. 5.2). The upper limestone layers have a fairly high storage coefficient (10%), which decreases quite rapidly with depth. The water storage capacity is around 10 billion m^3 (Le Coz, 2000).

The decision to manage this groundwater system as a single resource was based upon two key factors. Firstly, boreholes often pump water simultaneously from several different aquifers making it impossible to separately assess volumes pumped from each aquifer. Secondly, this complex hydrogeological system is drained by major rivers around the margins of the aquifer system (the Loire, Loing, Seine, Eure and Loir Rivers).

A few outflowing rivers originate in this geological area, namely, the Loing tributaries in the east, the Essone and its tributaries in the north and the Conie, the Aigre and the Cisse in the west (see Fig. 5.1). These natural outlets actually constitute the principal "outflows" of water from the area covered by the Beauce aquifer. There is no major discharge in much of the central area, which means that the aquifer has a considerable storage capacity which allows it to store excess winter rainfall for several years. This storage capacity lends itself to multi-annual management, which

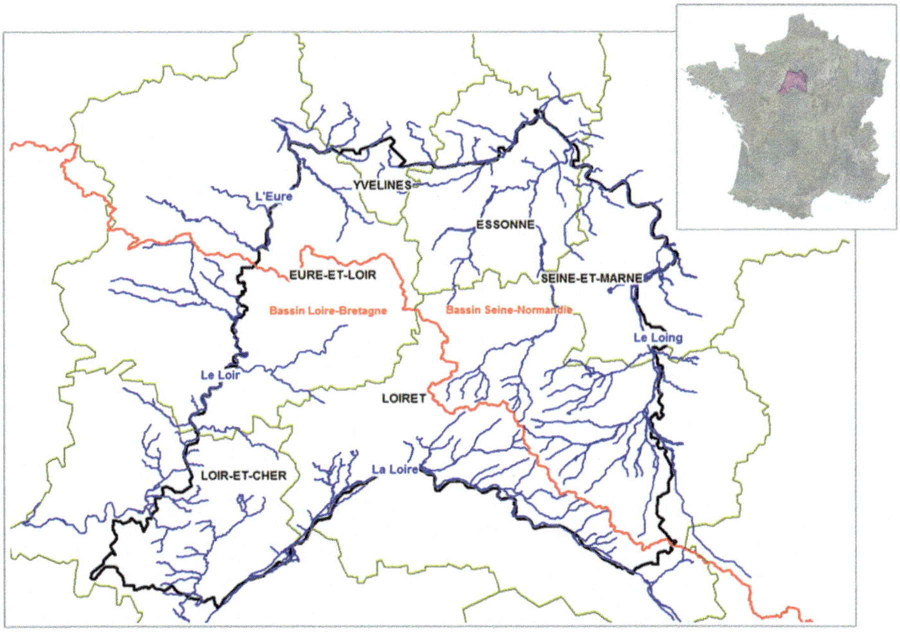

Fig. 5.1 Location of the Beauce aquifer

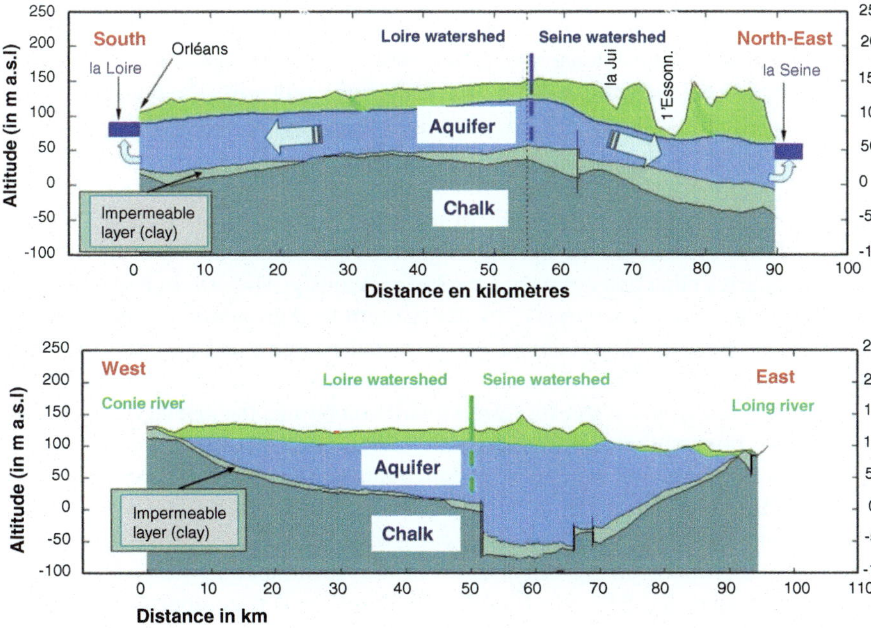

Fig. 5.2 Geological cross-section of the Beauce aquifer

would not be possible in other aquifers. In the eastern part where drainage is greater, the aquifer does not have the capacity to store water. Recharge and discharge are rapid, which makes the aquifer and the associated aquatic ecosystems far more fragile.

The average rainfall in Beauce is low (600 mm per year), is evenly distributed throughout the year and can vary greatly from year to year. This rainfall is comparable to that observed in the Mediterranean region, which makes it one of the driest sectors in France. The proportion of rainfall that contributes to recharge (i.e. the effective rainfall that generally occurs from October to April) is also modest, averaging around 140–150 mm/year with a high variability.

5.2.2 *Groundwater Uses and Their Development*

Beauce is a predominantly rural area located between two urban centres, Paris in the north and Orleans in the south. In order to provide drinking water to almost 1.5 million people, public water utilities abstract on average 100 million m^3 each year, the majority of which comes from groundwater. Industrial activities (agri-food, chemicals, computing, metallurgy, paper mills) abstract around 20 million m^3 per year. These volumes are much lower than the abstraction levels for agriculture.

Until the middle of the twentieth century, agriculture in Beauce was based on a three-year crop rotation: wheat or barley, root crops (potatoes and sugar beet) and forage crops (for sheep production). In the 1950s, the rural exodus led to a decline in livestock production and an increase in maize production because it can be mechanised and is not labour intensive (Désiré, 1972). This change was accompanied by the development of irrigation. From the late 1960s onwards, the consumption of groundwater for irrigation reached 30 million m^3, a figure that may have seemed large to observers at the time.

Subsequently, the area of irrigated land continued to expand with the installation of boreholes for coping with periods of drought which first occurred in 1976 and then later in the early 1990s. Spring irrigation also developed for crops (notably wheat and barley) that had not been irrigated previously. The development in irrigation was particularly encouraged by the Common Agricultural Policy (see the Chap. 24). As a result, abstraction levels for irrigation peaked in the early 1990s and later stabilised in the 2000s.

In Beauce today, irrigation for farming still constitutes the primary use of water in terms of consumption: up to 450 million m^3 was abstracted in the dry year of 1990. Current abstraction, which varies depending on the climate, is around 200 million m^3 on average (Fig. 5.3).

During the last agricultural census in 2010, 6543 km^2 of utilised agricultural land was surveyed, which corresponds to 68% of the region and a total of 6044 farms. The census found that 53% of utilised agricultural land is equipped for irrigation which is significantly higher than 22% which is the largest area recorded in the other four major regions in France where irrigation occurs, and 9% for the remainder of France.

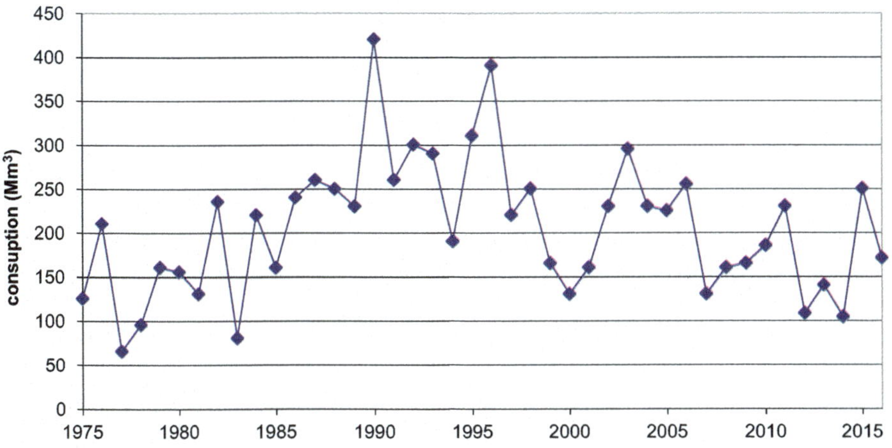

Fig. 5.3 Changes in abstraction for irrigation between 1975 and 2016 (in million cubic meters per year)

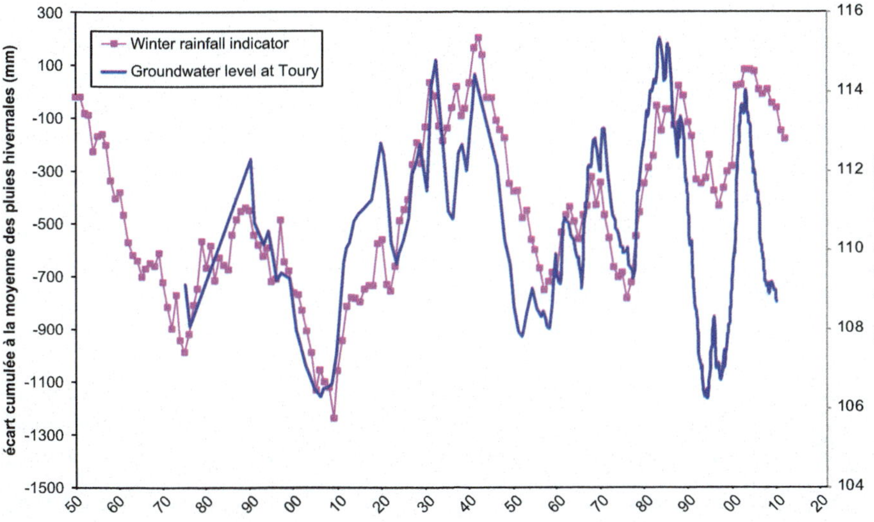

Fig. 5.4 Groundwater level trends in the Beauce aquifer at Toury from 1900 to the present day

5.2.3 The Onset of Overextraction

The groundwater level in the Beauce aquifer has always fluctuated, reflecting variations in the amount of water stored within the aquifer, which depends on the balance between recharge from rainfall, discharge into the rivers and abstraction by users (see Fig. 5.4). For years, the annual variation in groundwater level has shown a good correlation with winter rainfall. However, this relationship was disrupted by the 1976 drought and the shift became more apparent in the early 1990's when the fall in groundwater level then seemed to be decoupled from the decline in winter rainfall. This process reflects the development of abstraction for irrigation which accelerated the fall in successive dry years.

However at the start of the 1990s, there was a significant fall in the groundwater level to the lowest levels ever recorded. The level declined by 6 m over a period of 5 years which caused a general decline in the flow rates in the rivers draining the aquifer and resulted in several drying up for a sustained period. A similar decline occurred over a 16-year period in 1906.

Of all the rivers affected, the Conie River attracted attention and triggered conflicts between stakeholders. It normally flows continuously, but completely dried up between June 1992 and October 1993, affecting the condition of protected wetlands. Local residents set up a river protection group and requested the state to apply urgent measures to limit abstraction for irrigation, which was thought to be the source of the problem.

Given the size of the region and the large number of irrigators, the state could not react immediately. In response to what the group considered to be state inertia, the group referred the matter to the administrative courts. The state was condemned on the grounds that several of its decisions were judged inadequate (Petit, 2009). Ultimately, the group's action clearly raised awareness about the need to establish a joint management approach for water abstraction from the Beauce aquifer.

The 1992 Water Act gave the public authorities greater power to intervene administratively to regulate water use (see Chap. 3). In the Beauce region, the new regulatory tools made it possible to implement the first measures to limit irrigation. Initially, this involved a ban on irrigation for a set number of days, which was coordinated across six counties. These restriction measures had to be explained to the farming community. An "irrigation charter" was signed in 1995, laying out the rules for limiting abstraction depending on average groundwater levels in the aquifer. The average level was determined on the basis of a single groundwater level indicator based on the mean value of the groundwater levels recorded at nine monitoring wells across the territory. Three thresholds, T1, T2 and T3 were determined for this indicator and the following rules were established:

- when the level drops below T1, irrigation for all crops is banned for 24 h per week; but as soon as the level starts to rise, the restriction is lifted;
- when the level drops below T2, a ban is imposed for 48 h per week for cereals and 24 h for other crops; but if the level rises again, it is reduced to 24 h;
- when the level drops below T3, greater restrictions are necessary and must be negotiated between the state and the agricultural sector.

One of the key processes of this management approach was defining the groundwater level indicator. It was constructed as follows: (1) the aquifer was initially divided into geographic areas corresponding to the catchments for the main rivers; (2) a representative monitoring well was selected in each area. The indicator was then calculated by averaging the water levels of the nine monitoring wells, with values weighted as a function of the size of the catchment at each site. (3) The thresholds were established by taking into account the levels that were measured in several exceptional years in terms of climate, which were regarded as representative of the different crisis levels: April 1990 for T1; December 1976 for T2; and January 1994 for T3.

After the system was implemented, restrictions were repeatedly imposed between 1994 and 1997 due to drought. These restrictions were challenged in court by environmental campaigners, as well as irrigator organisations. The degree of conflict reflects the fact that the majority of irrigators were reluctant to accept the principle of state intervention, even though it was quite limited.

Therefore during this period, there was hardly any improvement to the status of the aquifer and its associated aquatic ecosystems. Abstraction remained too high because the restrictive measures only affected irrigators with limited pumping capacity. In fact, most irrigators had the capacity to abstract as much water as they needed over the restricted time period.

5.3 Second Stage: Introducing a Provisional Mechanism for Volumetric Management (1999–2005)

5.3.1 The SDAGE Sets the Guidelines

In 1996, the first two River Basin Management Master Plans (SDAGE) for the Seine-Normandy and Loire-Brittany catchments were approved. These plans which set the guidelines for managing water on the scale of river basins, also defined management goals on a more local level (see Chap. 4). In the case of the Beauce aquifer, they confirmed the definition of the groundwater level indicators and the three associated thresholds: a target level (T1), an alert level (T2) and a crisis level (T3). The SDAGE was thus giving the force of law to the system established at the local level. Both SDAGEs also set a target flow rate to be maintained in the rivers and streams that were dependent on groundwater baseflow. Alert and crisis flow rates thresholds were also determined which when exceeded, triggered restrictions on the use of both rivers and groundwater. Lastly, the establishment of a multi-party inter-basin commission was imposed with a view to designing a Local Water Management Plan (SAGE). The implementation of the SAGE was presented as a priority. Thus the SDAGE provisions outlined the regulatory framework for future action at a local level.

5.3.2 A Provisional Approach Prior to the Local Water Management Plan

In 1997, an inter-basin working group was established to ensure dialogue between the different stakeholders in the Beauce region. The group was the precursor of the future local water commission which would be responsible for drafting the Local Water Management Plan. The group decided to modify the system for regulating abstraction and validated the two following principles:

- the first involved determining an overall maximum authorized abstraction volume that could be extracted in a year of average rainfall. This volume was to be shared between authorized users. To ensure a degree of flexibility, irrigators were allowed to exceed their reference volume by up to 20%, but the volume in the following year was cut by the equivalent amount (penalty). They also had the option of saving up to 20% of their volume, which could be carried over to the following year (bonus);
- the second provision specified that the available volume would be estimated at the start of each year depending on recharge to the aquifer as measured by a unique groundwater level indicator which combined data from the groundwater monitoring network (the historic indicator from the 1995 charter). The volumes allocated to individual users were reduced by 10% if the indicator level in March dropped below T1, and by 20% if it was below T2.

The transition to volumetric management was facilitated by the installation of water meters on almost all of the boreholes pumping from the Beauce aquifer. The operation began in 1993, with 75% funding from the Loire-Brittany River Basin Agency, and continued until 1998/2000 in the Seine-Normandy Basin. Over 3500 volumetric meters were installed (mostly electromagnetic meters). The agricultural sector supported the action, convinced that it was a necessary transition for managing volumes.

5.3.3 Implementation of the First Volumetric Management System

The cap imposed for the maximum volume for abstraction was negotiated between the state and the agricultural sector. It was agreed that the volume allocated to agriculture must be at least 1000 m^3 per hectare of irrigable land, when the 20% restriction was imposed. This corresponds to a total volume of 360 Mm^3 (based on 360,000 ha of irrigable land). This meant authorising the allocation of 450 Mm^3 in a normal period (when the groundwater indicator is above T1). This volume corresponded almost exactly to the total volume that the state had previously authorised for abstraction (Fig. 5.5).

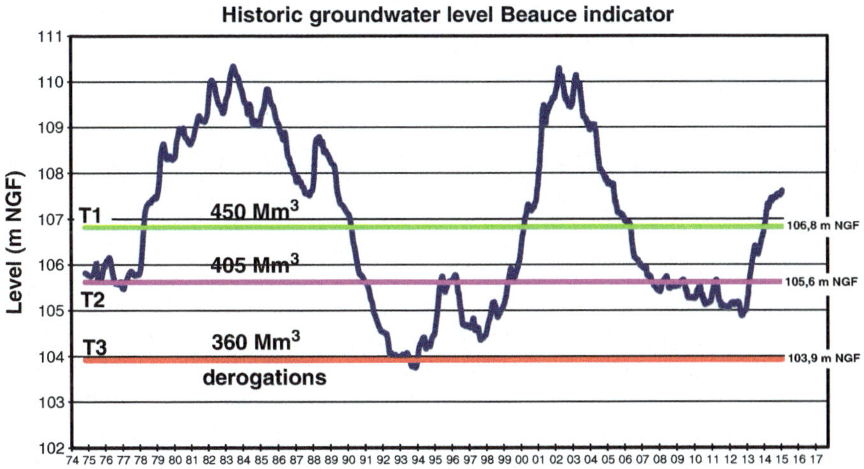

Fig. 5.5 Groundwater level indicator and volume authorised for irrigation as a function of the level on 1st March

After legalising formerly undeclared users, a total volume of 470 Mm^3 was authorised for pumping. Therefore, a reduction coefficient of 0.9554 was applied to all the previous pumping authorizations to meet the target of 450 Mm^3. An individual reference volume, corresponding to the maximum volume that could be pumped in a normal year, was then communicated to each user. In 2004, the management area was extended to include the adjacent region of Blois, which increased the total authorised volume for pumping from 450 to 525 Mm^3.

5.3.4 How the Agricultural Sector Accepted the Measure

The volumetric management system described above was introduced when the climate was extremely favourable, both in terms of groundwater recharge and streamflow. In the years from 1999 to 2002, the indicator remained above T1, and agriculture was authorised to abstract the maximum volume. The irrigation campaign progressed smoothly for all users (drinking water supply, industrial use and flow into shallow aquatic ecosystems).

The administrative operations were fine-tuned during this period. Every year, the state provided a pumping licence to each of the 3300 irrigators and read the meters at the end of the season. The favourable climatic conditions also meant that the farmers, who were initially reticent, accepted the approach more readily.

5.4 Third Stage: Revising the Volumetric Management System in the Framework of the Local Water Management Plan (2005–2013)

5.4.1 Volumetric Management Does Not Prevent Rivers from Drying Up

The 2003 drought began when the groundwater level indicator was near its highest ever levels, which meant that the irrigation season progressed unhindered. The allocated volume was approximately 500 Mm^3 and only 300 Mm^3 was actually pumped.

From 2003 onwards, while average groundwater levels remained high, extremely low water flows were recorded in several rivers draining the eastern periphery of the Beauce aquifer. Some actually dried up for prolonged periods. In the years that followed, the crisis situation continued. The situation revealed the limitations of a system based on a single groundwater level indicator located centrally in the region that did not adequately monitor levels close to the rivers and hence failed to guarantee the good ecological status of certain rivers on the periphery.

At the same time, the European Water Framework Directive strengthened the legal obligation for the state to restore a good ecological status (GES) for aquatic ecosystems and water resources. In 2005, the parties involved started discussions with the aim of improving the volumetric management system. Their findings also contributed to the development of the Beauce Local Water Management Plan (SAGE). Several studies were conducted which increased the understanding of the groundwater system due to more intensive groundwater monitoring and computer modelling. Discussions also contributed to the revision of the SDAGEs in Loire-Brittany and Seine-Normandy because one of the provisions was directly related to the Beauce aquifer. In-stream flow rate thresholds were defined for the main rivers, complementing the groundwater level threshold system. Last but not least, it was decided that irrigation would be totally banned if groundwater levels were to drop below the lowest T3 threshold.

5.4.2 Aquifer Modelling Indicates That Abstraction Must Be Reduced

The new management system was based on improved groundwater level monitoring and a better understanding of the interactions between the aquifer and rivers. The improved groundwater monitoring network, combined with improvements to the network of river hydrometric stations, contributed to a better knowledge of how the complex Beauce aquifer and the associated rivers interacted. The new approach recognised the fact that the different sectors of the Beauce aquifer have different storage capacities. Thus in the sectors with a low aquifer storage capacity, certain rivers have naturally low flow in summer (e.g. in the Fusain and Montargois sectors). In order to protect the aquatic ecosystems in these sectors, restrictions on abstraction must therefore be stronger than in other sectors.

At the same time, a groundwater model was developed to simulate the flow systems of the aquifer and the discharge to the rivers (Hydroexpert, 2004). The simulations revealed that under normal conditions (groundwater levels and recharge), the total volume pumped for irrigation should not exceed 200 Mm^3 per year. The simulations also show that if 400 Mm^3 is abstracted annually for several consecutive years, the groundwater level would drop to its lowest ever levels. This could drain the margins of the aquifer, causing rivers and streams to dry up rapidly and permanently.

5.4.3 The New Volumetric Management System

Individual reference volumes were initially allocated on the basis of historical irrigated area for each farm. It soon became apparent that actual use was less than the allocated reference volume (significantly lower, in some cases). Therefore the management system operated with large volumes of unused authorisations which farmers wanted to keep for possible future use.

In 2006, the state began discussions with the agricultural sector, with a view to reducing the volumes authorised for abstraction and aligning them with the volumes actually used. The initial goal was to ensure that pumping authorisations did not exceed 200 Mm^3, implying a reduction of the allocated volumes by over 50%. Farmers were fiercely opposed to it.

The negotiation that followed focused on three management principles that the state wanted to reform:

1. *reducing the authorised pumping volumes*: the stakeholders agreed on the principle of dividing the aquifer into four zones, which would be independently managed. A maximum authorised abstraction was agreed and fixed at 420 Mm^3 (state services initially proposed 310 Mm^3) for the entire aquifer (a 20% reduction compared to the previous 525 Mm^3). The volume was divided between the four zones, with the same county distribution as that established in 1999. The carry-over provision for unused water was withdrawn;
2. *better accounting for the interactions between groundwater and rivers*: the negotiation focused on defining new indicators for the groundwater status, based on the choice of new monitoring wells. The previous indicator was actually based on measurements of groundwater level in wells that had little correlation to river and stream flow rates, and therefore failed to anticipate the impact of abstraction on watercourses in dry periods. A study was then conducted to identify monitoring wells that were better correlated to the river flow rates. The research resulted in proposing a new indicator for central Beauce, which had far less inertia than the previous indicator and a better correlation to the flow of river outlets, particularly the Conie River;

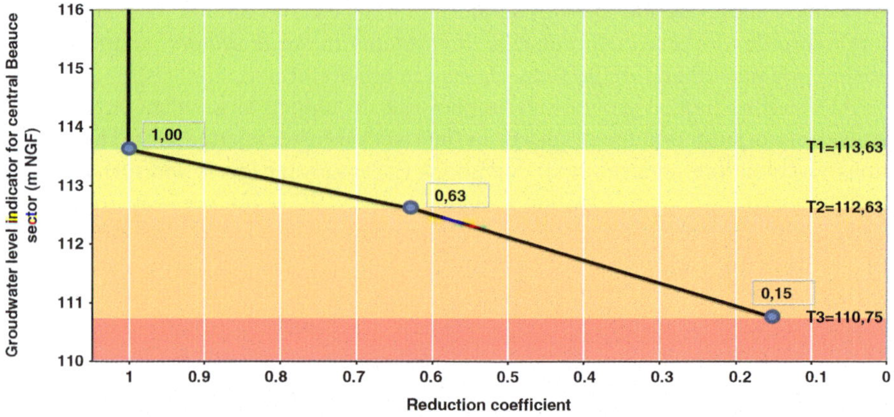

Fig. 5.6 Graph to determine the reduction coefficient for abstraction as a function of the groundwater level indicator at the start of the season in the central Beauce sector

3. *redefining the reduction coefficients associated with the alert and crisis thresholds (T1 and T2)*: lengthy negotiations took place between the different parties on this issue (Bouarfa et al., 2011). First, the decision was made to replace the existing system that was based on two set levels of restriction (10% and 20%), by a continuous function (Fig. 5.6) so the reduction coefficient could be adapted to the exact groundwater level measured at the start of the season. The negotiations almost failed as a result of the high tension surrounding the choice of the parameters for this function. The state defended the choice of a function capable of guaranteeing that there would be no more hydrological stress in the watercourses (in particular, the Conie River), while the agricultural sector wanted to maintain access to large volumes even in periods of low water flows. The outcome of the negotiations is illustrated in Fig. 5.6.

However the new management system could not totally prevent the occurrence of crisis situations, although it reduced their frequency. Measures to restrict pumping during the season (from 24 to 48 h per week) were maintained and applied when river flow rates fell below the alert thresholds. In the end, these rules were included in the SAGE and approved by all the stakeholders in the local water commission in June 2013. Since then, the rules have had a firm legal status.

5.4.4 The Implementation of the New Volumetric Management System

Table 5.1 shows the reduction coefficients that have been applied, sector by sector since 1999. In 2007, more restrictive coefficients for coping with low groundwater levels were applied in anticipation of the new management rules being discussed

Table 5.1 The coefficients applied from 1999 to 2017 in the different management sectors

Year	global volume	reduction coefficients				
		Beauce Blésoise	Beauce centrale	bassin du Fusain	Montargois	
1999	471		0.8			historic mechanism's rules
2000	471		0.86			
2001	471		0.955			
2002	471		0.955			
2003	471		0.955			
2004	525	0.955				
2005	525	0.955				
2006	525	0.86				
2007	525	0.5	0.65			rules before the SAGE was approved
2008	525	0.45				
2009	525	0.587				
2010	525	0.66	0.71	0.5625	0.5625	
2011	525	0.737	0.91	0.588	0.5625	
2012	525	0.594	0.66	0.594	0.594	
2013	525	1	1	0.9	1	
2014	420	1	1	0.95	1	after SAGE approval
2015	420	1	1	0.64	0.94	
2016	420	1	1	0.63	0.96	
2017	420	1	1	0.58	0.52	

in the framework of the future SAGE. This marked a break with the historic mechanism, which was advantageous for irrigators until then. From 2010, each sector was attributed its own coefficient. The practice of volume carry-over from one year to the next was not abandoned until 2010.

The first time the new rules defined by the SAGE were fully applied was in 2014. Time will tell if the measures are sufficient or if they need improving. The volume abstracted will probably have to be reduced further in order to bring it closer to the 200 Mm^3 estimated by the simulation model. At the moment, a reduction of this scale is politically unacceptable because it would involve a total.

5.5 Outlook

5.5.1 Additional Measures that Have Been Introduced or Envisaged

Additional measures have already been introduced or envisaged to improve the quantitative status of the Beauce aquifer and dependent aquatic ecosystems. The first measure involves relocating the boreholes that have a major impact on rivers because of their close proximity. A preliminary trial conducted in the Conie Basin in 1999, resulted in the transfer of four boreholes that severely affected the river's flow rate. A larger operation involving 11 boreholes, was conducted in the Fusain Basin in 2014. In this case, the boreholes near the river which had the greatest impact on the river flow, were moved over 800 m away from their original location. The operation was designed to produce an estimated gain of 130 L/s, the equivalent of the average monthly flow rate at low water level in the section of the river concerned. Specific provisions were set for the new boreholes (the flow and volume authorised), to ensure that the increased river flow rate did not deteriorate over time. The original boreholes were plug and abandoned. The project cost a total of 1,350,000 €, of which public subsidies financed 72% and the farmers the remaining 28%.

The second measure involves the development of winter water storage as a substitute for pumping at low groundwater levels. A project for a collective reservoir with a capacity of 400,000 m^3, fed by water pumped from the aquifer in winter, is currently being considered. This operation is extremely expensive (average cost of around 5 € per m^3 of storage), but would be subsidised like the borehole relocation project. The same strategy has been used in other regions in the Loire Basin (see Chap. 18).

The third measure explored involves artificially replenishing the aquifer by forcibly injecting water that is pumped from the river in the winter. The reinjected water could then be pumped over an annual or multi-annual period. In the case of multi-annual management, the reinjected water could be recovered at a rate of approximately 25% per year. This method has not yet been tested in a real situation.

5.5.2 Considering Climate Change

The Beauce aquifer is likely to be significantly affected by climate change. The scientific studies, RExHySS (Ducharnes, 2009) and Explore 2070 (Ministry of Ecology, Sustainable Development and Energy, 2012), have predicted a major and significant reduction in summer rainfall and, to a lesser but significant degree, in winter rainfall. The studies also show that potential evapotranspiration (PET) will increase significantly (+16% on average by 2050 and + 23% by 2100), which will lead to a decrease in recharge and a likely increase in abstraction for irrigation. There will probably be a very marked shift by the 2050s. The climate change will reduce the recharge by about 20% in 2050 and by nearly 30% in 2100. On a regional scale, the supply deficit for the Beauce aquifer will represent 180 Mm^3 by 2050, almost equivalent to the average volume now being abstracted from the aquifer for irrigation. Consequently, the groundwater levels could drop by several metres, reducing the baseflow to rivers and streams by up to 25%. Consequently, the current management approach used for irrigation abstraction will need to change.

5.5.3 The Collective Water Management Groups for Irrigation (OUGCs)

In the Beauce aquifer region covered by the SAGE, five collective management groups (OUGCs) were established in 2011 and 2012. They brought together all the stakeholders that use the aquifer for agricultural purposes. The OUGCs distribute the authorised volume for abstraction in the 10 areas that constitute the four groundwater management sectors (see Chap. 3). The OUGCs now have a single multi-annual water abstraction authorisation, and are responsible for managing and distributing the authorised volume that can be pumped for irrigation in their own jurisdictions for a 15 year period. Irrigators have lost their individual right to abstract, with their right to use water now dependent on the annual distribution plan which is presented annually by the OUGC to which they belong. Nevertheless, the OUGC has not changed the rules of distribution. Each user receives a share of the authorised volume which is proportional to historical use.

5.6 Conclusion

The Local Water Management Plan for managing abstraction for irrigation for the Beauce aquifer and its associated aquatic ecosystems, was adopted after a long drawn-out process that began after a severe drought and took over 20 years to complete. The long timeframe was helpful for improving the understanding of the complex aquifer through long term groundwater monitoring and groundwater flow modelling, and also for obtaining a more accurate assessment of groundwater use (borehole locations, widespread metering of volumes abstracted). Above all, it brought stakeholders together and enabled them to address regional issues jointly, despite their often conflicting interests. One salient feature which is specific to Beauce region, was the limited involvement of environmental protection groups. As a result, they had little influence on the outcome, although the process was actually initiated by an environmental protection group at the start of the 1990s.

During the process, farmers became more aware of the need to improve the management of a fragile resource and the fact that groundwater is not inexhaustible, contrary to some commonly held beliefs. At the same time, the regulatory framework changed and increasingly stringent environmental protection measures were enforced.

The stakeholders' commitment for the duration of the process was an undeniable advantage. The majority of the representatives from the state services and water agencies, as well as from the agricultural sector, participated to the negotiation from the beginning in the early 1990s until the end in 2013. They were able to develop a good understanding of the resource and the main issues and kept a memory of the compromises and concessions made. Importantly, they committed to deliver their promises. The same result would probably not have been reached with a high turnover of stakeholders involved in the design and implementation of the new Plan.

The outcome is a compromise between the interests defended by various parties, but nonetheless, represents a milestone in terms of water resource management in France. Obviously, the system will continue to evolve because it still does not fully satisfies all parties. Above all, the Plan will have to respond to the future challenge of climate change, an issue that it has not yet considered.

References

Bouarfa, S., Brunel, L., Granier, J., Mailhol, J. C., Morardet, S., & Ruelle, P. (2011). Évaluation en partenariat des stratégies d'irrigation en cas de restriction des prélèvements dans la nappe de Beauce (France). *Cahiers Agricultures, 20*(1–2), 124–129.

Désiré, E. (1972). L'extension récente de la culture du maïs en Beauce. *Norois, 73*(Janvier–Mars), 27–46.

Ducharne A. (2009). *GICC project final report*. Paris: University Pierre et Marie Curie. 61p. Available at http://temis.documentation.developpement-durable.gouv.fr/docs/Temis/0076/Temis-0076031/19589.pdf

Hydroexpert. (2004). *Reprise du modèle de gestion de la nappe de Beauce* (Rapport RPP04M026b). Orléans: Agence de l'eau Loire Bretagne.

Le Coz, D. (2000). Gestion durable d'une ressource en eaux souterraines. Cas de la nappe de Beauce. *La Houille Blanche*, (7–8), 116–121.

Ministry of Ecology, Sustainable Development and Energy. (2012). Explore 2070: hydrology souterraine. In *Available at: https://professionnels.afbiodiversite.fr/sites/default/files/pdf/RE_Explore2070_Eaux_Sout_Resume.pdf.*

Petit, O. (2009). Un regard rétrospectif sur l'évolution de la gouvernance de l'irrigation en Beauce (1993–2008). *Sécheresse 2009, 20*(3), 262–270.

Verley, F. (2014). Retour sur la construction des outils de pilotage de la gestion volumétrique des prélèvements agricoles pour l'irrigation en Beauce. DREAL Centre, Orléans. 101 p.

Chapter 6
Groundwater in Australia: Occurrence and Management Issues

Steve Barnett, Nikki Harrington, Peter Cook, and Craig T. Simmons

Abstract Groundwater is one of Australia's most important natural resources and is the only source of water available for many regions, supplying urban areas, agriculture, industry and mining developments. The economic value to the economy is estimated to be $A34 billion. Groundwater also sustains ecosystems, through baseflow discharges to surface water and artesian spring discharges.

Groundwater is found in both sedimentary and fractured rock aquifers, with most groundwater extraction occurring from the higher yielding sedimentary aquifers including unconsolidated Quaternary alluvial sediments, consolidated sandstones and limestones in large sedimentary basins. Low salinity groundwater is generally found in higher rainfall areas around the southern coastal areas. In the arid interior, high evaporation results in salinities up to 100,000 mg/L. Deeper confined aquifers may contain older low salinity groundwaters recharged thousands of years ago.

Groundwater resources have been rapidly developed over the past 40 years. Current extraction is about 5000 GL/year with 70% used for irrigation whereas in France, 60% of the total extraction of 34,000 GL/year was used for public water supplies. Early management intervention has resulted in only 2% of Australia's management areas being over-exploited.

Future challenges for groundwater management in Australia include potential impacts of climate change, impacts of mining and declining government funding.

S. Barnett (✉)
Adelaide, SA, Australia
e-mail: Steve.Barnett@sa.gov.au

N. Harrington
Innovative Groundwater Solutions Pty Ltd (IGS), Victor Harbour, SA, Australia
e-mail: nikki@innovativegroundwater.com.au

P. Cook
Flinders University, Adelaide, SA, Australia
e-mail: peter.cook@flinders.edu.au

C. T. Simmons
National Centre for Groundwater Research and Training, Flinders University, Adelaide, SA, Australia
e-mail: craig.simmons@flinders.edu.au

J.-D. Rinaudo et al. (eds.), *Sustainable Groundwater Management*, Global Issues in Water Policy 24, https://doi.org/10.1007/978-3-030-32766-8_6

Keywords Hydrogeology · Groundwater resources · Groundwater extraction · Climate change · Mining impacts · Groundwater management

6.1 Introduction

Australia is the driest inhabited continent on Earth with low rainfall, high evaporation and very limited surface water resources. Consequently, groundwater is one of Australia's most important natural resources and is the only reliable and cost-effective source of water available for 60% of the continent's total area, supplying urban areas, agriculture, industry and mining developments. It often plays a crucial role in sustaining ecosystems, through baseflow discharges to surface water and artesian spring discharges. Groundwater provides more than 30% of Australia's total water consumption and generates national economic activity worth $A34 billion a year across agriculture, mining, industry and water supplies in rural cities and towns (Marsden Jacob Associates, 2012).

This chapter provides an overview of Australia's groundwater resources and traces the history of their development. Section 6.2 provides details of groundwater resources in Australia, Sect. 6.3 discusses groundwater use, whilst Section 6.4 examines groundwater management issues, including the need for more timely policy responses (as will be explored in more detail in later chapters in this book). Section 6.5 briefly outlines future challenges.

6.2 Groundwater Resources in Australia

Australia lies between latitudes 10° and 44°S, and is covered by several climatic zones ranging from tropical with summer rainfall in the north, to temperate with winter rainfall in the south. Most of the interior has an arid climate, with some areas receiving less than 100 mm median annual rainfall and experiencing annual evaporation of more than 4000 mm. The rainfall averages 465 mm and results in a runoff of only 57 mm because of the high evaporation. The seasonal distribution of rainfall varies greatly over the continent which is generally of low relief. The tropical areas of northern Australia have high summer rainfall because of monsoonal conditions. Elsewhere higher rainfall due to topographic effects mainly occurs along the Great Dividing Range which runs parallel to the east coast, and in the south-western and south-eastern parts. These areas of higher rainfall are also where most of the highly urbanised population of 23.4 million resides. Figure 6.1 presents Australia's average annual rainfall, which in comparison by volume, represents only 1% of the annual rainfall falling on France.

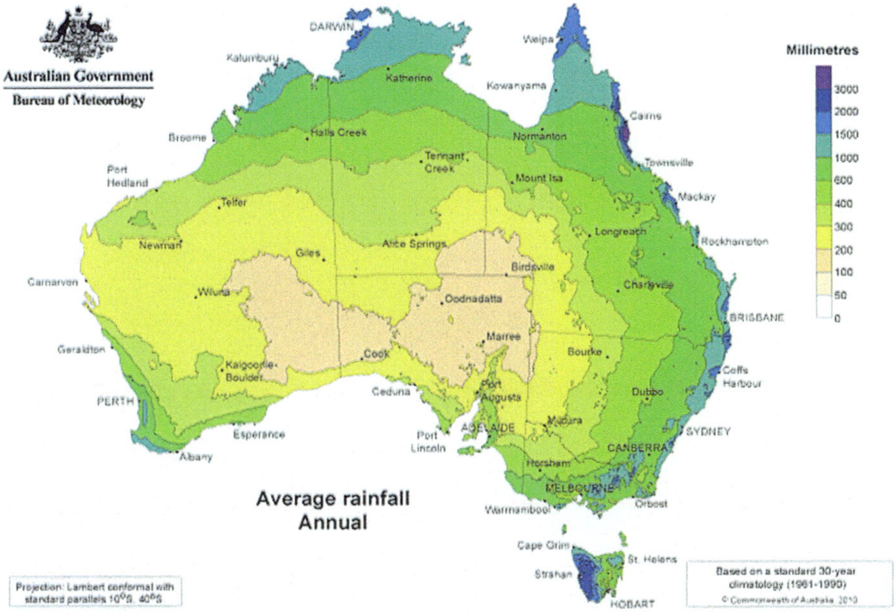

Fig. 6.1 Average annual rainfall of Australia

Although groundwater is found in both sedimentary and fractured rock aquifers, most groundwater extraction occurs from the higher yielding sedimentary aquifers which cover about 65% of the continent. Shallow unconsolidated alluvial sediments contain good quality groundwater in the higher rainfall highland areas along the eastern coastline, however further inland in drier areas, these sediments contain more saline groundwater due to lower rainfall and higher evaporation.

Aquifers in large sedimentary basins contain considerable volumes of groundwater in consolidated sandstones and limestones. Aeolian sediments and karstic limestones also form significant aquifers.

Figure 6.2 illustrates the major groundwater resources in Australia which include:

- the Great Artesian Basin, which covers one fifth of the continent;
- the major alluvial aquifers of the Murray–Darling Basin, which support Australia's major food bowl;
- the Perth Basin, which supplies much of Perth's water demands;
- the Canning Basin in northern Western Australia;
- the Daly River Basin of the Northern Territory; and
- the Otway Basin aquifers of south-east South Australia and south-west Victoria.

Fractured rock aquifers are prevalent throughout much of the Great Dividing Range of eastern Australia, Tasmania, the Mt Lofty and Flinders Ranges in South Australia, and the ancient hills and ranges of Western Australia and the Northern Territory (Fig. 6.2). Given the variability of water-bearing joints and fractures in these aquifers, well yields and salinities can vary greatly over short distances.

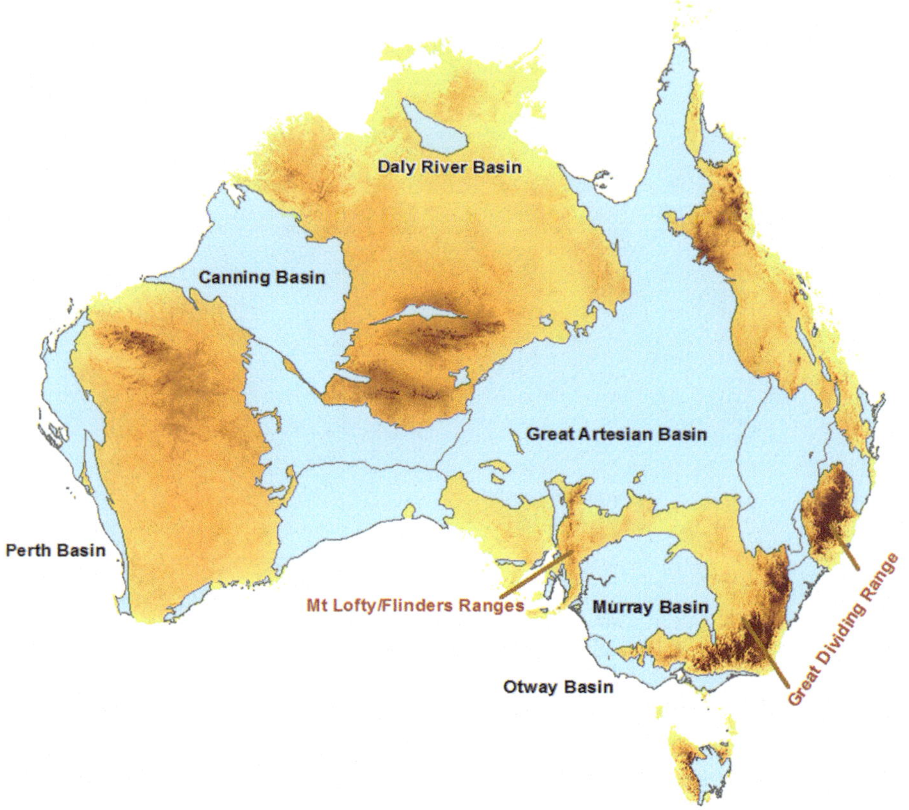

Fig. 6.2 Major groundwater resources of Australia

Typically, better quality groundwater is found in regions with higher rainfall where recharge to the aquifers can occur more readily. Sandstone, quartzite, siltstone and basalt aquifers tend to be productive because the joints and fractures are open and permeable, whereas in metamorphic and intrusive igneous rock aquifers, they are poorly developed or infilled by clayey weathered material.

A number of ecosystems that rely on groundwater discharge or access to groundwater, occur in a wide range of forms throughout Australia (Clifton, Cossens, & McAuley, 2007). There are three main types of these groundwater-dependent ecosystems (GDEs);

- ecosystems reliant on the surface expression of groundwater such as springs and wetlands
- ecosystems reliant on access to subsurface groundwater, for example vegetation that accesses groundwater through roots, drawing on moisture derived from the watertable
- aquifer and cave ecosystems that occur within the aquifer itself and support stygofauna

As the water requirements of GDEs must be taken into account in most groundwater management plans in Australia, a key advance has been the commissioning of a web-based GDE atlas for Australia (http://www.bom.gov.au/water/groundwater/gde/map.shtml). This atlas displays ecological and hydrogeological information on GDEs across Australia, collated from a number of sources, including published research and remote sensing data.

6.3 Groundwater Use

6.3.1 Historical Development of Groundwater

It is probable that groundwater has been used in Australia by aborigines for thousands of years according to the artefact evidence found adjacent to shallow wells and springs in several locations (AWRC, 1975). Although the first European settlements in the late eighteenth century were located according to the availability of surface water, shallow hand dug wells were also used when surface water was unavailable during drought, or of unsuitable quality due to contamination. As settlement spread from the well-watered coast into the more arid interior, the value of groundwater for providing reliable water supplies was realised. In 1857, the state of Victoria initiated the first known investigation into groundwater resources, and in 1871, an artesian well was drilled to a depth of 52 m near Perth in Western Australia (AWRC, 1975).

Development of the iconic Great Artesian Basin (GAB) began with the first well drilled in 1879 in north-western New South Wales. Similar wells were drilled in adjacent states over the next few years and by 1910, over 1500 artesian wells had been drilled throughout the GAB. Since then, over 30,000 wells have been drilled which have been essential for the expansion of the pastoral industry into the dry interior of Australia. This has resulted in declining pressure levels and in response, an extensive program has been undertaken to restore pressure levels by capping uncontrolled flowing wells (see Chap. 7).

After World War 2, the increased use of the rapid rotary drilling technique and downhole geophysical logging tools that could detect permeable layers, allowed more thorough investigations of groundwater resources. Demand for groundwater increased in the 1950s and 1960s to support agricultural development in the following regions;

- Lachlan Valley – alluvial sediments in tributary valleys of the Murray Basin in New South Wales
- Burdekin Delta – alluvial sediments deposited by the Burdekin River in northern Queensland
- Northern Adelaide Plains – Tertiary limestones of the St Vincent Basin in South Australia
- Western Port Basin – Tertiary sands, limestones and volcanics bordering the coastline in Victoria

During this period of rapid expansion in the 1970s and 1980s, a lack of resourcing for the management and monitoring of a number of groundwater systems in Australia, and occasionally some political interference, led to too many licences being issued and agricultural users being granted allocations in excess of the sustainable extraction limits (over-allocation) and in some cases, the volumes of groundwater extracted are also above the sustainable extraction limits (overuse) (NWC, 2012). These issues ae discussed in more detail in later chapters (see Chaps. 7 and 14).

In recognition of the huge task of assessing Australia's water resources, the Australian Water Resources Council was formed in 1962 by the federal and state governments (now no longer in existence). This led to a significant increase in investment into the investigation of major groundwater resources which continued through the 1980s and 1990s. Since 2000, the focus has been mainly on the monitoring and management of the known groundwater resources where groundwater extractions for irrigation purposes have generally stabilised.

6.3.2 Groundwater Usage

As detailed above, Australia has witnessed a rapid development of groundwater resources use over the past 40 years. Accurate statistics on groundwater abstraction and use in Australia has been difficult to obtain in the past due to limited monitoring infrastructure and inconsistencies in water accounting methods. However in recent years, more consistent national reporting procedures required by the National Water Initiative (see Chap. 7) has resulted in more accurate estimates.

For the 2015/16 water use year, an estimated 5000 GL was extracted, which represents about 63% of the total volume of entitlements (BoM, 2016). These volumes amount to approximately 30% of Australia's total water consumption. By contrast, the total consumption in France is about 34,000 GL/year. Because of the drier climate and smaller population in Australia, 70% of groundwater extraction is for irrigation whereas in France, 60% of extraction is for public water supplies. The mining industry in Australia is estimated to extract an additional 1500 GL/year. Figure 6.3 shows the regions of Australia and their dependence on groundwater.

The salinity of Australia's groundwater resources varies widely and is a major factor controlling groundwater use and development. In unconfined aquifers, low salinity groundwater which can be used for human consumption is generally found in higher rainfall areas around the northern and eastern coasts, as well as the south-eastern and south-western portions of the continent. In deeper confined aquifers, older low salinity groundwaters recharged thousands of years ago (possibly during wetter climates than that experienced today), may be found beneath arid areas after travelling along long flow paths e.g. in the Great Artesian and Murray-Darling Basins. These 'ancient' resources, sometimes over one million years old in the Great Artesian Basin, provide essential supplies for towns, livestock and irrigation. Over large areas of the arid interior where evaporation considerably exceeds rain-

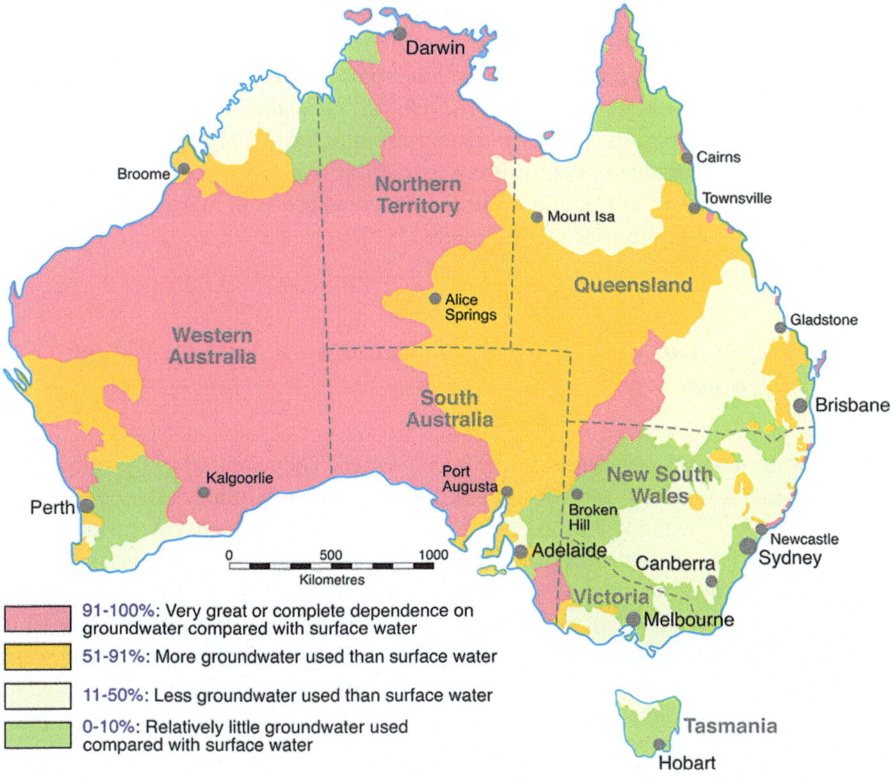

Fig. 6.3 Dependence on groundwater in Australia

fall, high salinities occur that are too high for drinking or agricultural purposes. However, the gold mining industry in Western Australia uses highly saline groundwater over 200,000 mg/L sourced from Tertiary palaeochannels for mineral processing.

6.4 Groundwater Management Issues

6.4.1 Overallocation and Overuse of Groundwater

The impacts of groundwater overuse are well documented (unacceptable decline in groundwater levels which could lead to loss of access to groundwater by users and ecosystems, increased infrastructure and pumping costs, seawater intrusion; and increased salinity of aquifers through inter-aquifer leakage).

The National Water Initiative (which is explained in more detail in Chap. 7) committed the States to address these issues. In 2016, there were 288 groundwater management areas (GMAs) throughout Australia. Of these, 136 GMAs had volumetric limits for extraction, with 25% of these classified as over-allocated, but more importantly, only 2% considered overused. While some of the GMAs that do not have volumetric limits could be considered 'at risk' or 'over-allocated', the management plans generally have rules that keep extractions at sustainable levels (BoM, 2016). Despite the uncertainties in determining extraction volumes in some GMAs, it can be concluded that over-extraction is not a widespread problem in Australia, although there may be localised issues in some areas.

6.4.2 Impacts of Groundwater Extraction on Surface-Water Systems

Groundwater and surface water are often connected. They are part of the one hydrologic cycle, and surface water resources can be significantly impacted by groundwater abstraction. Groundwater recharge can also be impacted by surface-water extraction or river regulation. Although hydrologists and hydrogeologists have long recognised the interconnection between surface-water and groundwater resources, and a study found that in some of Australia's more heavily developed catchments, reductions in baseflow to rivers and streams have caused reductions in surface water flows or complete drying out of streams (Evans, 2007). The impact of groundwater abstraction on surface-water systems has only been translated into policy and management to a limited extent so far (see Chap. 8).

The time lag from the commencement of groundwater pumping until the impact on a surface-water system is realised, can vary from days to decades or longer. Historically, groundwater and surface-water resources have been managed separately, however when increases in demand and low rainfall levels caused regulators to cap the use of many surface water resources, this had the unintentional result of shifting the extraction to nearby groundwater resources.

One of the best examples of this was in the Murray–Darling Basin, where there is usually a close hydraulic connection between the streams and the underlying alluvial aquifers. In 2008, groundwater use represented 16% of total water use in the Basin, and it was expected that this could increase to over 25% by 2030 under the prevailing water management arrangements that restrict surface water use. It was estimated that 25% of the groundwater use would eventually be sourced directly from induced streamflow leakage (CSIRO, 2008). The development of the Murray–Darling Basin Plan is Australia's most prominent example of government intervention to achieve effective conjunctive management of surface-water and groundwater resources (see Chap. 8).

6.4.3 Groundwater-Dependent Ecosystems

As mentioned earlier in this chapter, groundwater-dependent ecosystems (GDEs) occur in a wide range of forms throughout Australia. A great deal of progress has been made in the identification of GDEs. In 2007, a practical tool was developed to assist in the identification of GDEs and the management of their environmental water requirements (Clifton et al., 2007). This toolbox was updated in 2011 to include an assessment framework and assessment tools (Richardson et al., 2011a, 2011b).

6.4.4 Effect of Climate Change on Availability and Quality of Groundwater Resources

Over the past few decades, large areas of Australia have experienced a drier climate and reductions in surface water resources, causing increasing pressure on groundwater resources. Analysis of the climate over the past 80 years shows warming over most of Australia (except in the inland north-west); increasing rainfall over northern, central and north-western Australia; and decreasing rainfall in eastern, south-eastern and south-western Australia (Barron et al., 2011). From 1997 to 2009, large areas of southern Australia, particularly the southern Murray–Darling Basin, experienced the prolonged millennium drought which was the most severe drought in the 110 years of recorded rainfall history.

The south-west region of Western Australia has experienced a longer trend of gradually declining rainfall than the rest of the country. The average runoff to Perth reservoirs between 1975 and 2009 was 55% lower than prior to 1975, a result of a 16% fall in average annual rainfall. This led to the development of additional water supplies for Perth, resulting in a greater reliance on groundwater and commissioning of the first desalination plant to supply an Australian city (Bates, Chandler, Charles, & Campbell, 2010).

Most global climate models predict that southern Australia is likely to be drier in the future, which has significant implications for the future availability of groundwater resources. This is because variability in groundwater recharge can be two to four times greater than rainfall variability, with the effect of this being particularly obvious in areas of low recharge (Barron et al., 2011). Predictions of future changes in recharge have been made using 16 global climate models and these results have been scaled according to three global warming scenarios (low, medium and high) for both 2030 and 2050 (Barron et al., 2011). These predictions resulted in decreases in diffuse recharge across most of the west, centre and south-east of Australia, and increases across northern Australia. As well as reducing the availability of water resources, a drier and warmer climate may increase the demand for water resources from irrigated agriculture, cities, wetlands and other water-dependent ecosystems.

Hence, climate change intensifies the water scarcity challenge facing cities and rural catchments alike, but given the uncertainties in how and when the climate will change, an adaptive and flexible groundwater management approach is important to deal with these changes, together with comprehensive monitoring of the condition of the groundwater resources.

6.4.5 Impacts of Mining on Groundwater Systems

The mining sector is a large industrial user of water that is growing rapidly. Mining (including mineral, coal, petroleum and gas extraction) and quarrying tend to have a high gross value added per gigalitre of water consumed compared with agricultural uses. Despite an exponential increase in production, reported water use by the mining industry has historically been relatively steady, consuming within the range of 500–600 GL (Prosser, Wolf, & Littleboy, 2011). It is believed that the stable trend is due to improvements in water use efficiency in the mining industry, but also possibly to under-reporting of water use. Most water used for mining is in arid or semi-arid regions where water is scarce and often saline and there are few competing users. In these regions, the mining industry provides its own infrastructure, so water provisions tend to be part of the mining development approval process and are not always counted as a licenced extraction under a water management plan with other users.

There are exceptions to this, for example in the states of New South Wales and Victoria (and in Groundwater Management Areas in other states), all extraction for mining falls under the water planning process and requires a water access entitlement. Because mining is increasingly occurring in systems that are already developed for agriculture, the competition for resources and the potential impacts on existing water users is becoming more controversial.

A boom in coal seam gas (CSG) developments in Queensland and New South Wales presents major challenges in understanding and managing impacts of mining on other water users and the environment (Comino, Tan, & George, 2014). New technology to extract methane from deep lying coal beds has led to unprecedented CSG production in areas previously considered to be economically non-viable. Here, the gas extracted from coal seams is cooled and compressed to produce liquefied natural gas, which is ideal for export. Over 1000 gas production wells are in operation, requiring a peak extraction of about 95 GL to lower groundwater pressures which will allow the gas to be released from the coal deposits (DNRM, 2016).

Some key water management challenges in the current coal seam gas boom are (a) the effect of depressurisation on surrounding aquifers, (b) the likelihood and impacts of inter-aquifer leakage caused by aquifer depressurisation and hydraulic fracturing, and (c) chemical processes affecting the quality and safe disposal of the released water (Prosser et al., 2011). In Queensland, there are concerns over possible interactions of the CSG developments with usable aquifers (eg the Great Artesian Basin) which can occur above or below the coal seams. The locations of the potential development areas is shown in Fig. 6.4.

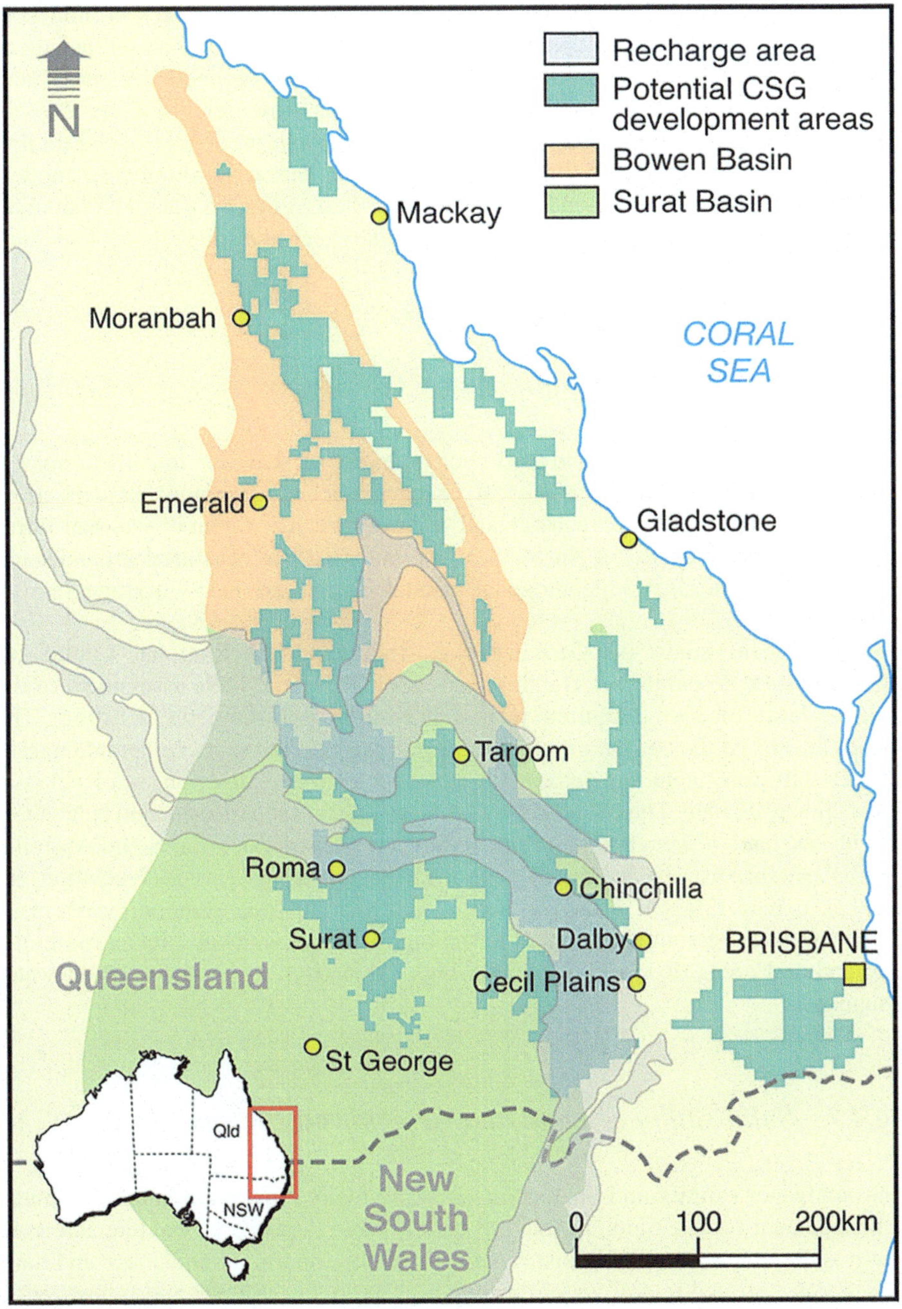

Fig. 6.4 Potential coal seam gas developments in Queensland

An overarching issue for the management of water resources around coal seam gas and mining developments in general is the uncertainty about the cumulative regional impacts of multiple developments on groundwater levels and pressures, and inter-aquifer leakage. Groundwater flow velocities are slow in many of the basins of interest, and any unforeseen consequences of the mining process can take decades or centuries to become apparent. Groundwater models are often desirable in this kind of analysis and these require a good characterisation of basin geology and how it controls groundwater pressures, flows, and quality (Prosser et al., 2011).

6.4.6 Seawater Intrusion

Seawater intrusion is the landward encroachment of seawater into fresh coastal aquifers in response to hydrological changes, such as groundwater extraction, reductions in groundwater recharge and sea level rise. The threat of seawater intrusion has been enhanced in Australia by an increased use of coastal groundwater, caused by increasing populations of coastal areas, and below-average rainfall (Werner, 2010).

One recent study by Geoscience Australia and the National Centre for Groundwater Research and Training, undertook a national-scale assessment of the vulnerability of coastal aquifers to seawater intrusion (Ivkovic et al., 2012). The study identified the coastal groundwater resources that are most vulnerable to seawater intrusion, including future consequences of over-extraction, sea-level rise, and climate change. The risk of seawater intrusion has been identified to be greatest in Queensland, although smaller but significant areas of Victoria, South Australia and Western Australia have also been identified as being at risk (Ivkovic et al., 2012). To date, targeted investigations into seawater intrusion to support water management have been mostly limited to the high value groundwater resources in the agricultural areas of Queensland, and high-value aquifers used for urban water supplies.

6.4.7 Salinisation of Land and Groundwater Resources

Groundwater salinity and land salinisation are major Australian natural resource management issues. Salts originating mainly from deposits of oceanic salt from rain and wind, are naturally distributed across the Australian landscape and have been concentrated in soil water through high evaporation rates and transpiration by plants for thousands of years. Salinisation of land and water resources occurs when comprehensive changes in land-use following European settlement changes the water balance and mobilises the naturally occurring salt. These land use changes involve dryland management systems (dryland salinity) and irrigation systems (irrigation salinity).

Dryland salinity occurs when European farming practices replaced deep-rooted native vegetation with shallow-rooted annual crops and pastures, resulting in a dramatic increase in groundwater recharge rates which can cause the watertable to rise, bringing salts into the root zones of plants and even to the surface where evaporation concentration occurs, impacting upon vegetation and rural infrastructure including buildings, roads and pipes. Increased recharge can also increase hydraulic gradients towards surface-water bodies which can increase the flow of saline groundwater into rivers and streams, thereby increasing river salinity. The National Land and Water Resources Audit defined the distribution and impacts of dryland salinity across Australia, with an estimation of 5.7 million hectares being affected or having a high potential for the development of dryland salinity (National Land and Water Resources Audit, 2001). The National Dryland Salinity Program which operated between 1993 and 2004, invested $40 million on research and development into the causes, costs, consequences, solutions and management of dryland salinity in Australia.

Since this audit, the previously discussed millennium drought from 1997 to 2009 affected large areas of southern Australia, reducing recharge and causing widespread declines in watertable levels which has reduced the risk of dryland salinity in many areas. It remains to be seen whether the predicted drying impacts of climate change will also reduce the risk of expansion of the current areas of dryland salinity.

Irrigation salinity (or recycling) occurs when groundwater is extracted from a shallow aquifer and is applied to a crop by irrigation, and as water is drawn up through the root system, most of the dissolved salt is not taken up by the plant and accumulates in the root zone. This salt then percolates back down into the shallow aquifer during subsequent irrigation applications or from rainfall recharge, resulting in increasing groundwater salinity over time due to the recycling of the irrigation drainage water.

6.5 Future Challenges

Australia has made good progress with water reform over the past decade or so. From a groundwater perspective, this included advances in groundwater science, education, management and policy reform (NWC, 2014). However, the process of water reform in Australia has now stalled. In 2007, the National Groundwater Action Plan was initiated with a $105 million investment from the federal government administered by the National Water Commission. This plan, which was the centrepiece of government investment in groundwater in Australia, terminated in 2012 with no continuation or replacement on the agenda.

Attention to groundwater and groundwater management from policy makers is deteriorating at an alarming rate, especially at a time when there are many groundwater related matters of national importance described previously (such as coal seam gas) that must be dealt with. Australia's future will increasingly depend on groundwater. It is expected that climate change and a potential doubling of the population in the next 50 years will be important future drivers of demand which will place further stress on already stretched groundwater resources (Simmons, 2016). The reform journey must continue if the many challenges are going to be overcome. But in order for this to occur, problematic institutional legacy and funding issues must be addressed (as discussed in other chapters in this book). The universal "hydro-illogical cycle" (Wilhite, 2012), whereby political attention and funding rains down during droughts but evaporates when water in plentiful, is being implemented in Australia.

National water reform must continue with the clear understanding that droughts are a fundamental component of the Australian climate. Recognising the need for an integrated, whole-of-water-cycle approach for effective water resources management, groundwater will be a critical part of any water reform strategy. An enduring, assertive, proactive, non-partisan approach to water reform in Australia is urgently needed which rejects short-term political and drought-driven interventions. The benefits for current and future generations will be enormous. The need for such reforms are discussed in later book chapters. For present purposes, this section will now focus on two emerging responses, namely managed aquifer recharge and general resourcing and education, that are otherwise not covered in detail in the remaining book chapters.

6.5.1 Managed Aquifer Recharge

Increasing pressure on water resources from climate change, population growth and increasing urbanisation means that Australia needs to diversify water sources to meet demand. Managed aquifer recharge (MAR) is a less established but growing alternative available to water resource managers. MAR is the purposeful recharge of water to aquifers through mechanisms such as injection wells, infiltration basins and galleries from sources such as rainwater, stormwater, reclaimed water, mains water and water from other aquifers. The recovered water may then be used for drinking water supplies, industrial water or irrigation, with the appropriate levels of pretreatment before recharge and post-treatment on recovery. Figure 6.5. presents a simple MAR operation.

Managed aquifer recharge can also be used to benefit the environment by leaving the stored water in the aquifer to sustain groundwater-dependent ecosystems or provide a barrier against seawater intrusion. The advantages of MAR are; it is less expensive than dam construction and uses far less land area, it minimises evaporation losses, and can make use of the natural attenuation properties of aquifer materials.

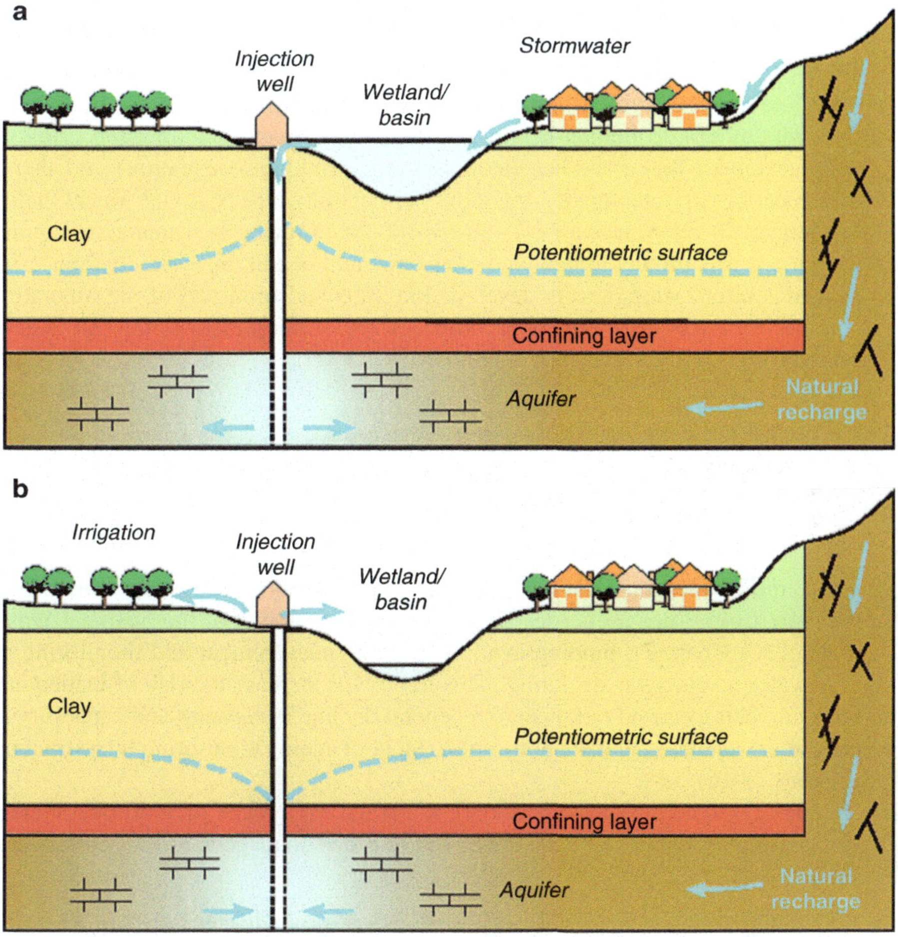

Fig. 6.5 Managed aquifer recharge (**a**) injection in winter (**b**) extraction in summer

A comprehensive overview of managed aquifer recharge is provided by Dillon, Pavelic, Page, Beringen, and Ward (2009). The occurrence and diversity of managed aquifer recharge in Australia has increased in recent years. In 2016, five states had operational managed aquifer recharge projects which contributed 410 GL/year to water supplies across Queensland, South Australia, Western Australia and the Northern Territory. The Western Australian Water Corporation has completed a trial of injecting more than 2.5 GL of highly treated recycled water into a deep confined aquifer over 3 years and it has been estimated that by 2023, groundwater replenishment of 28 GL/year could be achieved during winter for subsequent extraction during the dry summer months for potable supplies.

Currently growth in MAR in Australia is soundly based and is expected to make a greater contribution than sea water desalination in the longer term due to lower costs. However, there are some potential issues that need to be recognised and managed in the development of managed aquifer recharge schemes. These include water entitlement issues in water management plans (who owns the water), ensuring appropriate treatment of the source water (normally stormwater or recycled water) to prevent any risk to human health and the environment, including rendering an aquifer unsuitable for certain beneficial uses (e.g. drinking or irrigation), and also to prevent changes to water quality or aquifer permeability. The National Water Quality Management Strategy provides a framework for ensuring that managed aquifer recharge projects protect human and environmental health. Specific guidelines for managed aquifer recharge were developed in 2009 and form part of the Australian guidelines for water recycling along with other relevant guidelines for end uses of recycled water (NRMMC, EPHC and NHMRC, 2009).

6.5.2 Declining Resources for Understanding and Managing Groundwater

Because most states have not determinedly pursued full cost recovery of management costs from water users (despite being required as part of the National Water Initiative, see Chap. 7), funding available for the management and monitoring of groundwater resources is declining. This trend will increase the risk of inadequate monitoring being carried out, lead to a general decline in capacity and capability in management agencies, and the non-replacement of ageing monitoring infrastructure (Simmons, 2016).

In addition to addressing this decline in resources, a range of additional educational options to enhance better understanding and management of groundwater include;

- continuing to raise the profile and awareness of groundwater issues in Australia to politicians, policy makers and water users to combat ignorance and misunderstanding.
- building a national groundwater policy and planning forum where groundwater policy makers, managers, industry and scientists work together to define and solve important groundwater policy, management and technical issues.
- integrating all of the essential interdisciplinary subjects that relate to hydrogeology such as climate science, ecology, socioeconomics, public policy and law.
- quantifying and reducing uncertainty in analyses and explaining confidence levels to policy makers.

6.6 Conclusion

Australia's future will increasingly depend on groundwater due to the impacts of climate change and a growing population. Australia has made good progress with groundwater science, education, management, and policy reform, however there are many challenges ahead. Awareness of the importance of groundwater needs to be raised amongst politicians and policy makers so that current complacency and the resultant decline in budgets and resources within groundwater management agencies does not compromise their capability to address these issues in the future.

References

Australian Water Resources Council. (1975). *Groundwater resources of Australia*. Commonwealth of Australia, Department of Environment and Conservation.

Barron, O. V., Crosbie, R. S., Charles, S. P., Dawes, W. R., Ali, R., Evans, W. R., et al. (2011). *Climate change impact on groundwater resources in Australia* (Waterlines Report Series No 67). Canberra, Australia: National Water Commission.

Bates, B., Chandler, S. P., Charles, S. P., & Campbell, E. P. (2010). Assessment of apparent non-stationarity in time series of annual inflow, daily precipitation and atmospheric circulation indices: A case study from southwest Western Australia. *Water Resources Research, 46*, W00H02. https://doi.org/10.1029/2010WR009509

Bureau of Meteorology. (2016). Australian Groundwater Insight website: http://www.bom.gov.au/water/groundwater/insight

Clifton, C., Cossens, B., & McAuley, C. (2007). *A framework for assessing the environmental water requirements of groundwater dependent ecosystems*. Report prepared for Land & Water Australia, Canberra. http://lwa.gov.au/files/products/environmental-water-allocation/pn30042/pn30042.pdf. Accessed 26 Jan 2014.

Comino, M., Tan, P.-L., & George, D. (2014). Between the cracks: Water governance in Queensland, Australia and potential cumulative impacts from mining coal seam gas. *Journal of Water Law, 23*(6), 219–228.

CSIRO. (2008). *Water availability in the Murray–Darling Basin: a report to the Australian Government from the CSIRO Murray– Darling Basin Sustainable Yields Project*. Clayton South, Australia: CSIRO.

Department of Natural Resources and Mines. (2016). *Underground water impact report for theSurat Cumulative Management Area – 2016*. Brisbane, Australia: Government of Queensland.

Dillon, P., Pavelic, P., Page, D., Beringen, H., & Ward, J. (2009). *Managed aquifer recharge: An introduction* (Waterlines Report Series No. 13). Canberra, Australia: National Water Commission.

Evans, R. (2007). *The impact of groundwater use on Australia's rivers* (Land & Water Australia Technical Report). Braddon, Australia: Land & Water Australia, Canberra.

Ivkovic, K. M., Marshall, S. M., Morgan, L. K., Werner, A. D., Carey, H., Cook, S., et al. (2012). *National-scale vulnerability assessment of seawater intrusion: Summary report* (Waterlines Report No. 85). Canberra, Australia: National Water Commission.

Marsden Jacob Associates. (2012). *Assessing the value of groundwater* (Waterlines Report). Canberra, Australia: National Water Commission.

National Land and Water Resources Audit. (2001). *Australian water resources assessment 2000.* Commonwealth of Australia, Canberra. http://nrmonline.nrm.gov.au/catalog/mql:1674. Accessed 26 Jan 2014

National Water Commission. (2012). *Groundwater essentials*. Canberra, Australia: Commonwealth of Australia.

National Water Commission. (2014). *Australia's Water Blueprint: National reform assessment 2014*. Canberra, Australia: Commonwealth of Australia.

Natural Resource Management Ministerial Council, Environmental Protection and Heritage Council and National Health and Medical Research Council. (2009). *Australian guidelines for water recycling: Managing health and environmental risks (Phase 2) – Managed aquifer recharge*. Canberra, Australia: National Water Quality Management Strategy, NRMMC, EPHC and NHRMC.

Prosser, I., Wolf, L., & Littleboy, A. (2011). Water in mining and industry. In I. Prosser (Ed.), *Water: Science and solutions for Australia*. Collingwood, Australia: CSIRO.

Richardson, S., Irvine, E., Froend, R., Boon, P., Barber, S., & Bonneville, B. (2011a). *Australian groundwater-dependent ecosystems toolbox Part 1: Assessment framework* (Waterlines Report). Canberra, Australia: National Water Commission.

Richardson, S., Irvine, E., Froend, R., Boon, P., Barber, S., & Bonneville, B. (2011b). *Australian groundwater-dependent ecosystems toolbox Part 2: Assessment tools* (Waterlines Report). National Water Commission, Canberra, Australia.

Simmons, C. T. (2016). Groundwater down under. *Ground Water, 54*(4), 459–460.

Werner, A. D. (2010). A review of seawater intrusion and its management in Australia. *Hydrogeology Journal, 18*, 281–285.

Wilhite, D. A. (2012). Breaking the Hydro-Illogical Cycle: Changing the paradigm for drought management. *EARTH Magazine, 57*(7), 71–72.

Steve Barnett is Principal Hydrogeologist at the Water Science and Monitoring Branch of the Department for Environment and Water in South Australia. He has been involved in the investigation, monitoring and management of groundwater resources in SA for over 40 years, and has contributed technical and policy input into ten groundwater management plans which incorporate a variety of different aquifers and management issues. He is a past-president of Australian Chapter of the International Association of Hydrogeologists.

Dr. Nikki Harrington has over ten years of experience in the application of environmental tracer techniques and numerical modelling approaches to improve the conceptualization of groundwater flow systems. During her eight years working in State Government, she was involved in numerous projects applying environmental tracer and numerical modelling techniques to resource assessment investigations in the Murray Basin, Eyre Peninsula and the South East of South Australia. Over the past three years Nikki worked as a Research Associate at the National Centre for Groundwater Research and Training, where she was Project Leader for the Goyder Institute funded South East Regional Water Balance project. Nikki is now a Principal Hydrogeologist at Innovative Groundwater Solutions Pty Ltd (IGS).

Dr. Peter Cook is a Professorial Research Fellow with Flinders University, and is international recognized for his work on groundwater recharge and surface water – groundwater interactions. He has co-written books on environmental tracers and ecohydrology, and has published more than 100 research papers. He frequently advises the Australian government on water management, and was a member of the Victorian Government Technical Audit Panel for Water Resources from 2002–2008, and the National Groundwater Committee from 2002–2007. Between 2009 and 2014 he was Deputy Director of the National Centre for Groundwater Research and Training (NCGRT).

Prof. Craig T. Simmons, the Director of the National Centre for Groundwater Research and Training, headquartered at Flinders, is forging ahead to understand better ways to preserve and manage water both in Australia and around the world. His research applies computer modelling to identify ways to keep humans, livestock and crops alive, and protect the environment by identifying ways to better source and preserve fresh water supplies. Professor Simmons' team has also trained more than 80 postdoctoral fellows and 80 PhD students at the National Centre for Groundwater Research and Training since 2009, helping make Flinders University an international magnet for groundwater researchers.

Chapter 7
The Evolution of Groundwater Management Policy in the States of Australia

Rebecca Nelson, Steve Barnett, and Ann Kollmorgen

Abstract The isolated British colonies in Australia formed strongly independent jurisdictions that initiated management of their own resources. The states' power to manage water resources was enshrined in the Federal Constitution in 1900 and is still in operation today. This evolution of groundwater management in Australia contrasts with the French approach, whereby groundwater management approaches are determined at the national level. The colonies inherited the British riparian doctrine that gave landholders rights to water contiguous with and adjoining their land. In 1886, the state of Victoria enacted legislation that exclusively vested the right to the use of water in any watercourse in the state, and subordinated the rights of the individual. This approach forms the basis for the successful and efficient management of groundwater resources throughout Australia today. Two case studies present different approaches to the management and allocation of groundwater that are employed in Australian states, and how the level of entitlements is adjusted to meet sustainable extraction limits. Water reform efforts over the last two decades under federal frameworks and approaches are also discussed.

Keywords Water reform · Water entitlements · Management plans · Legislation · Extraction limits · Riparian rights

---

R. Nelson (✉)
Melbourne Law School, University of Melbourne, Melbourne, VIC, Australia
e-mail: rebecca.nelson@unimelb.edu.au

S. Barnett
Adelaide, SA, Australia
e-mail: Steve.Barnett@sa.gov.au

A. Kollmorgen
Department of Environment, Land, Water and Planning, Melbourne, VIC, Australia
e-mail: ann.kollmorgen@delwp.vic.gov.au

J.-D. Rinaudo et al. (eds.), *Sustainable Groundwater Management*,
Global Issues in Water Policy 24, https://doi.org/10.1007/978-3-030-32766-8_7

7.1 Introduction

The evolution of groundwater management in Australia is explored in this chapter. In a global historical context, the colonisation and development of Australia is a relatively recent phenomenon. This has allowed Australian decision makers to learn from the experience of other countries. Our short settler history also means there are few long standing traditions of settler water management, which, in recent years, has facilitated widespread reform of water resources management.

The original colonies were separated by vast distances and their isolation led to the development of strongly independent jurisdictions that initiated management of their own resources. The states' power to manage water resources was subsequently enshrined in the Federal Constitution in 1900 and is still in operation today. This situation contrasts with the French approach, whereby groundwater management approaches are generally determined at the national level.

Originally, the colonies of Australia inherited the 'riparian doctrine', which was enshrined in British common law and gave landholders conditional rights to water contiguous with and adjoining their land. Whilst this approach may have been adequate for the small dispersed villages of England, it soon became apparent that this doctrine was impeding the development of surface water supplies and infrastructure for the growing urban centres of the colonies. The English 'rule of capture' generally regulated early groundwater development, whereby landowners had the right to take all the water they could capture from under their property. While this was less problematic during an era of limited pumping power, technical advances would potentially allow almost limitless exploitation of groundwater resources.

In 1884, Alfred Deakin, the Victorian Minister of Public Works (he later became Prime Minister of Australia), travelled to California to investigate irrigation developments and how water rights were managed. In the western United States, 'prior appropriation' of water was the prevailing doctrine of water rights. This doctrine applies formal property rights to water, accrued to the user on a 'first in time, first in right' basis. Deakin's visit strongly influenced the introduction of the Victorian *Irrigation Act* of 1886. This ground-breaking legislation superseded the traditional English doctrine of 'riparian rights', and also rejected the western United States doctrine of 'prior appropriation'. The legislation:

- exclusively vested the right to the use of water in any watercourse in the state,
- subordinated private riparian rights to the rights of the state, and
- highlighted the need for the rights of the individual and the state to be fully defined.

Importantly, the Act also instituted a system whereby the state would administer the allocation of water rights to water users. Although designed to facilitate large scale surface water irrigation projects, the basic philosophy of the legislation has formed the basis for successful and efficient management of groundwater resources throughout Australia in later years.

After a long, drawn-out process, the federation of Australia occurred in 1900 when the six separate British self-governing colonies agreed to unite and form the

Commonwealth of Australia, which incorporated a three-tier system of federal, state and local government. Broadly, the division of powers between the federal and state governments follows the American model of federation. The constitution does not confer any direct powers for water resource management on the federal government and consequently, the states still have the primary constitutional powers and responsibilities for this management.

The first groundwater resource to literally 'flow' into national awareness was the Great Artesian Basin – the largest basin in the world, covering 1.7 million km^2 and underlying 22% of the Australian land mass. The first deep well was drilled in 1879 and by 1910, over 1500 artesian bores had been drilled throughout the Basin. Concerns about marked reductions in flows in many wells and potentially unsustainable extraction from the Basin led to a series of Interstate Conferences on Artesian Water, which were held between 1912 and 1928. These conferences helped clarify many of the issues of concern and instigated the systematic collection and interpretation of data from other artesian basins around the country. They also stressed the need for the controlled development and management of artesian groundwater resources into the future. This early focus on artesian resources was due to the lack of pumping technology which is required to lift water in significant quantities from unconfined aquifers.

Following the Second World War, the increased use of rotary drilling and improved pumping techniques allowed more development of groundwater resources. Demand for groundwater increased in the 1950s and 1960s to support agricultural and horticultural industries. In some areas, extractions reached unsustainable levels, which prompted the enactment of the first legislation that enabled management of the resources in those specific areas. Widespread investigations into new resources in the 1970s led to significantly increased extraction for irrigation.

In the 1980s, water management in Australia began to consider broader objectives. No longer did water authorities look solely to the construction of bigger dams to solve water issues; rather, they examined options of improving the allocation of existing water entitlements in conjunction with environmental and social policy objectives. Their objective was seen as promoting efficiency and equity of water allocation while protecting the environment.

This trend was triggered by the release of Brundtland report in 1987 by the United Nations, which signalled the need to adapt unrestrained economic growth by incorporating the principle of sustainable development that meets the needs of the present without compromising the ability of future generations to meet their own needs. Agenda 21, the working plan for action, was developed and ratified in Rio de Janeiro in 1992. As a signatory to Agenda 21, the federal government committed to the principles of ecologically sustainable development, which underpin the current state management approaches for groundwater resources in Australia.

The development of formal national-level policy about water began in the 1990s, motivated by these concerns about efficiency and sustainability. The 1994 National Water Reform Framework Agreement introduced the concept of comprehensive water allocation systems. This Framework was agreed by the federal government and all states and territories through the Council of Australian Governments (CoAG). This intergovernmental forum had been established in 1992 to coordinate government responses to common issues. The CoAG Agreement sought to reform water pricing, separate title to land from the right to use water and clarify legal rights to use water, ensure that the environment received allocations of water, attempt to deal with over-allocation of water, establish systems for water trading, and outline institutional reforms and increased public consultation. Although the CoAG Agreement was drafted primarily with surface water in mind, it specifically tasked the Agriculture and Resource Management Council of Australia and New Zealand (ARMCANZ) to report on management arrangements for groundwater and specifically recognised that groundwater basins had environmental requirements.

Even the relatively cursory way that groundwater was considered in the CoAG Agreement was significant for groundwater governance. It signalled the first notable involvement from the federal government in a resource management issue for which the states had direct and primary responsibility. The federal government had previously been involved in some groundwater-related measures. This was especially so where groundwater resources spanned state boundaries, as in the Great Artesian Basin, and in relation to measures to control groundwater salinity in the Murray-Darling Basin (MDB), Australia's most important agricultural basin (discussed later in Chap. 8). However, the CoAG Agreement represented its first sustained attempt to drive policy in relation to groundwater outside of the transboundary context. This federal influence was sweetened by a system of payments to states in return for them implementing the Agreement reforms, as part of overarching payments to the states to improve economic policies. The CoAG Agreement also established the use of a regular reporting mechanism for states to report on their performance in implementing agreed reform measures—an accountability and coordination mechanism that continues to characterise water management in Australia.

A subsequent national policy paper was developed in 1996 by ARMCANZ, entitled 'Allocation and Use of Groundwater: A National Framework for Improved Groundwater Management in Australia – Policy Position Paper for Advice to States and Territories'. This dealt with the concept of 'sustainable yield' in the groundwater context, and the idea of groundwater trading. Importantly, it also explicitly supported, for the first time at the national level, the need to protect the ecological values of groundwater in addition to sustainability concerns that centred on managing groundwater volumes for consumptive human uses.

The broadening of national-level groundwater policy to consider environmental sustainability was mirrored in large-scale regional policies. The Great Artesian Basin Sustainability Initiative (GABSI), established in 1999, aimed to increase artesian pressure by capping uncontrolled flowing bores and reduce evaporative losses by replacing open water delivery drains with pipelines in New South Wales, Queensland and South Australia. Under GABSI, the federal and state governments co-fund the works, with landholders also making a financial contribution. Although GABSI initially focused on the impacts of aquifer depletion on human groundwater users (particularly pastoral bore owners), it later adopted a broader focus that included maintaining or improving flows to springs for ecological purposes.

The Intergovernmental Agreement on a National Water Initiative (NWI), which all the federal and state governments signed between 2004 and 2006, aimed to build on the CoAG Agreement. It followed recognition that reforms had proven to be more difficult than initially anticipated, particularly in relation to ensuring environmental and public benefits from water management, and returning overallocated and overused water systems to sustainable levels of extraction. Unlike the CoAG Agreement, the NWI did not involve payments of incentives to the states to fulfil their commitments. Although the NWI does not frequently refer to groundwater specifically, it sets out reform and management principles that are intended to apply equally to surface water and groundwater.

The NWI's most fundamental governance reform was to commit states to a system of legally binding water plans, which most, though not all, states and territories have now implemented. Where they apply, these plans have changed the foundation of Australian systems for administering water rights. Traditionally, allocating groundwater depended on a case-by-case assessment of individual applications to take water. By contrast, plans now generally guide how groundwater withdrawals are licensed and managed in areas subject to relatively high levels of demand. This is a notably more strategic, holistic and proactive approach to water management that places much greater emphasis on public consultation and involvement. It seeks to avoid over-allocation and control impacts on the resource by imposing volumetric caps on aggregate licensed withdrawals. This means that if an application to take groundwater would result in a volumetric limit being exceeded, the application must be rejected. Water plans also offer an increased opportunity to manage groundwater and surface water in an integrated way, and can facilitate groundwater trading through rules that apply in the plan area or sub-areas.

In addition to dealing with licensed withdrawals, the NWI specifically dealt with other withdrawals that were not required to be licensed. The treatment of these 'interception activities' was particularly relevant to groundwater. The concept of interception activities included otherwise unregulated withdrawals of groundwater by stock and domestic bores where they are not required to be licensed and large-scale plantation forestry, though it explicitly excluded withdrawals of groundwater associated with minerals and petroleum. These latter extractions were considered to warrant tailored management due to their special circumstances and the difficulty of directly controlling them. The states committed to managing non-mining interception activities through monitoring and licensing for significant activities.

Like the CoAG Agreement, the NWI involved regular state reports on implementation in accordance with detailed timetables that applied to each reform commitment. The NWI also involved regular independent assessment of the jurisdictions' performance in implementing their commitments. Initially, this was undertaken by a specialist independent water agency, the National Water Commission. This agency was abolished in 2015 in an attempt to simplify water bureaucracy, and is indicative of the ongoing pressure to reduce regulatory burdens in the water sector. The ongoing responsibility to assess progress against the NWI is now carried out by the federal Productivity Commission, an independent research and advisory body of the federal government that deals with economic, social and environmental issues.

Taken together, the reports assessing progress against NWI reform commitments suggest that implementation of the commitments related to groundwater management tend to lag somewhat behind implementation of those that apply to surface water. Particular issues that have arisen over time include 'unbundling' (that is, separating) rights to use groundwater from ownership of land, considering the desirability and cost-effectiveness of developing systems to trade groundwater, protecting groundwater-dependent ecosystems, and recognising the connectivity between surface water and groundwater. The Productivity Commission's current assessment against the NWI (report in draft at the time of writing) notes several issues that especially affect groundwater. These include incorporating water use by extractive industries into the water rights frameworks that apply to other categories of uses; developing property rights frameworks for alternative water sources (including managed aquifer recharge) to encourage investment; and more integrated management of surface water and groundwater in certain areas.

Each of these issues is also highlighted as an area of particular need in the National Groundwater Strategic Framework (2016–2026), which was developed by Australian federal, state and territory governments. This most recent incarnation of national water coordination sets out 28 actions in three priority areas, but recognises that the states and territories will have different priorities and capacities to pursue the actions identified in the Framework. The three priorities are: ensuring sustainability and optimal use of groundwater, providing for 'confidence for investment through risk based, consistent and efficient regulation of groundwater resources', and 'developing integrated water supply planning' to increase water security. The Strategic Framework echoes most groundwater policies before it in calling for the development of better information to support groundwater management. Indeed, it sees groundwater information and effective regulation at the heart of its priorities and actions, and calls for greater linkages between groundwater decision-makers and managers, ongoing research, and recently developed groundwater information and capacity-building initiatives.

This emphasis on connecting research and knowledge to decision-making and increasing public awareness is important, given significant recent investment in groundwater information. A National Centre for Groundwater Research and Training was established in 2009; and a National Groundwater Action Plan (2007–2012) sought to improve knowledge and management in relation to groundwater. It recognised the historical under-investment in the area, and the need to correct this to more accurately reflect the economic, social and environmental significance of the resource. These initiatives have led to substantial nation-wide information platforms, including the Groundwater-Dependent Ecosystems Atlas, National Groundwater Information System, and Australian Water Resources Information System. Each is hosted by the Bureau of Meteorology under water information powers granted to it for the first time by the federal *Water Act 2007*.

Beyond policy measures and knowledge building, federal influence over groundwater has also increased through federal laws. In addition to a focus on groundwater information, the 2007 federal *Water Act* sought to limit the withdrawal of groundwater as well as surface water in the MDB to an 'environmentally sustainable level of take'. This occurred against the background of the long-established primary responsibility of the states for water management. On this basis, the legislation relied largely on the federal parliament's power to implement international treaties—specifically, environmental treaties related to water—as its foundation. While the definition of environmental sustainability adopted under the Act in the groundwater context is arguably relatively narrow, this law represents a serious attempt to implement the policy intent of the NWI in relation to water and the environment. The Act also adopts specific legal mechanisms to deal with low levels of knowledge about some groundwater systems. An adjustment mechanism allows for limits on groundwater withdrawals to be changed to account for new knowledge about the resource. The Act proposes that risks that arise from those adjustments, and other changes that arise from climate change, drought, and policy changes, are to be shared among groundwater users and governments.

It is important to recognise that the states and territories continue to be the primary managers of groundwater and are responsible for licensing processes and adopting legally enforceable plans to manage withdrawals. Table 7.1 below lists the state agencies, their enabling legislation and the plans they produce, whilst Fig. 7.1

Table 7.1 State agencies, legislation and management plans

State or Territory	Management agency	Legislation	Name of plan
Australian Capital Territory	Environment ACT	*Water Resources Act 2007* (universal)	Water Resources Management Plan
New South Wales	Department of Primary Industries	*Water Management Act 2000*; (universal) *Water Act 1912*	Water Sharing Plan
Northern Territory	Department of Natural Resources, Environment and Arts	*Water Act 1992* (risk-based)	Water Allocation Plan
Queensland	Department of Natural Resources, Mines and Energy	*Water Act 2000; Integrated Planning Act 1997* (risk-based)	Water Resource Plan
South Australia	Department for Environment and Water	*Natural Resources Management Act 2004* (risk-based)	Water Allocation Plan
Tasmania	Department of Primary Industries, Parks, Water and Environment	*Water Management Act 1999* (risk-based)	Water Management Plan
Victoria	Department of Environment, Land, Water and Planning	*Water Act 1989* (universal)	Water Supply Protection Area Management Plan
Western Australia	Department of Water and Environmental Regulation	*Rights in Water and Irrigation Act 1914* (risk-based)	Water Management Plan (non-statutory)

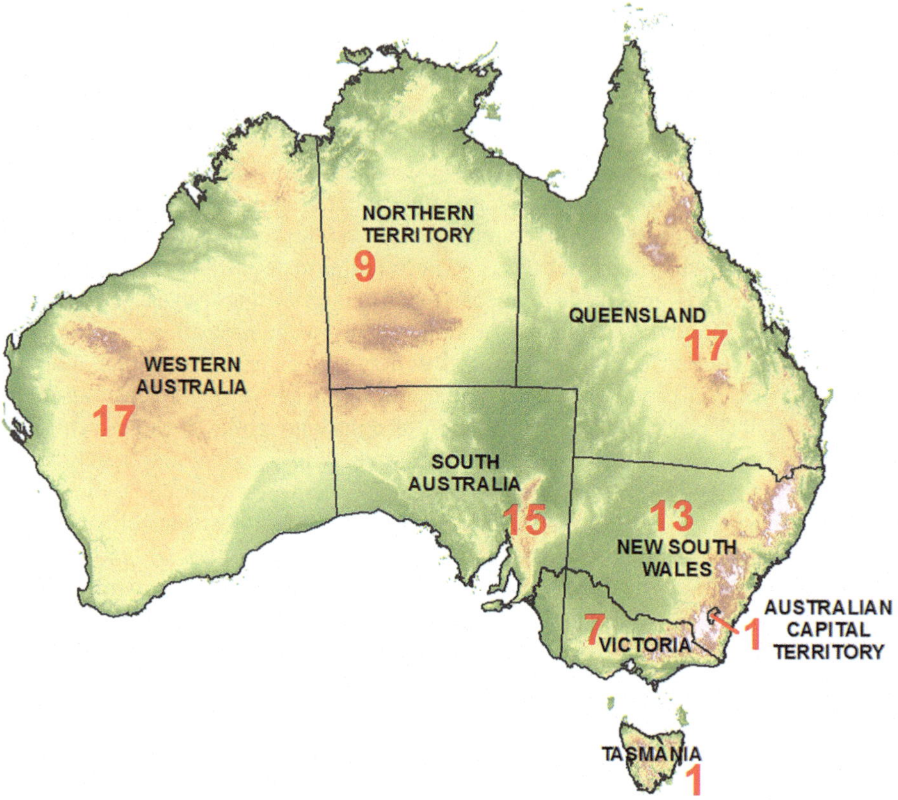

Fig. 7.1 Number of groundwater management plans in Australian states

presents the number of statutory plans produced by each state (with the exception of Western Australia, which, in practice, uses non-statutory water plans).

However, this new federal legal development in Australian groundwater governance requires that within Australia's most important agricultural area, the MDB, those plans meet requirements set by the federal government. In some respects, those requirements mirror commitments already agreed by states and territories under the NWI, many of which they were already pursuing on an individual basis. In others, they represent significant advances in areas that had not previously had the benefit of high-level attention, for example, consultation with Indigenous peoples about groundwater. These and other advances may drive developments in Australian groundwater management—directly in the MDB and perhaps indirectly outside it as well—for some time to come.

7.2 Case Studies

Although most states in Australia have the same broad approach to the management of groundwater resources, there is variation in implementation, which is illustrated in the two following case studies. Some states apply a 'risk-based approach', whereby a licensing and management regime is imposed only in areas where extractions pose a risk to the sustainability of the groundwater resource, whereas other states apply a 'universal approach', whereby a licensing regime applies across the whole state regardless of the intensity of extraction, but applies a stronger management regime in those areas with high extractions. Table 7.1 details the approach taken by each state and territory.

7.2.1 Risk-Based Approach – South Australia and Other Similar Jurisdictions

South Australia (SA) is one of the driest states and consequently has a low population, most of which resides in the capital city of Adelaide. Most of the state is arid with unusable groundwater resources. Common law access to groundwater prevailed in SA until the first legislation to control groundwater development came in the form of the *Water Resources Act 1976,* which operated to maintain common law rights but provided for limitations to these rights by declaring a water resource to be a prescribed resource in specified areas where the groundwater resources were considered at risk, and required a water user to hold a water licence in such areas. This Act also provided for Water Allocation Plans to manage the resource in these specified areas and also required a permit for the construction of a water well anywhere in the state.

The current legislation, the *Natural Resources Management Act 2004* ('NRM Act'), abolishes common law rights to water, meaning that nobody owns water in SA (not individuals and also not the state), and that water is a common or public commodity subject to use rights. The NRM Act provides the state with the right to control the management of water through Water Allocation Plans, authorising water-affecting activities and other means provided for in the legislation.

Under the Act, there is a variety of rights in relation to the ability of a person to take and use water. For example, in unprescribed resources, a land owner has a right on the basis of ownership or occupation of land to freely take and use the water on or under that land. Alternatively, in an area where the groundwater resource is prescribed under the Act, the right to take water is authorised by a water licence, which is an entitlement to access a designated part of the water resource. A water licence constitutes personal property and has the characteristics of personal property in that it can be owned, sold, leased, bequeathed, used as collateral, etc. What is owned then is the entitlement to access water rather than the water itself.

Once an area has been prescribed, the state issues existing groundwater users with a water entitlement that aims to meet their reasonable requirements, based on their use during a specified qualifying period, which usually extends over several years. In determining the reasonable requirements of existing users, the size and nature of the enterprise of each user (eg. 1.5 ha of grapevines or a 100 sow piggery) is examined and then converted to a volume using theoretical enterprise water requirements (TER). With regards to irrigated crops, water requirements have been determined using the internationally recognised Food and Agricultural Organisation (FAO) methodology, which determines the amount of water that needs to be applied to a crop in an average year. This methodology takes into account the average rainfall and evaporation over the area, and assumes that the crop is grown for maximum production under non-restricting conditions (for example, sufficient nutrients, non-limiting soil conditions, etc.) with exceptions for crops where irrigation levels are controlled for quality reasons, such as wine grapes.

A separate methodology to determine the reasonable requirements for non-irrigation enterprises (industrial/commercial water use) has been developed, which is based on industry standards and/or consultation with the licensee regarding their water needs.

If the total TER volume determined by this process is within the sustainable extraction limit, a user's reasonable requirements become their water allocation endorsed on their water licence. If however, the total TER volume exceeds the sustainable extraction limits, the state may reduce the existing user reasonable requirements so that resultant entitlements are within the sustainable extraction limits.

The advantages of this approach are that it can be seen to be transparent, consistent and fair to water users. As such, it is considered to be legally defensible if any water user chooses to appeal the entitlement issued to them. However, there are several factors that lead to many water users receiving an entitlement considerably higher than their actual use, which results in an over-allocation of the resource.

Firstly, because of the perceived monetary value of an entitlement (from both an increase in the value of the land on which the water is used, and also the value of the water itself if a water market exists for the selling or leasing of the entitlement), there is an incentive for users to exaggerate their reasonable requirements, sometimes in a dishonest manner. Secondly, the TER derived by the FAO (which assumes ideal growing conditions, maximum crop production and no constraints on water availability), is rarely applied by most irrigators due to ignorance, poor irrigation practices or constraints on water supply (well yields).

The State of Queensland adopts a similar risk-based approach to South Australia. In areas where usage is low and the groundwater resources are relatively unstressed, groundwater can be taken without a licence or other form of entitlement. However, where resources are considered at risk, the *Water Act 2000* provides for the creation of management areas, with licensing and management regimes implemented through a water plan.

In Western Australia, only groundwater in proclaimed areas, other than for domestic and stock watering, is licensed under the *Rights in Water and Irrigation*

Act 1914 and amendments, although as a matter of implementation, the vast majority of groundwater use occurs in proclaimed areas, which cover much of the state. The state is divided into groundwater management areas on the basis of hydrogeology, land use, or administrative divisions. Under non-statutory management plans, an allocation limit applies to each aquifer in each groundwater management area, and where an aquifer is not fully allocated, licensing officers may grant licences having regard to social and environmental considerations, and the impacts on the resource and other water users. A water use licence is issued to a landowner for a specific purpose, and is not transferable. In some fully allocated areas trading of allocations is allowed, with each trade requiring an assessment of the impact of pumping from the proposed point of extraction.

7.2.2 Universal Approach: Victoria and Other Similar Jurisdictions

Victoria has a population of over six million with the vast majority residing in the capital city of Melbourne. A significant portion of the state is mountainous with rainfall exceeding 1200 mm/year in alpine areas, but less than 350 mm/year is recorded in the semi-arid north-western region. The most populated southern coastal areas generally receive between 600 and 900 mm/year. Not surprisingly, surface water resources supply over 80% of water consumption (about 6300 GL/yr). Groundwater supplies about 400 GL/yr of Victoria's total consumption, including water for more than 70 towns that are dependent on groundwater either wholly or partly for supply.

Groundwater management throughout Victoria was initiated by the *Groundwater Act 1969,* which established licensing of groundwater wells and introduced conservation areas (groundwater management areas). This Act was superseded by the *Water Act 1989,* which provides the state with the right to control the management of groundwater through extraction limits and water management plans. The Act also includes provisions for private rights, whereby a person who has access to land also has the right to extract water for the purposes of stock and domestic use. It also requires well construction licences for all forms of take and take and use licences for extractions other than stock and domestic water use. Licensed extractions are metered in line with state policy.

If there is a risk of the existing licensed extractions causing undesirable impacts in any given area, the Act provides for the declaration of a Water Supply Protection Area (WSPA), a cap on licensed entitlements and the development of a groundwater management plan. Intensity zones may also be used to manage high demand areas within a WSPA to manage the risk of interference between groundwater users.

Under the *Water Act 1989*, a groundwater licence application from anywhere within Victoria must not exceed any entitlement limit applying to that area, and requires an assessment to determine whether the application is likely to impact on neighbouring users via an interference assessment (generally a maximum 10% additional drawdown is allowed at the site of another user), or on the aquifer or the environment under 'other matters to be taken into account' (S40, *Water Act 1989*). Any management plan local intensity rules and caps are also checked. In areas where groundwater is fully allocated up to the extraction limit, the only way to obtain a groundwater licence is via a trade with another licensee. Where there is water available and high demand, the remaining water allocations may be auctioned to interested parties.

Groundwater resources in the most populous state of New South Wales ('NSW') are managed in similar way to Victoria. Statutory water sharing plans, which have life of 10 years, cover the whole area of the state. These plans are established under the *Water Management Act 2000* which requires that water is provided to the environment as a priority and also provides licence holders with security through perpetual licences and greater opportunities to trade through the separation of water access rights from the land.

Most groundwater sources in NSW are fully committed and access licences can only be obtained through the water trading market. In areas that have unallocated groundwater available, applications for new access licences are considered through a competitive process such as auction or tender. A minimum price is set to prevent new allocations causing a devaluation in the value of existing licences.

7.3 Conclusion

Although there are differences in implementation, the groundwater management regimes established in the Australian states have been effective in controlling over-allocation to a large extent (as discussed in Chap. 6, 25% of management areas are considered over-allocated). However more importantly, they have prevented over-extraction in the vast majority of groundwater resources (only 2% of management areas are considered over-used). On-going refinement of the management plans is necessary to make them more efficient, flexible and adaptable to emerging threats of rising demand and climate change and recently recognised requirements of Indigenous peoples and groundwater-dependent ecosystems.

Dr. Rebecca Nelson is an Associate Professor, Melbourne Law School, University of Melbourne and a practising lawyer. Dr Nelson's research focuses on environmental and natural resources laws. She is an author of Water Resources Law (2nd ed., LexisNexis Australia), 8 book chapters and 32 articles and research reports. Dr Nelson was the IAH (Australia)/National Centre for Groundwater Research and Training Distinguished Lecturer (2016) and the Law Council of Australia's Young Environmental Lawyer of the Year (2014). She holds a JSD(Stanford), JSM(Stanford) and BE(Environmental)/LLB (University of Melbourne).

Steve Barnett is Principal Hydrogeologist at the Water Science and Monitoring Branch of the Department for Environment and Water in South Australia. He has been involved in the investigation, monitoring and management of groundwater resources in SA for over 40 years, and has contributed technical and policy input into ten groundwater management plans which incorporate a variety of different aquifers and management issues. He is a past-president of Australian Chapter of the International Association of Hydrogeologists.

Ann Kollmorgen is a Senior Hydrogeologist with the Water and Catchments group in the Department of Environment, Land, Water and Planning in Victoria, Australia. She has extensive experience in groundwater management and groundwater resource assessments across both government and private sectors. She has contributed technical and policy advice on a broad range of groundwater issues including salinity management, mine dewatering, urban water supplies, groundwater management plan development and groundwater – surface water interactions.

Chapter 8
Developing a Coordinated Groundwater Management Plan for the Interstate Murray-Darling Basin

Glen Walker, Steve Barnett, and Stuart Richardson

Abstract The Murray-Darling (MDB) Basin Plan is a strategic plan for the integrated and sustainable management of water resources, including groundwater. The MDB covers an area of more than 1 million km^2 across five states and territories in south-eastern Australia. The major proportion of the groundwater extraction occurs over relatively small areas of alluvial aquifers, while the rest of the land area is characterised by sparse extraction from a wide range of groundwater systems. The Basin Plan follows a 20 year period of water reform and a major drought. While there had been a cap on surface water diversions, it is only with the advent of the Basin Plan that a limit has been set on the level of groundwater extraction across the MDB. Consistent management arrangements have also been applied across the MDB. Within the regionalised limits on groundwater extraction in the Basin Plan, localised impacts on the groundwater resource (including water quality, baseflow and ecology) will be managed through water resource plans. These plans will be developed and implemented by the five states and territories and accredited by the MDBA. The groundwater elements of the Basin Plan will be fully implemented in mid-2019, with a review in 2026. Future challenges include compliance, and adapting to climate change.

Keywords Hydrogeology · Integrated management · Sustainable yield · Water plan · Legislation

G. Walker
Grounded in Water, Mount Osmond, SA, Australia

S. Barnett (✉)
Adelaide, SA, Australia
e-mail: Steve.Barnett@sa.gov.au

S. Richardson
CDM Smith, Adelaide, SA, Australia

J.-D. Rinaudo et al. (eds.), *Sustainable Groundwater Management*,
Global Issues in Water Policy 24, https://doi.org/10.1007/978-3-030-32766-8_8

8.1 Introduction

The Murray-Darling Basin (MDB) is defined by the catchment areas of the Murray and Darling Rivers and their many tributaries. The MDB is the largest and most complex river system in Australia and extends over 1 million km^2 of south-eastern Australia, covering three-quarters of New South Wales, more than half of Victoria, significant portions of Queensland and South Australia, and all of the Australian Capital Territory (Fig. 8.1). The MDB supports a population of 2.6 million people and is considered the food bowl of the nation, producing $A22 billion of food and agricultural products every year. The rivers and lakes support unique habitats for many waterbird and fish species and 16 internationally-recognised and protected wetlands.

The MDB is one of the flattest catchments on Earth. The low-lying topography of the Basin, warm to hot semi-arid conditions in most regions, and the meandering

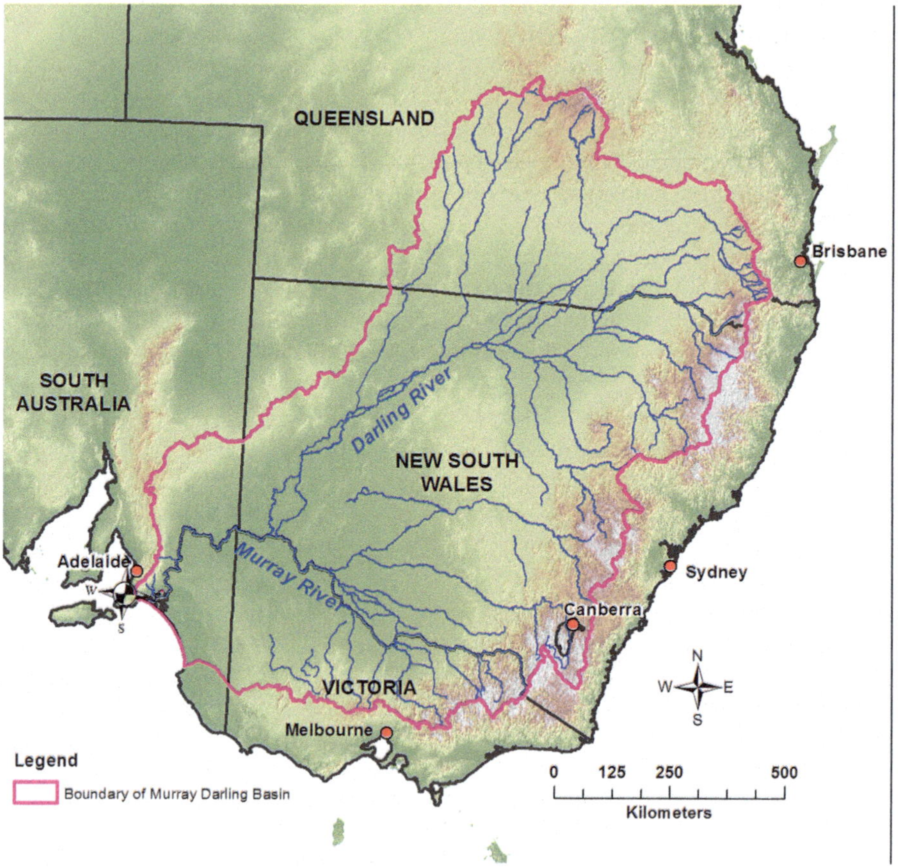

Fig. 8.1 Location of the Murray Darling Basin

and slow-flowing nature of the creeks and rivers, all combine to make an environment characterised by high evaporation. On average, the Basin receives about 530,000 GL of water as rainfall. Of this amount, 94% evaporates from waterways and floodplains or is transpired by plants. The system therefore carries one of the world's smallest volumes of surface water for its size (a mean annual discharge of only 0.4 ML/sec).

Because of the highly variable rainfall and runoff, the natural flow of surface water was historically altered with the construction of weirs and dams to support and foster large irrigated agriculture schemes as early as the 1880s. Approximately 42% of total surface water runoff is diverted from the river systems for irrigation which amounts to a long-term average of almost 11,000 GL/yr.

Groundwater use accounts for less than 20% of the water diverted in the MDB (CSIRO, 2008), but this use varies both spatially and temporally. A large number of water users in the MDB are totally dependent on groundwater, particularly in the northern MDB. Groundwater becomes an increasingly important source of water during drier periods when surface water becomes less available. Droughts are predicted to become more extended into the future (BoM & CSIRO, 2016) meaning that there will be a greater reliance on groundwater as a contingency measure. This, together with the high current use of surface water, means that groundwater will be needed to meet not only current needs, but also support new development opportunities.

This chapter describes how a coordinated joint management plan for the increasingly important groundwater resources of the MDB was developed using a consistent methodology to determine sustainable extraction limits across the five states and territories that comprise the basin. It also outlines some of the challenges arising from this joint management approach.

8.2 Groundwater Systems in the MDB

8.2.1 Hydrogeology

The Murray Darling Basin overlies a variety of geological units. The northern part of the Basin covers an area of 650,000 km^2 and is directly underlain by alluvial sediments up to 200 m thick which in turn overlie the Mesozoic deposits of the Great Artesian Basin (GAB) which comprise sandstones and mudstones of Jurassic to Cretaceous age (Fig. 8.2). Although the GAB is a vital resource for much of inland Australia, it is covered by separate management arrangement and will not be considered further in this chapter. The south-western portion of the Murray-Darling Basin is underlain by the Murray Geological Basin, a thin assemblage of flat-lying horizontally bedded fluvial and shallow marine sediments of Tertiary age which cover an area of approximately of 300,000 km^2. These sediments vary in thickness from less than 200 m in the north, east and south to 600 m in the west central part of the Basin (Brown & Stephenson, 1991).

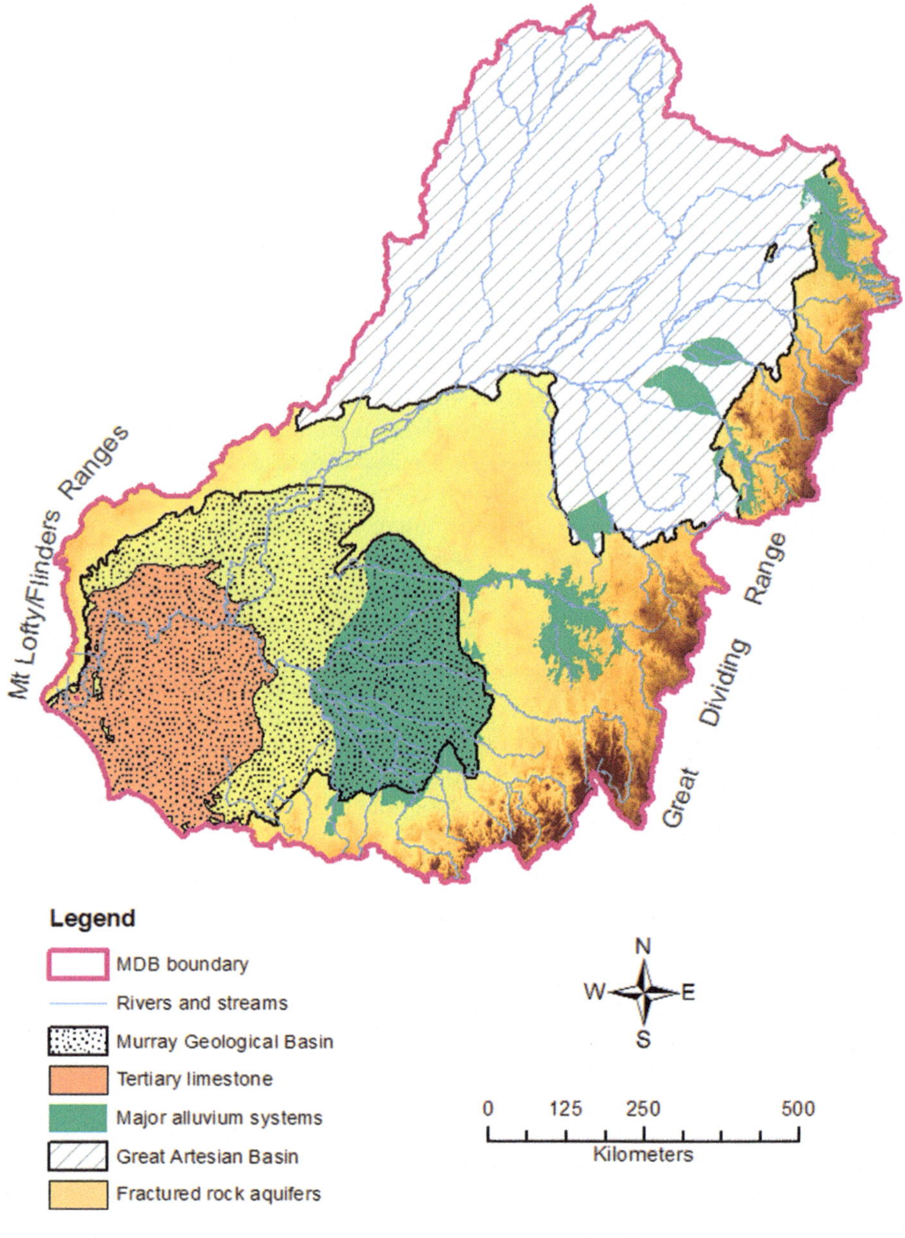

Fig. 8.2 Simplified hydrogeology of the Murray Darling Basin

The groundwater systems occur in a range of hydrogeological settings, but can be subdivided into three major provinces (Evans & Kellett, 1989) as shown in Fig. 8.2.

1. Fractured rock aquifers: The Mount Lofty/Flinders Ranges in South Australia and the Great Dividing Range through Victoria, NSW and Queensland form the margins of the southern part of the MDB. These highlands contain fractured rock aquifers of moderate productivity.
2. Major alluvial systems: Most groundwater extraction occurs from these systems which have been formed from the deposits of sand and gravel from the main river tributaries. River leakage and flooding are major sources of recharge which has formed areas of fresher groundwater in the otherwise brackish to saline regional groundwater. There is high connectivity between surface and groundwater, especially in the highland valleys.
3. Tertiary limestone of the western Murray Basin: Good quality groundwater in this aquifer was recharged tens of thousands of years ago during a wetter climate. Irrigation extractions are managed by a policy of controlled gradual depletion of the large volume of storage.

8.2.2 Groundwater Development

Over the last 30–40 years, groundwater extraction in the MDB has increased to around 1400 GL/yr. The increased extraction is causing some emerging problems especially in the highly connected alluvial systems. The increased drawdown of groundwater levels may cause not only increased losses from rivers and streams to the alluvial aquifer, but may also reduce baseflow discharge to them, resulting in reduced stream flow to downstream users. Another important issue is the over-allocation of entitlements within six major alluvial systems in New South Wales (Chap. 21 provides a case study of this issue). The remainder of this chapter covers the policy and management response to these issues.

In the past, a lack of reliable data on extraction volumes has hampered historic management, however greater investment (backed by regulation) has seem more reliable metering of groundwater extraction across the Basin.

8.2.3 Groundwater Salinity Issues

The low topography, slow groundwater movement, low rainfall (<300 mm/year) and high evaporation (>2000 mm/year) have together created large areas of highly saline groundwater in shallow aquifers in the western part of the Murray Geological Basin. Figure 8.3a shows that groundwater moves in a general west to north-westerly direction from the high rainfall recharge areas around the highland basin

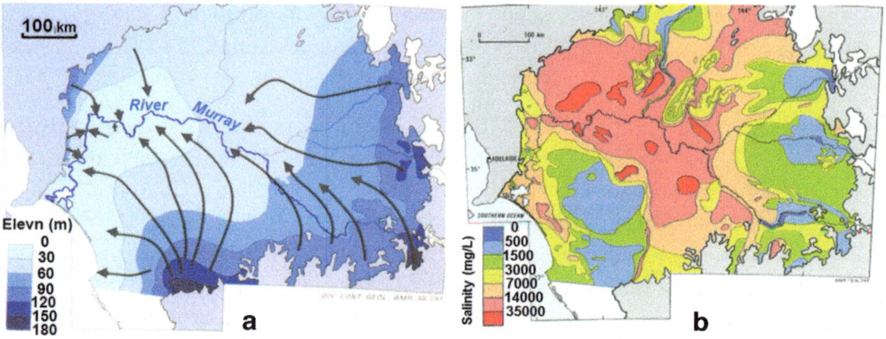

Fig. 8.3 (**a**) Groundwater flow directions and (**b**) Shallow groundwater salinity in the Murray Geological Basin

margins, to the downstream reaches of the River Murray where discharge occurs. Figure 8.3b presents the groundwater salinity. The low salinity resources of some of the major alluvial systems can be seen in the east, and the large volumes of fresh groundwater in the Tertiary limestone are located in the southwest. It also shows that where groundwater is discharging to the River Murray, salinities are over 20,000 mg/L. This natural discharge is exacerbated in some areas by drainage from irrigation areas established adjacent to the river. Before remedial measures were undertaken, salt loads entering the river were estimated to be about 1000 tonnes/day (MDBC, 1999). The increased irrigation development also led to land salinization and waterlogging on the low-lying riverine plains which have a shallow watertable.

These concerns led to the Salinity and Drainage Scheme, which later evolved to the Basin Salinity Management Strategy. The strategy involved the development of a series of salinity mitigation schemes. These included improvements in irrigation efficiency and water delivery systems, and diversion of saline irrigation returns. A number of salt interception schemes have been constructed that intercept saline groundwater before it enters the River Murray and its floodplain. About 175 wells extract over 10 GL/yr which is pumped to disposal basins distant from the river. A salinity offset scheme was developed to allow development and share costs for the salt mitigation schemes. This has successfully led to a considerable reduction in river salinity, especially in the lower reaches as shown in Fig. 8.4 which compares river salinity profiles from 2000 to 2015 (MDBA, 2018).

8.3 The History of Water Management in the MDB

There has been a long history of competition for water in the MDB. From the 1880s, diversions of water for irrigation caused tension between the interests of navigation and irrigation. In 1902, an Interstate Royal Commission was established to investigate the uses of the Murray and its waters. It attempted to strike a balance between

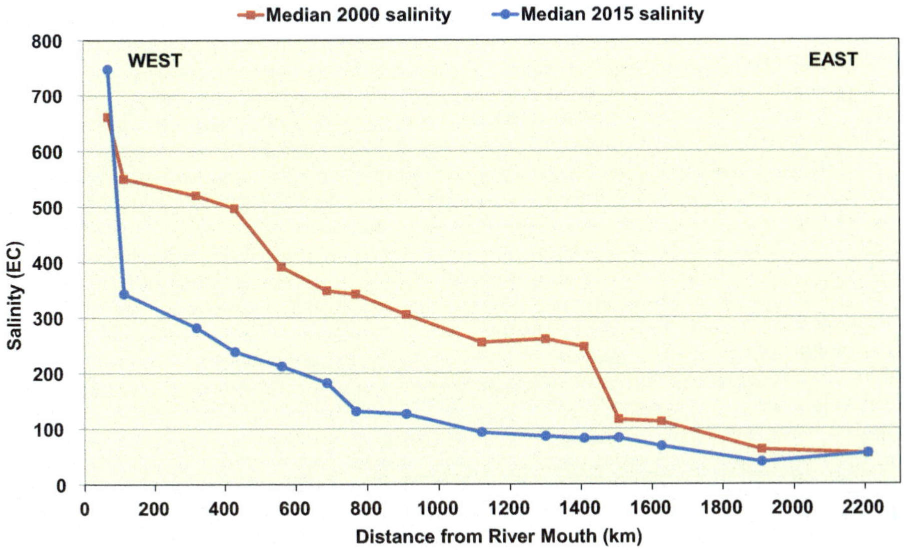

Fig. 8.4 Comparison of River Murray salinity profiles

irrigation and navigation interests by establishing the proportion of water that each of the states was allowed to take from the river. It was not until 1915 however, that the River Murray Waters Agreement was finalised and in 1917, the River Murray Commission was established to administer the Agreement. The Commission would manage the efficient sharing the Murray's waters between the three states and coordinate the construction of the locks and weirs by the state building authorities.

From that time to the 1940s, the rise of the railways and demise of river trade had caused a change in the major use of the river – the emphasis was now on irrigation rather than navigation, so only locks and weirs that aided irrigation diversion were constructed. Increasing demands on Murray waters for irrigation, human and industrial consumption were made in the following decades. Figure 8.5 shows the increase in surface water diversions from the late 1929 to 2007 for the individual states as well as the total for the whole basin. The increase is especially notable during the 1960–1990 period. In the late 1960s, severe drought coupled with these rising demands brought increasing public and political awareness of the escalating problems in the Murray-Darling Basin. Because the finite resource was being stressed, there was a need to change from a 'development' culture to a 'sustainability' culture. The closure of the Murray Mouth by sandbanks in 1981 due to lack of flow and reports about rising salinity levels in the Murray contributed to changes in the River Murray Waters Agreement in 1982, when its role was extended to address water quality, environmental and recreational issues.

A series of changes to the interstate agreement have occurred since 1980. The Murray-Darling Basin Commission was formed in 1987 to provide advice on natural resource management and water distribution throughout the whole Basin. By 1998, all of the Basin states had signed on to the agreement. A growing body of

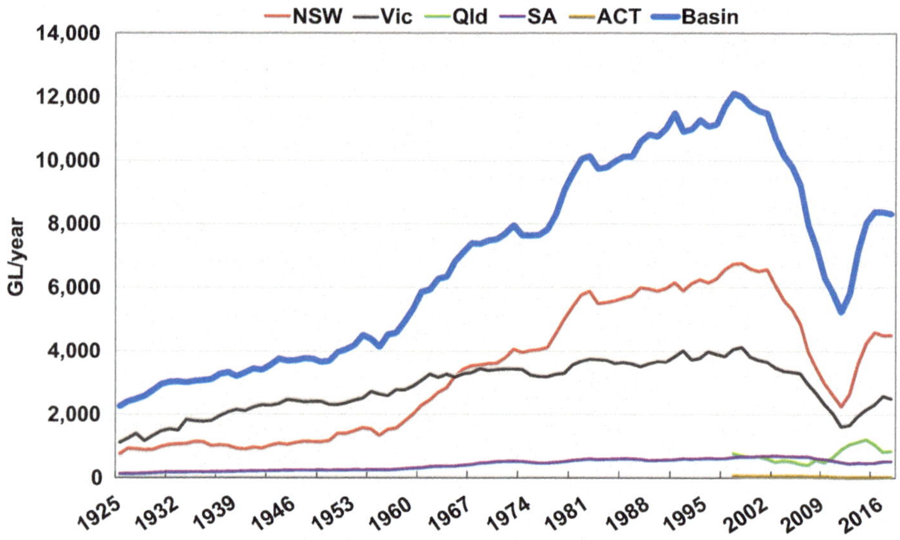

Fig. 8.5 Surface water diversions in the Murray-Darling Basin (1929–2007)

evidence indicated that the water resources of the Basin were being overextended, while the ecological health of the Basin was under increasing stress and degrading.

Following the review of the Operation of the MDB Cap in 2000, the MDB Ministerial Council agreed that groundwater would be managed on an integrated basis with surface water within the spirit of Cap, and that the MDB Groundwater Management Strategy be based on state jurisdictional management of groundwater through sustainable yields and include investigations clarifying how groundwater management practices may impact upon the integrity of Cap in future.

In response to such issues occurring in the MDB and elsewhere, the Council of Australian Governments in 1994 adopted a strategic water reform framework (see Chap. 7). The main objectives of the strategic framework were to establish an efficient and sustainable water industry, and to arrest widespread natural resource degradation partly caused by consumptive water use. In 1995, the Commonwealth and MDB states, as an early step in the water reform, agreed to cap the bulk of surface-water diversions in the MDB at 1993–1994 levels. The Council of Australian Governments (COAG) reinforced and extended the strategic water reforms in 2004 through the Intergovernmental Agreement on a National Water Initiative (NWI).

The NWI's most important governance reform was to commit the states, who have primary responsibility for water management, to a system of legally binding water management plans which would control impacts on the resource by imposing volumetric caps on licensed extraction (see Chap. 7). These plans were created for the good quality groundwater resources in the MDB that were developed for irrigation. However some over-allocation of some resources did occur (mainly in NSW) because in some irrigation areas that were fully allocated up the sustainable limit, actual extractions amounted to only about 40% of those allocations. In order to increase extractions closer to the limit, additional allocations were made which exceeded the limit. This over-allocation in major groundwater systems also led to the perception that groundwater extraction needed to be 'capped'.

The NWI also recognized the connectivity between surface and groundwater resources and the need for connected systems to be identified and to be managed and accounted as a single resource. It was increasingly being seen that the integrity of any cap on surface water diversions would be undermined without a similar cap on groundwater extractions.

The water reform process coincided with the Millenial Drought. The 1997–2006 decade not only was one of the driest ever recorded, but also led to the much lower surface water flows than previous droughts. This led to contingency measures with respect to water use and the death of some ecosystems. In response to this drought, the Prime Minister engaged CSIRO to review the water availability across the MDB, including explicitly accounting for risks of climate change, changed land use and increased groundwater extraction and farm dams (CSIRO, 2008). This study showed that in the event of these risks being realized, the main losers were the environment and downstream users.

In 2007, the then Prime Minister announced the National Plan for Water Security with $A10 billion funding. The Australian Government passed the Water Act (2007) and Addendum (2008). As a requirement of the Water Act, the Murray Darling Basin Authority (which took over the functions of the MDBC in 2008) was required to develop the Basin Plan, with the primary objective of determining a sustainable limit on water extraction in the Basin. The MDBA underpinned the development of the Basin Plan with scientific research, ecological response modelling and social and economic studies.

8.4 Groundwater and the Basin Plan

The 2007 Federal Water Act was the first to take precedence over the long-established states' primary responsibility for water management. It set a cap on groundwater extraction at current levels of extraction across the whole of the MDB for the first time because of the risks of increased extraction that could occur because of the over-allocation that occurred in key areas under the states' management regimes. The groundwater cap included areas with saline groundwater that were not included in existing state groundwater management plans. The application of such a cap meant that the MDB needed to be subdivided into spatial units (Fig. 8.6), and an extraction limit (Sustainable Diversion Limit: SDL) implemented for all such units (SDL resource units). The volume of entitlements issued prior to the Basin Plan was also assessed.

While there are stressed aquifers within the MDB, these tend to occur in 'hot-spots'. Much of the MDB is subject to a low density of groundwater extraction for a range of reasons, including high groundwater salinity, low aquifer transmissivity and a lack of economic drivers. This means that for some SDL resource units, it is

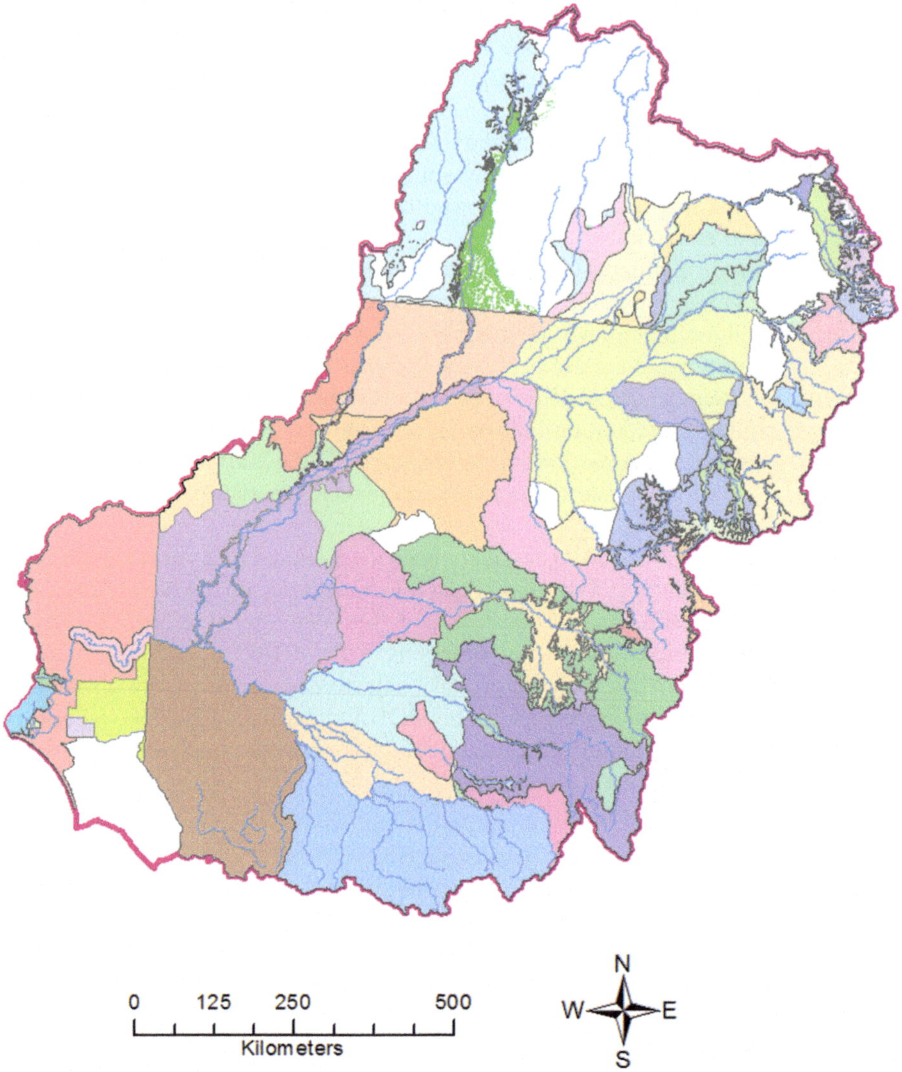

Fig. 8.6 SDL resource units in the Murray-Darling Basin

possible to increase groundwater extraction with 'acceptable' impacts. When aggregated across the MDB, average annual use from 2003 to 2017 (1401 GL/yr) is less than the Basin-wide entitlement volume (2380 GL/yr), which is less than the Basin-wide SDL (3494 GL/yr). This apparent potential increase in groundwater extraction has attracted criticism, as it seems at odds with and could possibly undermine, the surface water cap for which billions of dollars has been spent on reducing entitlements of more than 2000 GL/yr. In comparison, there is only a minor recovery of groundwater entitlements (40.4 GL/yr) from the northern MDB. In reality however, the potential for increases in groundwater extraction are low because of high salinities in the new SDL resource units which also have low or zero connectivity with surface water.

8.4.1 Determining Sustainable Diversion Limits

The SDL for each resource unit has been determined through a multi-step process (MDBA, 2012a, 2012b, 2012c) as follows:

1. A preliminary risk assessment was conducted across the whole MDB using a consistent methodology. The diffuse recharge was estimated for the SDL resource unit and a preliminary extraction limit (PEL) was determined as a fraction of the diffuse recharge which was dependent on the overall risk and the degree of uncertainty in the assessment of the impacts of extraction on;
   (a) aquifer integrity,
   (b) baseflow to streams,
   (c) groundwater-dependent ecosystems and
   (d) groundwater salinity

   The impacts were assessed using defined criteria. Where risk was assessed to be high, the PEL was set at 5% or less of recharge, and where risks were low, the PEL was up to 70% of recharge. Higher uncertainty would also lead to a reduction of the PEL. For many areas, data was sparse.
2. For units assessed to be of higher priority (generally areas of higher extraction), existing groundwater models were used to assess the impacts of extraction on the four characteristics described above. For example, aquifer integrity was assessed on whether modelled groundwater levels equilibrated within 50 years. The end result from these first two steps, is the determination of an extraction rate that should lead to acceptable impacts on key environmental values.
3. These preliminary estimates then underwent a filter process in which the PEL was compared to existing entitlement volumes. Also, policy overlays were taken into account for groundwater systems with high connectivity to streams, non-renewable groundwater resources (eg the Tertiary limestone aquifer), unassigned water (defined below) and areas already undergoing a reduction in entitlements. For example, for systems that are highly connected, the SDL was set at the existing entitlement volume or to the PEL, whichever was the lowest. This meant that extraction from highly connected systems would not further undermine streamflow than existing conditions, or extraction was limited to a small fraction of streamflow.
4. The cumulative impact on streamflow from groundwater extraction from different SDL units was determined and was considered unacceptable for some. This was used to further reduce the SDL for some highly connected systems.
5. For unassigned SDL resource units, ie those areas where the SDL is higher than the entitlement volumes, the SDL was further reduced as a precautionary

approach, so that the difference between the entitlements and the SDL was halved. This particularly affected areas where current use was very low, often less than the 10% of the determined SDL volume. For such areas, increases in extraction was unlikely to lead to the SDL being reached for several decades, if ever.

6. The value of the SDL was matched to the value determined by the existing state plan if the determination of extraction limits in these plans had adequate consideration of the four characteristics described above and there was no over-allocation of entitlements.

The average groundwater extraction from 2003 to 2017 is shown in Fig. 8.7 in comparison with the Basin-wide entitlement volume and SDL. Although there is no apparent trend over this period, there is a relationship with climate ie the use tends to be higher during drier periods and vice versa. However it should be noted that the methodology for assessing groundwater use has changed over time. Even if the groundwater extraction increased at a rate of 2–4%/yr in every groundwater unit up to the SDL, only low volumes of unassigned water would be used within the next 40 years. Rather, most increases would lead to current use being closer to the entitlement volume. In 24 SDL resource units where surface water connectivity is an issue, these increases in extraction would result in a reduction in streamflow of 56 GL/yr (MDBA, 2012b).

The extraction of groundwater for the purposes of coal seam gas production in the northern MDB has shown how extraction regimes can change quickly when economic and technological drivers are more prominent. Some mining processes can use brackish water and the economics of mining operations may result in desalination being feasible. Desalination has been used for horticultural purposes in the MDB during drought, but economic factors make this is a rare proposition. It is feasible that this situation may change with technological advances, and extraction

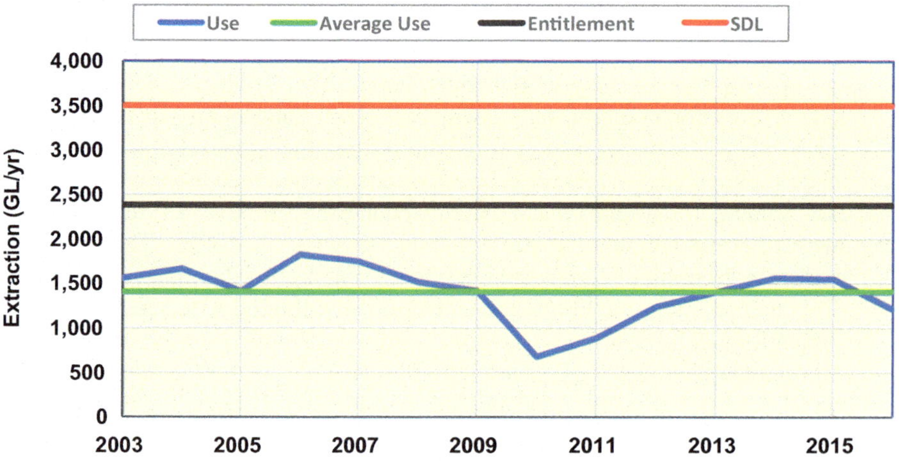

Fig. 8.7 Basin-wide groundwater use from 2003 to 2017

may approach the SDL in more areas in the coming decades. Groundwater itself is not a scarce quantity, but readily usable groundwater of appropriate quality is. This means that while there is a Basin-wide cap on groundwater use, that cap may never be reached.

The basin state jurisdictions are responsible for the groundwater management plans that keep extractions within the SDL. While the SDL is a limit to extraction, it is a very blunt instrument for groundwater management as it does not consider the spatial pattern of groundwater extraction which is important in determining whether adverse impacts occur. Highly focused extraction can lead to 'hot-spots' of groundwater drawdown while extraction close to GDEs or streams is likely to impact on those assets, even without any change in the SDL. The use of local management rules or management zones refines management to target higher priority impacts. Such refinement is not feasible at the Basin-scale and are generally included in state water management plans which requires communication with the local community and need to be accredited by the MDBA.

There are two examples of SDL resource units. The first example is the Lower Murrumbidgee, a floodplain alluvial groundwater system typical of those where most of the groundwater extraction occurs within the MDB. It had previously been previously over-allocated. Figure 8.8 shows a map of the groundwater extraction relative to the groundwater salinity and the groundwater drawdown response. It also

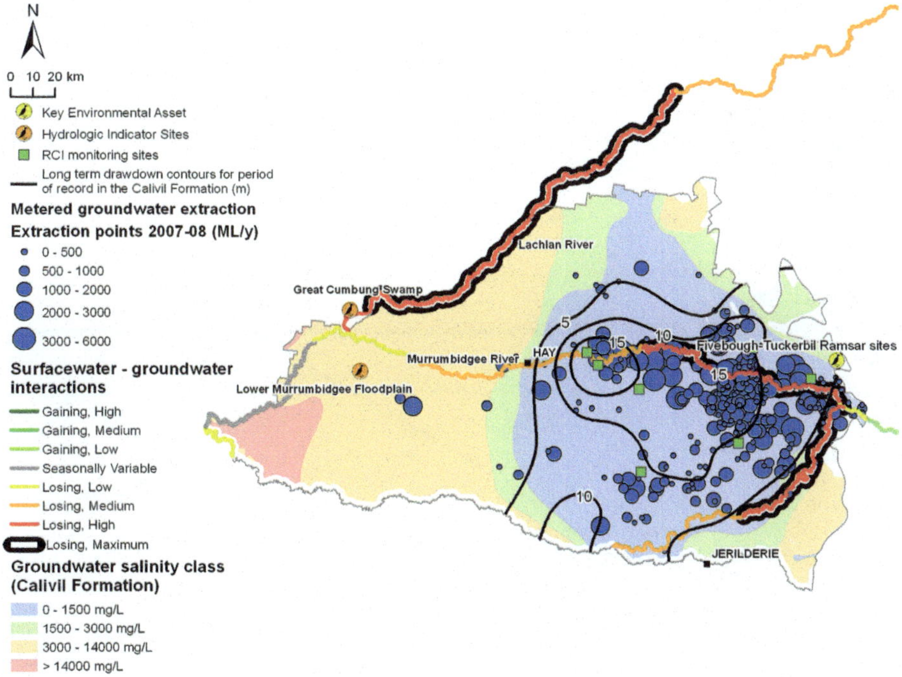

Fig. 8.8 Map of the Lower Murrumbidgee Alluvium SDL resource unit

shows the types of surface water – groundwater interaction. Characteristic features include:

(a) westerly flow with increasing groundwater salinity,
(b) groundwater extraction occurs mostly in the fresh groundwater zone, which is associated with a local cone of depression in the watertable,
(c) the main streams are losing and in some cases, are maximum losing (disconnected),
(d) the allocation had been previously reduced under the ASGE scheme, and
(e) most of the recharge occurs from drainage beneath surface water irrigation, which occurs over a much greater area than that irrigated from groundwater.

For the Basin Plan, the SDL has been set at the BDL (the reduced allocation), which is less than the PEL of 327 GL/yr. The historical groundwater use has fluctuated greatly in response to climate and floods (Fig. 8.9). In relation to the four key environmental characteristics:

1. Modelling had shown that water levels will equilibrate. From Fig. 8.8, it can be seen that a higher extraction rate may be feasible if the extraction had been more widespread.
2. The impact of increased groundwater extraction on streamflow is a 40 GL/yr reduction over a 50-year period,
3. groundwater extraction is not seen as a high risk to the key ecosystem site (Fig. 8.8),
4. increasing vertical gradients is likely to lead to increasing groundwater salinity in the deeper (main extraction) aquifer over a few decades.

The second example is the Western Porous Rock SDL resource unit (Fig. 8.10), which lies just to the west of the Lower Murrumbidgee Alluvium SDL resource

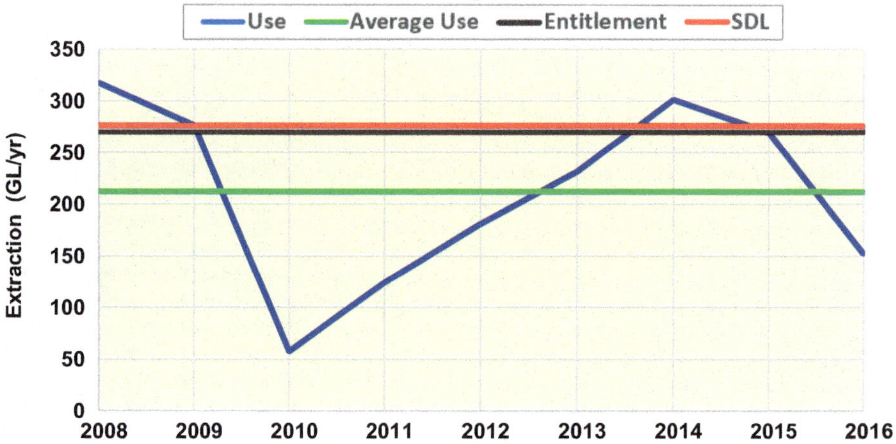

Fig. 8.9 Historical groundwater extraction in the Lower Murrumbidgee Alluvium SDL resource unit

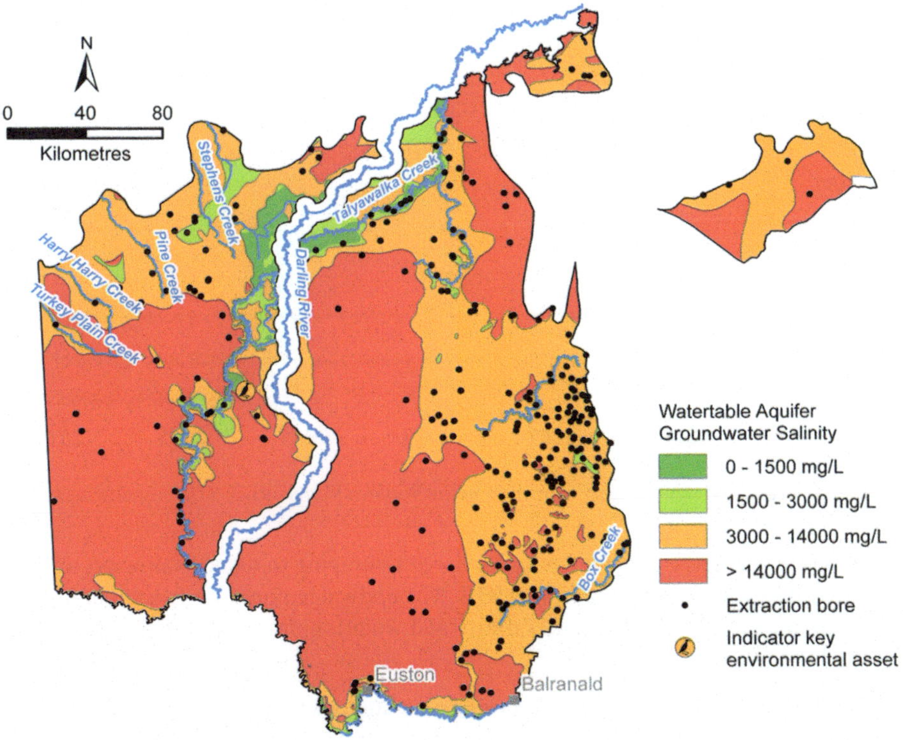

Fig. 8.10 Map of the Western Porous Rock SDL resource unit

unit. Current use is about half of the entitlement volume and only 15% of the SDL. Use is mostly for stock and domestic supply and can be seen to be mostly in the fresh to brackish groundwater salinity zones. There is no significant trend in groundwater use and without change are unlikely to reach the SDL in the next few decades (Fig. 8.11). However, if mining developments were to occur, this situation may change quickly.

One of the major reasons for including groundwater in the Basin cap is the connectivity between groundwater and surface water resources. There could be potentially an increase in groundwater extraction of 2000 GL/yr. The MDBA have estimated that the impact of this to be just less than 200 GL/yr, suggesting an average connectivity factor of only 0.1. Such a low connectivity is due to two main causes:

1. the potential increase in extraction from highly connected groundwater systems has been limited by the process used to determine SDLs,
2. the connectivity has been set to zero where groundwater extraction is seen to have environmental benefits. For example, where the groundwater is saline, discharge into the river can cause reduced water quality. Significant government

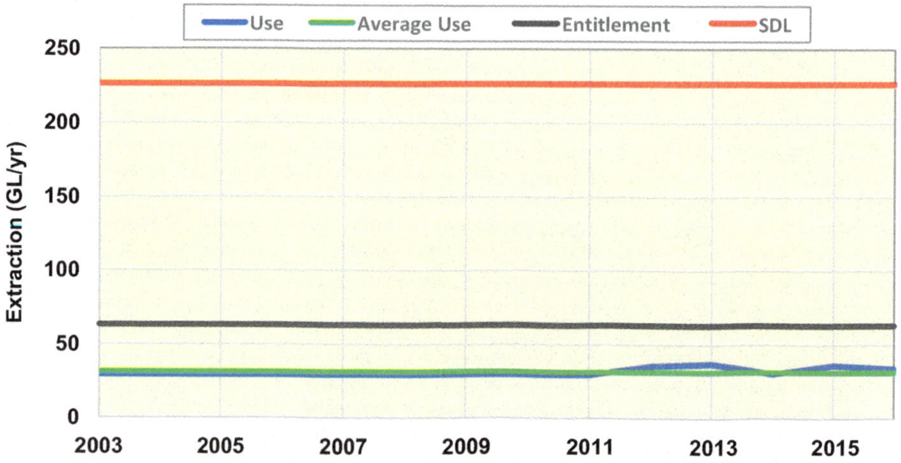

Fig. 8.11 Historical groundwater extraction in the Western Porous Rock SDL resource unit

funding is already allocated to groundwater pumping to reduce saline groundwater inflows to the rivers of the MDB. Groundwater pumping from the shallow alluvium can also reduce land salinity and waterlogging in the southern Riverine Plains region.

The Basin process addresses major basin-wide risks due to groundwater extraction, while Basin states are responsible for the water management plans that addresses threats within the plan area. Some adaptation of these plans is inevitable, as new information becomes available.

8.5 Future Issues

The long time-scale associated with groundwater processes means that adaptive management may be a practical way of dealing with future issues. Such issues include:

1. Climate change: The MDB is predicted to have more prolonged droughts (strong confidence) and less winter-spring rainfall (confident) in the southern MDB (BoM State of Climate, 2016). Climate change will affect groundwater in different ways:

   (a) surface water flow is likely to reduce along the southern connected system, putting more pressure on the groundwater as a source of water,
   (b) reduced surface water flow and increasing water use efficiency is likely to reduce irrigation recharge to some areas, such as the Lower Murrumbidgee,
   (c) lower winter-spring rainfall will reduce diffuse dryland recharge, especially in the southern MDB, where rainfall is winter-dominant,

(d) prolonged droughts will mean that groundwater will become increasingly important during these extreme events,
(e) there is likely to be an increase in use from climate-resilient water sources, such as deeper confined aquifers, desalination plants and managed aquifer recharge, especially for urban water supplies,
(f) climate change is likely to lead to land use change, such as a reduction in forest plantations in southern areas and shift from cropping to pastures in southern semi-arid areas. Where these changes are significant, there can be a change in recharge that dominates the change due to climate alone, and
(g) there will be shifts in irrigated crops due to higher temperatures and rainfall changes, that will change the demand for groundwater as a source.

Most of the above effects have a high uncertainty in timing, spatial distribution and magnitude. The response to this uncertainty is increased monitoring of stressed systems including land use; community education; planning for alternative water sources and drought contingency measures. The introduction of adaptive measures, such as water level response management and groundwater trade rules will be an important part of future management. Each jurisdiction has a climate change strategy. Climate change can be addressed without reducing the SDL through defining management zones around the 'hotspots', implementing extraction limits for management zones and using trade to shift extraction away. The stage may be reached however, where the SDL will need to be reduced.

2. Changing demand caused by factors other than climate change, such as demographic change, changes in irrigation crops, mining and gas industry and changes in technology. In particular, the mining and gas industry may be able to use brackish to saline water for some processes and use desalination for others. Should commodity prices change, or desalination becomes cheaper, saline groundwater could be increasingly used. Australia has an increasing population, with pressure to move people from major cities. It is difficult to predict the timing and spatial distribution of such changes. There is little alternative to using adaptive approaches and measures which are similar to those for climate change and are consistent with the current Basin Plan.
3. The use of groundwater trade has historically been much less than that for surface water. If there is greater pressure for groundwater use and metering becomes more widespread, trade of groundwater is likely to increase. New South Wales and South Australia are the only states in which trade widely occurs. Trade between groundwater and surface water is only applicable to two groundwater SDL units in the MDB. As climate change and variability becomes more important, it may become more beneficial to use groundwater in preference to surface water. Managed Aquifer Recharge (MAR) is only limited within the MDB due to the cost and the lack of sites where it can be implemented easily. Some arrangements for MAR need to be clarified, but there is no reason why it cannot be included within the Basin Plan. Conjunctive water use may be used to address climate variability and waterlogging caused by surface water irrigation.

The introduction of the cap has enforced a greater degree of consistency in the management approach across the Basin. The variability of hydrogeology, climate, land use, level of extraction and legislation mean that some difference in management arrangements is inevitable. However, consistency has benefits in efficiencies in oversight and reporting, community and industry understanding and bilateral arrangements across state borders. It is likely that some of the administrative processes in the Basin Plan will be adopted in areas outside of the MDB, at least in the Basin states.

The history of the development of the Basin Plan has resulted in it being designed for the management of surface water. The extrapolation of Plan processes to groundwater (due to the integrated nature of these resources) has not always made sense, and it is likely that some of these processes will be reviewed in the future.

The Basin Plan is a work in progress. In 2019, all State water resource plans will be revised to be consistent with the requirements of the Basin Plan. The Basin Plan itself will be reviewed in 2026. At that stage, it should be possible to review the effectiveness of the Basin Plan for groundwater.

References

Brown, C. M., & Stephenson, A. E. (1991). *Geology of the Murray Basin, southeastern Australia* (p. 235). Canberra, Australia: Bureau of Mineral Resources (Australian Bulletin).

Bureau of Meteorology and CSIRO. (2016). *State of the climate report* (23pp). Commonwealth of Australia.

CSIRO. (2008). *Water availability in the Murray-Darling Basin. A report to the Australian government from the CSIRO Basin Sustainable Yields Project* (p. 67). Canberra, Australia: CSIRO.

Evans, W. R., & Kellett, J. R. (1989). The hydrogeology of the Murray Basin, southeastern Australia. *BMR Journal of Australian Geology and Geophysics, 11*, 147–166.

Murray-Darling Basin Authority. (2012a). *The proposed groundwater baseline and sustainable diversion limits: Methods report* (MDBA publication no: 16/12, (49pp)). Canberra, Australia: MDBA.

Murray-Darling Basin Authority. (2012b). *Addendum to the proposed groundwater baseline and sustainable diversion limits: Methods report* (MDBA publication 62/12, (39pp)). Canberra, Australia: MDBA.

Murray-Darling Basin Authority. (2012c). *Groundwater sustainable diversion limit resource unit summary report cards* (MDBA publication no 49/12) (p. 103). Canberra, Australia: MDBA.

Murray-Darling Basin Authority. (2018). *BSM2030 biennial implementation report 2015–17.* Canberra, Australia: Murray–Darling Basin Authority.

Murray-Darling Basin Commission. (1999). *A salinity audit of the Murray Darling Basin.* Canberra, Australia: MDBC.

Glen Walker is a senior research scientist with interests in groundwater and salinity. Formally working at CSIRO, he is now a consultant (Grounded in Water) and member, Independent Expert Scientific Committee on Coal Seam gas and Large Coal Developments.

Steve Barnett is Principal Hydrogeologist at the Water Science and Monitoring Branch of the Department for Environment and Water in South Australia. He has been involved in the investigation, monitoring and management of groundwater resources in SA for over 40 years, and has contributed technical and policy input into ten groundwater management plans which incorporate a variety of different aquifers and management issues. He is a past-president of Australian Chapter of the International Association of Hydrogeologists.

Stuart Richardson has more than 25 years of experience in groundwater investigations, aquifer-scale impact assessments and regulatory liaison. Stuart's strong background in the development of water planning strategies for industry and government has allowed him to work on projects that include aquifer-scale assessment of sustainable yield, allocation planning and bringing together technical studies and policy development for many clients. Stuart is Principal Hydrogeologist and Managing Director of CDM Smith in Australia. He works with his team to provide innovative and lasting integrated solutions in water, environment, energy and facilities to public and private clients.

Chapter 9
Information Systems for Sustainable Management of Groundwater Extraction in France and Australia

John Sharples, Elisabetta Carrara, Lindsay Preece, Laurence Chery, Benjamin Lopez, and Jean-Daniel Rinaudo

Abstract Sustainable groundwater management relies on data to establish resource conditions and measure the effects of management intervention. As groundwater management grows in size and complexity so does the data needed to inform it, and the systems needed to manage this data. This chapter presents a discussion of groundwater information systems, their history, and examples of their application in France and Australia, including how these systems are used to inform and improve groundwater management. Examples are presented demonstrating the application of information systems in a range of agencies and legislative settings. These examples include systems used for local management, national data standardization, online data sharing, and environmental impact assessments. Finally, lessons learned and future developments are presented. This includes a comparison of the similarities and differences in the history and current state of groundwater management system development in each country.

Keywords Groundwater information systems · Data · Groundwater level

J. Sharples (✉) · E. Carrara
Bureau of Meteorology, Melbourne, VIC, Australia
e-mail: John.Sharples@bom.gov.au; elisabetta.carrara@bom.gov.au

L. Preece
Department of Water and Environmental Regulation, Perth, WA, Australia
e-mail: lindsay.preece@dwer.wa.gov.au

L. Chery · B. Lopez
BRGM (French Geological Survey), Orléans, France
e-mail: L.chery@brgm.fr; b.lopez@brgm.fr

J.-D. Rinaudo
BRGM, Montpellier University, Montpellier, France
e-mail: jd.rinaudo@brgm.fr

J.-D. Rinaudo et al. (eds.), *Sustainable Groundwater Management*,
Global Issues in Water Policy 24, https://doi.org/10.1007/978-3-030-32766-8_9

9.1 Introduction

Monitoring is an essential element of any effort to integrate groundwater science with water-management decisions (Holliday, Marin, & Vaux, 2007; Vaessen & Brentführer, 2014). Monitoring is especially important where anthropogenic impacts, such as pumping or pollution discharge, create stresses in an aquifer. Pumping without monitoring extraction or the state of the aquifer is like a business continually withdrawing money from a bank account without any bookkeeping system (Nelson, 2012). Hence, for a groundwater system to be managed effectively, the resource must be monitored to account for the "credits" and "debits". For groundwater resources this bookkeeping system is a Groundwater Information System (GWIS). These systems have become invaluable for groundwater resource management.

The focus of this book is quantitative management. As such, this chapter will only cover issues related to groundwater quantity monitoring, leaving aside the extremely important issue of quality (chemical) monitoring. Quantity monitoring focuses on the flows of water entering and leaving aquifers, on variation of water levels, and storage within the aquifers. Quantity monitoring systems are set up to provide technical and administrative information on (1) extraction points (wells and boreholes databases); (2) associated actual water extraction (pumping databases); and (3) water resources quantitative states, which can be assessed through water levels, e.g. water flows in springs, base flow to streams and rivers.

Other chapters have shown how quantitative groundwater data informs management decisions. In this chapter, the collection and management of that data is discussed. Monitoring systems, data management systems, and their relationship to resource management decisions, are also briefly described. Example GWIS in France and Australia are presented to highlight successes and challenges in those countries. Lastly, lessons learned and future challenges for these systems are discussed.

9.2 A Framework to Analyse the Development of Groundwater Monitoring and Information Systems

Groundwater Information Systems (GWIS) are the systems used to collect, store, and publish data relating to groundwater. These systems are ubiquitous with good management practices and have been developed by nations around the world to monitor groundwater resources (Klug & Kmoch, 2014; Lee & Kwon, 2016). Although many variations exist, reflecting local hydrogeology and management objectives, these systems are typically comprised of several distinct components; a monitoring network, a data store, and a data interpretation and publication system (Table 9.1).

Table 9.1 Groundwater Information System components

Monitoring network	Sites: Bores, piezometers, wells, springs
	Monitoring equipment: Data loggers, manual readings, telemetry, flow meters
	Data collection procedures: Sampling, handling, laboratory tests
Data store	Data entry process
	Quality checking and control
	Relational database
	Value add and contextual data
Publication and interpretation	GIS
	Reports
	Assessments
	Groundwater models
	Data sharing
	Internet data applications
	APIs

Adapted from UN-ECE Task Force on Groundwater Monitoring and Assessment (2000) and Tuinhof, Foster, Kemper, Garduno, and Nanni (2006)

9.2.1 Groundwater Monitoring Networks

The type of sites monitored, the number of sites, the kinds of data collected, and the frequency of monitoring will depend on the hydrogeology of the groundwater system as well as the desired management outcomes. However, in most systems the majority of monitoring occurs via piezometers, bores and wells. As such, data derived during bore construction and development typically forms the basis for a GWIS (Jousma, 2008). This data includes bore and site details, lithological and hydrogeological information, and bore construction details. This data is crucial in properly understanding and interpreting monitoring data from these bores.

Ubiquitous to all groundwater monitoring are measurements of groundwater level, or pressure head, measured in a bore. These data are the principal source of information about the hydrologic stresses acting on aquifers (Taylor & Alley, 2001; Tuinhof et al., 2006). Groundwater levels are used extensively to understand the hydraulic setting, being the primary way of estimating groundwater flow direction and magnitude. The value of groundwater level data increases with the length of ongoing monitoring. As groundwater typically responds slowly to changing stresses, long-term records of groundwater level are invaluable for evaluating the impact of these stresses. In more developed resources, where a greater degree of management is required, long-term level data is essential to develop groundwater models and for assessing the effectiveness of current and past management interventions (Taylor & Alley, 2001). As such, monitoring changes in groundwater level should be a key component of all GWIS.

Monitoring frequency of groundwater levels is an important factor to consider when setting up GWIS. In general, aquifers require more frequent monitoring if they are: shallow or unconfined, have a high through flow or recharge rate, have a higher level of extraction, or show a strong response to climate conditions or link to aquatic and related terrestrial ecosystem features. For extensive, confined aquifers changes in groundwater level typically occur very slowly. Adequate monitoring for such systems might be achieved with seasonal or annual records of water levels. Whereas monthly, weekly or continuous monitoring may be required in shallow, unconfined aquifers. For new, unknown resources, frequent or continuous water level monitoring should be considered to identify the magnitude and frequency of aquifer fluctuations (Taylor & Alley, 2001). The frequency can be appropriately adjusted once an understanding of the groundwater system is developed.

In areas where a licence or permit is required for the extraction of groundwater, compliance monitoring may be needed. This data should be part of the GWIS and link the licence to the physical resource, including the aquifer and bores used to extract groundwater. This is often not the case as the need for licensing administration systems typically arises long after monitoring data systems have been developed. Direct extraction monitoring, by fitting a meter to groundwater pumps, is the most accurate method, but costly and often difficult to implement, as it requires the cooperation of water users. Where meters are not feasible, surrogate measures may be employed to estimate use. For example, energy consumption from pumping, or hours of pump operation. In rural areas where agriculture is the dominant groundwater use, remote sensing can be used to infer groundwater use by estimates of evaporation or crop growth and coverage (Tuinhof et al., 2006; Vaessen & Brentführer, 2014). These estimation methods, however, do not typically form part of the core monitoring network but are added in the data interpretation phase of the GWIS.

In the practical implementation of a monitoring program, quality assurance and quality control must be carefully implemented to ensure the validity of the collected data. For detailed discussion of data validity see, for example, Jousma, 2008. Table 9.2 summarises the above discussion by giving broad groups of monitoring systems, and the types of management decisions they inform.

Once monitoring data has been collected, quality controlled and assured, it is imperative that it be systematically and securely stored for future use. Long-term records of groundwater data are invaluable in understanding, managing and forecasting resources. Ideally, this store should be a persistent relational database (Jousma, 2008; Tuinhof et al., 2006; Vaessen & Brentführer, 2014).

There are many database options for storing groundwater data. These range from generic, open source database applications through to commercial applications specifically designed for hydrologic data (Fitch, Brodaric, Stenson, & Booth, 2016). Commercial groundwater databases are often packaged with tools specially designed to view, interrogate, and publish hydrologic data (Fitch et al., 2016; Jousma, 2008). The choice of database application should be considered in the context of the data custodian's ability to use the application, its suitability to the data being collected, and the costs involved in initialising and maintain the application. In general, generic database applications are highly flexible but require significant effort and knowledge to build and customise for hydrogeological data. Commercial

Table 9.2 Examples of groundwater monitoring types and the management decisions they support

Type of monitoring	Ground water system	Parameters	Timescale	Management objective	Management decision
Resource investigation and monitoring	All	Levels and salinity	Ongoing – frequency appropriate to groundwater flow system	Improve system understanding and monitor changes in system	Does management approach need to be revised or modified?
Compliance monitoring	Highly developed aquifer	Abstraction volumes	Annually, or after pumping season	Understand volumes of water abstracted from aquifer and compare to issued permits	Should steps be taken to reduce abstractions, i.e. reduce allocations or issue fines?
	Highly developed aquifer	Levels	Daily to monthly	Maintain minimum levels in aquifer head, e.g. to protect flows to surface water or maintain levels in shallow bores	Have trigger levels been hit and do entitlements need to be reduces accordingly? Can more entitlements be issued without negative impacts?
Protection monitoring	Highly developed aquifer	Levels	Ongoing – continuous or high frequency monitoring	Protect public water supply well field from depletion	Do pumping schedules or spatial distributions need to be changed to protect access to the resource?
	Coastal aquifer	Salinity	Ongoing – monthly to seasonally	Protect aquifers from degradation by seawater intrusion	Do pumping locations and rates need to be adjusted to maintain groundwater heads near the coast?
	Aquifers connected to source of natural occurring pollutants	Arsenic, fluoride, etc.	Ongoing – weekly to monthly, or as required	Monitor natural occurring pollutant levels in groundwater resource	Should intervention be taken to prevent or minimise groundwater use? e.g. Public awareness campaign or to restrict certain uses?

applications typically work "off the shelf" but come with software-licensing fees and may attract further costs in maintenance and customisation.

More important than the choice of technology is the design and implementation of proper data management practices. This includes defined workflow and tools, roles and governance arrangements to ensure secure storage, ease of discovery, and access, as well as ensuring the quality and integrity of the data (WMO, 2008). The data management life cycle begins with the collection of samples and measurements in the field and extends through data handling, data entry, data validation, and publishing. Ensuring that data is accurate, trustworthy and available greatly increases the capacity to make informed management decisions. When establishing a Groundwater Information System, it is imperative that the data life cycle be considered in the design and planning phase. A significant amount of literature exists to guide the creation and review of information systems (e.g. Fitch et al., 2016; Jousma, 2008; WMO, 2008).

During development of a GWIS, it is important to consult widely. This includes engaging with a wide variety of stakeholders, such as water users, resource managers, government, interest groups, NGOs, and any other related groups. Understanding the current and future information needs of these groups will help to drive the content and structure of the information system. The success of groundwater management is dependent on communication and stakeholder investment, which are greatly affected by the availability of transparent, timely and relevant information (Global Water Partnership Technical Advisory Committee, 2000).

Publication via the internet has become the default method for distributing groundwater data in many nations. There is a proliferation of online applications for visualising, mapping, and downloading groundwater data from local catchments through to international coverage (e.g. waterdata.usgs.gov/nwis/gw; www.jejuwater.go.kr/; ggmn.un-igrac.org/). Typically, these applications will provide functionality to view bore locations on maps, plot water level and salinity, visualise bore hole logs and constructions details, and download data. In some cases, custom PDF reports and maps can be generated on the fly.

Many GWIS also use data management tools to expose and analyse the data. A common tool is to use a Geographical Information System (GIS) to view and analyse the data in a spatial context. This can be a highly beneficial way to view groundwater data as it can be overlain with other spatial data sets such as satellite imagery, terrain maps, and cadastre layers. Spatial data inquiry can be performed using specialised GIS software (e.g. ArcGIS, QGIS) or online services like Google Earth (Vaessen & Brentführer, 2014). Such tools are especially useful where surface features, such as land use, impact on groundwater resources. 3D visualisations are also becoming more readily available. These are particularly valuable for communicating the physical framework of aquifers, i.e. aquifer locations, confining layers extents, or faulting.

One of the most versatile tools for groundwater management is the use of numerical models to estimate current and future changes in a resource. In many regions, management decisions are driven by the outputs of a groundwater model. Models are data intensive to create, calibrate and run, requiring large volumes of input data.

This always includes hydrogeological and water level information and may include water quality data. Robust data systems can aid in model development by allowing quick and reliable access to groundwater data. Furthermore, as models and studies using models are becoming more ambitious in their scope, there is an increasing push to facilitate automatic data sharing via Application Programming Interface (API), web services and data sharing standards. The European Union's INSPIRE Directive is one example of a legislative requirement to share environmental data, including groundwater, via an agreed, well defined format. Major efforts have also been invested in creating international groundwater data sharing standards, such as the Open Geospatial Consortium (OGC) standard GroundWater Markup Language 2 (GWML2). Such efforts are invaluable in facilitating more complex, integrated studies in environmental and hydrological management (Alley et al., 2013; Fitch et al., 2016).

9.2.2 Challenges and Difficulties

The literature has many examples of groundwater monitoring being described as a trade-off between monitoring coverage (in terms of spatial extent, frequency and number of parameters monitored) and the cost and effort required (Bartram & Ballance, 1996; Taylor & Alley, 2001; Tuinhof et al., 2006). Given the slow movement of groundwater, monitoring can be protracted. The US Department of Energy estimates spending of US$5.5 billion on remediating polluted groundwater between 2000 and 2006, the majority of the cost going to long term monitoring. Long term stewardship is also expected to cost around US$100 million per year for 70 years. The US Navy expect to make similar expenditure on contamination monitoring (Minsker, 2003). While these examples are limited to pollution remediation they serve to demonstrate the magnitude of potential costs involved in groundwater monitoring.

In some situations, a "user pays; polluter pays" approach can be taken to recover the costs of monitoring and managing groundwater. Where the water manager issues licences or permits this can be achieved through collecting fees from groundwater users. This method has been successfully instituted in some countries, for example South Africa (DWA, 2010). However, a scale of economy comes into play, this method has mostly been applied where few extractors take large volumes of water. For regions such as Asia, which is typified by a large number of small volume extractors, such measures may not be feasible to enforce and administer (Shah et al., 2003).

New technologies, particularly in telecommunications and Internet of Things (IoT) sensors, are driving down the cost of collecting large volumes of data. Embedded sensors can now be installed inside bores and data collected via mobile networks or low flying satellite (e.g. Haley, Beck, Pollok, Grant, & McKilliam, 2017). This technology greatly increases the amount of data that can be collected, especially in remote and hard to access areas. While this technology can drive down the cost of data collection, the trade-off is the need for greater robustness in the data management components of the GWIS. Porter, Hanson, and Lin (2012), states that sensor technology is ahead of the information management and data storage technology typically used in water sciences. As such, adoption of new IoT and sensing technologies will require more robust and complicated data management and storage technology. Groundwater managers planning to move new or existing monitoring networks to an IoT based technology should be aware that it is an active, and fast moving, field of research and development.

In many countries, groundwater monitoring occurs for many purposes and is performed by multiple agencies. For example, one agency might monitor groundwater for resource extraction and supply, another for pollution monitoring and remediation. As these activities are typically legislative, the resulting data is often in inconsistent formats and held is separate databases (Dahlhaus et al., 2016; Fitch et al., 2016; Horsburgh et al., 2009). Similarly, where aquifers extend across administrative boundaries, data from a single resource might be collected and held by multiple agencies. This can occur internationally or within a country, e.g. across state borders. Typically, these situations are legally complicated; each agency will be operating under differing priorities, capacities, and legislation. In these cases, wide consultation can deliver great value to the design and implementation of information systems. Data sharing arrangements can provide opportunity to share monitoring and ensure best management practices can occur across borders (UN-ECE Task Force on Groundwater Monitoring and Assessment, 2000; Vaessen & Brentführer, 2014). This issue has been successfully addressed using data standards and API technology. The European Union's INSPIRE program is one such example.

9.3 Groundwater Information Systems in France

9.3.1 History of Groundwater Data and Metadata Collection in France

Monitoring of groundwater levels in France began in the middle of the nineteenth Century. The oldest known groundwater level data are from boreholes located in Albian aquifer (Paris sedimentary basin) in 1840. Overall, groundwater level data remained sparse until the end of the 1960s. The first networks for groundwater level monitoring were established in response to local needs, focusing on specific groundwater resources or on specific uses (e.g. drinking water) (Margat & Schneider, 1971).

In the 1960s and 1970s, the development of wider coverage groundwater monitoring networks was initiated by several ministries and by regional governments. This expansion occurred without real coordination between the different actors developing those networks. The first of these networks was set-up by the French Ministry for Health, under the drinking water regulatory framework. This first national network only collected information on water from boreholes used for drinking water abstraction (raw water sampling). During the same period, the

Ministry of Industry also funded the French Geological Survey to set-up a groundwater level monitoring network as part of the first "inventory of hydraulic resources". The main objective was to assess existing groundwater resources which were poorly known, to organise data collection and banking at a national scale and to draw the first hydrogeological maps.

These first uncoordinated actions were strengthened by the 1964 Water Law, which established the Water Agencies, and required them to set up water monitoring networks covering their entire territory. Each agency developed its own network, collected data and stored it in independent databases. From the late 1970's, Water Agencies also encouraged and financially supported county and regional governments to establish local groundwater monitoring networks. The objective of those local networks was mainly to monitor the increase of nitrate pollution (particularly in agricultural lands in the north and center of France) but most also monitored water levels. From 1970 to 1985, about 20 local or regional networks were created at various hydrological or administrative scales (catchments, counties, and water bodies). At the end of the 1980s, it became clear that all these independent monitoring networks should be coordinated and harmonized to improve both the geographical coverage and the consistency of data collected.

The new Water Law of 1992 provided the impetus for this reform (Blum et al., 2010, 2013). By imposing the elaboration of SDAGE - Water Resource Development and Management master plans (see Chap. 4), the law triggered the strengthening of existing water resource monitoring networks. Significant funding was provided to local governments by the Water agencies during the 1990s. As a result, a series of new networks were established and the number of monitoring points increased by a factor of four or five compared to 1970. The harmonization of existing networks started in 1999, after the six water agencies and the French ministry of environment signed a protocol establishing a "National Network for Groundwater Monitoring" (RNES in French). This protocol defined, amongst others, a minimum density of monitoring points, frequencies for water sampling (for quality measurement) and groundwater level measurements for each type of aquifer. A harmonized grid of sampling points was established, aggregating sites identified in each of the six French water basins. The first French national groundwater monitoring network was born.

In the early 2000s, this national groundwater monitoring network had to evolve again to comply the European Water Framework Directive (WFD) enacted in 2000 (WFD, 2000). According to the WFD, monitoring networks must provide data for conducting a reliable assessment of the qualitative and quantitative status of all groundwater bodies including assessment of the available groundwater resource (Grath et al., 2007; Blum et al., 2013). The WFD requires establishing two types of monitoring – surveillance and operational. Surveillance monitoring aims to supplement and validate the assessment of the status of water bodies and provide information for use in the assessment of long-term trends. Operational monitoring must be carried out for those groundwater bodies which are identified as being at risk of failing to meet the environmental objectives of WFD. Overall, the objectives of the WFD being quite similar to the former RNES, most of the former sampling sites were integrated into the WFD network which became operational on January 1st 2007.

9.3.2 The National Water Information System

The progressive integration of existing groundwater monitoring networks described above was supported by the development of a comprehensive Water Information System (WIS). The WIS (http://www.eaufrance.fr) collects, organizes and provides access to all water related data produced by 50 different organisations. The information, comprising 506 data sets, is regularly updated and published for each monitoring station, covering all catchments, regions, counties and aquifers. The system ensures data traceability (e.g. origin of the data, validation level). Data are produced, processed and stored according to standards defined by a network of institutions producing water data (http://www.sandre.eaufrance.fr/). These standards include technical specifications and code lists and describe how to exchange water data at the national scale. From an IT perspective, the Sandre guarantees interoperability of all French Water Information Systems. It ensures the creation and updating of detailed data dictionaries, the updating of common references, the development of data exchange standards (in accordance with European or international standards). The WIS comprises several modules which were progressively developed (see Table 9.3).

9.3.3 ADES: The National Portal for Groundwater

The development of the WIS was initiated with the creation of a groundwater portal, named ADES (Accès aux Données sur les Eaux Souterraines). ADES offers unique access to data from all groundwater networks in France (see Fig. 9.1) through a web portal (http://www.ades.eaufrance.fr). Data exchanges between participants implies certain rules defining both data content and format. The main data producers are the Ministry of Ecology, the six Water agencies, BRGM (French Geological Survey), the French Agency for Biodiversity (AFB, formally ONEMA), EDF (French Electricity company), Ifremer (Research institute for exploration of the sea), French institute for agricultural and environmental research, and the French Meteorological Institute.

In September 2018, the groundwater portal gives access to 15 million groundwater levels, from 4572 piezometers, and 76 million groundwater quality measurements, from 74,520 sampling sites. The main data users are groundwater local managers, water SMEs, drinking water producers, and environmental associations. The data is freely available to view and download. For groundwater levels, users may access historical data (see Fig. 9.2) but also to the results of a statistical analysis of water levels in the selected monitoring point as illustrated on Fig. 9.3. This figure shows the evolution of water levels from January to December for different climatic years, from very dry (ten years return period, in red) to very wet (ten years return period).

Table 9.3 Overview of selected water data portals part of the French Water Information System

Type of data	Portal	Data producers	Data volume (10^3 k)
Water levels in rivers (measured) and flows (calculated) – 3200 stations	HYDRO, http://www.hydro.eaufrance.fr/	MoE, MOA, WA, RI	20,000
Drinking water quality control (chemical and bacteriological)	SISE -EAUX, www.eaupotable.sante.aouv.fr	MOHA, PWUs	16,000
Water and ecosystem quality (physicochemical, hydro-biology, hydro) for rivers and lakes	NAIADES. http://www.naiades.eaufrance.fr	WAs, FBA	8000
Groundwater quality (chemical, physical)	ADES, http://www.ades.eaufrance.fr/	WAs, MoE, MoHA, LocGov, PWUs, NPs	6000
Groundwater levels			800
Performance of drinking water supply and sanitation utilities	SISPEA. http://www.services.eaufrance.fr/	DWUs	600
Coastal water quality (physical, chemical, biological)	OUADRIGE. http://quadrige.eaufrance.fr/	RI (Ifremer)	50
Bathing water quality (chemical, bacteriological) in rivers lakes and sea (3300 stations)	SISE-Baignades, http://baienades.sante.eouv.fr/baienades/home-Map.do#a	MoHA	200
Water abstraction (volume, per user, per year) in 85,000 abstraction points	BNPE. http://www.bnpe.eaufrance.fr/	Was, state services	80

Brgm French Geological Survey, *FBA* French Biodiversity Agency, *LocGov* Local Governments (municipal, couty, regional governments), *PWUs* Public Water Utilities, *MOA* Ministry of Agriculture, *MoE* Ministry of Environment, *MoHA* Ministry of Health Affairs, *RI* Research Institutes, *WAs* Water Agencies, *NPs* National and Regional Natural Parks, *Ind* Industries subject to Environmental Monitoring

ADES also collects data from the Subsurface Database (Banque de données du Sous-Sol – BSS http://infoterre.brgm.fr/page/banque-sol-bss) which contains information related to all underground works (deeper than 10 m), including all wells and boreholes used for groundwater extraction. Established in 1958 in application of the Mining Code, the BSS contains administrative information (name of the owner, location), it identifies the aquifer exploited and it provides the description of the geological levels encountered during drilling. When available, the drilling logs have been digitized and can be accessed online on the InfoTerre Portal. The BSS makes more than 700,000 descriptions of underground structures accompanied by a set of more than 2,000,000 digitized documents publicly available. Nearly half of the structures have a short geological section, and about 20% have an elaborate geological cross section verified by a professional.

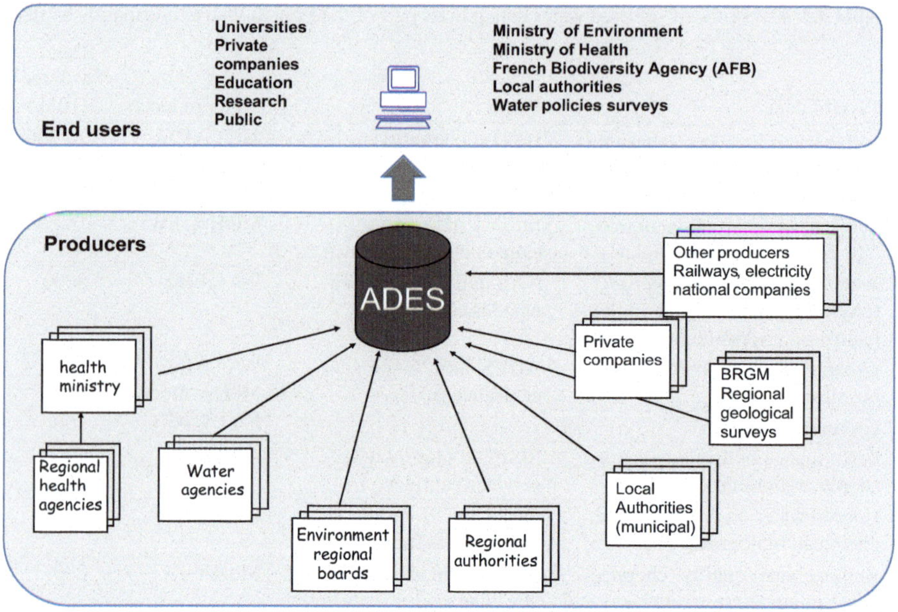

Fig. 9.1 Main data producers providing data to the Groundwater data portal (ADES)

Fig. 9.2 Groundwater level evolution in a selected monitoring point (screen shot of ADES)

9.3.4 BNPE: The National Water Abstraction Database

Obtaining accurate data on groundwater abstraction is essential for resource managers. Until 2010, this information was collected by several institutions and not consolidated into a national database. The main data producers are the following:

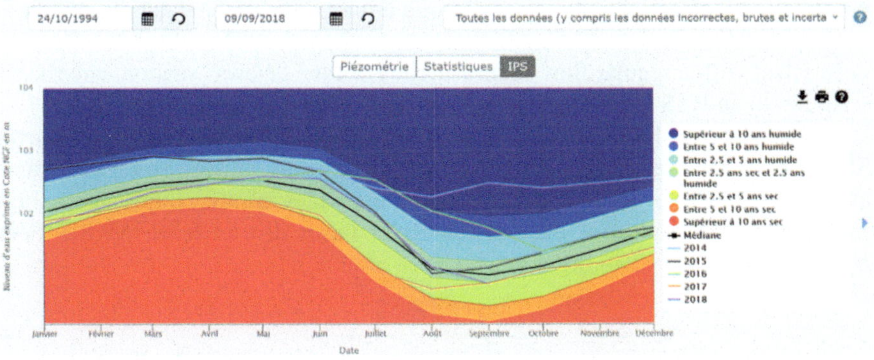

Fig. 9.3 Statistical groundwater level indicator (screen shot of ADES)

- Water agencies. By law, all users abstracting more than 10,000 m^3 (10 ML) per year, or 7000 m^3 (7 ML) in restriction zones, must declare the total yearly volume of water they abstracted to the water agency. This information is used to levy an abstraction tax, which is collected by each Water Agency.
- The regional Environmental Department also collects water abstraction data from all industries generating an environmental risk, and the corresponding data are stored in a database called GIDAF.
- Domestic wells are registered in a separate database (although househods rarely declare their wells and boreholes, see Rinaudo et al., 2015).
- And Government agencies in charge of water compliance and enforcement (See Chap. 23) also collect information related to actual water abstraction.

The National Water Abstraction database (BNPE in French) was set up to integrate these different sources of information. In 2018, it centralizes data from all water agencies. Further integration of other data sources is in progress. Information can be displayed for a single abstraction point or consolidated at different administrative levels (municipality, county, region) or hydrological scale (catchment, aquifer). The identity of abstractors is not disclosed, in accordance with the law.

9.3.5 *Other Information Systems on Groundwater at Local/Regional Scale*

From the 1990s, communication tools targeting the general public were developed in several French regions. Named SIGES (Systèmes d'Information pour la Gestion des Eaux Souterraines), they consist of a website which publishes information accessible to a wide public (expert, schools, and the general public). SIGES provides access to a large number of documents, maps and videos related to groundwater in a specific region. The user is offered access to different information and scientific material depending on their profile. Cross-sectional access also makes it possible to reach technical content through a map interface, a database search that links to ADES or a list of scientific literature references. The editorial information is enriched with a regular flow of information via the "News" section, and the subscription to an RSS feed.

Since the first SIGES was developed 20 years ago in Aquitaine region, several SIGES have been set up, most often at the level of an administrative region, but also at river catchment or aquifer level (upper Rhine valley aquifer: http://sigesar.brgm.fr/-La-nappe-d-Alsace-) or even at the River basin district level (SIGES Seine-Normandy; http://sigessn.brgm.fr/).

9.4 Groundwater Information Systems in Australia

9.4.1 Historical Development of Groundwater Information Systems

In Australia, collection and recording of groundwater information began in the late 1800s (Blake & Cook, 2006; Dahlhaus et al., 2016; NSW, 2012). These early activities predate the formation of the nation and were carried out by the self-governing colonies prior to their federation into the states and territories of Australia. As such, they were developed independently across the country, adapting to meet local needs. This arrangement continues to the present day; groundwater monitoring and data collection primarily remains the responsibility of state and territory governments.

In Victoria early drilling and bore data were published in Diamond Drills and Water Augers, and Diamond Drills and Other Boring Machinery reports dating back to 1884 (Dahlhaus et al., 2016; FedUni, 2015). In Western Australia, artesian bore drilling details were published in the annual reports of the Geological Survey between 1896 and 1911, and later a compiled dataset was presented at the Interstate Conferences on Artesian Water, 1912 et seq.

The tapping of the Great Artesian Basin (GAB) in New South Wales brought groundwater within the ambit of Government policy and administration (NSW, 2012). Two royal commissions in the late 1800's, the Lyne Royal Commission in New South Wales (1884–1887), and Deakin Royal Commission in Victoria (1884–1887), laid the foundation for water legislation reforms and, in the process, collected a vast amount of water data. In Queensland, increased exploitation of the GAB lead to extensive mapping from 1894.

In these early years, data collected was primarily concerned with exploitation of groundwater resources; bore location, construction, and yield. Pressures and levels were sometimes recorded to determine potentiometric heads and map regional resources in groundwater systems (Blake & Cook, 2006). These early data sets were recorded in hardcopy, as tabulated data, hand drawn maps, and periodically published reports. Over time some organisations developed a file system for storing bore data, typically on template cards (Fig. 9.4). This continued until the late 1960 and early 1970s when the use of computers revolutionised data management.

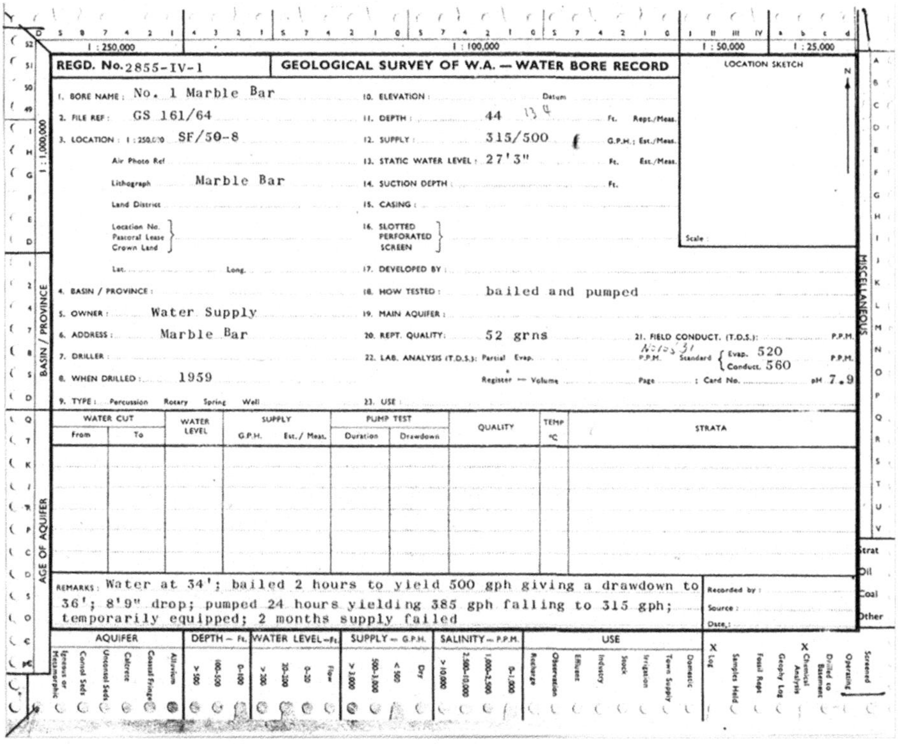

REGD. No. 2855-IV-1

GEOLOGICAL SURVEY OF W.A. — WATER BORE RECORD

LOCATION SKETCH

1. BORE NAME: No. 1 Marble Bar
2. FILE REF: GS 161/64
3. LOCATION: 1:250,000 SF/50-8
Air Photo Ref
Lithograph Marble Bar
Land District
Location No. / Pastoral Lease / Crown Land
Lat. Long.
4. BASIN / PROVINCE:
5. OWNER: Water Supply
6. ADDRESS: Marble Bar
7. DRILLER:
8. WHEN DRILLED: 1959
9. TYPE: Percussion Rotary Spring Well
10. ELEVATION: Datum
11. DEPTH: 44 134 Ft. Rept./Meas.
12. SUPPLY: 315/500 G.P.H.; Est./Meas.
13. STATIC WATER LEVEL: 27'3" Ft. Est./Meas.
14. SUCTION DEPTH: Ft.
15. CASING:
16. SLOTTED / PERFORATED / SCREEN
17. DEVELOPED BY:
18. HOW TESTED: bailed and pumped
19. MAIN AQUIFER:
20. REPT. QUALITY: 52 grns
21. FIELD CONDUCT. (T.D.S.): P.P.M.
22. LAB. ANALYSIS (T.D.S.): Partial Evap. P.P.M. Standard { Evap. 520 / Conduct. 560 } P.P.M.
Register — Volume Page : Card No. pH 7.9
23. USE:

WATER CUT From	WATER CUT To	WATER LEVEL	SUPPLY G.P.H.	SUPPLY Est./Meas.	PUMP TEST Duration	PUMP TEST Drawdown	QUALITY	TEMP °C	STRATA

REMARKS: Water at 34'; bailed 2 hours to yield 500 gph giving a drawdown to 36'; 8'9" drop; pumped 24 hours yielding 385 gph falling to 315 gph; temporarily equipped; 2 months supply failed

Recorded by:
Source:
Date:

AQUIFER: Igneous or Metamorphic, Consol Seds, Unconsol Seds, Calcrete, Coastal Fringe, Alluvium
DEPTH — Ft.: > 500, 100–500, 0–100
WATER LEVEL — Ft.: > 200, 20–200, 0–20, Flow
SUPPLY — G.P.H.: > 3,000, 500–3,000, < 500, Dry
SALINITY — P.P.M.: > 10,000, 2,500–10,000, 1,000–2,500, 0–1,000
USE: Recharge, Observation, Effluent, Industry, Stock, Irrigation, Town Supply, Domestic
Log, Samples Held, Fossil Rept, Geophy Log, Chemical Analysis, Drilled to Basement, Operating, Screened
BASIN / PROVINCE; AGE OF AQUIFER; MISCELLANEOUS; Strat, Oil, Coal, Other
1 : 250,000; 1 : 100,000; 1 : 50,000; 1 : 25,000; 1 : 1,000,000

Fig. 9.4 Example of card systems for bore data in Western Australia

Around this time, many state agencies established the ability to store groundwater data in digital formats. As well as storing newly collected data, historical data began to be digitised and ingested into these databases. This was the beginning of an ongoing process of storing and managing digital groundwater data (Blake & Cook, 2006; DoM, 1974; FedUni, 2015). In some cases, only the level and salinity measurements were digitised. For example, in the 1970s the Public Works Department in Western Australia created the State Water Resource Information System (SWRIS). Although the SWRIS was primarily used for surface water data, groundwater level and salinity data were also recorded. However, bore data, such as construction and geology logs data remained on a card system until 1993 when the Geological Survey received funding to computerise the bore data into the AQWABase. By then a separation between levels and salinity time series databases and bore logs and construction databases started creating future difficulties in relating these two datasets. This example is typical of hydrogeological data management in many states.

From 1994 onward, a series on national water reforms began additional collection of information regarding groundwater rights and allocation, including for the environment, and trading. The 2004 National Water Initiative supported the introduction of water registers at state level. This again created separated registers to store permits, use and trading data.

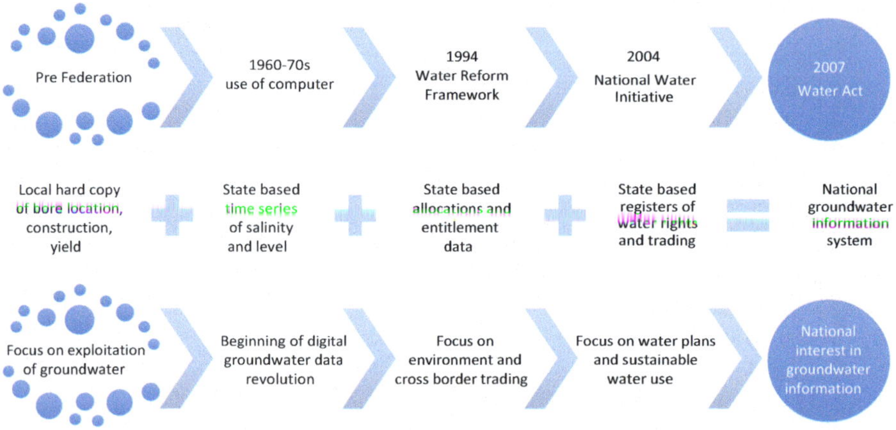

Fig. 9.5 Groundwater information systems timeline in Australia related to important events and reforms

While early groundwater data was collected primarily for developing groundwater resources, data was now collected for a variety of purposes, including: environmental; resource management and monitoring; resource investigation; contamination monitoring and compliance.

Significant effort went into collating data into consistent and complete datasets, however, results varied considerably for each state and territory and no standards were adopted nationally. This changed with the Water Act, 2007. The Federal government begun the task of establishing a consistent, national dataset for groundwater with a focus on promoting transparency and public data availability.

The history of water initiatives and changes to Groundwater Information System is shown below in Fig. 9.5.

9.4.2 *Organisation of Groundwater Information Systems*

In Australia, collection and recording of groundwater data and metadata is carried out by a variety of organisations both public and private. The vast majority of publicly available groundwater data is collected by State and Territory governments. However, other organisations and industries also collect and record groundwater data. These include:

- Environmental Protection Agencies;
- Other state government departments;
- Water corporations—State owned corporations responsible for water supply and licensing;
- Federal agencies—such as, Geoscience Australia and CSIRO;
- Research institutions;
- Mining and energy companies.

State and Territory governments remain the primary data custodians, due to their regulatory role in bore construction, groundwater management, and environmental management. However, even within a single organisation groundwater data is often found in disparate data management systems. For example, in New South Wales and Victoria high frequency data was stored in Hydstra, a specialised time-series database used for their surface water data, and the bore and manually read data was stored in a bespoke groundwater system. Similar arrangements exist in most other states. At time of writing, both Victoria and New South Wales are in the process of combining their groundwater data into single integrated systems, which demonstrates the ever changing nature of GWIS across Australia.

There are two main causes of this division in data stores are:

- Changes in the organisation of government departments, and corresponding responsibilities for water data management, have led to many departmental mergers, splits, and corresponding mergers and splits in GWIS.
- Ongoing developments in database technology and standards, along with increases in the volume of data collected, have led to almost constant changes in the technology. This process reflects the rapid growth in computer technology since data began to be digitised in the 1970s.

Through the Water Act, the Bureau of Meteorology (the Bureau) was given responsibilities to improve the integration, standardisation and dissemination of groundwater information across Australia. State agencies remain the primary data authority, but the Bureau is responsible for collating nationally consistent groundwater data. Below are two case studies of GWIS in Australia.

9.4.3 Case Study 1: Department of Water and Environmental Regulation, Western Australia

The State of Western Australia relies heavily on groundwater. The major population center around Perth sources two-thirds of its water needs from groundwater (BOM, 2017). Western Australia has invested in a network of groundwater monitoring bores, gauging stations and rainfall monitoring sites. This State Reference Network has provided a comprehensive set of scientific measurements. About 10,000 groundwater sites have measurements going back to the 1970/1980s, however, some measurements go back as far as the early 1900s. The valuable scientific data collected from the State Reference Network is maintained by the Department of Water and Environmental Regulation (DWER), who use the data extensively to manage Western Australia's water resources. It is a primary input into the department's groundwater and surface water models, which underpin the management of water resources (see Chap. 15).

There are 2500 groundwater bores which are currently monitored on a regular basis. Of these 500 sites have groundwater loggers, with the remaining being measured manually. The manually measured bores are typically dipped four times a year, with important sites being measured monthly. One important groundwater site on the Gnangara Mound groundwater system is logged and telemetered. DWER utilises the Hydstra time series data management application for surface water and for groundwater sites. This system is highly popular in Australia for storing both groundwater and surface water data. In addition to DWER, the Hydstra information system is used by other lead water agencies in the Northern Territory, New South Wales, and Victoria. It is also used by a variety of other organisation, including the Bureau of Meteorology.

DWER has chosen to combine all groundwater, surface water and water quality testing information into this single off-the-shelf database system for several reasons. It will simplify database management and reporting functions by allowing better integration with other departmental systems running on a uniform SQL server platform and lead to better reporting capability using Business Intelligence and other tools. This approach utilises existing knowledge and expertise in the Hydstra system and extends that to groundwater and water quality information previously stored in a bespoke Oracle system. As Hydstra is specially designed for time-series data management, it provides for the growing demand for the use of loggers and telemetry in groundwater bores.

Using an off-the-shelf system also provides a clear path for system updates and upgrades because the system suppliers provides support and maintenance for the system. This reduces overall operating costs by decommissioning the legacy bespoke systems and reducing the need for DWER to maintain and develop the application. The department is a long-time user of data loggers and telemetry systems utilising both cellular network and satellite communication systems. Western Australia is predominantly a sparsely populated, desert environment. Many monitoring bores are located in harsh and remote environments. Hence, durable, low power equipment, which allow remote administration, is a key factor for the department when choosing equipment for its monitoring systems.

The 500 groundwater sites where loggers are currently mounted down the bore inner casing are not telemetered due to current power requirements and the lack of a suitable low power telecommunications network in the South West of Western Australia. Emerging Internet of Things (IOT) technologies may enable these bores to also be telemetered in the future.

DWER has invested in an advanced self-service water information reporting (WIR) portal to make water data available online. It provides a one-stop-shop for groundwater, surface water and water quality information for Western Australia. The portal is based on a shopping cart design and is easy to use. The data is free to access and download, the user only needs to provide a valid e-mail address to get water data.

Before WIR was introduced all water data requests were handled manually with a minimum 10 business day turnaround. WIR now provides 99.5% of all water data with an average turnaround time of 43 s. Consultants and Universities are big users of WIR as are mining companies, farmers and the land and property development industry. Common use cases include; assessing drainage and land fill needs for property developments; evaluating potential environmental impacts; planning and design of new roads, bridges and other transport infrastructure; as well as supporting groundwater related research and management.

This information system underpins DWERs capacity to assess information and manage groundwater resources across Western Australia. It allows groundwater managers to understand the resource; understand the ecological, social and cultural needs; measure and estimate current and future demands and trends. The system also provides primary inputs to a suite of groundwater models that underpin many management decisions.

9.4.4 Case Study 2: The Bureau of Meteorology

Many organisations across Australia collect groundwater data for a range of purposes. The variety of methodologies employed in collecting, managing and transferring means that it can be difficult for other users to easily understand and interpret this data. The fractured nature of these datasets creates difficulties in producing nationally consistent information from data collected in different ways, and without reference to agreed or commonly applied standards and guidelines. The Bureau is actively working to develop water information datasets and standards, which support community understanding, comparison and sharing of water information.

The Millennium Drought (1997–2009) was a catalyst for unprecedented reforms in Australian water management, which were formalised through the National Water Initiative in 2004. As part of this reform, the Bureau was given a key role to improve the collection, standardisation and dissemination of water information, including groundwater, through the Water Act (2007). The Bureau is now responsible for publishing a standardised national dataset for groundwater. This is the first time such a dataset has been created and maintained for the whole of Australia.

The Water Act (2007), allowed for the creation of the Water Regulations (2008) which legislated the detailed requirements of the water information that must be given to the Bureau. The Regulations define the type of data that needs to be supplied to the Bureau and who needs to provide it, as well as the delivery frequency and format. The preferred format for time series data, such as of groundwater level and salinity data, is the Water Data Transfer Format (WDTF) (Walker, Taylor, Cox, & Sheahan, 2009), an XML file format for transferring water information.

Information about bore location, construction and bore log details are also required through the Water Regulations. The preferred format is in an ESRI geodatabase using the National Groundwater Information System (NGIS) data model, which is derived from ESRI's ArcHydro for Groundwater. Each State and Territory water agency produces an NGIS database for their jurisdiction, which is integrated into a national dataset by the Bureau.

The NGIS contains data for more than 870,000 bores, and is growing larger every year. Detailed information is provided about each bore, including (where available) purpose, lithology, construction and hydrostratigraphy logs. Aquifer geometry is available for some areas in 2D or 3D, including 3D hydrostratigraphy models for the Murray Basin and the Great Artesian Basin.

A major challenge for the Bureau in building a national groundwater dataset is that each State and Territory uses local terminology to describe, among other things, aquifers, aquitards, boreholes and bore pipe identification systems. These differences are problematic, particularly when examining aquifers that span multiple States and Territories. The Bureau, in collaboration with each State and Territory, developed a National Aquifer Framework (NAF). Hydrogeologic data in the NGIS is standardised across the nation using this Framework.

Groundwater data held by the Bureau can be viewed, analysed, and downloaded through the Australian Groundwater Explorer (http://www.bom.gov.au/water/groundwater/explorer). The Explorer now contains data for more than 220,000 bores with water level or salinity data as provided through the Water Regulations (Fig. 9.6). The Explorer provides a truly national picture of groundwater data, makes this data readily available at a national scale and puts local, State and Territory groundwater information into an Australia-wide context.

In addition to the above-mentioned data, the Bureau also collect groundwater data relating to groundwater extraction, and licences for extraction, through the Water Regulations. This data can be visualised through the interactive Australian Groundwater Insight (http://www.bom.gov.au/water/groundwater/insight). The Insight shows maps of hydrogeological information such as aquifer types, alongside information about licences, entitlements and extractions by groundwater management areas, providing background to the analysis of groundwater salinity and trends in levels presented in the application (Fig. 9.7). This significantly increases the capacity to provide a consistent analysis of groundwater resources across the nation.

The Bureau's suite of groundwater products is based on a common format and terminology for groundwater, resulting in a standardisation of groundwater data across Australia. For the first time, decision-makers have easy access to comprehensive, nationally consistent information on groundwater to support sustainable use of the groundwater resource across the nation (Fig. 9.8).

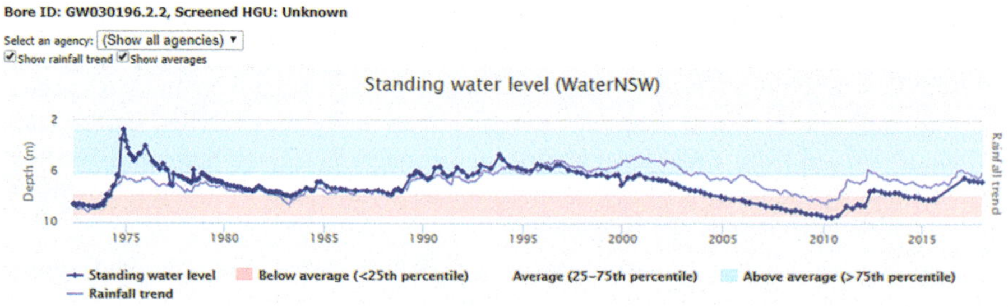

Fig. 9.6 Example of hydrograph from the Australian Groundwater Explorer

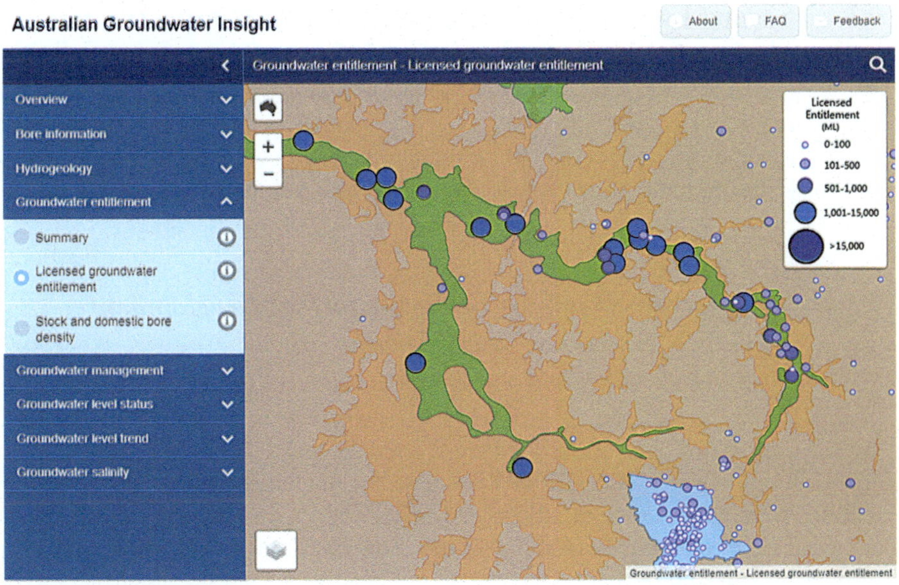

Fig. 9.7 Locations and indicative size of extraction licences and aquifers extents for the Upper Lachlan Alluvial Aquifer

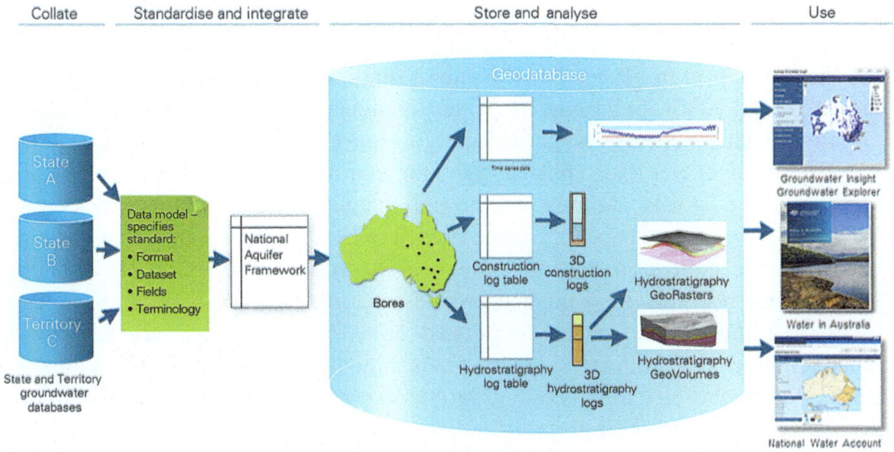

Fig. 9.8 Ingestion, standardisation, analysis and publication of groundwater information at the Bureau of Meteorology

9.5 Lessons Learned, Future Challenges and Opportunities

The history of GWIS development in France and Australia provides valuable lessons for countries currently engaging in developing a GWIS. This section summarises the differences and similarities in GWIS development in both France and Australia, including lessons learned and future developments.

9.5.1 Comparative Analysis of the Historical Development of GWIS in France and Australia

In both countries, the need for groundwater information has emerged locally, leading local actors to design and implement independent GWIS. In France, diverse organizations have invested in GW monitoring, including Public Water Supply Utilities, various government ministries (agriculture, environmental, health affairs), and local government (county and regional councils). In Australia, data has historically been collected by state government and local water resource managers. Over time the focus of groundwater monitoring has varied, from resource exploitation, to dryland salinity management, to environmental protection and maintenance. Information produced by these early GWIS was not consistent in spatial coverage, monitoring frequency, measurement protocol, and data organisation and processing.

In both countries, the first challenge was to improve the geographical coverage of GWIS. Early systems were developing based on local initiatives, but public agencies had to step-in to fill gaps, using public funding. This mainly happened during the 1970s and 1980s. In France, the Water Agencies played a key role in developing GW monitoring network, sometimes relying on county and regional councils or Government agencies to establish and run the monitoring networks and information systems. The cost was paid by users through the water abstraction fees collected by Water Agencies (see Chap. 4). Unlike France, where groundwater monitoring covers the entire nation, in Australia monitoring programs focus on areas of high groundwater use and good quality groundwater resources. Many aquifers, especially in remote and sparsely populated areas have little, or no, monitoring. Abstraction fees are collected in many management areas across Australia, however state governments also fund GWIS programs as part of their responsibilities to manage water resources for all users, including the environment.

Once the coverage of GWIS was satisfactory, the second challenge faced in both France and Australia was to standardise the existing heterogeneous GWIS. In France, the ministry established and imposed formats and protocol to all data producers. Conversely, the Australian Water Act of 2007 did not mandate any change in state GWIS, instead it implemented mandatory transfer formats, requiring water agencies to send data in these formats. The data was then standardised once ingested into the national dataset.

The third challenge was to facilitate access to data collected and stored by many organisations. In France, this was facilitated by technological innovation in computer sciences such as Web Services and APIs. By developing automated data exchange between the many organisations that hold groundwater data, nationally consistent GWIS were created by federating these existing systems into a coherent network. Conversely, in Australia, each state and territory continues to maintain its own, independent GWIS. The vast majority of this data is published via the internet on data access portals specific to each state or territory (see example for Western Australia above). Nationally coherent groundwater datasets are produced by the Bureau of Meteorology who receive data from water agencies across the country and ingest into the national GWIS (see above, Groundwater Information Systems in Australia).

9.5.2 Lessons Learnt

The development of independent GWIS in separate jurisdictions is most likely unavoidable. No single agency is able to develop a tool that meets the information requirements of all interested parties, e.g. resource managers, public water utilities, environmental protection agencies, abstraction compliance agents, among others. However, what can be learned from GWIS development in France and Australia is that the State should define, as early as possible, technical specifications so that the data and the independent GWIS are compatible. To reach this objective, a combination of economic incentives and regulation can be used. Also, significant resources should be devoted to the development of tools that can federate/integrate the data and make them available to users via the internet. This is because the cost of collecting this information is large and making these datasets publicly available is good practice and good use of public resources.

The responsibility for collecting, storing, and managing groundwater data is typically tied to a legislative requirement to manage groundwater resources. However, changes in groundwater systems typically occur at a much slower rate than changes in legislation and governments. As such, meaningful groundwater monitoring and data collection efforts occur across multiple iterations of governments, departments, and legislative changes. Both France and Australia have a long history of water data systems undergoing change as departments split and merge. Responsibility and funding can vary greatly over the monitoring history of a single resource. As such, when planning new information systems, it is important to plan for future management and maintenance of these systems. Are these systems extensible? Can extra functionality be added to meet new legislative requirements? For example, introduction of licensing information, where this was not previously enforced. Planning for a long-term system can greatly improve the functional lifespan of an information system.

Effective data sharing across state borders has been, and remains, an issue within Australia. The Bureau of Meteorology has developed a standard to share water data,

WDTF. The development of this standard has greatly increased the ability of the Bureau to manage the transfer and ingestion of large volumes of data. However, due to its complexity, and being dissimilar to existing formats, adoption of the standard was slow. Furthermore, the standard was developed to align with legislative requirements set out in the Water Regulations (2008). This did not include a holistic approach to groundwater data and does not cater for some commonly collected data, e.g. hydrogeochemistry. Where cross boarder data sharing is likely, adoption of such a standard is recommended, as it facilitates easy data sharing. However, to reduce the complexity and cost of implementation an existing standard can be adopted, for example GWML2 (Brodaric et al., 2018).

9.5.3 Future Developments

Fifty years after the French and Australian GWIS started to be developed, managers have access to sophisticated technologies for data acquisition, transmission, and publication. These technologies are bringing about huge transformations in GWIS, including changes of infrastructure, operational process, volume and currency of data.

The availability of new technology is driving changes in monitoring devices and how they record and transmit data. More and more bores are being equipped with electronic monitoring devices as low powered IOT sensors, along with new transmission networks (GSM, low orbit satellite), reduce the cost and footprint of monitoring equipment. This is particularly attractive in Australia where monitoring networks are often spread over vast distances and cannot be monitored using existing communication networks.

Another transformation of GWIS may come from a greater demand of citizens to participate in the monitoring of the environment. Developments in communication technology, data processing and visualization will increasingly allow the general public to participate in the collection of data (crowed sourcing) and, more generally, the production of knowledge (citizen science). While such data have a significant potential to create increase spatial coverage, in particular in remote regions, their integration with traditional monitoring network is challenging (Grieef & Hayashi, 2007).

Publication of real-time groundwater data is a current, and ongoing, development in both France and Australia. Real-time data gives complete data transparency to managers, users, and the public. For example, the Méteau-Nappe application is currently being developed by Brgm to provide real time access to groundwater levels and to prediction of groundwater level evolution, updated at a monthly time step, based on realtime groundwater level data (Mougin, Nicolas, Bessiere, Vigier, & Loigerot, 2017). The state of New South Wales in Australia publishes extensive real-time groundwater level data. Their web portal (https://realtimedata.waternsw.com.au/) provides data for 488 bores, covering the major groundwater resources across the state.

In both France and Australia, developments in water information systems are now directed towards the development of APIs. In addition to data provisioning APIs, new programming interfaces are being developed to allow data processing and complex querying. This will eventually make it possible to call multiple remote environmental data sources and apply automated statistical processing. Spatially enabled APIs will also allow GIS users to make these aggregations based on spatial summaries and queries. These APIs will facilitate environmental management by making available not just raw data, but indicators that aggregate and draw inferences from multiple data sources.

References

Alley, W. M., Evenson, E. J., Barber, N. L., Bruce, B. W., Dennehy, K. F., Freeman, M. C., et al. (2013). *Progress toward establishing a national assessment of water availability and use* (U.S. Geological Survey Circular 1384) (p. 34). Reston, VA: U.S. Geological Survey. Available at http://pubs.usgs.gov/circ/1384

Bartram, J., & Ballance, R. (Eds.). (1996). *Water quality monitoring – A practical guide to the design and implementation of fresh water quality studies and monitoring programmes*. London: Chapman & Hall.

Blake, T., & Cook, M. (2006). *Historical overview: Great Artesian Basin*. Brisbane, Australia: Natural Resources Mines and Water, Queensland Government.

Blum, A., Lopez, B., Guilbert, M., Noël, Y., & Chery, L. (2010). *Bilan national des efforts de surveillance des eaux souterraines à partir d'ADES. Indicateurs d'évaluation de l'état qualitatif et quantitatif des eaux souterraines: exemples de réalisations et recommandations* (Rapport BRGM/RP-57943-FR). Paris: Onema. 166 p.

Blum, A., Winckel, A., Laurent, A., & Michon, J. (2013). *Les efforts de surveillance des eaux souterraines*. Paris: ONEMA. 16 p. Accessible at http://www.eaufrance.fr/IMG/pdf/surveillanceeauxsouterraines_201301.pdf

BoM. (2017). *National groundwater information system, data product specification*. Bureau of Meteorology. ISBN 978-0642-706-430.

Brodaric, B., Boisvert, E., Chery, L., Dahlhaus, P., Grellet, S., Kmoch, A., et al. (2018). Enabling global exchange of groundwater data: GroundWaterML2 (GWML2). *Hydrogeology Journal, 26*(3), 733–741.

Dahlhaus, P., Murphy, A., MacLeod, A., Thompson, H., McKenna, K., & Ollerenshaw, A. (2016). Making the invisible visible: The impact of federating groundwater data in Victoria, Australia. *Journal of Hydroinformatics, 18*(2), 238–255.

DoM. (1974). Department of Mines, South Australia, a data base management system for the storage and retrieval of geological data. Rept.Bk.No. 74/76, G.S. No. 5395, D.M. No. 105/73, Eng. Geol. No. G34.

DWA. (2010). *National Groundwater Strategy* (p. 2010). South Africa: Department of Water Affairs, Pretoria.

FedUni. (2015). Visualising Victoria's groundwater. (Internet data portal). Centre for eResearch and digital innovation, Federation University Australia, Mt Helen, Ballarat, Victoria. Retrieved December 20, 2017, from: http://www.vvg.org.au

Fitch, P., Brodaric, B., Stenson, M., & Booth, N. (2016). Integrated groundwater data management. In *Integrated groundwater management* (pp. 667–692). Cham, Switzerland: Springer.

Global Water Partnership Technical Advisory Committee. (2000). *Integrated water resources management*. Stockholm: Global Water Partnership. ISBN: 91-630-9229-8.

Grath, J., Ward, R., Scheidleder, A., & Quevauviller, P. (2007). Report on EU guidance on groundwater monitoring developed under the common implementation strategy of the water framework directive. *Journal of Environmental Monitoring, 9*(11), 1162–1175.

Grieef, L. A., & Hayashi, M. (2007). Establishing a rural groundwater monitoring network using existing wells: West Nose Creek pilot study, Alberta. *Canadian Water Resources Journal, 32*(4), 303–314.

Haley D., Beck A., Pollok A., Grant A., & McKilliam R. (2017). Global reach for the Internet of Things. 68th International Astronautical Congress (IAC), Adelaide, Australia. IAC-17-B2.4.3.

Holliday, L., Marin, L., & Vaux, H. (2007). *Sustainable management of groundwater in Mexico*. Washington, DC: National Academies Press.

Horsburgh, J. S., Tarboton, D. G., Piasecki, M., Maidment, D. R., Zaslavsky, I., Valentine, D., et al. (2009). An integrated system for publishing environmental observations data. *Environmental Modelling & Software, 24*(8), 879–888.

Jousma G. (2008). Guideline on: Groundwater monitoring for general reference purposes – Revision 2008. IGRAC, The Netherlands. Report nr. GP 2008-1.

Klug, H., & Kmoch, A. (2014). A SMART groundwater portal: An OGC web services orchestration framework for hydrology to improve data access and visualisation in New Zealand. *Computational Geosciences, 69*, 78–86.

Lee, J. Y., & Kwon, K. D. (2016). Current status of groundwater monitoring networks in Korea. *Water, 8*(4), 168. https://doi.org/10.3390/w8040168

Margat, J., & Schneider, J. L. (1971). *Dispositif de contrôle des fluctuations de niveau des nappes d'eaux souterraines en France. Etat en 1969–1970*. Report 70-SGN-204-HYD (p. 22). Orléans: BRGM.

Minsker, B. (2003). *Long-term groundwater monitoring–the state of the art*. Reston, VA: American Society of Civil Engineers.

Mougin B., Nicolas J., Bessiere H., Vigier Y., & Loigerot S. (2017). Vers une prévision en temps réel du niveau des nappes françaises ? L'outil d'aide à la décision « MétéEau des nappes ». Revue *Géologues* n°195.

Nelson, R. L. (2012). Assessing local planning to control groundwater depletion: California as a microcosm of global issues. *Water Resources Research, 48*(1).

NSW. (2012). NSW strategic investment plan for water monitoring systems, New South Wales Office of Water. http://www.bom.gov.au/water/regulations/fundingProgram/document/NSW_Strategic_Investment_Plan.pdf, cited 6th March 2018.

Porter, J. H., Hanson, P. C., & Lin, C.-C. (2012). Staying afloat in the sensor deluge. *Trends in Ecology & Evolution, 27*(2), 121–129.

Rinaudo, J. D., Montginoul, M., & Desprats, J. F. (2015). The Development of Private Bore-Wells as Independent Water Supplies: Challenges for Water Utilities in France and Australia. *In Grafton et al. Understanding and Managing Urban Water in Transition. Springer. pp*, 155–174.

Shah, T., Roy, A. D., Qureshi, A. S., & Wang, J. (2003). Sustaining Asia's groundwater boom: An overview of issues and evidence. *Natural Resources Forum, 27*(2), 130–141.

Taylor, C. J., & Alley, W. M. (2001). *Ground-water level monitoring and the importance of long-term water-level data* (U.S. Geological Survey Circular 1217) (p. 68). Reston, VA: U.S. Geological Survey.

Tuinhof, A., Foster, S., Kemper, K., Garduno, H., & Nanni, M. (2006). *Groundwater monitoring requirements – For managing aquifer response and quality threats*. https://doi.org/10.13140/RG.2.1.4530.9529.

UN-ECE Task Force on Groundwater Monitoring & Assessment. (2000). *Guidelines on monitoring and assessment of transboundary groundwater*. Lelystad, The Netherlands: RIZA Publication. ISBN 9036953154.

Vaessen, V., & Brentführer, R. (2014). *Integration of groundwater management: Into transboundary basin organizations in Africa* – A training manual by AGW-net, BGR, IWMI, CapNet, ANBO, & IGRAC.

Walker, G., Taylor, P., Cox, S., & Sheahan, P. (2009). Water Data Transfer Format (WDTF): Guiding principles, technical challenges and the future. In *Proceedings of 18th World IMACS Congress and MODSIM09 International Congress on Modelling and Simulation* (pp. 4381–4387).

Water Act. (2007). Act No. 137, C2007A00137. Retrieved from https://www.legislation.gov.au/Series/C2007A00137, 9th March 2018.

Water Regulations. (2008). F2008L02170 SLI 2008 No. 106. Retrieved from https://www.legislation.gov.au/Series/F2008L02170/, 9th March 2018.

WFD. (2000). Directive 2000/60/EC of the European Parliament and of the Council of the 23 October 2000 establishing a framework for Community action in the field of water policy. *Official Journal L 327*, (22 Dec. 2000).

World Meteorological Organization (WMO). (2008). *Guide to hydrological practices*. Geneva, Switzerland: WMO.

John Sharples is a senior hydrogeologist with the Bureau of Meteorology, where he led the creation of the first ongoing national groundwater timeseries database. He also led major developments in the Bureau's Groundwater Information Suite, publishing national datasets on aquifer boundaries, groundwater entitlement and use, and assessing groundwater trends across Australia. John is a champion for the Bureau's mission to provide accurate, up to date, freely available, groundwater data to support decision making across Australia. Prior to joining the Bureau, John worked in the Department of Sustainability and Environment, Victoria, assessing groundwater management and licencing application.

Dr. Elisabetta Carrara is the Water Insight Unit Head at the Bureau of Meteorology. In the last 11 years, she has been responsible for shaping the Bureau's suite of groundwater products and ensuring widespread access to nationally-consistent groundwater information across Australia. Elisabetta believes that transparent high-quality information will improve decision making about our water resources. Previously, she was a physical limnologist at the Institute for Ecosystem Studies part of the Italian National Research Council. Elisabetta holds a PhD from the University of Melbourne in Hydrogeochemistry. That study focused on understanding the chemical and physical controls on groundwater evolution in a semiarid basin for sustainable groundwater management. She is also a Fellow of the Peter Cullen Water and Environment Trust and part of the Bureau's 2018 Winnovators in support of WaterAid.

Lindsay Preece is working for Department of Water and Environmental Regulation of Western Australia as manager of Manager Science Data. He has a long working experience in the government administration industry. Skilled in Managerial Finance, IT Consulting, Stakeholder Management, Land Use Planning, and IT Management. Strong information technology professional with a Bachelor of Commerce – BCom focused in Accounting, Finance and Information Systems from UNSW.

Laurence Chery heads up a team of 20 persons, a unit of Assessment and Evaluation of Water Knowledge, in the Water, Environment and Ecotechnologies division in BRGM. She has more than 25 years experience in hydrogeology and management of water resources projects. Her main interests at the beginning are isotope hydrology, hydrogeochemical studies with policy application, origin and transfer time of waters, hydrogeochemistry and particularly natural background levels. Since 10 years, occupational fields are, in applying the Water Framework Directive, monitoring of groundwater (networks design), water information system, Decision Support Systems. She was in charge of the national groundwater database project: quality and quantity national database for groundwater, ADES. She ensures also animation with the Ministry of Environment and ONEMA of different databases at the national level (training, conceptual modelling, Usage Scenarios for a Knowledge Exchange Network at national scale but also European scale).

Benjamin Lopez After 8 years of experience in project leading in the domain of groundwater protection management, particularly in fate and mobility of pollutants through unsaturated and saturated zones of aquifers, Benjamin Lopez took the scientific coordination of the water policy studies at BRGM, the French Geological Survey, in 2017. Both expert in diffuse pollutions and compounds of emerging concern in groundwater, and in statistical data analysis of groundwater quality, Benjamin leads and actively participates in European sub-groups that link EU governments, other EU bodies, international organisations, medical professionals, academics, researchers, NGOs and the business community. In these groups, Benjamin focuses on skill and knowledge transfer from research domain to water management, regulations and European policies.

Dr. Jean-Daniel Rinaudo is researcher at Brgm, Montpellier University, where he coordinates the scientific program on environmental and risk economics. Initially trained as an agricultural engineer (Montpellier SupAgro), he specialized in agricultural and resource economics (PhD University of Auvergne). Prior to joining Brgm, he worked for the International Water Management Institute in Pakistan where his research focused on the political economy of irrigation management reforms. His current research mainly focuses on the institutional economic dimension of groundwater management. Most of his research is conducted in France but he also works in Morocco and Chile. He is currently developing new research activities in the field of natural disasters economics, focusing on the methods to assess economic vulnerability and resilience. Dr. Jean Daniel Rinaudo is also member of the Scientific Council of the Adour Garonne River basin agency.

Chapter 10
The Challenge of Making Groundwater Visible: A Review of Communication Approaches and Tools in France

Audrey Richard-Ferroudji and Gaïa Lassaube

Abstract Groundwater specialists strive to make groundwater issues visible. They face a dual challenge: first to develop knowledge on groundwater and secondly to share this knowledge with other stakeholders who should be included in knowledge development, groundwater management and protection policy. Questioning communication is all the more interesting as groundwater is a quasi-invisible resource. How groundwater and issues can be made more visible? In the field of sociology, with a pragmatist stance, our chapter questions how instruments frame interactions and represent groundwater. Indeed, the groundwater is made visible by tables, indicators, maps, photographs, videos, games, stories in newspaper and spokespersons such as hydrogeologists. Within a project funded by AFB (French Agency for Biodiversity), we reported on a number of communication approaches and activities implemented in 11 case studies in France. The inventory is based on web mining, grey literature review and interviews. The chapter develops a transversal analysis of the use of the instruments, and identifies assets and limits across the cases according to the following categories: public targeted; content, issues brought to the fore and normative stance adopted; type/format. Finally, concrete recommendations are made.

Keywords Mediation · Representation · Policy instruments · Format · Participation · Spokespersons

A. Richard-Ferroudji (✉)
French Institute of Pondicherry, Pondicherry, India

UMR G-EAU, Montpellier, France

G. Lassaube
Centre Emile Durkheim, Institut Français de Pondichery, Pondicherry, India

J.-D. Rinaudo et al. (eds.), *Sustainable Groundwater Management*,
Global Issues in Water Policy 24, https://doi.org/10.1007/978-3-030-32766-8_10

10.1 Exploring the Social Depth of Groundwater and Issues of Communication

Over the last 50 years, the development of access to groundwater has increased the pressure on these resources. There is wide recognition today that groundwater over-exploitation urgently needs to be curtailed but there is little consensus on how this can be best achieved (Jakeman et al., 2016). In the 1970s and 1980s, groundwater specialists were mainly asked to provide technical support for groundwater prospecting and resource development. Growing concerns over groundwater depletion have challenged the historical mandate of water management institutions. In France, the 1964 Water Act promoted monitoring of groundwater and the 1992 Water Act promoted planning and local management. However, groundwater specialists strive to make groundwater issues visible to policy makers, water users and civil society. They face the challenge of shedding light on groundwater while eyes are focused on surface water. In many situations, there is no shared representation of aquifers (in particular their boundaries as management units) between experts, actors involved in land or water management nor the numerous dispersed users. Hydrogeologists face a dual challenge: first to develop knowledge on groundwater and secondly to share this knowledge with other stakeholders (Baldwin, Tan, White, Hoverman, & Burry, 2012; Van Der, 2017) who should be included in knowledge development, groundwater management and protection policy. Meeting this second challenge requires different skills and methods. How should be shared the already available data? How can this knowledge be turned into standardized indicators? How can awareness be raised at the local and national level? How can an enabling environment be created for effective communication? Communication is understood here in its broad meaning as the action of making groundwater visible and common with crafting institutions and a body of shared knowledge.

Questioning communication is all the more interesting as groundwater is a quasi-invisible resource. It is mostly hidden from view. It can be seen in broad daylight only when it gushes out from a bore well or when it lies at the bottom of an open well. In contrast to waters flowing in rivers and channels, underground water streams circulate and create hidden interdependencies between human beings and communities. These interdependencies can be shown with maps representing the ground water perimeter at the surface. The quantities stored are materialized in the productions of experts employing instruments: piezometers, satellites images, tables, etc. The groundwater is made visible by photographs, by stories in newspapers or by spokespersons such as hydrogeologists. The users also produce their own representations and instruments. This chapter focuses on objects, artefacts, settings and persons which represent groundwater. Tool is understood here as any means used to communicate.

Within a project funded by AFB (The French Agency for Biodiversity), we reported on a number of communication approaches and activities implemented in field projects related to groundwater resources (Richard-Ferroudji, Lassaube, Bernard, Daly, &

Table 10.1 Description of the 11 cases

Name of the aquifer(s)	Management structure (2017)	Team Nbpers. (2016–2017)	Procedure	Starting	Area. km^2	Nbhab. (2016)	N° (map)
Ill Rhin Alsace	Region	1	SAGE	1954 1st management structure	3596	1,300,000	1
Astien	Syndicat	4	SAGE and *contract*	1990 syndicat creation	540	110,000	2
Beauce	Syndicat	2	SAGE and *contract*	1994 charter on irrigation	9500	1,400,000	3
Breuchin	EPTB	0,6 54 in the hosting structure	SAGE	2011 emergence of the SAGE	380	28,673	4
Champigny	Association	10	*Contract*	1971st contract	2600	800,000 (2013)	5
Crau	Syndicat	4	*Contract*	2010 emergence of the *contract*	550	270,000	6
Gironde	Syndicat mixte	5	SAGE	1999 SAGE	10,138	1,400,000	7
Roussillon	Syndicat mixte	4	SAGE	2003 framework agreement	900	455,000	8
Stand stone of the early Triassic	Departmental Council	2	SAGE	80s protection of Vittel spring 2009 SAGE	1497	60,642	9
Lower valley of the Var	Syndicat Mixte	2 FTE 20 in the hosting structure	SAGE and *contract*	1995 monitoring of the aquifer	346	400,000	10
Vistrenque	Syndicat	4	SAGE	1986 Syndicat creation	785	250,000	11

Sources: www.gesteau.fr, June 2016, SAGE documents, *contracts*, technical reports and interviews

Latusek, 2018).[1] This chapter reports our findings on the way tools are used to make groundwater visible toward different publics: general public, farmers, elected representatives, etc. Tools were inventoried in 11 case studies in France (See Sect. 10.2.1.1 and Table 10.1). Concrete recommendations are made to improve the same.

[1] The challenge of raising groundwater visibility is shared by many countries. In this project, a comparative stance with India was developed.

This chapter develops a sociological approach to contribute to the exploration of the social depth of groundwater complementary to the physical one. Groundwater practices are indeed deeply rooted in societies and cultures. Achieving more sustainable management requires a comprehensive understanding of socio-economic, political and institutional structures which is complementary to the technical ones. Such an understanding has significant relevance for better governance of groundwater, which has been of increasing concern since the 90s (Ostrom, 1990; Shah, 2009; Villholth, Lopez-Gunn, Conti, Garrido, & Van Der Gun, 2017). There is a need to develop interdisciplinary approaches that integrate the diversity of scientific knowledge on groundwater resources (ranging from hydrogeology to social sciences). However, interdisciplinary projects are still rare (Bouarfa & Kuper, 2012) and the social depth of groundwater deserves to be explored on a more systematic basis. There is a growing body of literature that studies the social aspects of groundwater resources but with a broad scope of development (Curtis, Mitchell, & Mendham, 2016; Faysse & Petit, 2012; Mitchell, Curtis, Sharp, & Mendham, 2012). Mitchell et al. posit the literature on the topic can be grouped in five broad themes: power and influence, social impact assessment, self-regulation, stakeholder engagement and farmer decision making. Faysse and Petit point that the approaches differ in the content of governance systems recommended to achieve sustainable groundwater use, and especially in the benefits of involving water users in the implementation of governance. Therefore, they also differ on what should be the focus of academic analyses.

In the field of sociology, with a pragmatist stance, our chapter question how tools frame interactions and represent groundwater, considering a plurality of values, interests and attachments to the environment (Richard-Ferroudji & Barreteau, 2012; Thévenot, Moody, & Lafaye, 2000). Indeed, groundwater can be represented in various ways. Plural interests but also plural social values are associated to groundwater. For example, through the analysis of 5 years publications in *The Hindu,* one of the leading newspapers in India, we identified four typical qualifications of groundwater associated with best management measures: (a) endangered heritage whose access must be regulated, (b) limited resource that must be optimized, (c) issue of survival whose access must be ensured (d) source of emancipation that must be acknowledged (Richard-Ferroudji, 2019). The two last ones condone the overexploitation of aquifers. This led us to advocate for a careful consideration of the multiple normative perspective toward groundwater management and emphasizes the importance of compromises between conservation and consumption.

The chapter is organized as follows. Section 10.2 describes the methodology and introduces the framework used to analyse the communication approaches and tools deployed in each case (public targeted; content, issues brought to the fore and normative stance adopted; type/format). Section 10.3 develops a transversal analysis of the use of the tools, and identifies assets and limits across the 11 cases. Section 10.4 discusses the transversal results and concludes with recommendations.

10.2 Learning from Pioneering Experiences

10.2.1 Methodology

10.2.1.1 Eleven Cases of Policy Instruments Dedicated to Aquifers

During the past 45 years in France, water policy has evolved from a sector-based and centralized form of management to a more local and integrated one. French water policy promotes tools and procedures that consider hydro-territories[2] as the relevant areas for integrated management. At a local level SAGEs (local sub-basin plans for water development and management),[3] *contracts* (for coordinating agency and other government investment in local public action)[4] and management structures (which support the making and the implementation of SAGE and *contracts*)[5] completed the apparatus. Our study focuses on SAGE and *contracts* that are dedicated to aquifers. We consider them to be pioneering in making groundwater more visible. Focusing on these cases, we aimed at identifying some original activities and tools to capitalize on the experiences. 11 case studies were selected to illustrate the variety of forms which those initiatives may take (See Table 10.1 and Fig. 10.1). They strongly vary in terms of policy instrument (SAGE or *contract*), management structure (state body, association, etc.), size (from 1 to 10 employees), duration and maturity (one goes as far back as 1954, another one was launched in 2011), area (from 346 km^2 to 10,138 km^2) and number of inhabitants (28,673 to 1.4 million).

10.2.1.2 Inventory and Analysis of the Uses

The tools used in each case study were inventoried. The inventory is based on web mining, grey literature review and interviews. We explored (in June 2016) the websites dedicated to groundwater and the management structures of the 11 cases. We used a search engine (Google) to explore the web pages dealing with each aquifer and also the illustrations used on the web. These explorations were completed by research targeted on the use of specific tools in each case, with the following keywords: "scale model", "3D model", "Facebook", "Twitter ", "film", "video", "game", "exhibition" and "observatory". Besides, more than 40 interviews were

[2] Area of land delimited by interdependence to a waterbody (river, lake, wetlands, aquifer, etc.) and draining ultimately to this particular body.

[3] They were founded by the 1992 Water Act to define the management and restoration strategies at the local scale. In 2018, 184 SAGE were implemented, in areas that range from 300 km^2 to 10,000 km^2, more on www.gesteau.fr

[4] Contract between funders (e.g. a Water Agency, French State, municipalities) instituted by memorandum in 1984.

[5] Territorial bodies tend to associate municipalities at the basin scale in the frame of Syndicat, EPTB or Syndicat Mixte. One should consider that the Water Framework Directive 2000/60/EC strengthened the role of territorial body in water management.

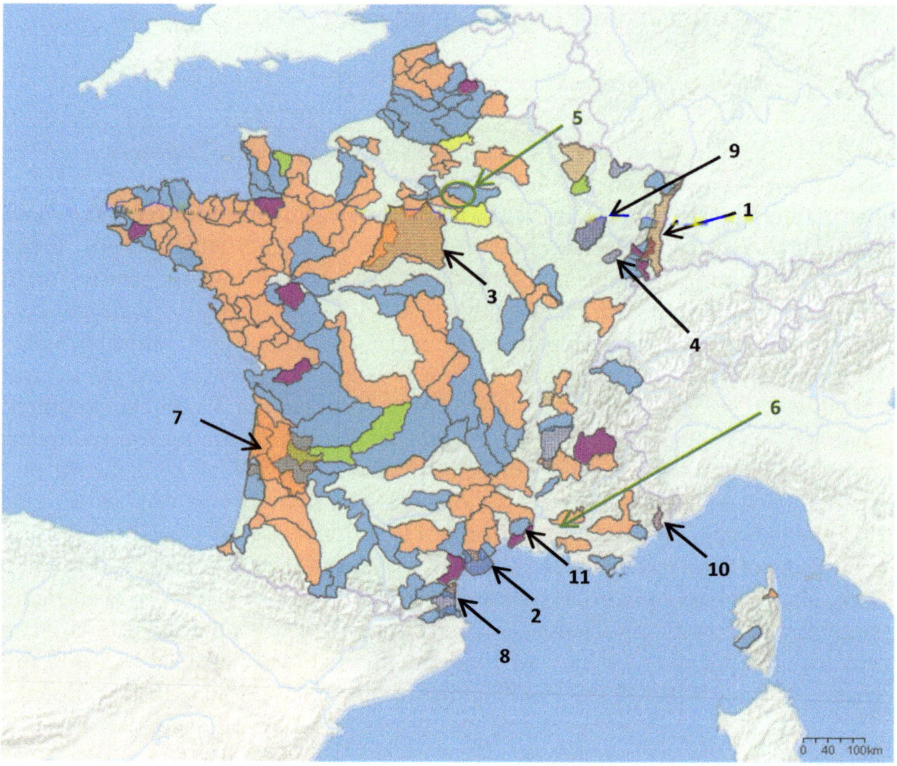

Fig. 10.1 Situation of the 11 cases in France. (Source of the background map: progress of the SAGE procedures, Gest'eau, www.gesteau.fr, April 15th 2016, Yellow: emerging, green and blue: drafting, orange and pink: implementation)

conducted with transcripts or detailed reports.[6] We first interviewed SAGE facilitators and directors of the management structures. Interviews were then conducted with other stakeholders to gather different points of view (representatives of associations or the administration, elected, teachers, researchers and consultants). Documents (technical studies, guides, reports on school programs, booklets, posters, etc.), web pages, maps, photos, movies and games were collected and analysed. The fact of having 11 cases of study favoured the gathering of a diversity of experience.

[6] Two to five interviews per case with some collective interviews and some people concerned by several cases (consultant, civil servant).

10.2.2 Framework of Analyse

10.2.2.1 Who Participates?

For each tool are identified who promoted them to which audience, and who eventually participated or used it (public engagement). Numerous categorizations are used in the field of water to designate participants. The "Water Parliament" which gives its vote on the SAGE document is for example made up of three committees: elected representatives, users representatives (farmers, industrialists, landowners, etc.), and State representatives. With a different perspective, the theory of communication (Shannon-Weaver's model) distinguishes sender and receiver to define a strategy for effective communication through a channel and that can be affected by noise. Communication is then intended as transmitting information to target groups from the general public to specific users. In doing so, it is based on a linear approach to communication. Our approach of communication leads to distinguish participants by their connection to groundwater: their interests, attachments or knowledge. For example, an article in a special issue of the journal *Géologues*[7] on Communication and Mediation distinguishes "outsiders" from "insiders" (Marjolet & Normand, 2006).[8] According to these authors, with insiders there is no problem of communication. They share a common scientific or administrative culture and language. The outsiders, by far the most numerous group, are not part of this circle of "common culture" as they use a different idiom and frame problems in a different way.[9] The circle is however a restricted one. It includes, besides experts, some elected representatives, civil servants and members of NGO. Outsiders also include elected officials, civil servants, as well as many members of the civil society. However, lay people may well know the aquifer but not be familiar with the technical language.

10.2.2.2 Which Issues Are Made Visible and According to What Normative Stance?

The following sections examine each tool to identify the issues they tackle regarding groundwater and their normative perspective. Indeed specific issues were at the origin of each SAGE or *contract* studied, and these may be considered for communication to various interested or affected publics. Issues include groundwater depletion and pollution (e.g. salt intrusion and fluoride), etc. The tools are also underpinned by normative conceptions on groundwater resources management. Many tools promote the principle of resource conservation. However, different objectives can potentially be assigned:

[7] Geologists in English.

[8] "Initiés" vs "non initiés" in the French original version.

[9] Marjolet et al. observe during meetings a gap between those who speak of nitrogen and other participants who mention the issue of nitrate which has received much more media attention.

- Develop scientific knowledge and create indicators.
- Make people understand the specificities of groundwater in general or the local resource in particular.
- Change practices: save water, reduce pollution, increase available resources.
- Develop governance and participation of the concerned people

The last objectives focus on the participative nature of the communication tools. Over the last years, participatory groundwater management has been much commented upon but remains a bone of contention between the proponents of expert management and those who advocate the principle of letting water users shape their management institutions (Ostrom, 1990). In this regard, special attention should be paid to distinguish participatory tools (Callon, Lascoumes, & Barthe, 2009).

10.2.2.3 Which Format of Interaction?

There are different types of communication tools:

- Traditional media (Press, Radio, Television, posters, booklet, mail, etc.).
- Digital media (Online Press, Online advertising, social networks, blogs, groups, forums, websites, emailing, newsletters, videoconferencing, mobile applications, SMS, shared videos).
- Events (stands, fair, conferences, etc.).
- Direct contact (in the premises of the structure, by telephone, meeting, training sessions, etc.).

In addition, the given information may be of different formats: texts, numbers, images, diagrams, videos, etc. Special attention will be paid to these different formats across the above four type of tools, as they frame interaction and can affect communication.

10.3 A Wide Range of Activities and Tools to Make Groundwater Visible

10.3.1 Increase in the Available Information

The first result of the study is a strengthening of the visibility of aquifers in the studied cases. We have observed a growing production of information over time[10] and a gradual widening of the range of tools. Every year new documents are released. Web site or Facebook pages are created. Many documents we examined were intended primarily for specialists, while other communication tools were tailored specifically for awareness campaigns for the lay persons and water users.

[10] Our study provides a benchmark for quantitative evaluation.

Numerous documents are available, not only technical reports for specialists, but also documents for policy makers and the lay public. Booklets and newsletters are produced for a targeted or mass audience. They address a wide range of topics such as wetlands, chlorine pollutions, SAGE procedures, and practical guides to save water or drill borewells. Most of the analysed documents are prepared by the staff of the management structure and technical consultants. Communication consultants are rarely hired. The documents are made available on the internet and distributed during events or through targeted mailing etc. In none of the cases did we identify a systematic mailing to all inhabitants of the groundwater management area, as this is considered too expensive and inefficient.

10.3.1.1 Internet Used to Share Information, but Rare Use of Social Networks

Many documents and related information are made available online in public or private spaces. All the management structures have web pages, either their own website or a webpage hosted by a larger structure. However, the use of the Internet is often limited to information sharing, with little use of the potentialities of this support (interactivity and live communication), with the exception of interactive mapping tools or Facebook pages.[11] Interviewees tend to be sceptical about using social networks. Some people point out that they do not need to communicate quickly on the news. They feel that people are already over solicited and may not be interested in the topic. Others argue in favour of using social networks, recognising that digital media is increasingly used among participants, including elected representatives. However, a lack of time for posting and updating was brought up by all as a critical challenge limiting social media and internet uses.

10.3.1.2 Traditional Media: Visibility in the Regional Press

When it comes to traditional media, articles appear in regional and local press, television or radio stations, for example when signing a *contract* or for a particular event (e.g. a Science Festival). In the national press, the topic is rare, with articles tending to be limited to reporting extreme events (e.g. drought or pollution) or in the case of public controversies (e.g. exploitation by private companies). Groundwater professionals rarely inform the media on a regular basis. Mass communication is perceived as expensive and inefficient. In two cases, however, we noted the use of billboards to promote water savings (Roussillon and Gironde). In such instances of broad dissemination, the campaign benefited from the support of partner organisations (e.g. technical support for the communication services, free access to municipal board journals or district billboards).

[11] In two cases: the Breuchin SAGE and the Crau Aquifer.

10.3.2 Toward Conventional Representations

10.3.2.1 Indicators for Information, Alerts and Regulation

SAGEs and *contracts* procedures provide policy settings suited for gathering data on groundwater resources and implementing new studies. These procedures aim at building a common representation of the groundwater systems. In the water field, issues are usually divided into two categories: quantitative (related to volume of water) and qualitative (related to water quality). All structures rely on both quantitative and qualitative monitoring networks. Yet, there is still an issue of knowledge development (e.g. on groundwater recharge). Besides, there is less harmonization and formalization of indicators for groundwater than for surface water because groundwater monitoring is younger and the monitoring network less dense (but with territorial variability). Hydrogeology is a relatively young discipline, in which measurement units used for aquifers representation are sometimes yet to be standardized. As a consequence, different indicators for groundwater conditions are in circulation. While an indicator such as the piezometric level is common to all cases, others are more specific (e.g. salinization). The use of these indicators is deeply embedded in the history of local territories. There is a path dependency in the choice of indicator in each case, but to the benefit of adaptation to local issues (e.g. monitoring salinization in the case of coastal Astien Aquifer).

Indicators such as piezometric level are used to objectify groundwater and issues. In all the cases studied, information is conveyed about groundwater levels but at different scales (from annual average, monthly measurement, to real-time information). Information is presented with curves, maps or with a clepsydra as an illustration. Groundwater level data is represented with other information such as rainfall or water consumption to understand their dependency. We found that piezometric records are used for different purposes. In some cases, those records were used to inform and/or alert stakeholders about groundwater trends and potential implications for management decision. They are also becoming increasingly instrumental in regulating groundwater uses. SAGE documents can define threshold levels to be used to regulate extraction. The definition of such thresholds is subject to debate and results from negotiations.

Over time, maps and indicators have been refined in terms of spatial and temporal scales. In the cases studied, groundwater professionals now benefit from a range of tools providing shared representations of the local aquifers. In five cases, observatories or dash boards are set up to gather data sets and offer an integrated approach to understanding groundwater conditions. Modelling is also developed to explore management scenarios. Most of the indicators used are biophysical ones. Indicators of socio-economic dimensions are rarely used, with the notable exception of the SAGE of Gironde aquifers which set progress indicators for task completions along with an annual opinion survey entitled "Gironde people and water". Finally, we observed that data production is entrusted to experts and consultants, with rare use of experience with citizens. One rare example of citizen science was found in the

Crau case. The managing structure called for volunteers to participate in the monitoring of groundwater levels. This kind of approach has proven to be effective in complementing existing government-run monitoring programs in other regions (Little, Hayashi, & Liang, 2016).

10.3.2.2 Maps: Essential Tools

All the organisations in our study produce maps and use them in their documents. In SAGE and *contracts* processes, it is common practice to collect maps in a booklet. Maps are abundant. One interviewee goes as far as to say: "there are never too many maps!" This medium has been used by hydrogeologists since the beginning of the discipline to show these hidden resources while representing their borders at the surface. Today, cartographic methods are used to represent a wide array of topics: aquifer perimeters, piezometric networks or socioeconomic issues (tourism or farming in the area, institutions, etc.). We emphasize the fact that maps can be used to cross aquifers representations with other issues (e.g. groundwater resources and population increase). Maps are produced in different formats and for different audiences. The maps produced in the SAGE documents and in the *contracts* are mainly used by "insiders" (elected representatives, NGO representatives, state services officers) and by consulting firms working on groundwater related projects. A certain level of knowledge is required to understand these maps, as well as technical references (e.g. concentration thresholds of pollutants). Many maps are thus difficult to understand by lay people. While considering the purposes for which maps are made, all the interviewees recognize the ability of maps to synthesize information and simplify technical aspects. During public hearings, maps are instrumental in fostering discussion with stakeholders. Users confront their own spatial landmarks and their field knowledge with them. Some maps are designed to alert users and convince them to change their practices by highlighting management issues and depletion. We are witnessing the growing use of maps in a regulatory perspective (protection of catchment perimeters, definition of Strategic Zones for drinking water supply or zones vulnerable to nitrates, definition of threshold volumes, etc.). In a few pilot projects, the building of such maps is participatory and proved to be instrumental in involving users to promote common pool resources management. Maps are to play an increasing role in the consultation and the regulation of the uses.

Yet, during interviews, several people also pointed out that map proved at times to be unnecessary or mere decoration. They report low usage and little discussion of SAGE maps that are accepted as technical data. Some deplore the systematic and unavoidable nature of the production of maps without questioning their relevance. Besides, for several respondents, the mapping must remain the responsibility of the expert. In short, although maps appear to be essential to groundwater management, it is necessary to keep a critical stance on their production and uses.

There was consensus among the people interviewed that miniature models are relevant: from the rough (and low cost) ones made by the teams[12] to detailed representation of the territory and the aquifer (e.g. the upper Rhin Aquifer miniature model). There is a growing use of such models and 3D mapping as this medium can meet a wide array of needs including raising awareness among lay people. 3D makes it possible to represent the superposition of aquifers and to introduce users to the complexity of aquifer dynamics. It is a tool that deploys its potential when used in a digital and interactive form, with the user exploring the 3D view from multiple angles. The advent of web 2.0 technologies is seen as an opportunity to increase the potential of cartography with interactive mapping platforms. These tools, however, remain difficult to apprehend for people unfamiliar with GIS software. Besides, the cost of 3D technologies or viewers makes this media difficult to access for most organizations. In our study, some of them resented investing in a tool such as 3D modelling that does not necessarily provide added value (compared to maps) to management and collective discussion.

10.3.3 The Potential of Arts, Field Visits and Intermediaries

10.3.3.1 Groundwater Is Not Photogenic but Inspires Fictions

We were interested in the use of art and illustrations to make groundwater visible. How can a hidden resource be captured in an image? Illustrations could be photographs, numerical data (tables or graphic illustrations) or drawings.[13] The analysis of the websites showed that photos are barely used. Groundwater is obviously not photogenic, with the exception of some karsts which can be misleading to the public as they represent only one type of aquifers. Groundwater can only be captured in a traditional photographic image in caves or when it gushes from a pipe, percolates on the surface, or lies at the bottom of a well. Stored in sand or pebbles, it is difficult to photograph it. In most cases, groundwater resources are represented by proxies, such as photographs of (A) surface water that interacts with groundwater (tank, lake and river), (B) infrastructures (pumps, motors, pipe) or measurement equipments (piezometers) (C) the users and their practices (a farmer in a field, children drinking, etc.), (D) events concerning groundwater or groundwater professionals (water parliament meeting, exhibition, the team of the management structure), or (E) generic photos on the theme of water (a drop of water, flowing water). Photos may show the social or political dimension of groundwater when capturing groundwater uses or meetings. Some structures have developed photo libraries. This is for example the case of the Symcrau, whose website presents a participatory photo library. Interestingly, the photo library is part of their observatory.

---

[12] E.g. one crafted with an aquarium, layers of sand and stones, and straws.

[13] Cartoons are used in the national press, but not in our cases.

In a number of cases, short documentaries were produced about the local aquifer dynamics and/or its management. Often, these films dealt with water more generally than the specifics of local aquifers. Such videos were considered as necessary for raising awareness by the interviewee, but expensive. They were able to capture the social and political dimensions via people's testimonies. Animated movies were also used to assist peoples' understanding of phenomenon such as groundwater recharge (e.g. in one case, a dinosaur was used to remind the old age of groundwater). Yet, the potential of fictions and the presence of groundwater in culture are under-exploited. Feature films and novels are largely untapped formats for increasing the general public's awareness and understanding of groundwater. Interviewees confirm the very low use of fictions, stories or myths despite their relevance to regain a "culture of water" which is fading. However, some interviewees were wary of fictional material because it may convey and perpetuate misconceptions (according to them) of complex groundwater dynamics and management policy.

From the perspective of visibility and participatory management, it would be interesting to develop the use of popular culture, graphic arts and games. We observed that some management structures produced games (e.g. The "game of the camel" on the Astien or the game Gaspido on the Roussillon which are combining goose game, quiz and challenges) that are used mainly for schools. Drawing from the innovative use of serious games (Meinzen-Dick et al., 2018), this kind of tools can be used to the general public to invigorate awareness campaigns or with insiders to foster collective discussion and explore scenarios.

10.3.3.2 Rallying Around Aquifers

Various meetings are organized involving groundwater. The SAGE and the *Aquifer contract* processes may include meetings of the "Groundwater parliament", of consultative meeting, of thematic groups, of advisory groups (See Chap. 4). Consultation bodies are set up on a permanent or ad hoc basis. Events, conferences and exhibitions are organized to promote a knowledge-awareness and to transmit knowledge to a broad audience whether temporary, travelling or permanent in government or other groundwater manager offices. They are organized by the management structure but more often by partners (e.g. environmentalist associations, Water Agencies, municipalities, universities). Groundwater professionals are invited to share their experience and knowledge.

Our research identified activities dedicated to schoolchildren in all 11 cases. In most of the cases, environmental education associations were mandated by the management structure to implement these activities. Educational activities benefit from funding from French Water Agencies and Ministry of Education. The activities carried out with school children often focus on water saving. They include field trips which play important roles in raising awareness or sharing experiences.

Practitioners also organize field trips for the newcomers in the Water Parliament, to get them acquainted with the issues of the territory. Field trips are activities implemented regularly in some cases but more often once off. They could be further expanded for general public or targeted ones (e.g. bore well owners). Some practitioners shared with us their ideas about how to manifest the physical presence of aquifers at the surface, with boards or art settings as a symbol to represent and map the water under our feet. This is promising area, and would benefit from further exploration to understand how to mark the boundaries and features of groundwater on the surface, so as to raise awareness of its otherwise hidden presence.[14] Yet, face-to-face events and on-the-ground communication (e.g. information stalls, field trips) are still too rarely used. This is because existing management structures lack of investment in staff time and supporting budgets to organize events on a regular basis.

10.4 Discussion and Conclusion: How to Make Groundwater More Visible?

10.4.1 Diversify the Format of Communication: From Scientific Reports to Art

The analysis conducted in the 11 case studies showed that over the last two decades[15] a variety of communication tools has been developed in the field of groundwater at the local scale. More and more documents are available, mainly for specialists but also for decision makers and the lay public. A lot of information is available online. The potentialities of the digital media could still be developed to favour interaction and participation, but this would require more human resource for facilitation. When it comes to traditional media (e.g. press, TV, etc.), some stories concerning groundwater are covered by the local media but rarely by national ones. Groundwater managers are not inclined to mass communication, which is perceived as costly and inefficient. Their focus is on local appropriation. Documents and websites are illustrated with maps, photographs, numerical data (tables or graphic illustrations) or drawings which represent groundwater. Maps are essential to represent groundwater at the surface. They are abounding. Then a critical stance is needed on their objectives and uses. Maps and indicators have been refined in terms of spatial scale and time scale but also with a legal perspective to regulate the extractions (e.g. "Strategic Resources studies"). They have become conventional representations that support groundwater management and are shared among "observatories". If groundwater is not photogenic, it can be shown indirectly (connected surface water, pumps, pipes, users, etc.) and narrative fiction offers a promising area to share knowledge and explore multiple points of view. Beyond scientific representations, artistic representations deserve to be used to reach a broad audience and represent social and political dimension of groundwater. Moreover, there is potential for development of face-to-face events and field trips as well as landmarks and land art works to

[14] Facing the same issue of oblivion, flood markers remind the possibility of flood.

[15] Several management structures were created in the 90s and the SAGE were set up in 1992.

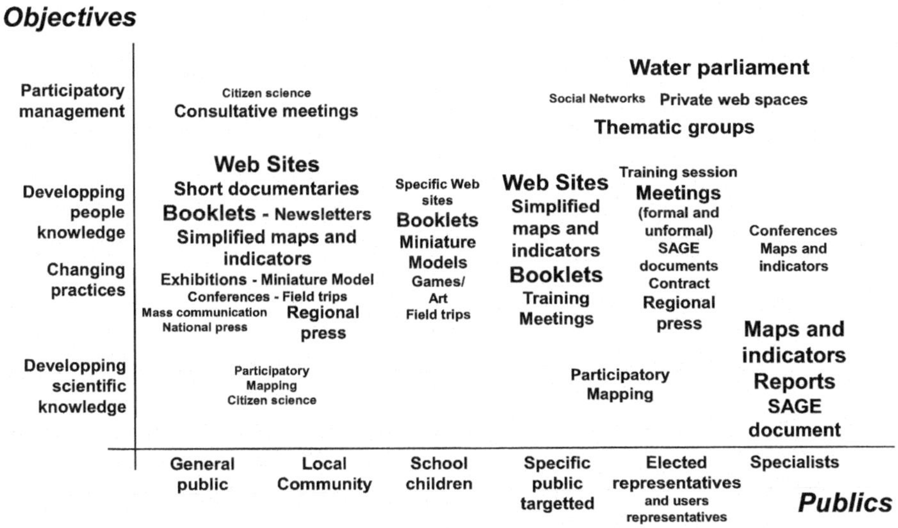

Fig. 10.2 Preferred tools for different objectives and publics with font size representing the extent of their use (the biggest for those used in all the studied cases)

materialize the presence of water beneath our feet. Different approaches are required for different publics and to develop capacities, as it is observed in other cases (Re & Misstear, 2017). Figure 10.2 illustrates the diversity and predominance of some tools.

10.4.2 Foster the Unconfining[16] of Groundwater Management

Four types of objectives assigned to the tools that we have inventoried were identified and discussed:

10.4.2.1 Develop Scientific Knowledge and Create Indicators

A first objective is to develop scientific knowledge of the resource and conventional monitoring equipment with a management perspective. This leads to the production and circulation of indicators and maps. This study confirms the hypothesis that nowadays, quantitative issues are more visible than qualitative ones with few exceptions. The quantitative stakes are more emphasized and rely on important equipment from the piezometric maps to "volumetric groundwater management" process. The

[16] If some aquifers are confined, management can also be. Sociology of science distinguishes participatory tools from the ones that are "confined" (Callon et al., 2009) within the restricted spaces of secluded research and representatives designated by ordinary citizens.

interviews show that significant investments have been made and are still made in the production of knowledge, which is an important work of visibility. However, efforts are still needed to share this knowledge and to involve people concerned in the production.

10.4.2.2 Make People Understand the Specificities of Groundwater in General or the Local Resource in Particular

With the former perspective, a second objective assigned to the tools is to make people understand the groundwater systems and issues, raise awareness and capacity. However, we distinguished two approaches. Either communication is about groundwater in general or it insists on the local resource as a common heritage. It is then a matter of developing groundwater knowledge by highlighting local issues and the neighbouring environment related to everyday life and people own experiences.

10.4.2.3 Change Practices: Save Water, Reduce Pollution, Increase Available Resources

A third goal is to change practices. Some expected changes are general in the field of water: save water, improve sanitation, reduce the use of pesticides, etc. Other messages are specific to groundwater management: protection of catchments, promoting maintenance of the bore wells and good drilling practices, etc. Even if water mining practices are justified for some people, nowadays, in France, claiming publicly the relevance of groundwater overexploitation can no longer be deemed reasonable. Practically, tools often tackle both this objective and the previous one (See Fig. 10.2). We distinguished them as this one is oriented toward convincing while the other one is more oriented toward capacity building.

10.4.2.4 Develop Governance and Participation of Concerned People

The fourth objective focuses on governance. Groundwater related participatory practices are little developed and social mobilization is weak. Studies and data production are entrusted to experts. The lack of resources and the reluctance of technicians or elected representatives are also obstacles to the implementation of participatory approaches. Opponents fear that outsiders would pollute the debate if it is unconfined, while the tenants of participatory approach expect that they will recharge it. Among our pioneering case studies, participatory practices are developing with original initiatives and positive feedbacks (e.g. participatory cartography, citizen science). This fourth objective needs to be fostered.

10.4.3 Build on Local Communities

As budgets dedicated to communication are limited, there is a tendency for groundwater managers to focus on specific themes and target audiences. The "general public" appears to be a fuzzy notion too difficult to reach. The promoters we interviewed prefer to develop tools targeting specific publics such as tourists or socio-professionals (e.g. farmers or camps managers in the Mediterranean coastal area), municipalities (e.g. Campaign "Stadium without pesticides" on the aquifer of Vistrenque), well owners (e.g. to inform them of good practices in the construction or maintenance of a borewell) or urban planners (e.g. to inform them about water constraints). Communication toward elected representatives is considered as a priority. Meetings, documents or training sessions are specifically tailored for them. However, this public remains heterogeneous, with variable levels of knowledge, involvements and scales of action (from the municipality to the Region). Elected representatives also face a challenging dilemma between territory development and water resources protection. Schoolchildren are considered by interviewees as a multiplier group because they are an investment in the future, as well as transmitters to their family, relatives and neighbours. Initiatives brought to schools are numerous. Yet there is little follow up study of the effectiveness of school education on groundwater. Besides, in the field of groundwater, we found that the associations were mainly involved in an educational perspective with partnership with the management structure. Exceptionally, they are involved as activists and contest projects. Public administration representatives are involved in the SAGE or *contract* procedures. They are from the sector of water, agriculture or urban planning. It is often a captive public whose participation is linked to their position. The challenge is then to involve them more in local issues. An asset of SAGE and *aquifer contracts* is their territorial approach. Indeed the objective is not to make groundwater visible but to get people to take care of a specific aquifer that is a common heritage.

10.4.4 Recognize and Promote Spokespersons for the Aquifers

In France, the employees of management structures play a key role in making groundwater visible. Communication activities often depend on their commitment. Most of them are willing to develop communication. Only one interviewee stated that communication does not relate to his area of work and that groundwater professionals should not venture beyond technical management. They can conceive their role in different ways, from an expert role to that of facilitator, with a dimension of taking care of water bodies and participants such as family doctors (Richard-Ferroudji, 2014). Groundwater professional and specialists dedicate a major part of their time to groundwater. They can be considered as spokespersons for groundwater. This can also be the case of elected representatives, NGO representatives or users which have a thorough knowledge of the subject from different perspectives. Several interviewees also stress the importance of relying on local intermediaries to reach users. In the process, groundwater specialists are asked to expend the gamut of their activities. While groundwater professionals used to be focused on the supply side of groundwater, they now deal with activities meant to curb groundwater uses, ranging from public sensitization to facilitation. Those activities deserve better recognition. Yet, time resources and financial means are missing in most cases. In the management structures, budgets allocated to communication are low. Most of the time, promoters seize opportunities to communicate. In two cases only a communication consultant was hired for advice and drafting a communication plan. In order to make groundwater more visible, some support is requested, not only financial but also institutional. Communication activities should be better recognised and supported by public funds for livening up groundwater policy.

In short, from this study, we recommend (1) to continue the development of tool with a diversity of formats including artistic ones and field trips, (2) to develop participatory approaches, building on local communities, (3) to recognize and promote spokespersons for the aquifers. Specific budget and public support are needed to create an environment for effective communication and sustainable groundwater management.

Acknowledgements This work has been supported by AFB (French Agency for Biodiversity). The authors would like to thank Bénédicte Augeard, Delphine Loupsans and Claire Magand for stimulating discussions; Xavier Bernard, James Daly and Jeanne Latusek (IFP) for their fruitful contribution to the study; the editors for reviewing the first draft of this chapter and providing valuable comments and all the interviewees who shared their point of view and experiences.

References

Baldwin, C., Tan, P.-L., White, I., Hoverman, S., & Burry, K. (2012). How scientific knowledge informs community understanding of groundwater. *Journal of Hydrology*. https://doi.org/10.1016/j.jhydrol.2012.06.006

Bouarfa S, & Kuper M (Eds.) (2012). Groundwater governance: Learning from local experiences. *Irrigation and Drainage, 61*(Supplement S1).

Callon, M., Lascoumes, P., & Barthe, Y. (2009). *Acting in an Uncertain World*. Cambridge, MA: MIT Press.

Curtis, A., Mitchell, M., & Mendham, E. (2016). Social science contributions to groundwater governance. In A. J. Jakeman, O. Barreteau, R. J. Hunt, J.-D. Rinaudo, & A. Ross (Eds.), *Integrated groundwater management: Concepts, approaches and challenges* (pp. 477–492). Cham, Switzerland: Springer International Publishing.

Faysse, N., & Petit, O. (2012). Convergent readings of groundwater governance? Engaging exchanges between different research perspectives. *Irrigation and Drainage, 61*(S1), 106–114.

Jakeman, A. J., Barreteau, O., Hunt, R. J., Rinaudo, J.-D., Ross, A., Arshad, M., et al. (2016). Integrated groundwater management: An overview of concepts and challenges. In A. J. Jakeman, O. Barreteau, R. J. Hunt, J.-D. Rinaudo, & A. Ross (Eds.), *Integrated groundwater management: Concepts, approaches and challenges* (pp. 3–20). Cham, Switzerland: Springer International Publishing.

Little, K. E., Hayashi, M., & Liang, S. (2016). Community-based groundwater monitoring network using a citizen-science approach. *Groundwater, 54*, 317–324.

Marjolet, G., & Normand, D. (2006). *Le Conseil général des Cotes d'Armor: communication et médiation à la Direction* 'Agriculture et Environnement'. Geol-PARIS- 151:50.

Meinzen-Dick, R., Janssen, M. A., Kandikuppa, S., Chaturvedi, R., Rao, K., & Theis, S. (2018). Playing games to save water: Collective action games for groundwater management in Andhra Pradesh, India. *World Development, 107*, 40–53. https://doi.org/10.1016/j.worlddev.2018.02.006

Mitchell, M., Curtis, A., Sharp, E., & Mendham, E. (2012). Directions for social research to underpin improved groundwater management. *Journal of Hydrology, 448–449*, 223–231. https://doi.org/10.1016/j.jhydrol.2012.04.056

Ostrom, E. (1990). *Governing the commons: The evolutions of institutions for collective action*. New York: Cambridge University Press.

Re, V., & Misstear, B. (2017). How scientific knowledge informs community understanding of groundwater. In *Advances in groundwater governance* (pp. 215–230). Boca Raton: CRC Press.

Richard-Ferroudji, A. (2014). Rare birds for fuzzy jobs: A new type of water professional at the watershed scale in France. *Journal of Hydrology, 519*(Part C), 2468–2474. https://doi.org/10.1016/j.jhydrol.2014.09.070

Richard-Ferroudji, A. (2019). Ambivalence about groundwater: promoting conservation while justifying over-exploitation in an Indian newspaper. *RILE, Revista Interdisciplinar de Literatura e Ecocrítica*, 1–2 (pp. 64–92)

Richard-Ferroudji, A., & Barreteau, O. (2012). Assembling different forms of knowledge for participative water management – Insights from the Concert'eau game. In C. Claeys & M. Jacqué (Eds.), *Environmental democracy facing uncertainty* (pp. 97–120). Bruxelles, Belgium: Peter Lang.

Richard-Ferroudji, A., Lassaube, G., Bernard, X., Daly, J., & Latusek, J. (2018). *Mise en visibilité des eaux souterraines et de leurs enjeux de gestion- Expériences des syndicats de nappes et structures porteuses de SAGE en France et perspectives indiennes*. Paris: AFB/CNRS/IFP.

Shah, T. (2009). *Taming the anarchy groundwater governance in South Asia*. New Delhi, India: Routledge.

Thévenot, L., Moody, M., & Lafaye, C. (2000). Forms of valuing nature: Arguments and modes of justification in French and American environmental disputes. In M. Lamont & L. Thévenot (Eds.), *Rethinking comparative cultural sociology : repertoires of evaluation in France and the United States* (pp. 229–272). Cambridge, UK: Cambridge University Press.

Van Der Gun, J. (2017). Data, information, knowledge and diagnostics on groundwater. In *Advances in groundwater governance* (pp. 193–214). Boca Raton: CRC Press.

Villholth, K. G., Lopez-Gunn, E., Conti, K., Garrido, A., & Van Der Gun, J. (2017). *Advances in groundwater governance*. Leiden, The Netherlands: CRC Press.

Dr. Audrey Richard-Ferroudji's trajectory is transdisciplinary. She holds a diploma of engineer and a PhD in Sociology. She worked for 11 years at IRSTEA (2002–2013), a French research institute on environmental and agricultural engineering, within the joint research unit G-EAU dedicated to water management and water uses. In October 2013, she moved to India, as head of the Social Sciences Department of the French Institute of Pondicherry/CNRS USR3330. Since 2018, as a consultant, her work focuses on water governance and participative management. She looks at the evolution of institutions, policies and adaptation to global changes and risks at the local scale. She is dedicated to promoting the Social Sciences in the water field.

Gaïa Lassaube is Ph.D student at Centre Emile Durkheim, University of Bordeaux and affiliated to Institut Français de Pondichery in India. In her Ph.D research, she conducts a comparative socio-historical analysis of groundwater policies in France and India. She explores the professional practices and knowledge of French and Indian hydrogeologists working in public administration.

Chapter 11
Conceptual Approaches, Methods and Models Used to Assess Abstraction Limits for Unconfined Aquifers in France

Luc Arnaud

Abstract This chapter presents a review of methods an tools used in France to assess groundwater abstraction limits in unconfined aquifers. The experience gained from over 30 studies shows that the estimation of Maximum Permissible Volume (MPV) is complicated by numerous uncertainties. The first prerequisite is a good knowledge of the dynamics of the hydrosystems and abstraction volumes, but unfortunately this is rarely achieved. Moreover, both the calculation methods and modelling tools that aim to conceptualize these complex systems have limitations due to the simplifying assumptions required for their application. Technical recommendations are proposed for a proper assessment of such uncertainties. In many cases, the calculated maximum permissible volumes were much lower than the previously authorized volumes. Therefore, many of the results were contested by affected users. Such disputes concerned not only the economic consequences of reduced abstraction, but also the scientific basis of the studies in view of the known uncertainties and limitations. The last section of this chapter discusses this phase of negotiations, specifically based on examples from the Adour-Garonne water basin in southwest France.

Keywords Abstraction limits · Calculation methods · Hydrogeological models · Uncertainties · Unconfined aquifers

---

L. Arnaud (✉)
BRGM (French Geological Survey), Orléans, France
e-mail: l.arnaud@brgm.fr

J.-D. Rinaudo et al. (eds.), *Sustainable Groundwater Management*, Global Issues in Water Policy 24, https://doi.org/10.1007/978-3-030-32766-8_11

11.1 Introduction

The French law on water and aquatic environments of 30 December 2006 requires the implementation of volumetric water-abstraction management in all river basins that are considered as being water deficient. Such volumetric management is mainly based on the definition of an abstraction limit, specified in volume, that the State must respect when delivering the yearly abstraction authorizations (see Chap. 3).

In its circular of 30 June 2008, the French Ministry for Ecology defines a maximum permissible volume (MPV) as "*the water volume that the environment can supply under satisfying ecological conditions*" (Ministry for Ecology, Energy, Sustainable Development and Territorial Development, 2008). In order to consider climatic variability, the MPV is statistically calculated so as to guarantee that, for 8 out of 10 years, such abstraction does not jeopardize the good quantitative, qualitative and ecological status of the water resources and their associated aquatic environments. Contrary to Australia, this French approach is purely environmental, and does not consider actual water use and its economic importance.

An MPV calculation considers where the abstraction is located. Rather than an absolute value, it is a value associated to the spatial distribution of wells and boreholes, and the distribution of water abstraction between them. A significant modification of the spatial distribution thus will result in a modification of the MPV.

An MPV calculation thus requires in-depth hydrogeological understanding. In this chapter, we present lessons learnt from recent French studies for determining the MPVs in unconfined aquifers, where the possible hydraulic connection between groundwater and surface water renders such evaluation particularly delicate. Our analysis is based on a review of over 30 studies, carried out by various organizations between 2008 and 2015 (Arnaud, 2016).

The first section of the chapter reviews the methods and tools used in the studies, and describes the criteria that determined which method was chosen. The second section describes the main difficulties and limits of the various methods employed, as well as the uncertainty associated with produced results. Technical recommendations are proposed for a proper assessment of such uncertainties. The third section discusses the phase of negotiations between stakeholders, experts and government agencies that generally follows after the technical studies, which eventually leads to a regulatory definition of the MPVs. This last section is specifically based on examples from the Adour-Garonne water basin in southwest France, a region where the assessment of MPVs has generated significant conflict and disputes.

11.2 Review of Methods and Tools for MPV Calculation

Following the publication of the 2006 Water Law, the Ministry for Ecology in charge of implementing the new legal framework did not impose a specific method for calculating maximum permissible volumes. The reason was that a single methodology, regardless of its relevance, cannot cover the great variety of hydrogeological settings found in France. This decision has allowed experts and managers responsible for determining the MPVs to use a diverse range of methods that are presented hereafter.

11.2.1 General Approach for Assessing MPV in the Context of an Unconfined Aquifer

Even though tools and methods can vary greatly from one study to the next, MPV studies are generally organized into nine main stages. This framework was first adopted by the Rhône-Méditerranée-Corse Water Agency in order to integrate the various studies carried out in its region (Agence de l'Eau RMC, DIRENs of the RMC Basin, ONEMA, 2009).

1. A steering committee is appointed before each study, with the objective of including all stakeholders affected by the study, involving them in the technical decisions during the study and facilitating the adoption of the results. The committee generally consists of representatives of national and local government services, the Water Agency, water managers and other stakeholders.
2. The second stage consists of creating a conceptual model describing the main characteristics of the aquifer and its flow systems. In the case of unconfined aquifers, particular attention should be paid to defining the interaction between groundwater and surface water, and to the evaluation of the recharge (Healy, 2010). This is because the final selection of the method for determining maximum permissible volumes is in large part governed by the hydrogeological setting.
3. Stage 3 consists of conducting a complete inventory of abstractions and discharges. Given that the data collected by government agencies generally are incomplete, further work will often be needed. For example, water abstraction declared by irrigators can be compared to the theoretical irrigation-water requirements corresponding to the irrigated areas, which can be measured using satellite imagery.
4. The steering committee then defines the environmental objectives that must be respected. They will differ according to their context, such as: maintaining minimal flow in a stream connected to an aquifer; limiting saline intrusion into coastal aquifers; avoiding flow reversal between different aquifers near a wetland; respecting the long-term equilibrated groundwater budget, etc. In some cases, the environmental objectives can be defined beforehand as part of a Water Management Plan established at the local or river catchment level (SAGE and SDAGE respectively, see Chap. 4).
5. The next stage consists in selecting the scientific methods and tools which will be used for calculating the MPV. At a minimum, retrospective analysis of climatic, hydrological and hydrogeological observations must be carried out. Depending on the quality of the available data, the use of groundwater flow models may be possible.

6. Stage 6 consists in defining appropriate management areas on the basis of hydrological and hydrogeological criteria. The MPVs will then be determined at the scale of these management areas.
7. For each zone, indicators are defined for measuring how the environmental objectives are going to be met. Generally, these are groundwater levels in specific monitoring wells, associated with river flows measured at specific gauging stations during low water periods. Other indicators can be added, such as the frequency and duration of periods during which the riverbed becomes dry, the salinity of water, etc.
8. The MPVs can then be estimated for each management area, ideally with monthly time steps and being appropriate for dry seasons.
9. Finally, the MPV study should highlight the limitations of the analysis and propose recommendations for future improvement (collecting further data, etc.). An MPV estimate should not remain fixed in time and regular updates are required by law. The aim is to progressively incorporate new hydro-climatic datasets, modelling updates, a better understanding of local conditions, etc.

11.2.2 Calculating Maximum Permissible Volume Without Using a Groundwater Model

Once the environmental objectives have been defined, three methods can be used for evaluating the MPVs.

The simplest approach consists of a retrospective analysis of climatic, hydrologic and hydrogeologic datasets. By examining historical trends, the maximum volume that has been abstracted in the past without jeopardizing environmental objectives can be identified over a range of climatic conditions. Unfortunately, this highly pragmatic approach is rarely used, even though it has the advantage of very easy implementation and enables a comparison of the estimated MPVs with the observed responses in the field. It seems particularly suitable in the case of systems that are or were exposed to known overexploitation.

A good example is the case of the alluvial Gapeau aquifer in the Var department, where chronic over-use led to saltwater intrusion from the Mediterranean Sea in the past. A cross-analysis of available datasets was carried out by the consulting firm Grontmij (2014). This allowed identification of those volumes abstracted in the past that not only maintained groundwater levels above sea level, but also prevented saline intrusion, under different precipitation conditions ranging from below to above average rainfall.

Based on the quantity of available data and their existing correlations, more in-depth data processing may be possible. The work by the Calligée consulting firm in 2008 on the aquifers of South Vendée (western France) is an interesting example. Their study showed the existence of a linear relation between static water levels and abstracted volumes during summer. This allowed deduction of a mathematical equation for estimating the permissible volume that could be abstracted from the aquifer at the start of the growing season. This equation then was adjusted to be statistically valid for four out of 5 years. Use of this method requires detailed knowledge of abstractions (bi-monthly frequency in this instance), and is only suitable for low-storage aquifers with a seasonal response to recharge and discharge.

In short, even though such analyses may not always result in determining the MPVs that can be abstracted, they should be seen as an important preliminary step for any modelling exercise. They not only allow a first evaluation of the available data, but also provide an understanding of aquifer behaviour under pumping, and an indication of data gaps and recommendations for further investigations.

11.2.3 Calculating Maximum Permissible Volume Using Global Models

Figure 11.1 shows the two main types of global models used in hydrogeology. The first type is based on a schematic and very simplified representation of hydrological systems and processes. It generally consists in a series of reservoirs representing

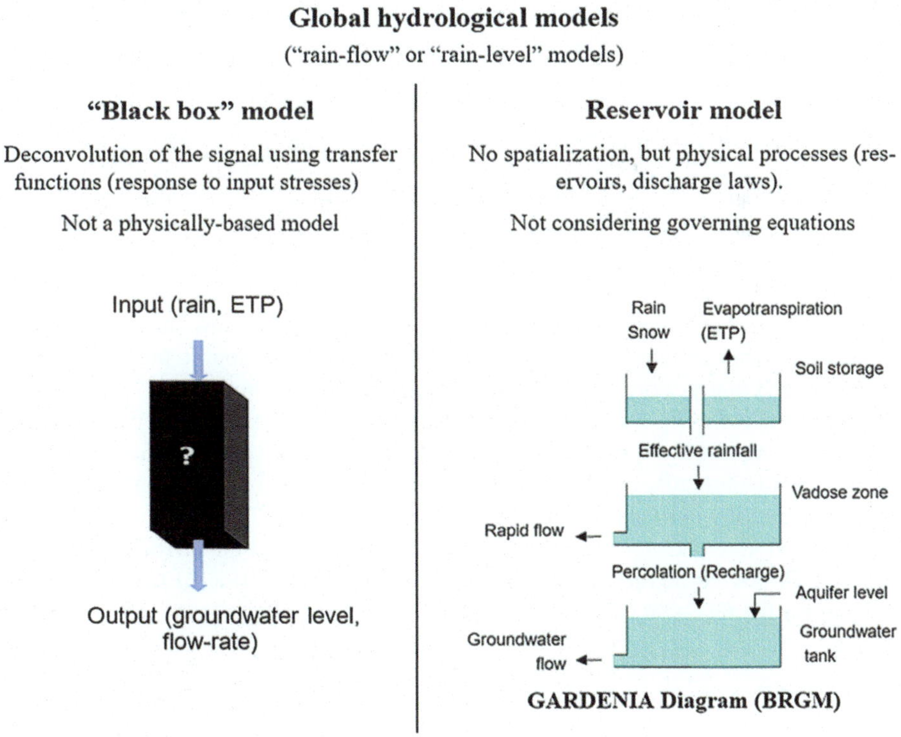

Fig. 11.1 The two main global models used in hydrogeology

aquifers and river stretches connected by simple hydrological functions. The second, "black box", type is based on establishing a mathematical relationship between a output variable describing the condition of the hydrosystem being studied (e.g. the water level in the aquifer, or the flow-rate in a river depending on the aquifer, in Fig. 11.1), and one or more input variables determining this condition (rainfall, evapo-transpiration, abstraction). This mathematical relationship is called a transfer function.

As global models are not spatialized and do not describe the physical environment, they require few data inputs (precipitation, ETP, static water level, flow-rate) and have the advantage of rapid implementation and short calculating times. Such tools should be used when timeframes are short, or the budget is restricted.

From a hydrogeological viewpoint, a global model is to be preferred over a spatialized groundwater flow model (see later) when studying highly heterogeneous (e.g. karst) aquifers, as such a model can show a hydrosystem as a whole, independent of its internal structure which may be complex and difficult to characterise.

However, as global models cannot show spatial differentiation, they are difficult to use in the following configurations: irregular distribution of pumping sites in a catchment; areas with high recharge; variable aquifer-river exchanges between upstream and downstream, etc. The method will thus be more suitable for single aquifers which are pumped in a spatially regular pattern.

Moreover, from a viewpoint of water-resource management, a global model cannot be used to identify an optimal spatial distribution of abstraction which could take into account the environmental constraints related to the impacts of extraction on surface aquatic environments, such as wetlands and streams.

Finally, if the hydrogeological synthesis shows a spatial heterogeneity of the hydrosystem to be modelled, it may be possible to use a semi-global model. This is based on assembling several interconnected global models representing sub-basins of similar characteristics, and allows the assigning of specific abstraction rates and suitable parameters to each sub-basin.

Some software packages offer the possibility of considering two underground reservoirs, thus distinguishing between two flow types, e.g. "slow" and "very slow". This option can be interesting for modelling aquifers with double porosity (fissure porosity and matrix porosity in carbonate aquifers, for example).

In all studies using a global model that were examined, the first step consisted in reconstituting the flow-rates and/or the natural water levels, i.e. the values that should be theoretically observed in the absence of pumped abstraction. This can be done in several ways (see Box, below).

The software packages used in the studied cases are Tempo (BRGM, 2011) for the "black box" type models, and GARDENIA (Thiéry, 2014; BRGM, 2013) and NAM[1] for the reservoir models.

---

[1] https://www.mikepoweredbydhi.com/products/mike-11

Box: Reconstituting Natural Flow-Rates or Static Water Levels

Strictly speaking, the natural flow-rates (or water levels) should be modelled with a model that was earlier calibrated over a period predating the development of water abstraction. However, lacking historical pre-1980s data, this is rarely possible. Among the work studied, this approach was used on only one occasion for the karstic Mosson aquifer in the Hérault department (BRGM, 2011).

In practice, natural flow rates (or water levels) must be modelled with known abstraction figures. Two options are then possible depending on the software used:

1. The software considers abstraction from the aquifer, in which case the precipitation/flow-rate calibration can be directly used for the observed flow rates. The natural flow rates are then reproduced by a simulation without abstraction. The abstractable (permissible) volume can then be calculated from the simulation of different abstraction scenarios that result in a minimum flow rate objective for four out of 5 years.
2. If the software does not consider abstraction from the aquifer, it will be necessary to improve the existing flow rate dataset with figures from streamflow rates. In that case, the abstractable volume can be calculated from the number of natural flow rates that exceed the minimum flow rate objective, similar to what is done for surface waters. The limitations of this approach, commonly used for unconfined aquifers, are discussed below.

11.2.4 Calculating Maximum Permissible Volume by Spatialized Modelling

The third methodological approach uses a spatialized, or distributed groundwater flow model (Anderson, Woessner, & Hunt, 2015; Bear & Cheng, 2010). Supported by the equations of subsurface-flow physics, spatialized models are the most complete modelling approach for showing a complex reality, offering the widest range of applications. Examples of such applications can be found in Chaps. 13 and 18 of this volume.

From a hydrogeological viewpoint, using a distributed model is mandatory where multi-layer aquifer systems are concerned. In such a setting, each aquifer layer can exchange water through vertical leakage with over- and underlying layers. However, though well-suited to a sedimentary porous environment, distributed models are generally unsuitable for strongly discontinuous environments, such as fissured or karstic aquifers.

Such models also allow showing the hydraulic exchanges between aquifers and streams by mobilizing different approaches. In the context of an unconfined aquifer connected to the surface drainage system, it is recommended to use explicit coupling of underground and surface flow, an increasingly common option in modelling software. In this configuration, the model calculates the flow exchanged between the stream and aquifer in both directions (drainage and infiltration) over time, and for each aquifer cell located below surface water. Software used in the work studied includes Modflow,[2] MARTHE,[3] Feflow[4] and Talisman.[5]

The construction and calibration of a distributed groundwater flow model requires a large quantity of data:

- Input data: Three-dimensional geological description of the aquifer geometry; structure of the river-drainage network; spatialized description of aquifer hydraulic parameters and stream sampling; rainfall and ETP datasets (or recharge data); and initial water level conditions.
- Calibration data: These correspond to data that must be adjusted or determined during model calibration: hydrogeological properties of modelled formations (permeability, storage and hydraulic boundary conditions) and of streams (thickness and permeability of streambed); parameters involved in hydro-climatic calculations (storage capacity of water in soil, distribution between runoff and infiltration; dephasing caused by the unsaturated zone).
- Observed data (potentiometric levels and gauged flows) that should be reproduced as well as possible during model calibration.

Compared to global modelling, creating a spatialized model takes much longer and requires a far higher budget.

Once calibrated, the model can be used to simulate the effects of various climatic and abstraction scenarios. Determination of the MPVs is done by trial-and-error *via* the simulation of different abstraction-reduction scenarios. The model is used for determining the maximum abstraction for safeguarding the earlier-set environmental objectives.

In addition to such planning simulations, the model can be used for exploratory simulations, showing what happens if the spatial distribution of abstractions is modified. This capability inherent in distributed models can be very useful when preparing for dialogue between stakeholders. The studies carried out in France commonly review the following alternatives:

- Modification of the timing of agricultural groundwater abstraction with a carry-over of part of the summer abstraction into the winter period; for instance in the case when substitution reservoirs are constructed to store this water (see Chap. 18).
- Modification of the spatial distribution of aquifer abstractions: for instance evaluating the impact of increasing the distance of certain wells from a stream (see

---

[2] https://water.usgs.gov/ogw/modflow/

[3] http://www.brgm.fr/production-scientifique/logiciels-scientifiques/marthe-logiciel-modelisation-ecoulements

[4] https://www.mikepoweredbydhi.com/products/feflow

[5] https://who.rocq.inria.fr/Martin.Vohralik/Files/Doc_Talisman.pdf

Chap. 5 on the Beauce region), or of changing the upstream/downstream position in the catchment basin.
- Modification of the distribution of abstraction between different resources: changing from a shallow to a deep aquifer, changing from a stream to an aquifer, etc.

11.3 Main Difficulties Encountered and Limits of the Studies

Most of the studies examined for this chapter suffered from the widespread difficulties of obtaining sufficient data to derive reliable estimates of MPVs. The two main data gaps encountered were abstraction volumes and monitoring data, especially for water levels and streamflow.

11.3.1 Lack of Abstraction Data

With regard to groundwater abstractions, many studies mention the difficulty of accurately quantifying abstractions for agriculture and also those from domestic wells (Rinaudo, Montginoul, & Desprats, 2015). Most studies ignored domestic abstraction because of the lack of data. In the case of agricultural use, some studies rely on authorized volumes data rather than actual abstraction data which is largely unknown. Although over 90% of agricultural wells are equipped meters, they are not systematically monitored by Government agencies (see Chap. 25). In addition, the seasonal distribution of such abstractions is rarely known. However, the studies for determining MPVs require data for at least monthly intervals which must be extrapolated using simplifying assumptions that inevitably create uncertainty.

Similarly, there is little reliable data on the discharge of treated wastewater from water treatment plants into surface or groundwater. Recharge to groundwater due to infiltration from gravity irrigation systems is also poorly estimated. All these uncertainties have a direct impact on the accuracy of the estimated MPVs.

11.3.2 Insufficient Resource Monitoring Data

The construction of observation wells for monitoring the water levels in unconfined aquifers is relatively recent in France, occurring during the 1990s and 2000s. Consequently, long hydrogeological time-series data are rare, which obviously hinders the ability to determine MPVs, regardless of whether or not modelling is used in the process. Long-term datasets are indispensable for the development and calibration of models, as they should cover a variety of climatic conditions. This is a particularly critical point when studying minimum river flow conditions, or robust aquifers with large storage that have delayed responses to climatic influences. This prerequisite was not always fulfilled in the studies that were examined.

11.3.3 Differences in Definition of Flow-Rate/Discharge Objectives

In circumstances where an unconfined aquifer is connected to a stream, the MPV determination often directly depends on the minimum river flow objective defined by the stakeholders before commencement of the technical study. Definition of this minimum flow target thus is essential: the higher it is, the lower the abstraction volume will be for the unconfined aquifer connected with the stream.

In the studies examined, different discharge values are used, mostly based on the Biological Discharge[6] and Minimum-Flow Discharge[7] (DOE in French) as found in planning documents such as SDAGE (see Chaps. 4 and 5), but also based on analyses of statistically observed discharge rates (QMNA5,[8] VCN30,[9] etc.). The use of different definitions of target river flow-rates leads to the question of whether or not the results from different catchments can be compared.

Moreover, there is significant uncertainty concerning physical flow-rate measurements in streams, especially for low-water periods when the uncertainty can be more than 20% (Rhône-Mediterranean-Corsica [RMC] Water Agency, 2011). Once more, this uncertainty directly affects the planned flow-rate and the calculated MPV.

Strong uncertainties also affect the evaluation of biological flow-rates that must be respected in streams, and which are imposed by the regulations (AERMC et al., 2013). In the RMC Basin, it is recommended to propose a range of values for such biological flow-rates and their derived MPV values. It should be noted that, even though the recommendation is commonly followed for biological flow-rates, this is rarely the case for the MPVs.

Finally, the stations where biological flow-rates are evaluated are not necessarily the same as the gauging stations on which the modelling is based. In that case, the downstream or upstream biological flow-rate value must be extrapolated, introducing further uncertainty into the hydrological conditions.

[6] Minimum discharge into a stream for safeguarding the life, movements and reproduction of the species living in it. Its estimation is commonly based on using a modelling tool of the habitats of the various fish species.

[7] Value of the minimum discharge at a (nodal) point, above which it is considered that all upstream uses (activities, abstraction, discharge) are in equilibrium with the proper functioning of the aquatic environment. This structural objective is laid down in the SDAGE, SAGE and equivalent documents (www.eaufrance.fr).

[8] Minimum monthly discharge over a 5-year return period.

[9] Average minimum annual discharge calculated over 30 consecutive years.

11.3.4 Limitations Associated with the Global Models

Three problems were identified concerning the application of global models. Firstly as seen before, a global model must respect certain application conditions because of its globalizing and non-spatialized character. The problem is that some studies applied global models to heterogeneous aquifer systems (multi-layered aquifers), or to aquifers with an irregular spatial distribution of abstraction points. In both cases, the modelling results may be erroneous, but this limitation was not discussed in the studies.

The second problem is related to considering the aquifer abstraction at the scale of a catchment area. Because of its global character, the model cannot distinguish between abstraction near a stream—which will have an immediate impact on discharge—and those far from streams that will affect discharge only after several days, or even weeks or months. Numerous global modelling studies thus consider the abstraction from an unconfined aquifer as direct abstraction from a stream, without attenuation or any lag time. If low water flow occurs in a stream after an irrigation period, this assumption may lead to an optimistic evaluation of the MPV to the extent that it ignores the delayed impact of the wells farther away from the stream.

The third problem encountered in the global modelling work is related to the extrapolation of rainfall/flow rate relationships from one basin to a neighbouring basin. This practice occurs when the data for any basin is insufficient for developing or calibrating a model. In view of the probable differences in abstraction and hydrological functioning between two basins, such extrapolation appears to be particularly uncertain for reconstituting a "natural" flow-rate dataset. However although this option should obviously be used with caution, it could be envisaged if at a minimum, low water flow gauging data is available.

11.3.5 Limitations Associated with the Spatialized Models

Analysis of studies based on the use of spatialized models revealed several specific problems. The first is related to the re-use of existing models that were originally developed for a purpose different from calculating an MPV. Here, the initial objectives determined the model type, in particular the extent of the modelled domain. Whether the model is "fit for purpose" for determining an MPV must therefore be verified, and if necessary, the model should be modified and recalibrated.

The second problem is related to the definition of the conceptual flow model that underlies the construction of a distributed model. The studies analysed commonly mention the uncertainties related to the assumptions concerning the geometry of the different aquifer layers to be modelled, the hydraulic conditions at the model boundaries or the type of groundwater-surface water interactions. Imposed flow-rates are quite commonly applied to the model limits without validation being possible. The uncertainty related to such flow can be quite high. Even though such methods may be able to reproduce the potentiometric values of the aquifer, they are not without risk when running the model.

The third problem concerns the often imperfect understanding of the spatial distribution of hydraulic parameters. The adjustment of calibration parameters is effectively equivocal: several combinations of parameters may apparently satisfy the re-transcription of aquifer levels and streamflow. Where aquifers are connected to streams, it is important to test the sensitivity of the results to variations in hydraulic parameters such as streambed permeability (due to clogging) or the aquifer characteristics in a valley bottom (permeability and storage). Such sensitivity analyses are almost never carried out, even though they allow testing the representativeness of the model and understanding the uncertainties associated with the calculated MPVs.

11.4 From Technical Evaluation to Decision Making: The Example of the Adour Garonne Basin

For about half of the studies that were examined, the calculated MPVs turned out to be lower than the volumes actually abstracted (Table 11.1).

When the imposed reduction in extraction is large, the users may challenge the scientifically based MPV, using both technical arguments that show the limitations of the study methodology and economic arguments stressing the impact of recommended reductions. The technical objections commonly refer to the poor quality of the basic data and thus the uncertainties associated to the results.

In same cases, the resulting negotiations produced a higher MPV figure than that initially calculated. The objective here is to illustrate this phase of negotiation using the example of the Adour Garonne Basin.

11.4.1 Economic Consequences of Reducing the MPV

In the Adour-Garonne Basin, a strict application of the initially calculated MPVs would have required an average 10% reduction in the authorized volume at basin scale. Reductions could be as much as 50% in specific aquifers and negligible in others (Hébert et al., 2012). After a very strong backlash against the MPVs, the Adour Garonne Water Agency commissioned a study to quantify the economic impact of MPVs on the agricultural sector (Hébert et al., 2012). This study covered six sub-basins which were considered representative of the diversity of the agricultural economy in the Basin. All six required reductions in abstraction from 28% to 90% compared to the existing authorizations for agricultural use (Table 11.1).

The study was based on a micro-economic modelling of farms, carried out with the stakeholders who systematically validated the choices made for the economic modelling. The results showed that a reduction of MPVs would lead to a 9–34% loss

Table 11.1 Examples of study results on determining MPVs (Arnaud, 2016)

Water Agency district	River catchment	Calculated recommendations for reducing abstraction
Adour Garonne	Seudre	90% reduction in agricultural abstraction
Artois Picardie	Somme catchment	20% reduction in abstraction (reference year 2005)
Loire Bretagne	South Vendée	Reduction of summer abstraction between 20% and 50%
	Around the Poitevin marsh	50% reduction in agricultural abstraction: 70–80% in spring and 30–50% in summer
	Upstream Cher SAGE	6% reduction in agricultural abstraction compared to the maximum 1996–2008 annual abstraction for one sub-basin and 32% for another
Rhône Méditerranée Corse (RMC)	East Lyons region	Reduction of 2.2 Mm^3/year after discussions on Meyzieu management area
	Lower Ain valley	Period 2003–2007: reduction of summer abstraction (June to August) of between 30% (2004/2005) and 50% (2003)
	Alluvial aquifer of the Garon	Period 2002–2009: reduction of abstraction between 6% and 43% for the prudential scenario
	Drôme hills	Overall reduction between 20% and 45%
	Galaure	Overall reduction of 40%
	Roussillon toll	Reduction between 30% and 87% in terms of connecting habitat to aquifer
	Véore-Barberolle	40% reduction of present abstraction during low-water periods
	Lez Basin	July: reduction of 17% on the Lez and 40% on the Hérin; August and October: Free of actual abstraction; September: 40% reduction on the Lez and 30% on the Hérin
	Eygues Basin	40% reduction in abstraction on the entire basin from July to September
Seine- Normandy	Caen plain	Reduction of 2010 abstraction between 8% and 82% in different management units

in the economic value of agricultural production[10] for farms for the various sub-basins. The most strongly affected farms would be grain and cattle farmers, whose land commonly is characterized by soils with low soil moisture reserves. Under such conditions, the viability of some farms might even be jeopardized. These results assumed average climatic conditions and average agricultural prices and also showed that, if the water-resource allocation was optimized via a re-allocation of available water volumes to crops with a higher added value, this economic impact would be almost nil. Such re-allocation could take place by proposing compensation,

[10]The indicator used in the study was the gross operating surplus (GOS). This allows estimating the profitability of the farming system by neutralizing the effects related to differences in patrimony or investment strategy between the farmers.

or by establishing a mechanism inspired by a water "market", even though this type of mechanism at present is theoretically impossible in France (see Chap. 3).

11.4.2 Opposition from the Farming Profession and First Political Concessions by the State

Since 2008, the farming profession has been strongly opposed to the principle of calculating MPVs based on hydro-meteorological data corresponding to a dry year occurring every 5 years. This theoretically ensures that in 4 years out of five, the environmental objectives will be met and the volume allocated to farmers will be available without constraints. However, it also implies that any surplus of water available during these 4 years will be granted to the environment, which for farmers, represents an unacceptable loss of income.

In order to show their opposition, the farmers' union representatives decided to boycott the consultation meetings from June 2010 (CGEDD & CGAAER, 2015).

Following this first protest to the reform of MPVs, the Ministry for Ecology softened its position (Circular of 3 August 2010, Ministry for Ecology, Energy, Sustainable Development and the Sea, 2010). Without calling into question the definition of the MPVs, the Circular planned for financial aid to farmers to assist them to minimise the impact of reducing the MPVs. Such measures are applicable in basins where the difference between present water use and the MPV is over 30%. The first measure proposed prolonging the delay in applying the reduction in abstraction by 2 years to the end of 2017. The second measure proposed the reduction should take place in a progressive manner up to 2017. The third concession by the State was to accept the principle of a yearly revision of the MPV, considering the volume effectively available at the start of each year, and the actual climatic and hydrologic conditions, in order to avoid an over-restrictive limit in times of abundant resources. In addition, the State proposed more substantial financial assistance through the Water Agencies for the creation of private storage reservoirs (subsidies of up to 70% compared to the initial 50%), which enable the storage of available winter precipitation to compensate for the reduction in authorized abstraction in summer. Finally, the State showed further flexibility by attributing an additional volume of up to 20% of the scientifically calculated MPV to take into account the uncertainties associated with evaluating the MPVs.

11.4.3 The Conflict Reached the Presidency

The concessions made by the State in its Circular of 2010 were still considered to be insufficient by the agricultural profession, which then elevated the debate to the national level in November 2010. This resulted in additional adaptive measures, now arbitrated by the French presidency, for implementation in the Basin. Two

Table 11.2 Maximum permissible volumes notified for the Seudre catchment (Charente-Maritime department)

Catchment unit	Authorized volume (Mm^3)	Notified MPV 2009 (Mm^3)	Refined MPV (Mm^3)	Final MPV (Mm^3)
Upstream Seudre	2.6	0	1.74	1.74
Middle Seudre	6.5	0	0.5	0.6
Downstream Seudre	2.5	2.2	0.5	0.6
Total	11.6	2.2	2.74	2.94

memoranda of understanding were signed between the State and regional Chambers of Agriculture, the first in June 2011 for the Poitou-Charentes region and a second in November 2011 for the Midi-Pyrénées and Aquitaine regions.

The two memoranda have different contents and conditions of application, but both contain the main concessions that were previously negotiated. The application of the MPVs was delayed by a further 4 years to 2021, instead of 2015 initially. The State finally authorized additional abstraction during spring, depending on the state of the water resource.

Furthermore through these memoranda, the State committed to improving the rigour of the MPV studies (by verifying the relevance of the minimum river flow rates targets), as well as implementing compensatory measures (in particular financial assistance). For their part, the Chambers of Agriculture committed to ensuring the sharing of the available abstractable volumes between agricultural users, as part of the creation of Water Users' Associations (Organisme Unique de Gestion Collective) (see Chap. 3).

Following several years of negotiation, the MPVs for irrigation were notified by the regional Prefects, distinguishing the initial MPVs (derived from scientific studies), the refined initial MPVs (from local consultations before 2011), and the final MPVs which integrate the corrections and flexibility allowed by the memoranda of understanding (Préfet de la Région Midi-Pyrénées, 2011). Table 11.2 provides a quantitative illustration of the negotiation process carried out for the Seudre catchment. Here, a 34% increase was granted compared to the initially planned MPV for agricultural use. To help reduce the abstraction to the MPV, this catchment was the subject of a territorial development project (see Chap. 24).

The negotiation phase is, however, not yet complete at the time of writing for the Adour-Garonne Basin, as the November 2011 memorandum of understanding for the Midi-Pyrénées and Aquitaine regions was appealed in the administrative court by environmental protection associations. The judgement was to be rendered in 2018.

11.5 Conclusion

In France, quantitative groundwater management policy is mainly driven by environmental objectives. Groundwater abstraction limits must be set to ensure that water use will not lead to any deterioration of aquifers, groundwater dependent streams and rivers and other aquatic ecosystems. This policy requires the calculation of an abstraction limit, expressed as a volume that if respected, guarantees a good state of aquatic environments for 4 years out of five. The calculation of this volume requires an in-depth understanding of the hydrogeology of the aquifers being exploited.

For unconfined aquifers, the experience gained from over 30 studies shows that the estimation of MPVs is complicated by numerous uncertainties. The first prerequisite is a good knowledge of the dynamics of the hydrosystems and abstraction volumes, but unfortunately this is rarely achieved. Moreover, both the calculation methods and modelling tools that aim to conceptualize these complex systems have limitations due to the simplifying assumptions required for their application.

In many cases, the calculated maximum permissible volumes were much lower than the previously authorized volumes. Therefore, many of the results were contested by affected users. Such disputes concerned not only the economic consequences of reduced abstraction, but also the scientific basis of the studies in view of the known uncertainties and limitations.

A negotiation phase at both national and local levels was thus started between stakeholders, experts and government services. Several compromise measures were agreed upon by the State for catchments with large over-allocations, including authorizing an increase in the initially planned abstractable volume. The negotiation process to resolve local disputes is still underway in some areas.

References

Agence de l'Eau RMC. (2011). – Bureau du comité de bassin Rhône Méditerranée. Séance du 10 juin 2011. Point IV: méthodes et études des volumes prélevables.

Agence de l'Eau RMC, DIRENs of the RMC Basin, ONEMA. (2009). – Études de détermination des volumes maximum prélevables. "Standard" Terms of Reference.

Agence de l'Eau RMC, IRSTEA, ONEMA. (2013). – Note du secrétariat technique du SDAGE. Mieux gérer les prélèvements d'eau. L'évaluation préalable des débits biologiques dans les cours d'eau.

Anderson, M. P., Woessner, W., & Hunt, R. (2015). *Applied groundwater modeling* (2nd ed.). New York: Academic Press-Elsevier.

Arnaud, L. (2016) – Estimation des volumes prélevables dans les aquifères à nappe libre: retour d'expériences sur les méthodes utilisées, identification des problems rencontrés, recommandations. Rapport final, BRGM/RP-64615-FR. 107 p., 42 fig., 1 app.

Bear, J., & Cheng, A. H. D. (2010). *Modeling groundwater flow and contaminant transport*. Dordrecht, The Netherlands: Springer.

BRGM. (2011). – Caractérisation du comportement d'un indicateur piézométrique et définition des volumes prélevables sur les compartiments carbonatés Nord et Sud de l'entité MOSSON de la masse d'eau FR_DO_124. Report BRGM/RP-59658-FR.

BRGM. (2013). – Connaissances des ressources réellement disponibles sur l'ensemble des bassins versants crayeux de Champagne-Ardenne. Report BRGM/RP-61371-FR.

Calligée Consultants. (2008). Expertise et objectifs du projet de SDAGE de la ressource en eau sur la bordure Nord du Marais Poitevin vendéen, Bassin du Lay, de la Vendée et des Autizes, et propositions de principes de gestion. Phase 3: étude d'un cadre de gestion des nappes du Sud-Vendée compatible avec les enjeux écologiques et économiques. Report N08-85138C.

CGEDD, CGAAER. (2015). – Evaluation de la mise en œuvre des protocoles Etat-profession agricole conclus en 2011 dans le bassin Adour-Garonne pour la gestion quantitative de l'eau.

Grontmij, Rivages Environnement. (2014). – Identification et préservation des ressources majeures en eau souterraine pour l'AEP. Lot n°4: Etude des alluvions du Gapeau, et des alluvions et formations du Muschelkalk de la plaine de l'Eygoutier. Rapport de phase 2. Etape 3 – Complément à l'étude des volumes maximums prélevables des alluvions du Gapeau.

Healy, R. W. (2010). *Estimating groundwater recharge*. Cambridge, UK: Cambridge University Press.

Hébert, N., Grandmougin, B., Loubier, S., Graveline, N., Marsac, S., Amen, J.F., & Brunel, L. (2012). – Réforme des autorisations de prélèvement dans le bassin Adour-Garonne: impacts sur l'économie agricole. Agronomie Environnement et Sociétés, Vol. 2, n°2.

Ministry for Ecology, Energy, Sustainable Development and Territorial Development. (2008). – Circular (in French) of 30 June 2008 on the reduction of quantitative disequilibria during water sampling and collective management of irrigation sampling.

Ministry for Ecology, Energy, Sustainable Development and the Sea. (2010). – Circular (in French) of 3 August 2010 on the reduction of quantitative disequilibria during water sampling and collective management of irrigation sampling in basins where the difference between abstracted volume during a five-yearly dry year and the volume that can be abstracted is over 30%.

Préfet de la Région Midi-Pyrénées. (2011). – Notification des volumes prélevables sur la région Poitou-Charentes.

Rinaudo, J. D., Montginoul, M., & Desprats, J. F. (2015). The development of private borewells as independent water supplies: Challenges for water utilities in France and Australia. In Q. Grafton et al. (Eds.), *Understanding and managing urban water in transition* (pp. 155–174). Dordrecht: Springer.

Thiéry, D. (2014). GARDÉNIA software, version 8.2. User Guide. BRGM/RP-62797-FR report, 126 p., 65 fig., 2 app.

Luc Arnaud is Senior Hydrogeologist at Brgm (French Geological Survey). He has 17 years-experience in the characterization and management of different aquifer systems: sedimentary basins, crystalline and volcanic rocks. He successively held positions in Madagascar, in Normandy, in the French West Indies before joining in 2014 the scientific and technical center of BRGM in Orleans. He was the project leader of hydrogeological prospections for drinking water supply in the French West Indies (volcanic islands): delineation of favourable zones, field, geophysics and drilling supervision (2010–2014). He is presently involved in hydrogeological modelling for unconfined aquifers connected to rivers and for deep groundwater resource management of the two main French sedimentary basins (Parisian basin, Aquitaine basin) (2014–2016). He recently led a project aiming to consider the studies relating to the determination of sustainable yields in French unconfined aquifers (2015/2016).

Chapter 12
Setting Sustainable Abstraction Limits in Confined Aquifers: Example from Deep Confined Aquifers in the Bordeaux Region, France

Frédéric Lapuyade, Marc Saltel, and Bruno de Grissac

Abstract This chapter describes the management of the deep aquifers in the Gironde Department of south-western France, which supply drinking water to the City of Bordeaux and almost all the1.5 million inhabitants of the Department from about 400 wells. These deep aquifers are a strategic resource for the Gironde area because of their accessibility and excellent water quality. Already in the 1950s, the risk of overexploitation of these resources was recognised, in particular for the Eocene aquifer whose groundwater levels showed a clear decline. The resulting awareness of this risk led to the implementation of specific regulations, before implementation of management policies as set down in the Law on Water of 1992. Major investigations were carried out to improve knowledge of the aquifers, monitor the groundwater levels, and develop ground-water flow models. The local stakeholders involved in aquifer management employed these modelling tools to create the principles and policies for controlling groundwater-abstraction. The current water management regime in the Gironde Department is the result of a long scientific and technical policy evolution, which has led to an operating process that supports consultation and regulation within the legal framework of a Water Development and Management Plan (SAGE in French).

Keywords Confined aquifer · Consultation · Groundwater model · Management · Regulation

---

F. Lapuyade · B. de Grissac
SMEGREG, Bordeaux, France
e-mail: bruno.degrissac@smegreg.org

M. Saltel (✉)
BRGM (French Heological Survey), Bordeaux, France
e-mail: m.saltel@brgm.fr

J.-D. Rinaudo et al. (eds.), *Sustainable Groundwater Management*,
Global Issues in Water Policy 24, https://doi.org/10.1007/978-3-030-32766-8_12

12.1 Introduction

This chapter describes the management of the deep aquifers in the Gironde Department of south-western France, which supply drinking water to the City of Bordeaux and almost all the1.5 million inhabitants of the Department from about 400 wells. These deep aquifers are a strategic resource for the Gironde area because of their accessibility and excellent water quality. Already in the 1950s, the risk of overexploitation of these resources was recognised, in particular for the Eocene aquifer whose groundwater levels showed a clear decline. The resulting awareness of this risk led to the implementation of specific regulations, before implementation of management policies as set down in the Law on Water of 1992. Major investigations were carried out to improve knowledge of the aquifers, monitor the groundwater levels, and develop groundwater flow models. The local stakeholders involved in aquifer management employed these modelling tools to create the principles and policies for controlling groundwater-abstraction.

In addition to retracing the history of managing the deep aquifers of Gironde, this chapter describes the concepts of overpumping confined aquifers and of maximum permissible volumes.

12.2 Groundwater Resources: Usage and Management Stakes

12.2.1 Hydrogeological Context

The Gironde Department is located in the north of the Aquitaine Basin, the second largest sedimentary basin of France (Fig. 12.1) which contains alternating permeable and impermeable layers that were deposited from the Jurassic (200 million years ago) to the Pliocene-Quaternary periods (±10,000 years ago). Because of this configuration, the basin contains abundant groundwater resources of great quality within a multi-layered aquifer system with inter-aquifer leakage occurring over very long time frames. The average residence time is as high as 35,000 years (Saltel, Lavielle, Thomas, Rebeix, & Franceschi, 2016). The Northern Aquitaine Basin is largely open to the Atlantic Ocean, but is limited to the east by the foothills of the Massif Central and by two major structural features, the Jonzac anticline in the north and the Villagrains-Landiras one in the south (Fig. 12.2).

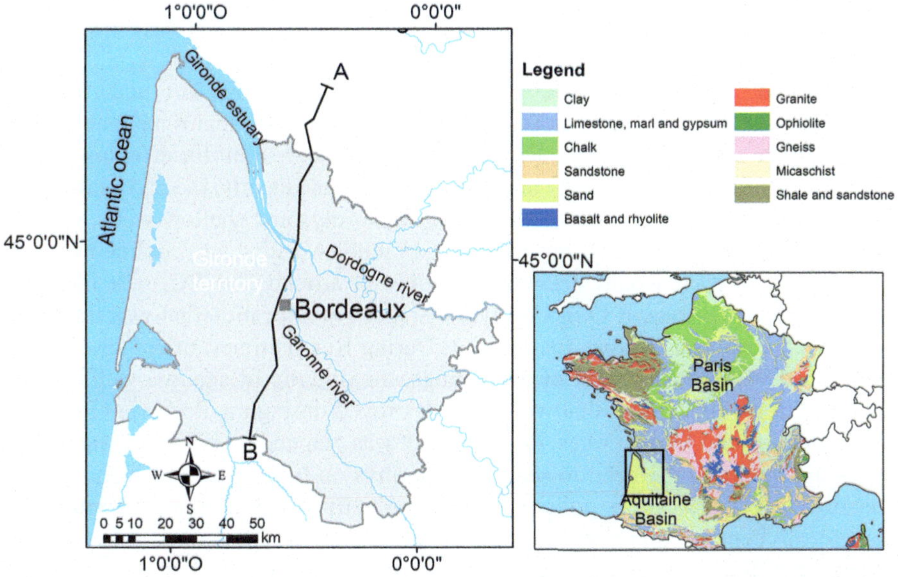

Fig. 12.1 Location of the Gironde Department and of trace A-B of the cross-section on Fig. 12.2

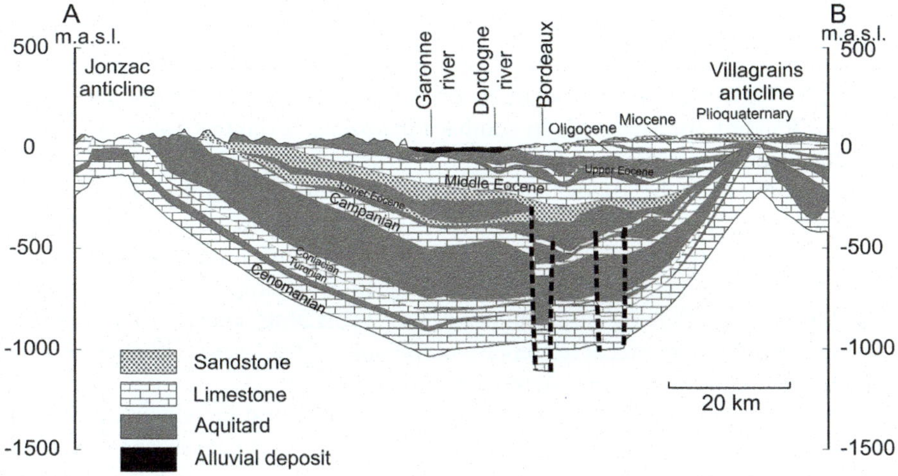

Fig. 12.2 Geological cross-section through the Aquitaine Basin in the Gironde department. (Trace A-B on Fig. 12.1)

12.2.2 Groundwater Uses

The Gironde Department has a surface area of nearly 10,000 km^2 and supports a population of about 1.5 million people. It is drained by two major navigable rivers, the Garonne and the Dordogne Rivers, whose waters are unfit for drinking because of marine influence (tidal bores and mud-flats). Consequently, the early inhabitants had to rely totally on groundwater, first using springs and shallow wells, and then from the nineteenth century onward, using boreholes drilled into confined aquifers. The siting of what would become Bordeaux at around 300 BC, was intimately linked to the presence of Oligocene karst springs in a location that was favourable for river transport as well as for defence. During Roman times, urban development continued with development of the springs, the digging of shallow wells and the construction of aqueducts that were later destroyed during the Barbarian Invasions.

Starting in the Middle Ages, urban growth was plagued by recurrent water shortages that imposed the need to search for springs up to over 40 km away, as those nearby were increasingly polluted by human activities and urban growth. In the nineteenth century, major works were undertaken to create the first modern water supply for Bordeaux, by tapping various springs in the surrounding area and delivering water to the centre of town via aqueducts (Le Taillan's aqueduct built in 1850 and Budos's dating from 1880). However the risk of shortages was not alleviated, even though all springs in a radius of 30 km from Bordeaux were now tapped. In 1830, an attempt was made to drill a bore in the centre of Bordeaux, but the relatively high elevation of the hole and its small diameter resulted in failure. Even though by the end of the nineteenth century, the Eocene aquifer already supplied over 100,000 m^3/day from about thirty artesian boreholes for industrial purposes as well as flooding the vineyards to combat *Phylloxera*,[1] drinking water still was not supplied from this resource.

It was not until the 1940s that the exploitation of deep aquifers for producing drinking water commenced. These aquifers contain excellent quality groundwater and are effectively accessible everywhere in the Department. After WW2, the number of boreholes increased very rapidly (Fig. 12.3) to support rural water supply pipeline networks, having the advantage of low access costs, not requiring significant transportation and treatment (except for eliminating excess iron), and negligible risk of contamination from human activities.

Today, this resource is used for supplying the 7th largest French urban conurbation in population numbers, as well as an extensive rural landscape—including some of the largest French municipalities—with little network interconnection because of easy access to the deep groundwater resources.

[1] A parasite that ravaged French vineyards, whose proliferation was controlled by submersion of the vines.

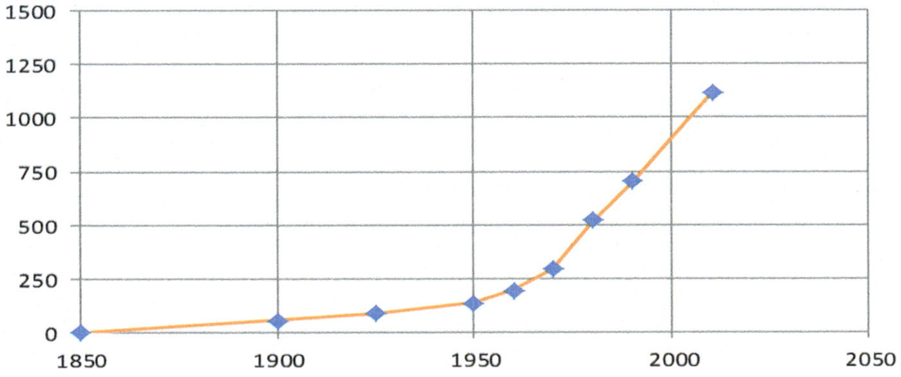

Fig. 12.3 Number of boreholes in the deep aquifers pumped in the Gironde Department

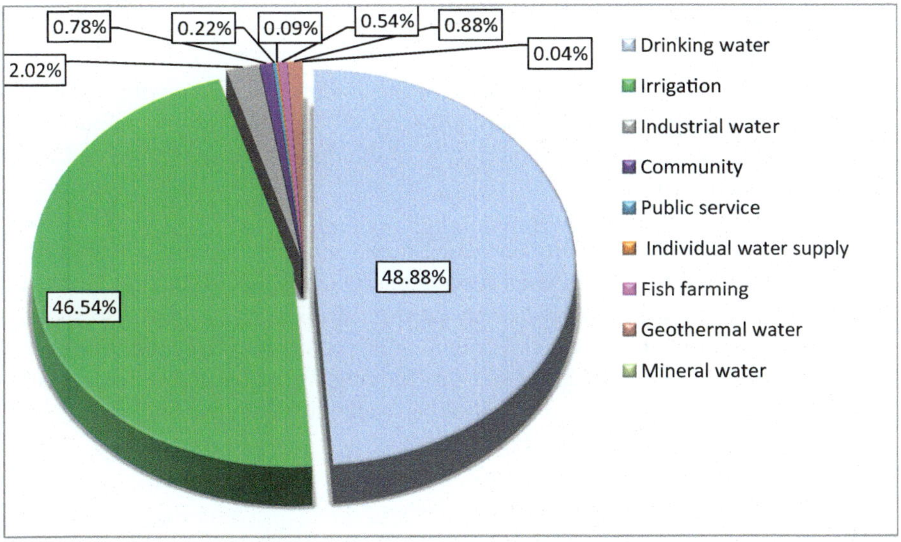

Fig. 12.4 Distribution of abstraction compared to water use. (From Douez et al., 2017)

12.2.3 Which Resources Are Used Today and for What Purpose?

In 2015, 246 million m^3 were abstracted from aquifers in the Gironde Department (Douez, Abou, & Bourgine, 2017), including 99 million m^3 from the phreatic Pliocene-Quaternary aquifer and 147 million m^3 from confined aquifers. In terms of use, drinking-water supply was the most important (~49%) followed by agriculture (46.5%). Other uses are minor: industry is the third largest user but represents only 2% of groundwater abstraction (Fig. 12.4).

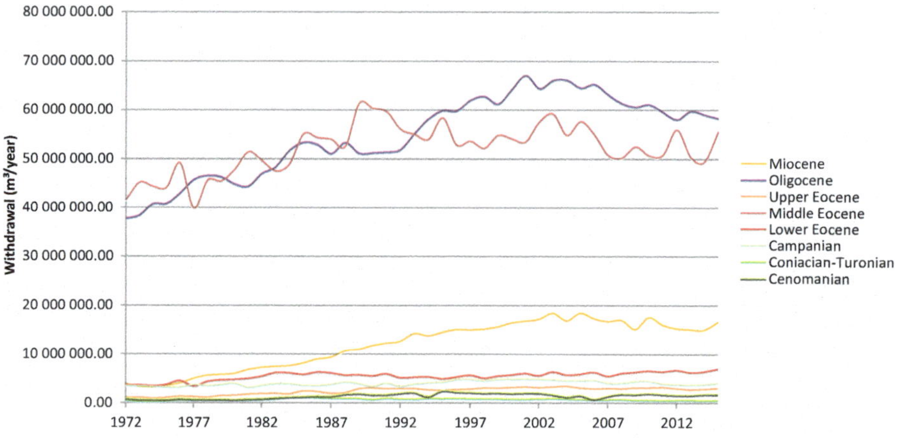

Fig. 12.5 Evolution of abstractions from aquifers in the Gironde department, 1972–2015. (Modified from Saltel et al., 2017)

Drinking water is by far the main use of the confined aquifers,.with extractions of about 120 million m^3/year accounting for over 80% of the volume pumped from these aquifers. This supply comprises 97% of the drinking water consumed in the Department from all sources.

The Middle Eocene and Oligocene aquifers contain the main drinking-water resources of the Gironde Department. The Eocene aquifer is also used for industrial purposes, though the volumes used have declined since the 1970s (Corbier et al., 2012). The Oligocene aquifer is heavily used for drinking water, but also supplies water to certain agricultural sectors, such as wine production. Only the (shallowest) Miocene aquifer is not primarily used for drinking water, but rather for irrigation supplies. Overall, aquifers used for drinking water supply have been increasingly exploited since the early 1970's (Fig. 12.5).

12.2.4 The Management Issues for the Confined Aquifers of the Gironde

To the extent that confined aquifers supply 97% of drinking water in the Department, the main priority for their management is to perpetuate this low cost and secure supply which has a high quality. These resources being naturally protected from human activities, their actual quality allows distribution without prior treatment, except for eliminating excess iron.

Schoeller (1956) observed a decline in water levels coinciding with a major increase in the number of boreholes and raised questions about the long term sustainability of the resource, and the risk of overexploitation of these aquifers.

12.3 From Understanding Aquifers to Groundwater Dynamics

12.3.1 Knowledge of Groundwater System

Following the questions raised on the risks of overexploitation in the mid 1950s, the first hydogeological studies were carried out by Bellegarde, Bourgeois, Camus, Camus, and Schoeller (1964). This then led to a dynamic that federated the Faculty of Sciences of Bordeaux, BRGM (the French geological survey), and State services, before integrating local and regional authorities, and the Adour-Garonne Water Agency. This work led to defining a first geometry of the various aquifer units and construction of the first piezometric maps, which providing an understanding of the groundwater flow systems and exchanges between aquifers of this extensive system (Astié, Bellegarde, & Bourgeois, 1967). The ever-improving knowledge of these aquifers was in part made possible by the increasing number of boreholes being drilled with each contributing new data on the local geology, and the purpose of these wells (to produce water, or for oil-and-gas exploration (Seronie-Vivien, 2001).

12.3.2 Piezometric Level Monitoring

As soon as the first inventory of water resources in Aquitaine Basin was completed in late 1958, regular monitoring of the hydraulic heads in these aquifers was required. Initially, a piezometric network was designed for the Eocene aquifer which resulted water levels being measured manually in 217 wells (Bellegarde, 1969). However, the need for a network for on-going monitoring was not recognised until later (Bellegarde, 1975), starting with seven wells monitoring the Eocene aquifer and another six monitoring the Oligocene aquifer (Astié, 1978). This piezometric network was a national first, and it was not until the Law on Water of 1992 that the general need was confirmed to dispose over piezometric-monitoring networks. In 2015, the network covered eight aquifers, comprising 138 continuous-monitoring points and 135 points for regular manual measurements (Douez et al., 2017). This network allows a real-time evaluation of local trends (Fig. 12.6), as well as the construction of large-scale regional maps (Fig. 12.7).

The piezometric monitoring has shown that the hydraulic heads in some areas of the Middle Eocene aquifer are showing a constant decline (Fig. 12.6). Even though this decline was recognised in some areas because of the drying up of artesian wells, the monitoring has shown that this decline was occurring regionally and was clearly linked to the increasing abstraction.

The piezometric maps drawn from these measurements have shown the existence of an extensive cone of depression, whose size and depth have increased over recent years (Fig. 12.7).

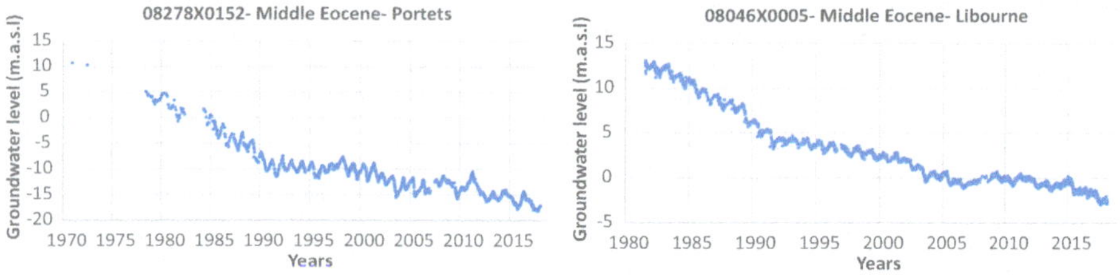

Fig. 12.6 Observed lowering of the hydraulic head in the Eocene aquifer

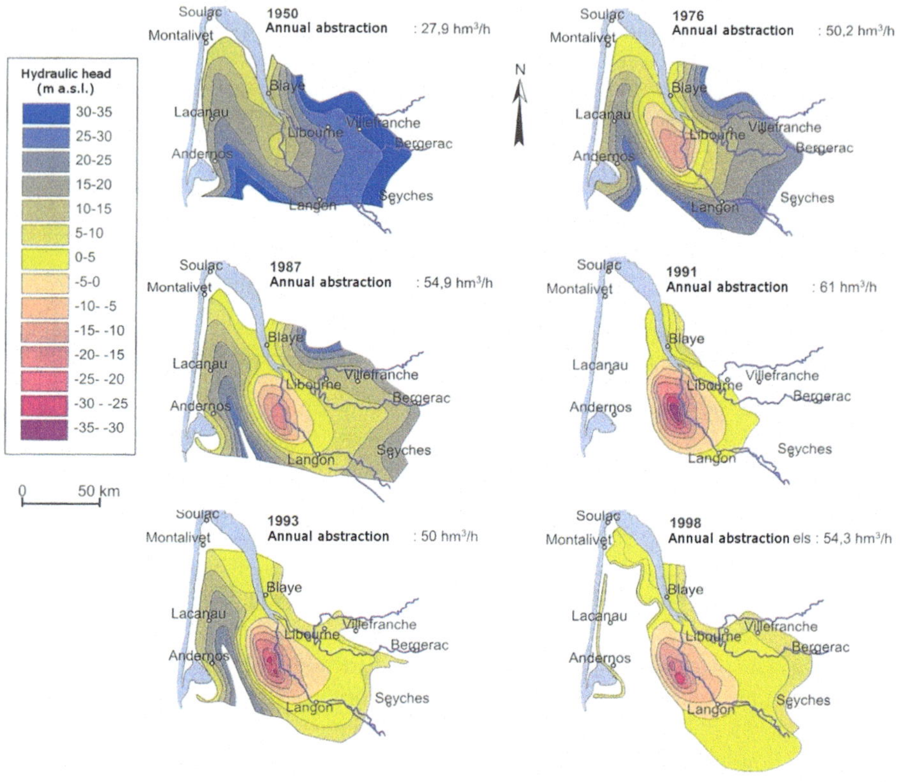

Fig. 12.7 Piezometric evolution of the Eocene aquifer from 1950 to 1998. (Mauroux, Sourisseau, & Bonnery, 1999)

As a consequence of this continuous decline in hydraulic heads in the Eocene aquifer, saline intrusion from the Gironde River estuary is now considered to be a major risk for the sustainability of the groundwater resources due to a possible reversal of groundwater flow. Instead of the aquifer discharging to the estuary, the lowering of the watertable could result in the aquifer being recharged from the estuary (Fig. 12.8).

Fig. 12.8 Overall condition and local risks – the SAGE concepts for evaluating the good state of an aquifer. (SMEGREG, 2013)

Evidence of intrusion adjacent to the estuary was noticed as early as 1964 (Bellegarde et al., 1964), especially in the North Medoc region with the chloride content in groundwater rising above 250 mg/L at St-Vivien-de-Médoc, which prompted several studies and the installation of a specific observation network.

Even though the risk of intrusion has been the driving force for management action for decades, it is now considered with less apprehension. In fact, the chloride contents found in observation wells show that the Eocene rocks are contain natural salt. The almost complete absence of tritium in these saline waters, and higher salt

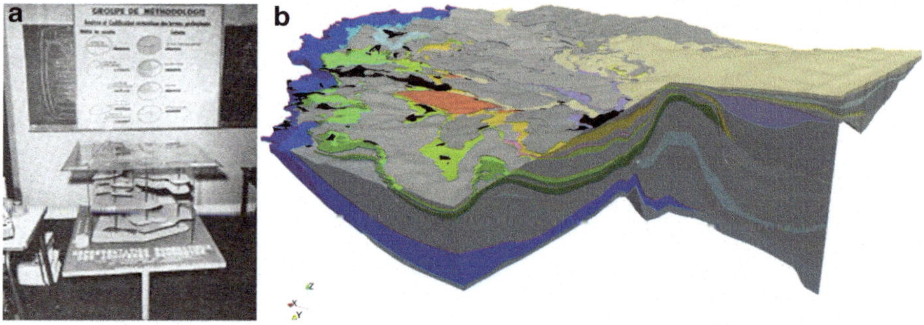

Fig. 12.9 Examples of how models visualize water-management problems: (**a**) Electric analogue model, 1960s; (**b**) Digital MONA model (MOdèle Nord Aquitaine)

concentrations than those in estuary waters, indicate that such salt may be contained in fossil waters within the sub-Flandrian terraces (Platel et al., 1999). Another sustainability risk recently identified is the dewatering of the Oligocene aquifer south of Bordeaux. Dewatering occurs when the lowering of the water level within an aquifer causes it to become unsaturated. This notion applies in particular to an initially confined aquifer, which has become unconfined through the lowering of the pressure level below the top of the aquifer. Dewatering of an aquifer may have adverse impacts on the physico-chemical, microbiological and hydraulic properties of the resource (Fig. 12.8).

Such local risks must be carefully considered. However at a larger scale, a confined aquifer should not be considered over-exploited based only on the observation of declining hydraulic heads in wells, especially if the aquifer is thick, well confined and covers a large area. In such a large robust aquifer, it may take several decades for a new equilibrium to be established between inflows and outflows that will result in stable pressure levels. Any assessment on the potential of over- exploitation of a large-scale aquifer is in fact impossible without groundwater modelling tools.

12.3.3 Development of Groundwater Modelling Tools

In addition to monitoring and investigations, groundwater flow modelling approaches became an indispensable tool very early on for understanding how a multi-layer aquifer system functions. Such modelling tools, successively developed over the past four decades, were conceived by the University of Bordeaux, the Paris School of Mines, and BRGM. The first such model developed in the late 1960s, used an analogue approach (Astié et al. 1969), but very rapidly, the use of mathematical models became evident (Besbes, De Marsily, & Plaud, 1976; Douez et al., 2016; Larroque, Treichel, & Dupuy, 2008; Pédron & Gomez, 2010; Pedron, Platel, & Marchet, 2012; Saltel, Pédron, & de Grissac 2010; Saltel, Picart, & Lousteau, 2016; Thiery, Amraoui, Gomez, Pédron, & Seguin, 2011). These different tools

(Fig. 12.9) allowed synthesizing both geological and hydrogeological data. Each model required a conceptualization phase for retranscribing the complexity of the system being studied with the technical means available at the time. The ever increasing power of calculation with time allowed integrating ever more complexity in the model, thus transcribing ever more faithfully the studied processes, such as a finer geometry, shorter time steps, interactions with surface water, or the role of aquitards. Where the first models incorporated 1612 cells and 8 model layers (Besbes et al., 1976), the most recent include over a million cells and 15 layers (Saltel, Picart, et al., 2016).

The advantage of modelling tools is that they determine the cause of observed piezometric variations (whether local or regional) based on assessments of the water balance which can determine whether the resource is overexploited, at least on a large scale. Such tools were used for prediction simulations to guide management policies (Cabaret & Saltel, 2012), in particular for:

- Establishing acceptable large-scale abstraction limits for each aquifer (Cabaret & Saltel, 2012);
- Evaluating at a more local scale, the impact of infrastructure projects in a context of climate change, at the same time searching for an optimal distribution of abstraction between the different areas (Saltel, Picart, et al., 2016).

12.4 From the First Precautionary Measures to the Current Management Approach

In parallel to data acquisition, a management policy was formulated for the deep Gironde aquifers. The initial precautionary measures have been replaced by regulated access and use of the aquifers, with the objective of restoring and guaranteeing a good state of these resources.

12.4.1 The First Regulations for Using Deep Aquifers: 1956–1990

In 1959, a specific regulation called the "Decree-Law of 1935" was enforced in the Gironde Department. Initially drawn up for the Albian aquifer in the Paris Region (see Chap. 3), this regulation controls the access to groundwater resources and stipulates that, "*because of the public interest in the conservation and rational use of groundwater resources, no well or borehole of more than 80 m depth can be drilled without prior authorization.*" For the Gironde Department, the authorization depth was changed to 60 m; all existing wells and boreholes had to be declared, and the abstraction from them could be limited.

Despite the fact that the most heavily pumped aquifer (Eocene Aquifer) is extending into neighbouring departments as well (in particular,the Dordogne area where it is also used for producing drinking water)the application of Decree-Law of 1935 concerned only the Gironde Department. This limitation can be explained by the strategic importance of this deep aquifer for the Gironde,whereas the other departments had access to alternative water-supply sources.

Even though the drilling of wells and abstraction now had to be authorized, no limit was fixed on the extraction from the aquifers. Data from 2003 indicated total authorizations of 180 million m^3/year, total abstraction of 60 million m^3/year, and an estimated maximum permissible volume without risk for the aquifer of only around 40 million m^3/year.

12.4.2 Start of a Planned Water-Abstraction Management Regime (1991–1998)

In response to the observed decline in hydraulic heads (Fig. 12.6), a Groundwater Management Committee for the Gironde Department was created in 1991 on the initiative to central and local government services, bringing together government services, drinking-water distribution companies and scientists. Its function was that of a 'think tank', proposing solutions for all questions related to the groundwater resources (Servat & Gaillard, 2000).

In 1994, the deep aquifers of the Gironde were classified into Zones of Water Distribution. This regulatory provision (see Chap. 3) allowed the lowering of the thresholds for authorization and required the declaration of abstraction from surface waters as well as groundwater.

Investigations into water supplies necessary to sustain the future development of the Bordeaux metropolitan also began in 1994 with the Gironde Department drawing up a master plan for drinking-water supply. This planning document was based on an inventory of the existing resources and infrastructure, and proposed optimized and where possible, shared facilities for securing a safe and clean long-term drinking-water supply. The preparation of this plan required the identification of aquifers that could be confirmed as being overexploited. The Department ordered simulations of the changes in the deep aquifer pressure levels over 20 years to 2015 using the new MONA (MOdèle Nord Aquitain) groundwater model developed by BRGM (Fig. 12.9).

In 1996, these investigations for the departmental master plan confirmed that certain deep aquifers in the Gironde area were overexploited, 40 years after the risk was identified by Henri Schoeller in 1956.

In view of the significance of this overexploitation which was estimated at about 15 million m^3 for the Eocene aquifer (or more than 10% of the total volume abstracted for drinking water from all resources), the Gironde Department and the Urban Community of Bordeaux (now called Bordeaux Métropole, see below) decided to collaborate in order to:

- Build expert capacity independent of pressure groups, in the form of a specialized public institution tasked with finding alternative water resources;
- Request the creation of a Water Development and Management Plan (SAGE in French) under the Water Law of 1992 (see Chap. 3)

12.4.3 Creation and Implementation of the Deep Aquifers SAGE (1998–2017)

During several months, the area from application of the measure of the Water Development and Management Plan (SAGE) is in discussion to determine if it will applies to physical limits (aquifers extension) or administrative limit. It has finally been decided to limit the area to the most impacted zone, the Gironde Department.

In 1998, SMEGREG (Joint Association for Study and Management of Water Resources in the Gironde Department) was set up. Today, this public institution for cooperation between the Gironde Department, Bordeaux Métropole and about 20 drinking water providers covers almost 70% of the drinking water volume supplied.

In 1999, the CLE (Local Water Commission) of the Deep Aquifers SAGE of Gironde was formed. It has 24 members in 3 colleges: local politicians, water-resource users and the State. The CLE's task is to establish the SAGE and to ensure its strict application.

The Deep Aquifers SAGE was adopted unanimously by the CLE in 2003, and was the first SAGE in France that only concerned groundwater. It was revised in 2013 and now contains almost 100 specific provisions for the deep aquifers of the Gironde that impose legal requirements on public decisions as well as a regulation applicable to third parties.

Initially, the objective of the SAGE was to provide a sustainable guarantee for the good state of the groundwater resources within its perimeter, however the revised 2013 version also covers the resources outside its perimeter by integrating the downstream environments in its definition of a 'good state', as these could be affected by outflow from the deep aquifers.

12.5 Overexploitation and Maximum Permissible Volumes Objectives

12.5.1 How to Define the Overexploitation of a Confined Aquifer

The first methods to evaluate groundwater resources in a quantitative sense were derived from the approaches used for surface water. The European and French texts that mention equilibrium between abstraction and recharge, are unsuited to confined aquifers without stating the duration over which such equilibrium should be ensured. Though it is conceivable that winter precipitation will restore an unconfined aquifer to the same level each year, an even partial compensation of abstraction levels over a short period is inconceivable for a large confined aquifer. The fact that pressure levels decline over long periods in a confined aquifer, is not the sole criteria for judging whether or not is overexploited. It is not easy to answer the question "Is this confined aquifer overpumped or not?"

It is therefore clear that the decision of whether or not an aquifer is overpumped cannot be taken without long-term simulations by a suitable groundwater model. Another difficulty is that an aquifer is not only sensitive to the total amount of water abstracted, but also to the spatial distribution of those withdrawals.

Owing to a lack of a definition of a 'good state' for a confined aquifer in the technical literature or in legislative or regulatory texts, the CLE has formulated its own definition of 'good state' (with the advice from several expert hydrogeologists), which is now included in the SAGE. The objective is to guarantee, under acceptable socio-economic conditions, the 'good state' of the groundwater resources within its perimeter of application, which refers:

- For a 'good *qualitative* state' to directives 2000/60/CE and 2006/118/CE of the European Parliament and Council, including a list of polluting substances and their threshold values;
- For a 'good *quantitative* state' to the definition adopted by the CLE on July 4th, 2011, which combines a general approach for evaluation of the groundwater balance and local approaches for pressure, all attainable within time frames compatible with SDAGE deadlines.

The 'good *quantitative* state' definition adopted by the CLE on July 4th, 2011, stipulates that a confined aquifer is in 'good state' when:

- A storage decrease does not endanger the sustainability of the resource, as indicated the annual groundwater-balance calculations taken over the medium- and long-term (at least several decades),
- The piezometric levels in areas with identified risks will not result in:
  - Permanent and extensive dewatering of the reservoir;
  - Flow directions and patterns causing the inflow of extraneous water;
  - Insufficient outflow into downstream environments to allow maintaining or reaching their good state.

To evaluate the state of an aquifer using these principles, the SAGE defines a Maximum Permissible Volume (MPV) objective for each aquifer which is the annual abstracted volume that will not endanger the sustainability of the resource, provided that the spatial distribution of abstraction points is appropriate.

Initially, it was planned to define an MPV for each of the four major aquifers covered by the SAGE, but it soon became clear to the CLE that a finer subdivision of the aquifers would offer a greater degree of flexibilty for management of the

Table 12.1 Comparison of maximum permissible volumes and abstraction in 2013 (in million m³ per year)

Management unit	MPV	USE	MPV-USE (Mm³)	USE/MPV (in %)
Centre				
Miocene	12	7.9	4.1	66%
Oligocene	48	44.2	3.8	**92%**
Eocene	38.3	47.2	**−8.9**	**123%**
Campanian-Maastr.	2.5	2	0.5	**80%**
Cenomanian	4	1.4	2.6	35%
Médoc-estuary				
Miocene	3	0.2	2.8	7%
Oligocene	7	4.5	2.5	64%
Eocene	7.5	5.2	2.3	69%
Campanian-Maastr.	1	0.1	0.9	10%
Cenomanian	1	0.2	0.8	20%
Littoral				
Miocene	12	2	10	17%
Oligocene	22	8.1	13.9	38%
Eocene	6.6	6.3	0.3	**95%**
Campanian-Maastr.	2.5	1.2	1.3	48%
Cenomanian	–	–	–	–
North				
Miocene	–	–	–	–
Oligocene	–	–	–	–
Eocene	7	6.1	0.9	**87%**
Campanian-Maastr.	2	0.4	1.6	20%
Cenomanian	–	–	–	–
South				
Miocene	12	4.4	7.6	36%
Oligocene	2	0.2	1.8	10%
Eocene	–	–	–	–
Campanian-Maastr.	0.5	0	0.5	0%
Cenomanian	15	0	15	0%

groundwater resources. The MPVs were thus defined for Management Units (MU), which are spatial subdivisions of the aquifers based on hydrogeological criteria.

The MUs listed in columns of Table 12.1 each correspond to the combination of a geographic area (North, South, Centre, Médoc Estuary and Littoral) and an aquifer (Miocene, Oligocene, Eocene, Campanian-Maastrichtian and Cenomanian). Comparison of the abstracted volume in an MU with its MPV allows an assessment its state: surplus, in equilibrium (i.e. at the limit of overexploitation), or deficit (overpumped).

12.5.2 Calculation of Maximum Permissible Volumes

The MPVs presented in Table 12.1: Comparison of maximum permissible volumes and abstraction in 2013 (in million m^3 per year) Table 12.1 were defined by the CLE on the basis of expert advice and simulations by the MONA groundwater model. In version 3 (Fig. 12.10), used for this work, the model can simulate flow in 15 aquifer layers, with regular 2 × 2 km meshes and nearly 67,000 cells. MONA is a pseudo-3D model, which means that the capacitive role of the confining beds is not considered and that no head is calculated for these layers. However, any exchanges through leakage are well reproduced in the model, as the permeability values of the confining beds are integrated.

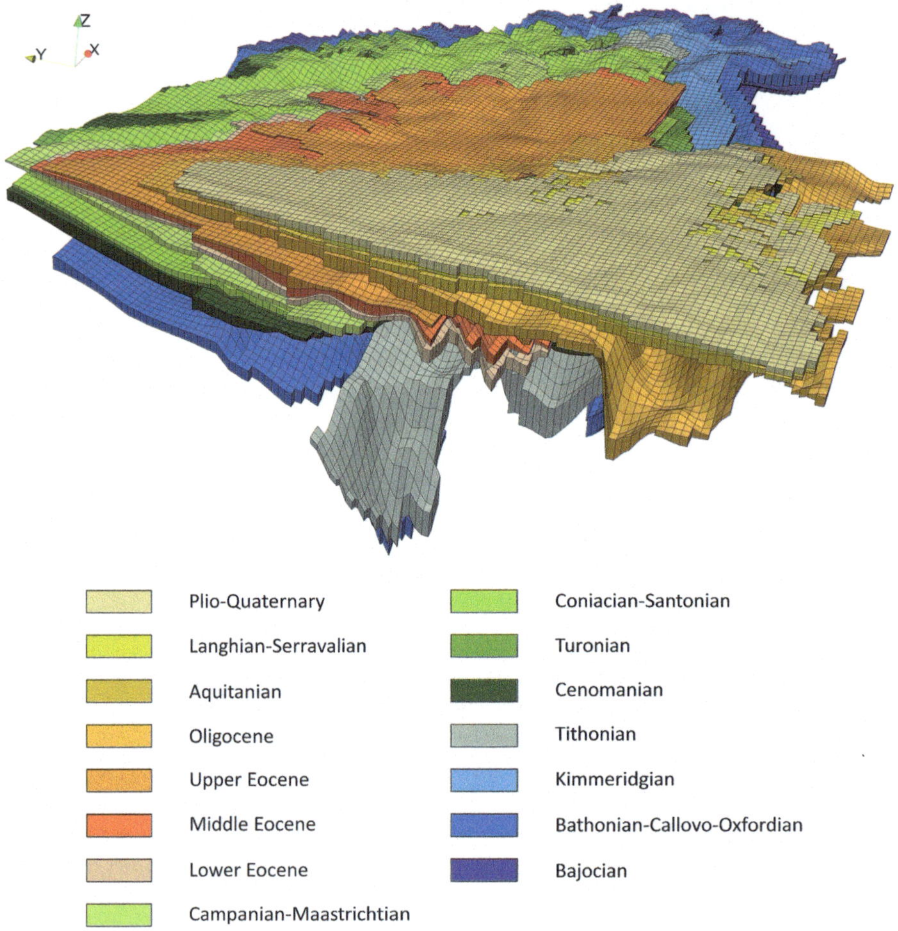

Fig. 12.10 3D view of the geometric structure of the aquifer reservoir as shown in the MONA model

Definition of the MPVs has to consider physical factors, such as aquifer properties, recharge conditions, the spatial distribution of withdrawal points, increases in water demand, etc.

Groundwater modelling enables the integration of all these different parameters. Several simulations were run, using credible scenarios in terms of growth in demand based on population increases, household consumption patterns, performance of the water supply networks and the creation of new well fields, as well as on changes in climatic conditions.

Only two climate conditions were considered in the simulated scenarios. The first corresponds to the average climate observed over the period 1978–2007; the second covers the average precipitation for the years 1998–2007, corresponding to a period with lower rainfall.

As there are multiple variables, creation of the scenarios must ensure that the results are easy to interpret. For this reason, it was decided during scenario preparation to use a reference scenario in which only one parameter would be varied at the time, thus determining the relative weight of each variable in the results (Table 12.2).

Each simulation of a scenario produced the following outputs:

- Piezometric maps that can allow comparison between scenarios;
- Piezometric hydrographs that allow the identification of risks related to a decline of the pressure levels (dewatering, salt-water intrusion, etc.);
- Changes in the storage volumes within the aquifers.

The results (Fig. 12.11) were examined by a group of expert hydrogeologists who evaluated whether or not the storage variation is acceptable on the basis of various criteria of acceptability: variability, trends and. The final judgement is therefore based on expert assessment.

Box: Are Declining Pressures Acceptable?

In a confined aquifer, all pumping will generate a decline in water pressure levels. This decline indicates a lowering of the storage that in the long term, will reach a new state of equilibrium where the rate of inflows into the area equals the rate of extraction, and stabilization of the pressure levels occurs. A fall in pressure level therefore does not mean that the aquifer is overpumped, and a declining trend should not be the sole criterion for judging whether the the aquifer is in a 'good' or 'bad' state as the pressure levels are likely to stabilize at a lower level. The water balance approach comparing recharge to extraction which is used to determine extraction limits for unconfined aquifers, should be replaced by the notion of "acceptable impacts" for confined aquifers, for which the storage reduction over time is not zero, but is low in comparison to the flow volumes involved and can be controlled over the medium term (10–20 years) with a possibility of returning to the initial aquifer condition when pumping stops.

This approach for confined aquifers is also applied in Australia with the use of groundwater models. An example is presented in Chap. 16.

Table 12.2 Description of the various MPV scenarios

Scenario	Effect
Status quo scenario	Population rise – no improvement in efficiency ratio of the water distribution network – no decrease of domestic water consumption
Optimized scenario	Population rise – construction of new pumping and distribution infrastructure to improve the spatial distribution of pumping for minimizing groundwater depletion in specific sites – reduction of leaks in the water distribution network – reduction of per capita domestic water consumption
Impact of water savings	Population rise – constructing of new facilities for improving the spatial distribution of pumping areas to minimize groundwater-mining impact in specific sites – no improvement in efficiency ratio of water distribution network – no decrease of domestic water consumption
Impact of low recharge condition	Population rise – realization of new facilities to improve the spatial distribution of pumping areas in order to minimize groundwater mining impacts on specific sites – improvement of the efficiency ratio of the water distribution network – reduction of domestic water consumption – dry climate
Increased pumping	Increase of pumping in areas currently underexploited
All pumping stops	End of groundwater mining

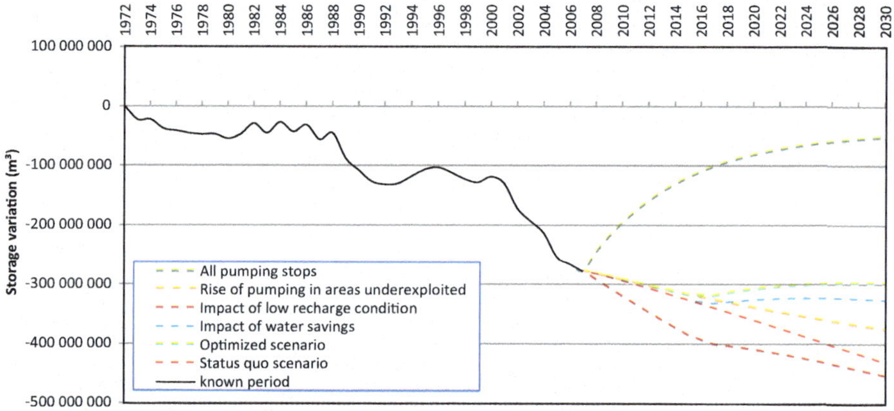

Fig. 12.11 Evolution of storage in the middle to lower Eocene aquifer by 2030

The Status-quo scenario prediction (Table 12.2) did not lead to equilibrium conditions in the Middle to Lower Eocene aquifer (Fig. 12.11). However, the optimized scenario for this aquifer shows a stabilization of storage levels over time and was judged suitable for defining the MPV for various Management Units. However for some Management Units, all simulated scenarios give acceptable results and the planned MPVs are the known Status-quo extraction volumes.

The results also show that the impact of water savings policies (decreased leakage from water-distribution networks, incentives for reduced individual consumption) is not neutral.

Such imposed abstraction limits do not guarantee sustainable development of a confined aquifer forever, but are values based on the best information currently available. They should be revised if new data, improved models or new scenarios result in different outcomes from the modelled predictions.

In conclusion, setting an MPV is a technically complex and laborious process that involves some arbitrary assumptions. This is why MPV values must be regularly reviewed by the Local Water Commission for their suitability if new information becomes available or pressure level trends change.

12.6 Management Rules for the Deep Gironde Aquifers

12.6.1 Water Conservation Policy

In order to reach the SAGE objectives i.e. a 'good state' of the deep aquifers, reduced abstraction was required,from the overexploited aquifers (representing more than 92% of drinking water supplies). The strategy adopted by the CLE to achieve this end included a priority policy of water con-servation and demand management, followed by resource substitutions (Grissac, 2008).

The public water utilities, as the primary users of the deep aquifer resources, were the first affected by these measures. Specific regulations imposed on them required a full examination of their distribution networks (leak detection, repair and renewal in order to obtain an acceptable efficiency of drinking water distribution networks); permanent monitoring and modelling of these networks; a yearly calculation of standardized performance indicators; and the adjustment of their pumping licences to the minimum volumes that would meet their needs.

The water conservation policy also involved the public consumers through awareness campaigns to reduce consumption. This comprised:

(i) Accredited associations for education programmes on water resources and their use in schools [the "*L'eau un enjeu majeur*" ("*Water is a Big Issue*") programme];
(ii) Classic printed communication tools such as brochures and posters; and
(iii) The internet site "jeconomiseleau.org" (i.e. ISaveWater.org) that has become a national reference for communication on this subject.

The effectiveness of this policy has become clear through the reduction in abstracted drinking-water volumes observed between 2005 and 2013 (Fig. 12.12), a reduction of almost ten million m^3, mainly due to fewer distribution-network leaks, even though the population increased by about 120,000 over the same period.

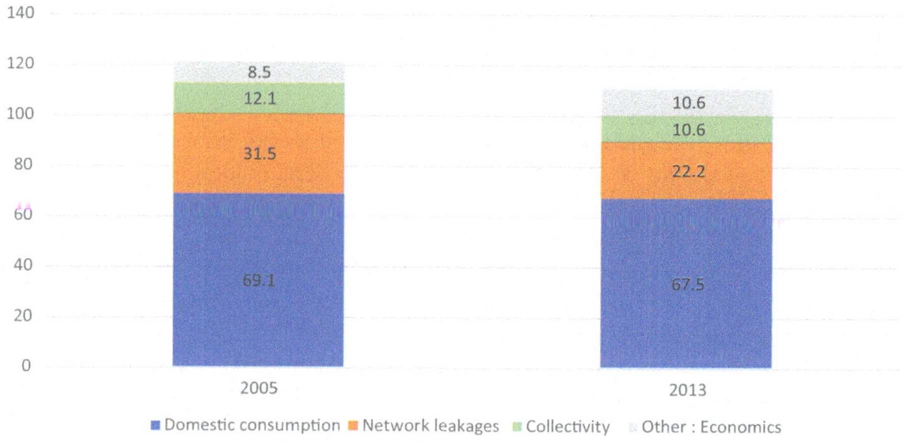

Fig. 12.12 Abstractions for drinking water from 2005 to 2013, in million m^3

12.6.2 Resource Substitutions

Evaluation of the regularly updated future drinking water demands of an ever increasing population, indicates that about 20 million m^3/year (almost 20% of the current abstracted volume in the Department) will need to be obtained from alternative sources due to the requirement to reduce pumping from the already overexploited aquifers. The search for resource substitution is the task for SMEGREG, which must evaluate the technical, legal, administrative and financial feasibility of possible solutions. Bordeaux Métropole is presently building infrastructure costing around 60 million Euros to supply 10 million m^3/year from 14 wells completed at a depth of 250 m in a deep aquifer that is not over-exploited and can support this level of abstraction.

12.6.3 Eco Conditions, Technical Specifications and Master Plans

In addition to the above-mentioned technical aspects, the management of the deep aquifers in the Gironde Department is also based on eco-conditions that must be respected for obtaining the necessary authorizations. For instance, water-withdrawals are only authorized if it can demonstrated that:

- Water use will be optimized;
- Outside the SAGE, no alternative resource exists that can fulfil the water requirements taking into account technical, economic and quality considerations.

In addition, the SAGE imposes technical specifications for the drilling and/or pumping from the wells, such as setting a maximum permissible drawdown in areas where the risk of a water level decline has been identified.

Finally, local development master plans drawn up for certain areas, define the conditions for access to the resource, for instance stipulating that drinking water supplies should be distributed from south to north, and not the other way around.

12.7 Conclusions – The Lessons Learned

The current water management regime in the Gironde Department is the result of a long scientific and technical policy evolution, which has led to an operating process that supports consultation and regulation within the legal framework of the SAGE.

The management of regional confined aquifers which are robust and may take decades to equilibriate to increases in extraction, presupposes that criteria must be set for judging the state of such aquifers. The decline of pressure levels cannot alone indicate the overexploitation of such an aquifer. It is thus necessary to calculate specific assessments for assessing whether the impacts of long term exploitation are acceptable. The question of whether or not the deep aquifers of the Gironde Department can be sustainably developed, could not be answered until sufficient knowledge and technical capability were available for the representative flow modelling of the groundwater system as a whole. Even then, it must be understood that this answer was partially based on expert evaluation and was only valid for the assumptions used in the abstraction scenarios that were modelled.

It is therefore highly desirable that strategic confined aquifers have an adaptive management approach which would be reliant on:

- Groundwater flow models that are constantly updated to improve their accuracy (finer geometry, better consideration of recharge and surface-water exchange, etc.), and reduce the uncertainties inherent in this type of model as far as possible (Delottier, Pryet, & Dupuy, 2017).
- Future abstraction scenarios that are as realistic as possible in terms of future water demand.

These requirements make close cooperation between groundwater specialists and urban-country planning specialists essential.

References

Astié, H. (1978). Surveillance et gestion des nappes d'eau souterraine dans le département de la Gironde. BRGM Techn. Report 78 SGN 481 AQI, 12 p.

Astié, H., Aubic, J., Bellegarde, R., Bourgeois, M., Cazal, A., Emsellem, Y., et al. (1969). Synthèse des études sur le système multicouche des aquifers tertiaires du nord de l'Aquitaine. *La Houille Blanche*, 907–918.

Astié, H., Bellegarde, R., & Bourgeois, M. (1967). Contribution à l'étude des différences piézométriques entre plusieurs aquifères superposés. *Chronique d'hydrogéologie, 12*, 49–59.

Bellegarde, R. (1969). Historique de la piézométrie de la nappe des sables éocènes en Gironde -BRGM Technical Report 69 SGN 144 AQI, 23 p.

Bellegarde, R. (1975). Nappe des sables éocènes en Gironde (situation en 1975), BRGM Technical Report 75 SGN 293 AQI, 14 p.

Bellegarde, R., Bourgeois, M., Camus, G., Camus, R., & Schoeller, H. (1964). Étude de la nappe des sables éocènes en Gironde – compte rendu des observations effectuées en 1963 et début 1964, rapport DSGR. 64. A38, 41 p., BRGM Technical Report.

Besbes, M., De Marsily, G., & Plaud, M. (1976). Bilan des eaux souterraines dans le Bassin Aquitain, Hydrogéologie des grands bassins sédimentaires, Mém. *Conférence de Budapest, XI*, 294–304.

Cabaret, O., & Saltel, M. (2012). Contribution à la révision du SAGE Nappes Profondes de Gironde: Utilisation du Modèle Nord-Aquitain – phase « Tendances et Scenarios » – Simulations complémentaires – BRGM Technical Report RP-61479-FR, 46 p.

Corbier, P., Abou Akar, A., Mazurier, C., Bourgine B., Capéran, F., Grabenstaetter, L., et al., (2012). Que retenir de l'évolution de l'état des nappes de Gironde en 2011? BRGM Technical Report RP-61771-FR, 116 p.

De Grissac, B. (2008). Gestion de la demande en eau: une alternative aux resources conventionnelles. L'expérience du département de la Gironde. *La Houille Blanche, 4*, 22–28.

Delottier, H., Pryet, A., & Dupuy, A. (2017). Why should practitioners be concerned about predictive uncertainty of groundwater management models? *Water Resources Management, 31*, 61–73.

Douez, O., Abou, A. A., & Bourgine, B. (2017). Réseau départemental de suivi des nappes de Gironde – Bilan de l'état des nappes en 2015. BRGM Technical Report RP-66626-FR, 165 p.

Douez, O., Cabaret, O., Saltel, M., Abasq, L., Wuilleumier, A., & Pédron, N. (2016). Groundwater management of large aquifers in southwestern France by regional hydrodynamic models – Poster 43rd IAH CONGRESS.

Larroque, F., Treichel, W., & Dupuy, A. (2008). Use of unit response functions for management of regional multilayered aquifers: Application to the North Aquitaine tertiary system (France). *Hydrogeology Journal, 16*, 215–233.

Mauroux, B., Sourisseau, B., & Bonnery, H. (1999). Schéma d'aménagement et de gestion des eaux – Nappes profondes en Gironde – Commission locale de l'eau – État des connaissances des nappes profondes à fin 1998. (Etat des lieux techniques) -BRGM Technical Report RR-40844-FR.

Pédron, N., & Gomez, E. (2010). Utilisation du MOdèle hydrodynamique Nord Aquitain (MONA) pour appuyer à la définition des volumes prélevables dans les aquifères profonds du nord du Bassin aquitain, 12 p.

Pedron, N., Platel, J. P., & Marchet, P. (2012). Les modèles hydrogéologiques multicouches développés en Aquitaine. *Géologues, 156*, 24–26.

Platel, J. P., Mauroux, B., Bonnery, H., Sourisseau, B., Seguin, J. J., Chery, L., et al., (1999). Gestion des eaux souterraines en Aquitaine. Année 3. Opération sectorielle – Relations entre l'estuaire de la Gironde et la nappe de l'Éocène – Réalisation d'un piézomètre complémentaire – Mise en place du réseau d'alerte et protocole d'actions BRGM Technical Report BRGM/RR-40729-FR, 88 p.

Saltel, M., Picart, D., & Lousteau, D. (2016). Picart Delphine and Lousteau Denis Multi-model approach to evaluate the impact of a future well field on forest production (South-West of France) -43rd IAH CONGRESS [Conférence]. - Montpellier, France: [s.n.].

Saltel, M, Lavielle B, Rebeix B, Thomas, R and Franceschi M. (2016). Projet Eaurigine. Étude de la paléo-recharge des aquifères du nord du Bassin aquitain par l'utilisation de traceurs isotopiques et des gaz rares. Application à la paléoclimatologie. Colloque Q10 AFEQ CNF-INQUA, 2016.

Saltel, M., Compère, F., & Cabaret, O. (2017). Gestion des eaux souterraines en Région Aquitaine -Développements et maintenance du Modèle Nord-Aquitain de gestion des nappes – Module 1.1 – Année 1 – Convention 2015 – 2020. report BRGM/RP-66832-FR. In *43 p.*

Saltel, M., Pédron, N., & de Grissac, B. (2010). Caractérisation par modélisation hydrodynamique du dénoyage d'un aquifère: impact des facteurs climatiques et anthropiques – exemple du reservoir Oligocene au sud de l'agglomération bordelaise – 23ème Réunion des Sciences de la Terre Bordeaux 25 au 29 Octobre 2010 [Conférence].

Schoeller, H. (1956). Méthode de détermination du rayon d'appel des forages et du coefficient de Darcy, application à l'étude de l'épuisement de la nappe des Sables paléocènes de l'Aquitaine. *Ass. Intern. d'Hydrologie Scientifique, 41*, 67–75.

Seronie-Vivien, M. R. (2001). Contributions apportées par la Société Essorep à la connaissance stratigraphique du Bassin d'Aquitaine, Ressources minérales du sol et du sous-sol des Landes de Gascogne. *Actes du colloque de Brocas, 3*, 35–40.

Servat, H., & Gaillard, B. (2000). Le SAGE nappes profondes de la Gironde. In *La Houille Blanche* (Vol. 7–8, pp. 127–130).

SMEGREG. (2013). *SAGE Nappes profondes de Gironde - Règlement Commission Locale de l'Eau*. SMEGREG: Bordeaux.

Thiéry, D., Amraoui, N., Gomez, E., Pédron, N., & Seguin, J. J (2011). Regional model of groundwater management in North Aquitania aquifer system: Water resources optimization and implementation of prospective scenarios taking into account climate change. Water Security in the Mediterranean Region, NATO Science for Peace and Security Series C: Environmental Security. pp. 275–290, eds. Scozzari A. & El Mansouri B. https://doi.org/10.1007/798-94-007-1623-0.

Frédéric Lapuyade is an expert hydrogeologist at Sondalp (Suez Group), working on digital systems and technical assistance on wells and water re-source worlwide. Trained at the National School of Geology of Nancy; Frédéric completed his training with engineering degrees in industrial systems management and Mines of Nancy schools. He began his career in the water police of Meurthe et Moselle, then joined the Rhine Meuse Water Agency where he ensured rising of measurement networks of groundwater, national water in-formation systems, international concertation. His paper here deals with his experience at SMEGREG, public institution in charge of the management of deep aquifers, the creation of new water resources and the implementation of a water saving policy.

Marc Saltel is Senior Hydrogeologist at Brgm (French Geological Survey). PhD in hydrogeology, he has 10 years of experience in groundwater management from wells to large multilayered aquifer systems. He has developed several models to guide local stakeholders in-volved in aquifer management. These models were used for public service is-sues: assessment of the volumes that can be pumped to prevent over-exploitation, evaluation of the impact of a future well field (10 millions m^3/year) on a multilayered aquifer system, wells fields optimization, climate changes impact. Currently, he is involved in a research program to evaluate more precisely uncertainties in groundwater models. He has also worked on many topics of hydrogeology, particularly geochemistry and past-climate changes and their impact on groundwater.

Bruno de Grissac is the director of SMEGREG, an public institution in charge of managing groundwater in the Gironde county, since its creation almost 20 years ago. Bruno holds a Ph.D in was hydrogeology. He previously in charge of water management in the Charente County, then at the entire Charente river basin. President of the Association of Hydrogeologists of Public Utilities, he is also serving as a member of the Scientific Council of the Water Agency since its creation in 2010.

Chapter 13
A Tool to Determine Annual Ground-Water Allocations in the Tarn-et-Garonne Alluvial Aquifer (France)

Pierre Le Cointe, Vorlette Nuttinck, and Jean-Daniel Rinaudo

Abstract The Tarn-et-Garonne department is crossed by three main rivers (Garonne, Tarn and Aveyron) whose alluvial plain covers an area of almost 1000 km^2. Since 1996, the "Direction Départementale des Territoires" (Gov. Administration at county level), with the technical help of the French geological survey (BRGM), initiated the development of a groundwater model and a decision support tool to define annual groundwater abstraction allocations. As the field data and the computing capacities increased and the law evolved, three versions of the groundwater model were successively developed to better assess the Maximum Permissible Volumes (MPV) of groundwater abstraction on a yearly basis. The last transient state version takes into account the annual fluctuations in groundwater recharge and the water exchanges between the aquifer and the rivers. The MPVs are calculated each year in 21 management zones outside of the previously defined riverside aquifer. These zones are now managed by five agricultural users' associations, known as collective management agencies (or OUGCs). Further improvement should lead to the decision support tool being available online, to encourage OUGCs and farmers to be more proactive in managing the groundwater resource.

Keywords Alluvial aquifer · Annual allocation · Collective management · Groundwater model · Management tool · Riverside aquifer

P. Le Cointe (✉)
BRGM, Ramonville-Saint-Agne, France
e-mail: p.lecointe@brgm.fr

V. Nuttinck
Direction Départementale des Territoires de Tarn-et-Garonne, French Government, Montauban, France
e-mail: vorlette.nuttinck@tarn-et-garonne.gouv.fr

J.-D. Rinaudo
BRGM, Montpellier University, Montpellier, France
e-mail: jd.rinaudo@brgm.fr

J.-D. Rinaudo et al. (eds.), *Sustainable Groundwater Management*,
Global Issues in Water Policy 24, https://doi.org/10.1007/978-3-030-32766-8_13

13.1 Introduction

This chapter presents the quantitative management system for groundwater resources that was set up in the 1990s in the Tarn-et-Garonne department in southwestern France. In this region, crop irrigation constitutes the principal use of groundwater. The groundwater resource consists of shallow alluvial aquifers with limited storage capacity. This makes them very sensitive to annual climatic fluctuations, which can be considerable in the department. In dry years, the decline in water levels causes certain boreholes to dry up, generating localised conflicts over water use. In addition, when groundwater is abstracted from the aquifer, river baseflow is reduced, exacerbating problems at low water levels for the three major rivers flowing through the department.

At the end of the 1990s, the Adour-Garonne Water Authority, the region and the state launched a programme designed to improve knowledge of the groundwater resource and to model how it functions. This led to the development of a water allocation decision support tool used by the state for issuing water use authorisations to farmers on an annual basis. The most original feature of the tool is its capacity to adjust the volumes allocated to users at the start of each year, by taking into account the climatic conditions and aquifer recharge. Initially, the allocation process was the exclusive responsibility of the state, but gradually the users became involved.

This chapter presents the principal stages of implementing the groundwater management system. Section 13.1 describes the main characteristics of the study area, the groundwater resources and its uses. Section 13.2 presents the first management mechanism applied between 1996 and 2006, based on the allocation of authorisations expressed in terms of pumping flow rate. Section 13.3 describes the implementation of a volumetric management system, based on the calculation of a maximum permissible volume shared between users in the form of individual quotas. Section 13.4 recounts how a collective management approach emerged based on the creation of water user associations. The state transfers the responsibility of allocating volumes of water for abstraction to the user associations.

13.2 Presentation of the Case Study

13.2.1 Geographic and Climatic Context

Tarn-et-Garonne is a French department (county) located in the Aquitaine Basin, in the southwest of the country. The alluvial plain covers an area of almost 1000 km^2, which represents 30% of the department's area. It is located at the confluence of three major rivers: the Tarn, the Garonne and the Aveyron. The altitude in the region varies between 50 and 210 m. The plain is surrounded by hills composed of the Lomagne and White Quercy Tertiary marly sandstones and, on its eastern border, by the karstic plateaux of Caylus Causse, which is a part of Quercy Causses (Fig. 13.1).

The climate is characterised by mild wet winters and hot, generally dry summers. Annual rainfall is around 700 mm and is relatively homogenous across the region. However, there is high inter-annual variation (ranging from 426 mm in 1967 to 1007 mm in 1959 at the station in Montauban), which leads to variations in aquifer recharge.

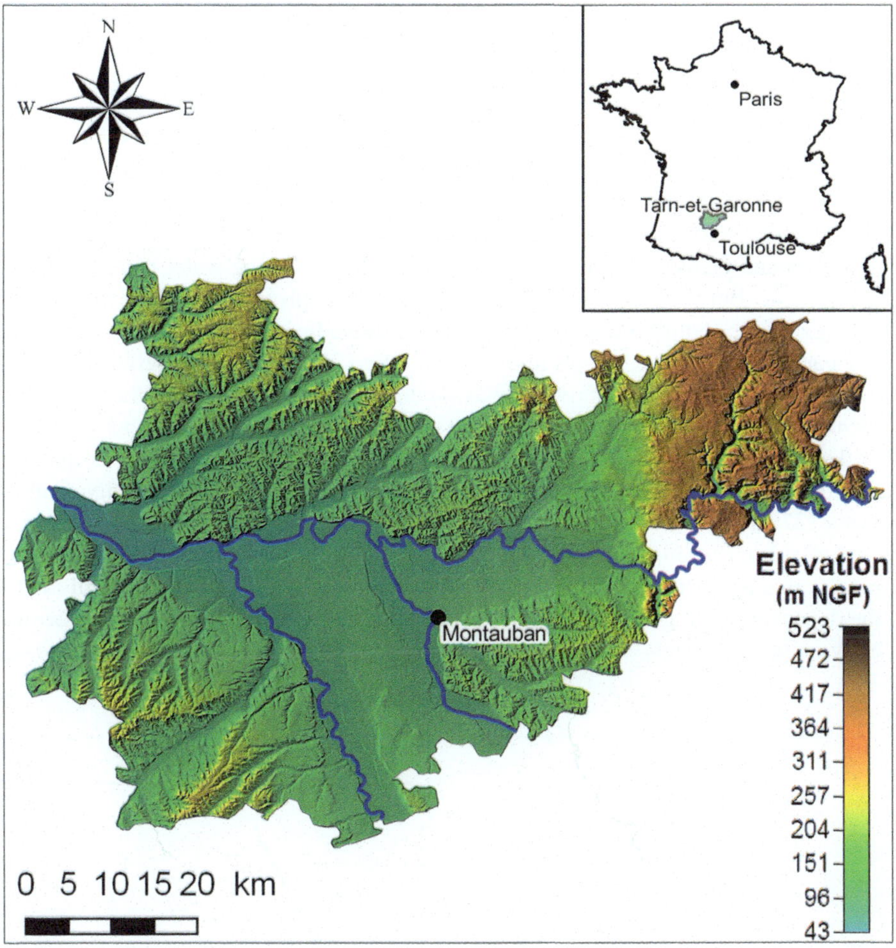

Fig. 13.1 Location and topography of the Tarn-et-Garonne department

13.2.2 Geological Context and the Groundwater Resource

The confluence of the Garonne, Tarn and Aveyron Rivers consists of an extensive series of Quaternary alluvial deposits composed of sand and gravel deposited on Tertiary clay-limestone molasses formations, which are thought to be fairly impermeable (Bouroullec, 2013). The alluvial system comprises tiered terraces, created during a succession of glacial and interglacial phases. Because of erosion, the terraces are frequently separated by molasse banks, from which springs emerge (Fig. 13.2).

The alluvium on the different terrace levels averages 5–8 m thick and form unconfined aquifers. These aquifers are mainly replenished by rainfall infiltration and contribute discharge to the watercourses by baseflow. The alluvial aquifer in Tarn-et-Garonne thus contributes on average to 3% of the total flow of rivers in the department, a contribution that rises to 8% during low-water periods.

13.2.3 The Uses of Surface and Groundwater

The Tarn and Garonne Rivers represent a significant surface water resource for the department. Between 2003 and 2012, surface water provided 75% of the region's average water requirements (48 million m^3, excluding extraction for cooling the nuclear power plant), while groundwater supplied 25% of the requirements (16 million m^3). The use of groundwater has declined significantly since the end of the 1990s, when the total volume abstracted reached 35 million m^3 per year.

The alluvial aquifers are primarily used by the agricultural sector for crop irrigation (67% of volume abstracted). The remaining volume is used to supply drinking water (27%) and several industries (6%) (Table 13.1; Fig. 13.3).

Agriculture is important for the economy in Tarn-et-Garonne. The fruit sector, predominantly located on the alluvial plain, represents 11% of agricultural land in the department but generates 33% of production in value. Fruit crops are systematically irrigated, largely with water pumped from the main rivers (the Garonne, the

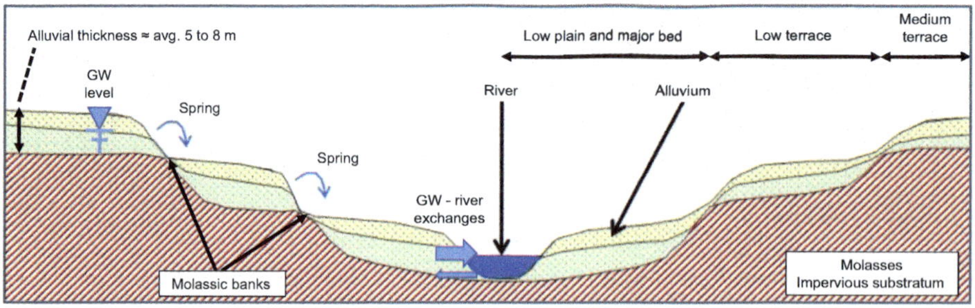

Fig. 13.2 Diagrammatic cross-section of the tiered terrace system

Table 13.1 Volumes of surface and groundwater abstracted in Tarn-et-Garonne Plain, per use (in thousands of m3 per year)

Sector	Groundwater	Surface water
Drinking water	4,370	10,968
Industry	1,016	550
Agriculture (irrigation)	10,848	36,289
Total	16,234	47,807

Average volume abstracted from 2006 to 2012. (Source: Bardeau & Le Cointe, 2016)

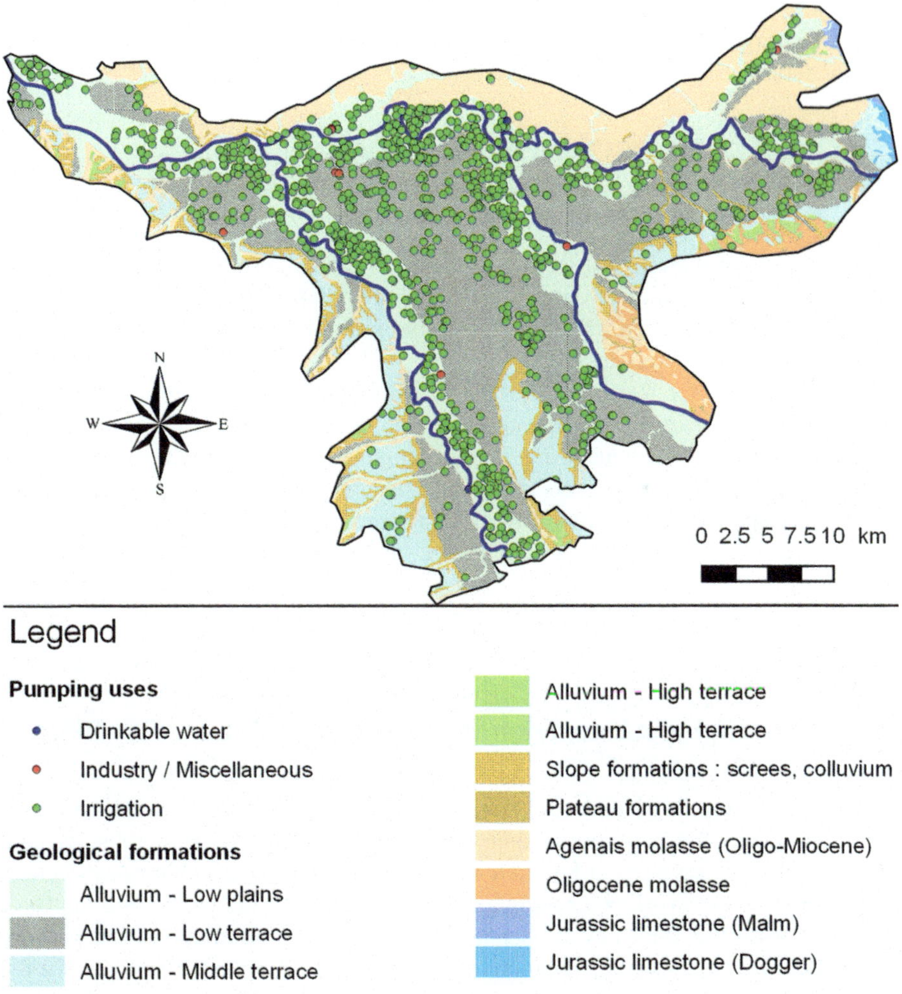

Fig. 13.3 Geographic distribution and use of groundwater abstracted from the alluvial aquifers in Tarn-et-Garonne in 2015

Tarn and the Aveyron) or from the aquifer in the sectors with no access to surface water. The water is used for irrigation in the summer, but also to reduce frost damage in the spring.

The large arable farms (cereals, maize) comprise 40% of the farmland in the department, but only produce 22% of the output in value. Irrigation for arable crops is highly developed, although dry cultivation is still practiced. Irrigation primarily concerns maize, but also cereals (irrigated once or twice in spring for seedling emergence), sunflower and soya. Cereals and maize can be grown without irrigation in soils with sufficient available water.

As in most regions in France, farms are still family businesses. This equally applies to the largest farms, which may have over 350 ha of fruit trees. In Tarn-et-Garonne, large and average-sized farms cover 86% of the utilised agricultural land and generate 95% of the output in value.

13.2.4 The Challenge of Managing Groundwater

The alluvial aquifers in Tarn-et-Garonne have a limited storage capacity and react quickly to climatic fluctuations. In dry years, the combination of low recharge and high abstraction for crop irrigation lowers the groundwater level, potentially causing some wells or boreholes to dry up.

In addition, when a volume of water is abstracted from the aquifer, baseflow is reduced and the rivers lack that water, accentuating the problem of low water levels in watercourses. Nonetheless, this impact remains inconspicuous because the discharge in Tarn-et-Garonne rivers is largely determined by the dams located upstream of the area studied. The inflow from the aquifers to the rivers is marginal compared to the streamflow from upstream. Thus, the situation is very different from Beauce region (Chap. 5), where the over-exploitation of groundwater has caused some small watercourses to dry up completely.

13.3 Managing Abstraction Based on Flow Rate: 1996–2006

For years, the use of groundwater was far less regulated than that of surface water. This situation actually encouraged groundwater use. Until the mid-1960s, the construction and use of wells or boreholes for irrigation were only subject to a declaration, in the case of works exceeding a depth of 10 m, according to the Mining Code of 1951. With the 1964 law, a licence was required for installations that pumped over 80 m^3/h, while other installations simply had to be declared. Therefore, the use of groundwater for agricultural purposes developed in a context where there was virtually free access to the resource.

The gap between the regulations for using surface water and groundwater was further widened when the 1984 law came into force (the so-called "fishing" law).

The law actually created a regulatory mechanism allowing state services to restrict surface water abstraction in the event of severely low water levels in order to protect aquatic habitats and fishery resources in particular. As the regulation did not apply to groundwater, many farmers replaced (or supplemented) their river water intake with a borehole in the alluvial aquifer, sometimes located only a few metres away from the riverbank. By relocating their point of abstraction, they evaded the temporary restrictions applied to the watercourses and continued pumping the same resource at little extra cost. This situation did not change until the 1992 Water Act came into force.

13.3.1 The 1992 Water Act

The 1992 Water Act radically changed how surface water and groundwater were managed in Tarn-et-Garonne. In general, the law set out to restore a balance between abstraction and available resources, by taking the health of aquatic habitats into account (see Chap. 3). Four of the provisions in the Water Act had a direct impact on water management in Tarn-et-Garonne.

The first provision introduced the concept of "riverside aquifer", which was designed to take into account the impact of groundwater abstraction on the surrounding watercourses. The concept of riverside aquifer refers to an aquifer that is hydraulically connected to a watercourse, meaning that any pumping from this aquifer will negatively impacts the flow of the watercourse. From a regulatory perspective, the law states that abstraction from the riverside aquifer should be considered as surface water abstraction. Therefore, it should be subject to the same regulatory restrictions where appropriate.

The second change was linked to the creation of water use restriction zones ("Zones de Répartition des Eaux", ZRE). These zones included the basins, subbasins or aquifers characterised by a structural shortage of resources in relation to requirements. The regulatory zoning allowed the state to tighten restrictions. It became mandatory to declare abstraction points with a flow rate of 8 m^3/h or more and if warranted, any new abstraction could be banned. The whole Tarn-et-Garonne department was declared a water use restriction zone in 1994.

The third major change was the obligation to install a volumetric meter at all abstraction points within 5 years. In Tarn-et-Garonne and more generally throughout southwestern France, the farming community was fiercely opposed to installing meters. Thus in 2005, only 40% of water abstraction points were equipped with meters, but by the 2010s, meters had been installed in 90% of abstraction points.

In Tarn-et-Garonne, the 1992 Water Act was only genuinely implemented in 1995. The state took the initiative, by appointing an inter-ministerial water commission (MISE) to issue water use permits. The MISE conducted the first survey of water abstraction (surface and groundwater) and developed a network of groundwater monitoring wells in the alluvial aquifers. A study was also launched to define the extent of the riverside aquifer and develop a groundwater flow model to simulate how the alluvial aquifer functions for different abstraction scenarios. In 1996, based on the newly acquired knowledge, the MISE issued the first individual annual permits for abstraction. The permits stipulated an authorised flow rate. The installation of the volumetric meters had only just begun for both surface and groundwater at that time, which ruled out the possibility of issuing volumetric authorisations for abstraction.

13.3.2 Acquiring a Knowledge Base for Alluvial Aquifers

When the state started introducing a management system for water abstraction, very little information was available on groundwater and its uses. The available knowledge was collated in the comprehensive hydrogeological study of the Tarn-et-Garonne department, conducted in the framework of the assessment of the hydraulic resources of France (Soulé, 1978). The study contained an inventory of the department's wells, boreholes and springs, which was conducted over a 2-year period, but had incomplete information about the aquifers' geometry and hydrogeological properties. There was insufficient data to establish a synchronous piezometric map.

In 1995, the first step involved acquiring supplementary data. The state conducted a census of the water abstraction points (surface and groundwater) which resulted in a large number of undeclared wells and boreholes being registered. In addition, a field survey was carried out and measurements obtained from the 387 wells were used to establish a piezometric map for the alluvial system in Tarn-et-Garonne. The field survey also provided the opportunity to verify the abstraction data collected by the MISE and the Adour-Garonne Water Authority.

Since 1982, the water level in the alluvial aquifer was constantly monitored by a state service in an unused borehole in the lower Garonne Plain. This was the only long-term piezometric record available for the entire alluvial system in Tarn-et-Garonne at the time.

13.3.3 Defining the Riverside Aquifer

The available knowledge was applied to defining the boundary of the riverside aquifer. It was defined as "the aquifer(s) hydraulically connected to the watercourse and where abstraction is likely to have an impact (direct or indirect) on the river flow rate before the end of the low water period" (Collin & Daum, 1995). Therefore, the extent of the riverside aquifer depends on the choice of (i) the impact threshold for pumping groundwater on the river flow rate and (ii) the time t, after which the impact becomes apparent. In the case of the alluvial aquifer in Tarn-et-Garonne, the riverside aquifer is initially defined using time $t = 90$ days, which corresponds to the duration of the irrigation season. The impact threshold is calculated for two values,

1% and 5%. This percentage is the ratio between the flow that is directly or indirectly deducted from the river, and the pumping flow rate.

Theis and Darcy's formulae were used to define the boundary of the riverside aquifer for the Garonne, the Tarn and the Aveyron and Rivers (Gandolfi, Danneville, Petit, & Tilloloy, 1997). The formulae presume that the values for the aquifer's transmissivity and storage coefficient are known. In the absence of accurate data, a single value was used for the entire lower plain for both parameters ($T = 2.5 \times 10^{-3}\ m^2/s$; $S = 4\%$). Identifying the boundaries of the riverside aquifer in this way revealed the zone was often 3 or 4 km wide and was hence limited to the lower plain of the main watercourses. It is considered the only alluvial terrace with groundwater-surface water connectivity (Fig. 13.4).

13.3.4 Developing the Groundwater Flow Model

The next step consisted in developing a groundwater flow model for the whole alluvial aquifer (Gandolfi et al., 1997). The aim was to develop a simulation tool to provide the MISE with the elements it required to issue abstraction permits for irrigation. The software MARTHE (Thiéry, 1990, 2010b, 2015) was used for this model. It resolves the flow equation in porous media, which link flow rates to the groundwater levels (Darcy's law), on a rectangular grid using the finite difference

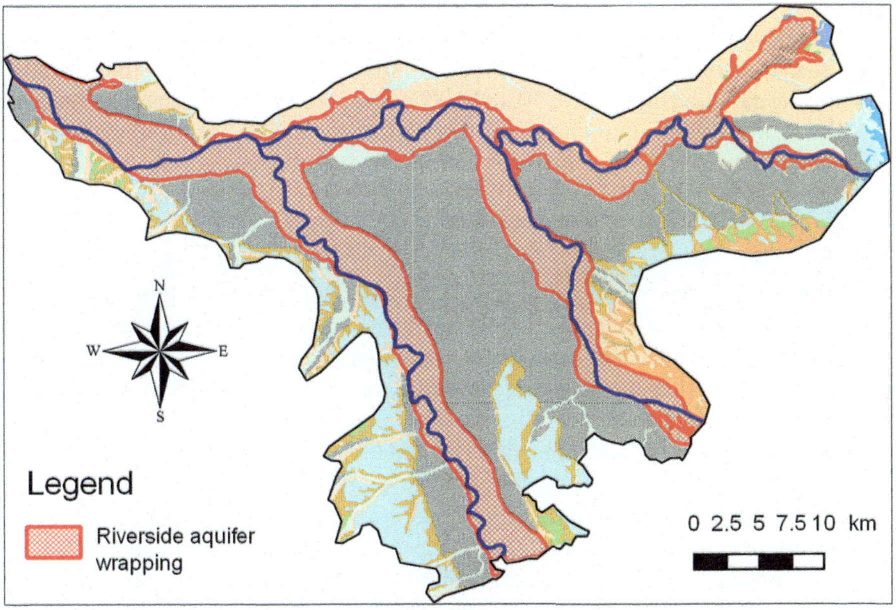

Fig. 13.4 The extent of the riverside aquifer in 1996. (Source: Gandolfi et al., 1997)

method. A square grid (1 km^2) was applied on the alluvial plain for a total of 895 computational grid cells. The presence of numerous boreholes in the zone meant that it was possible to describe correctly the aquifer's geometry.

The model's development was severely limited by the lack of data. Only one piezometric record was available when the first model was built, which ruled out the possibility of developing a transient state model that could reproduce the seasonal fluctuations in the groundwater level (as a function of rainfall and pumping). The first model (a steady state model) shows the average state of the aquifer during the low water period.

The model was adjusted for the low water period in 1996, using data collected during the field survey. The volumes abstracted in 1996 were estimated as annual average pumping rates, calculated to ensure that the abstracted volume corresponded to the whole year. An average recharge of 80 mm was estimated based on the rainfall record and the piezometric history of the only well monitored since 1982. The model was calibrated using permeability values, the parameter that had the least field measurements. These were based on the range of values used for the different terraces. Once calibrated, the model was capable of accurately reproducing the piezometric baseline (the low water period in 1996).

13.3.5 From the Model to the Management Tool for Abstraction

The mathematical model was then used to simulate the impact of different abstraction scenarios on the groundwater levels. In the simulations, the abstraction levels are increased uniformly by 20, 50 and 100% compared to the estimated 1996 level.

The results of the simulation show that a growing number of model grid cells are dewatered when pumping increases. Although aquifer dewatering is unlikely to occur in the field given the scale of the model, the findings indicate that the resources are limited or over-exploited in several sections of the alluvial aquifer.

The model was divided into 58 management zones (Fig. 13.5). These zones were defined in relation to the geology (distinction between different alluvial terraces), flow lines (distinction between the different hydrogeological catchment areas), the density of abstraction points (calculated by subdividing the heavily exploited zones) and the model's grid cells (the construction of zones with too few cells was avoided).

Based on the simulation results, the maximum permissible flow rate for irrigation was calculated for each zone that ensured that the aquifer was not dewatered in the simulations.

The results of the modelling were transposed to a simplified decision support tool that can be run using Excel. Once the tool has integrated the data produced with the groundwater model, the abstraction authorisations can be updated and compared with the available resources to determine whether or not new abstraction authorisations can be issued. When first used in 1996, the tool revealed that in 10 out of 58

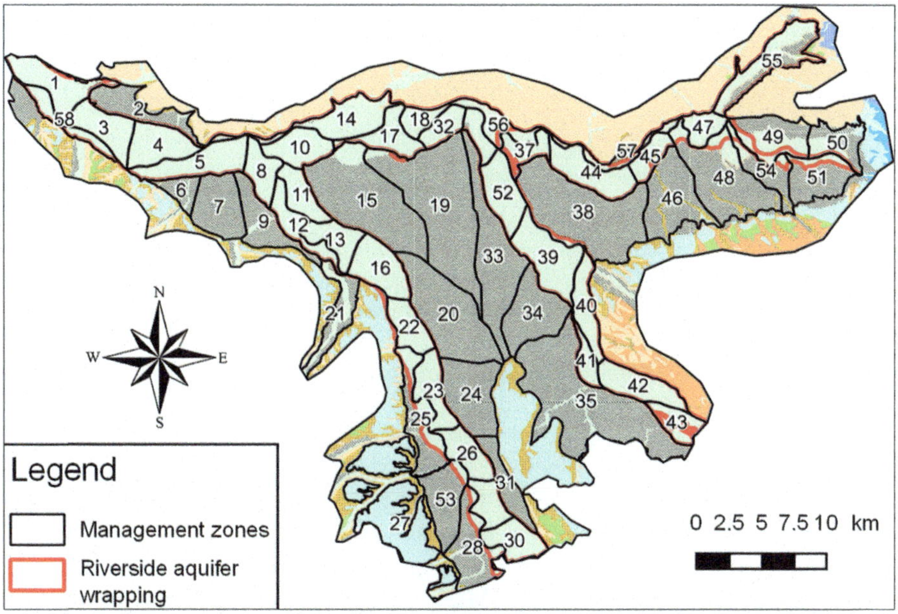

Fig. 13.5 Map of the 58 management zones for the alluvial aquifers in Tarn-et-Garonne. (Source: Gandolfi et al., 1997)

management zones, the authorised abstraction rate should have been reduced by a total of about 1000 m^3/h. On the other hand, the abstraction rates in 46 zones could have been increased to a total rate of 10,000 m^3/h. These findings show that if abstraction were better managed in terms of spatial distribution, pumping rates could be increased by a further 40%.

13.3.6 The Procedure for Allocating Annual Authorisations for Abstraction

Since 1996, this tool has been used by the MISE to determine the annual abstraction authorisations. The procedure was as follows (Fig. 13.6):

1. In February before the irrigation season begins, each farmer submits their irrigation plan to the MISE, with details of the crops and the area of land that the farmer wants to irrigate, the pumping rate and the required volume. The MISE checks whether the plan is consistent with the license (maximum authorised abstraction flow rate) issued for the borehole. It also ascertains whether the pumping rates and volumes requested are consistent with the crops and the acreage concerned.

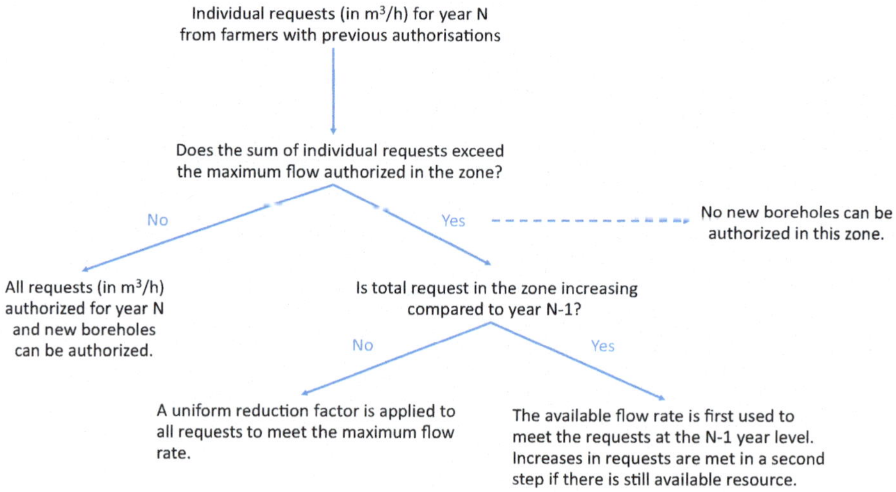

Fig. 13.6 Procedure for allocating the annual authorisations for abstraction, based on flow rate

2. For each management zone, the MISE compares the total flow rate requested with the maximum permissible abstraction rate (estimated using the model). If the total request is less than the maximum rate, all the annual applications are accepted. The MISE can also accept new applications to install boreholes in these management zones.
3. However if the total request exceeds the maximum abstraction rate, the resource in the management zone is considered to be over-allocated. No new boreholes can be authorised in the zone and the annual applications cannot be fully satisfied. First, the MISE strives to meet the individual requests, by allocating the abstraction rate attributed in previous years. Then, it shares out what is left of the available resource if any, in proportion to the extra request. The MISE notifies each farmer of the authorised pumping rate before the start of the season.

This procedure operated from 1996 to 2006. From the late 1990s, the Chamber of Agriculture processed the applications, which involved reception, compilation and verification (stage 1), and acted as the intermediary between farmers and the administration. In cases where the maximum abstraction rate allocated to certain management zones was exceeded, the Chamber of Agriculture negotiated with the applicants to reduce their requests. The negotiations involved stakeholders from the agricultural sector, not the administration, which meant there was more chance of the negotiations being successful.

Overall, during this period, few management zones experienced conflict situations where authorisations were refused.

13.3.7 Limitations of the Management Procedure

One of the procedure's limitations was that it failed to take into account the actual groundwater situation at the start of the season, when the farmers' annual allocation of resources was calculated. In fact, the model was not used each year to estimate the volumes genuinely available at the start of the season. Given that the alluvial aquifer is shallow and unconfined, the groundwater level is vulnerable to major fluctuations due to seasonal and inter-annual rainfall, which significantly modify the volume of water available for irrigation from year to year. As the management procedure was based on a steady state model, it could induce over-exploitation of the aquifer in some years if the volumes allocated were too high. On the contrary, it could unnecessarily restrict users in wet years. This observation, combined with regulatory changes and the acquisition of relevant data, led to a revision of the management model and the entire procedure for allocating annual pumping authorisations.

13.4 Volumetric Management of Abstraction: 2007–2015

After the 2006 Water Act came into force, the state was obliged to implement a volumetric management system for abstraction in all the basins designated as restricted zones. The management tool used in Tarn-et-Garonne for allocating abstraction authorisations (expressed as flow rate) failed to meet these new requirements. As a result, the model and the management tool were overhauled, which involved collecting new data.

13.4.1 Setting Up a Groundwater Monitoring System

Until 1996, the groundwater level was monitored at a single point. This was totally inadequate for measuring seasonal and inter-annual fluctuations in the groundwater level and especially for developing a transient state model.

To overcome this shortfall, two groundwater level monitoring networks were gradually set up (Ricard, 1998; Ricard & Tilloloy, 1999). The first was established between 1996 and 2005, which continuously measures and records the level in 10 monitoring wells equipped with telemetry. The second network comprises 26 other wells (16 were monitored regularly), where the state took manual measurements every 2 months (Fig. 13.7). In addition to groundwater monitoring, the flow rate and water levels were monitored in the main watercourses, using automatic hydrometric stations, combined with manual measurements taken in the three main rivers (the Garonne, the Tarn and the Aveyron).

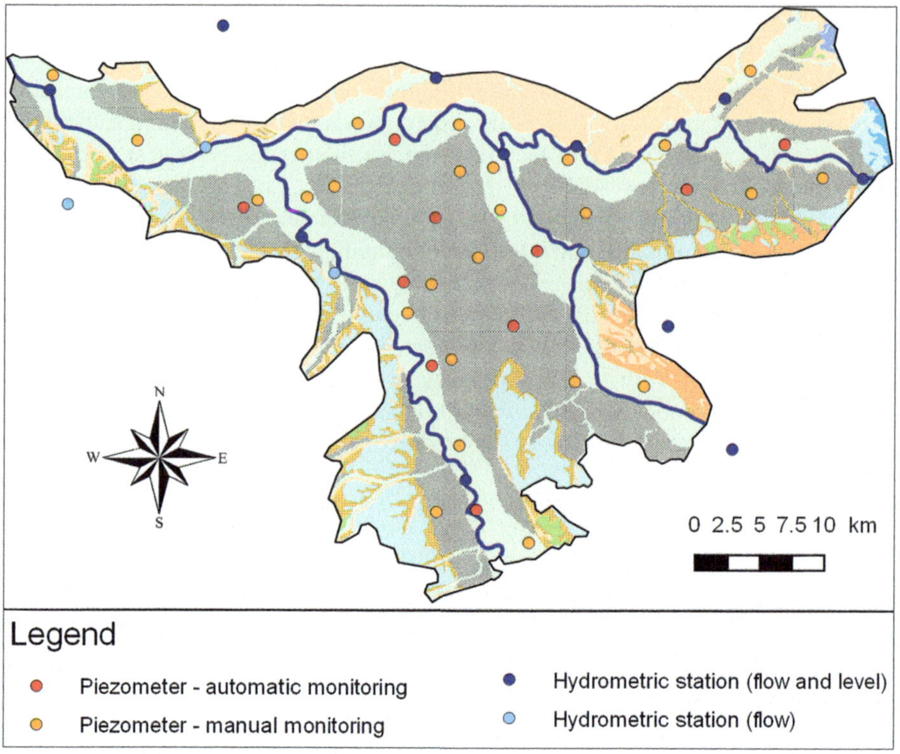

Fig. 13.7 Map of the groundwater and river monitoring network. (Source: Bardeau & Le Cointe, 2016)

13.4.2 The Revision of the Groundwater Model

These data were used to develop a new transient state model[1] to estimate the water resource available each year from 2007 onwards (Ghyselinck-Bardeau, 2004a, 2004b, 2007). The model was capable of reproducing seasonal and inter-annual fluctuations in water levels by taking into account the variability in rainfall and the observed abstraction. The model's spatial resolution was also improved with a smaller grid of 250 m.

Twelve scenarios of annual recharge were then designed using climatic data from the station in Montauban, including rainfall measurements since 1949. Seven scenarios of abstraction based on pumping rates for the year 2005 (±0 to 30%) were also used. By combining these scenarios, 84 simulations were conducted to predict the groundwater level at the end of the irrigation period. The simulations were used

[1]The model was calibrated based on the piezometric history for the period 1996–2005, with a bimonthly interval, i.e. the groundwater level was simulated every 2 months and abstraction for irrigation was allocated to June–July and August–September.

to determine the volumes and rates of abstraction for 12 recharge scenarios. The new abstraction volumes and rates estimated with this model are higher on average than those derived from the tool used since 1996.

13.4.3 A Tool to Assess the Annual Water Allocation

The results of all the simulations were integrated into a decision-support tool used by the MISE for allocating the annual authorisations for abstraction. The tool was developed using Excel and contains two modules. The MISE applies the first module to issue administrative approval for abstraction (in January) and the second to inform users of the actual volumes available in June (Saplairoles, 2005).

In January, when the administration has to issue authorisations for use, the aquifer's annual recharge is still underway. However, at this stage, farmers want some information to prepare their crop plan and decide on the area to plant. The MISE feeds data into the management tool for the total groundwater recharge between October and January, based on the recorded rainfall data, plus a hypothetical expected recharge for the next 3 months. The sum of the two provides a hypothetical, but probable, annual recharge. This value is compared to the 12 recharge scenarios simulated previously. The scenario that is the most similar to the current year (year N) is used to determine the maximum volume for abstraction. The MISE compares this volume to the requests made by farmers and the drinking water services. It then applies the rules (described above) to determine the volumes allocated to each applicant when the authorisations are issued. The volume specified in the authorisation is provisional and subject to revision if spring rainfall is lower than expected.

The second module is used at the start of June, when the irrigation period begins. At this stage, the effective rainfall for the whole hydrological year is known (apart from a few exceptions, summer rainfall does not recharge aquifers). An estimate of the annual recharge can now be fed into the management tool. As previously, the scenario that is most comparable to year N is used to provide the volume that can be abstracted in each of the 58 management zones. If spring recharge is lower than expected, the volumes actually allocated are lower than those specified in the annual authorisation established in advance in January.

13.4.4 Differentiating Between Water Entitlements and Allocation

When the volume for abstraction has been estimated, it is shared between the users in proportion to the flow rate that they are authorised to abstract. A parallel can be drawn with the Australian management model, in which the authorisation (flow rate) allocated to users amounts to an "entitlement". In fact, it constitutes a long-term right of use, which gives access to a fraction of the available resource. Whereas the allocation is the annual volume that is estimated each year, after the volume for abstraction has been calculated (as presented above).

Table 13.2 shows how individual allocations are assessed each year. At the start of year N, the users inform the manager of their irrigation plan. They declare the share of their entitlement that they wish to activate (1) and the share that they do not intend to use (dormant entitlement) (2). The manager then calculates the total maximum request for year N (5), which is capped and allocated on a per unit basis set for each hydrographic basin (4). If total request exceeds the maximum permissible volume (6), a reduction coefficient must be applied to determine the actual allocation (9). It is important to note that users, who have deferred the use of a share of their entitlement 1 year, can activate it freely the following year.

Table 13.2 presents two examples to illustrate the rationale applied in 2015: Zone 13, in the Garonne basin, where the resource was over-allocated; and Zone 14, in the Tarn basin, where the resource was under-allocated. In Zone 13, all the activated entitlements amounted to 1048 m^3/h. The maximum allocation (833 $m^3/m^3/h$) could not be granted to users because the total corresponding volume (872,984 m^3) significantly exceeded the Maximum Permissible Volume (MPV), estimated at 284,954 m^3 in 2015 with the management tool. The actual allocation to the users was reduced to 358 $m^3/m^3/h$. This corresponds to the MPV divided by the activated entitlement for the year 2015. On the other hand, in Zone 14, the users received the maximum allocation, which is 686 $m^3/m^3/h$, because the zone was not suffering from a problem of historic over-allocation.

Table 13.2 Rationale to assess yearly allocation

		Zone 13 Garonne	Zone 14 Tarn	Unit
1	Activated entitlements for year N	797	235	m^3/h
2	Dormant entitlements (possibly activated year > N)	251	27	m^3/h
3 = 1 + 2	Total entitlement	1048	262	m^3/h
4	Max. allocation in the basin	833	686	$m^3/m^3/h$
5 = 3 × 4	Total maximum request (year N)	872,984	179,732	m^3
6	Max. Permissible Volume (MPV) for year N	284,954	249,637	m^3
7 = Max(5–6)	Excess request in volume	588,030	0	m^3
8 = 5/6	Groundwater exploitation rate	306%	72%	% of MPV
9 = Min(6/1, 4)	Allocation in year N	**358**	**686**	$m^3/m^3/h$

Source: (DDT Tarn-et-Garonne, 2015)

13.5 The Emergence of Collective Management

The implementation of the 2006 Water Act led to the creation of agricultural users' associations, known as collective management agencies (or OUGCs). As explained in Chap. 3, when the OUGCs were set up, all the individual entitlements were cancelled and replaced by a single authorisation, which was attributed to the OUGC and corresponded to the sum of the former individual entitlements. The OUGC was then responsible for designing its own rules to share the resource between its members.

In Tarn-et-Garonne, the alluvial plain was split between five OUGCs after the creation of the collective management agencies. The division corresponded to the main rivers catchment areas in the department. The number of management zones was reduced to 21 and their geographic perimeter was adjusted to match that of the OUGCs (Fig. 13.8). Water pumped from the riverside aquifers was still considered as surface water.

The hydrogeological model was upgraded[2], which improved its accuracy in terms of calculating the volumes for abstraction. One of the main changes was to ensure that a minimum groundwater level limit was built into the calculation. This level, set in each of the model's grid cells, corresponds to a baseline dry year.

The result of the new simulations involving years with very contrasting climates, shows that the volumes available for irrigation vary from 14 (in 2001–2002) to 130 million m^3 per year (in 1992–1993) for all of the alluvial aquifers in the department. For the period 2006–2012, when the total requests for agricultural abstraction per management zone are compared to the permissible volumes, the model shows that the three driest climatic scenarios (minimum recharge, 20 dry years and 10 dry years) would trigger drastic reductions (up to 83%) in one to five management zones. Management zones 17 and 21 are the most vulnerable to a low annual recharge. For the other wetter years of recharge (from 5 dry years to a year with maximum recharge ever observed), the current request for irrigation can be met in all of the management zones (Fig. 13.9).

The management tool has also been improved with the calculation for recharge now built in. The user simply has to provide the data for rainfall and evapotranspiration. The new version is also capable of estimating the volumes that can be abstracted at three distinct dates: end of January (when the irrigation plans are submitted), end of March (before authorisations are allocated) and the start of June (to check whether the authorised volumes are consistent with the resource available).

Initially, the tool was only used by the state services. Later, it was transferred to the five OUGCs and used for allocating volumes of water to their members. The

[2] The improvements include: the aquifer geometry, calculation of recharge, accounting for the fluctuations in the water levels in the rivers and reducing the modelling timestep from 2 months to 10 days (Bardeau & Le Cointe, 2016; Thiéry, 2010a, 2014). The model was recalibrated, using data from groundwater monitoring for the period 2005–2015 and validated for the period 1995–2015.

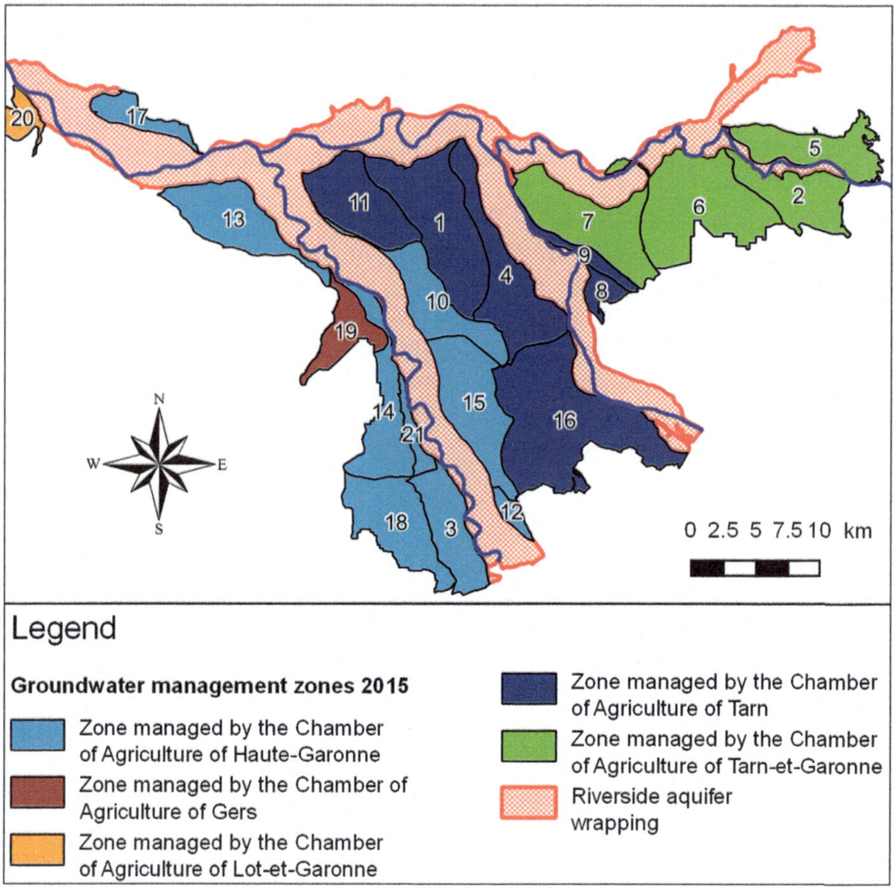

Fig. 13.8 Map of the 21 zones managed by the OUGCs, here, the departmental Chambers of Agriculture. (Source: Bardeau & Le Cointe, 2016)

OUGCs adopted the rules for sharing the volume allocated, which had previously been applied by the state.

13.6 Conclusion

The decision-support tool implemented for allocating water in the Tarn-et-Garonne department is the only one of its kind in France. Its main innovative feature is decoupling the management of entitlements (which define long-term access to a resource) from annual allocations. In theory, this means that crisis situations can be avoided because the volumes allocated are reduced in years when the groundwater

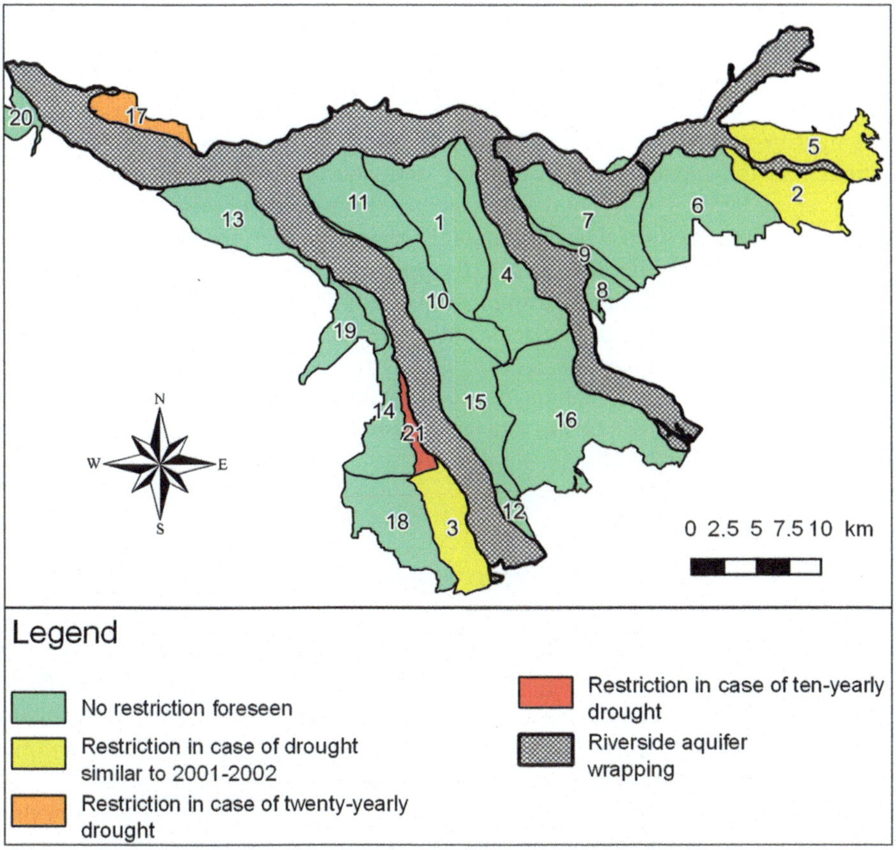

Fig. 13.9 Map of likely restrictions, based on abstraction levels that are similar to those for the period 2006–2012. (Source: Bardeau & Le Cointe, 2016)

levels are low at the start of the season. It also means that the volumes allocated can be increased when groundwater levels are extremely high.

13.6.1 Opportunities for Improvement

The quantitative management set up in Tarn-et-Garonne can be described as a process of continuous improvement over a period of 20 years. The stakeholders did not wait to acquire perfect knowledge of the aquifer before developing a model and a management tool. Gradually, both tools have been improved and this will continue as knowledge advances.

From a technical point of view, one possible improvement involves taking into account the real groundwater level at the start of the season when calculating the volume that can be abstracted (at the moment, the calculation relies on observed recharge alone). This would require installing a piezometer in each zone to determine more precisely the groundwater level in each zone.

From a societal point of view, one of the major issues is raising awareness amongst users about the tool – how it functions, the assumptions used for the calculations and the input data. This type of outreach is essential when it comes to encouraging farmers to accept the management rules and minimising the likelihood of conflicts in a dry year when volumes for abstraction are reduced. The aim is to provide farmers with Internet access to the groundwater records, supplemented with projected changes in the groundwater levels (as a function of the weather predictions). This would help each irrigator anticipate the amount of water available for the year and make an informed decision about crop choice and area.

Before this is achieved, making the tool available online is a first step. This would encourage OUGCs to adopt the tool and facilitate the management of the water resource available in the alluvial aquifer.

13.6.2 Compliance and Enforcement

This chapter would be incomplete if the issue of compliance and enforcement was not addressed. In fact, although the state and its partners invested significant resources to develop knowledge and produce innovating management tools, it was far less effective in fulfilling its responsibility to monitor compliance and enforce regulations. The water policing services only monitor 1% of water meters each year. Their inspection is limited to checking whether the device is in place and working properly. The water policing services lack the means to carry out regular meter readings, which would allow them to determine whether the volumes declared by farmers correspond to the volumes actually pumped. Their operations are sometimes hindered by political interference. The lack of resources for effective enforcement is inconsistent with the highly sophisticated technical tools available.

Implementing a quantitative management plan for water resources takes time. This is one of the difficulties when it comes to applying regulations. In fact, the transition to volumetric management constitutes a revolution for most users who have always considered water as a freely available resource. A change in mentalities cannot be decreed, but must depend on training and raising awareness, which requires education. The state services are striving to achieve this, particularly with regard to the water policing services' mission. Most agents consider that their action should be gradual and that full compliance can only be achieved after a period of social learning, which may take 10–20 years. Thus in Tarn-et-Garonne when users default, the water policing services systematically apply sanctions progressively. The first time there is an infraction, the water policing services inform the offender of the regulations and offer to help them comply. For the second time, an official warning is issued (there is an administrative record of this in the case of recidivism). For the third time, and only then, legal proceedings are initiated.

References

Bardeau, M., & Le Cointe, P. (2016). *Gestion des systèmes aquifères alluviaux dans le bassin Adour-Garonne – Résultats de la modélisation et outil de gestion des prélèvements dans le Tarn-et-Garonne*. BRGM report/RP-65583-FR, BRGM, Ramonville-Saint-Agne, France.

Bouroullec, I. (2013). *Carte géologique harmonisée du département du Tarn-et-Garonne*. BRGM report/RP-61974-FR, BRGM, Ramonville-Saint-Agne, France.

Collin, J. J., & Daum, J. R. (1995). *Méthodes d'évaluation de l'impact des prélèvements réalisés dans les nappes d'accompagnement sur les eaux superficielles associées*. BRGM report/R-38586, BRGM, Orléans, France.

Gandolfi, J. M., Danneville, L., Petit, V., & Tilloloy, F. (1997). *Connaissance, évaluation et protection des aquifères alluviaux de Tarn-et-Garonne (82)*. BRGM Report/R-39543, BRGM, Toulouse, France.

Ghyselinck-Bardeau, M. (2004a). *Etude des aquifères alluviaux du Tarn-et-Garonne – Modélisation nappe-rivière en régime transitoire (Garonne, Tarn et Aveyron) – Intermediary report– Year 2*. BRGM report/RP-53207-FR, BRGM, Ramonville-Saint-Agne, France.

Ghyselinck-Bardeau, M. (2004b). *Etude des aquifères alluviaux du Tarn-et-Garonne – Modélisation nappe-rivière en régime transitoire (Garonne, Tarn et Aveyron) – Final report – Year 2*. BRGM report/RP-53406-FR, BRGM, Ramonville-Saint-Agne, France.

Ghyselinck-Bardeau, M. (2007). *Gestion des systèmes alluviaux dans le bassin Adour-Garonne – Modélisation de la nappe alluviale de la Garonne, du Tarn et de l'Aveyron dans le département de Tarn-et-Garonne – Final report*. BRGM report/RP-55315-FR, BRGM, Ramonville-Saint-Agne, France.

Ricard, J. (1998). *Gestion des systèmes alluviaux dans le bassin Adour-Garonne – Appui à la Police de l'Eau en Midi-Pyrénées – Mise en place des réseaux d'acquisition de données sur les aquifères alluviaux du Tarn-et-Garonne en vue de la réalisation d'un modèle en régime transitoire (phase 2 – tranche 1)*. BRGM report/R-40364, BRGM, Toulouse, France.

Ricard, J., & Tilloloy, F. (1999). *Gestion des systèmes alluviaux dans le bassin Adour-Garonne – Appui à la Police de l'Eau en Midi-Pyrénées – Réseaux d'acquisition des données sur les aquifères alluviaux du Tarn-et-Garonne (phase 2 – tranche 2)*. BRGM report/R-40731, BRGM, Toulouse, France.

Saplairoles, M. (2005). *Etude des aquifères alluviaux du Tarn-et-Garonne – Présentation de l'outil de gestion issu de la modélisation en régime transitoire*. BRGM report/RP-54309-FR, BRGM, Ramonville-Saint-Agne, France.

Soulé, J. C. (1978). *Evaluation des Ressources Hydrauliques de la France*. Etat des connaissances et synthèse hydrogéologique du département de Tarn-et-Garonne. BRGM report/78-SGN-070-MPY, BRGM, Toulouse, France.

Thiéry, D. (1990). *Logiciel MARTHE – Modélisation d'Aquifère par un maillage Rectangulaire en régime Transitoire pour le calcul Hydrodynamique des Ecoulements*. BRGM report/R-32210, BRGM, Orléans, France.

Thiéry, D. (2010a). *Reservoir Models in Hydrogeology* in *Mathematical Models Volume 2*, chapter 13, pp. 409–418. Environmental Hydraulics Series, Tanguy J.M. (Ed.), Wiley/ISTE Editions, London, United Kingdom.

Thiéry, D. (2010b). *Groundwater Flow Modeling in Porous Media Using MARTHE* in *Modeling Software Volume 5*, chapter 4, pp. 45–60. Environmental Hydraulics Series, Tanguy J.M. (Ed.), Wiley/ISTE Editions, London, United Kingdom.

Thiéry, D. (2014). *Logiciel GARDÉNIA, version v8.2. Mise à jour version v8.7. Guide d'utilisation*. BRGM reoprt/RP-62797-FR, update: February 2019, BRGM, Orléans, France.

Thiéry, D. (2015). *Code de calcul MARTHE – Modélisation 3D des écoulements dans les hydrosystèmes – Notice d'utilisation de la version 7.5*. BRGM report/RP-64554-FR, BRGM, Orléans, France.

Pierre Le Cointe is hydrogeologist at BRGM (French Geological Survey). After an initial formation as geological engineer (2005) and hydrogeologist (2006), he completed his doctorate in quantitative hydrology and hydrogeology (2011), working for the French nuclear waste management agency (Andra), to assess the mobility of radionuclides in groundwater at a surface radioactive waste disposal through laboratory and field experiments combined with geochemical and hydrodynamic modeling. After several experiences in environmental consulting, to characterize groundwater quality on contaminated sites and define remediation processes, Pierre joined BRGM in 2015. He contributed to the hydrodynamic modelling of the alluvial aquifer in Tarn-et-Garonne and the conception of the spreadsheet tool designed to manage groundwater irrigation, in partnership with DDT 82.

Vorlette Nuttinck (DDT 82, Direction des Territoires de Tarn-et-Garonne) has been an agricultural technician with the Ministry of Agriculture since 1991. Since 2009, she has been involved in quantitative water management in Tarn-et-Garonne, covering the different uses (agricultural – drinking water – industry) and resources (surface water, rivers and lakes, and groundwater). Since 1996, the French geological survey (BRGM) has worked on groundwater resources in Tarn-et-Garonne, with three successive studies. Vorlette implemented the second phase of groundwater management (2008–2015) and participated in the third study (2013–2016). She has also contributed in the setting up of the agricultural users' associations, known as collective management agencies (OUGC), the elaboration of the quantitative management procedure and the multi-annual single authorization.

Jean-Daniel Rinaudo is researcher at BRGM, Montpellier University, where he coordinates the scientific program on environmental and risk economics. Initially trained as an agricultural engineer (Montpellier SupAgro 1994), he specialized in agricultural and resource economics (PhD University of Auvergne, 2000). Prior to joining BRGM, he worked for the International Water Management Institute in Pakistan where his research focused on the political economy of irrigation management reforms. His current research mainly focuses on the institutional economic dimension of groundwater management. Most of his research is conducted in France but he also works in Morocco and Chile. He is currently developing new research activities in the field of natural disasters economics, focusing on the methods to assess economic vulnerability and resilience. Dr. Jean Daniel Rinaudo is also member of the Scientific Council of the Adour Garonne River basin agency.

Chapter 14
Conceptual Approaches, Methods and Models Used to Assess Extraction Limits in Australia: From Sustainable to Acceptable Yield

Daniel Pierce and Peter Cook

Abstract The establishment of a limit for extraction is a fundamental requirement for the long-term sustainable development of groundwater resources. As in many countries, the context and methodology for determining these limits in Australia has evolved over time. The instigation of the National Water Initiative (NWI) in 2004 was a major milestone in Australia which enabled the development of a nationally consistent framework for water management. A key component of this major reform process has been a commitment across the States and Territories to the concept of establishing a sustainable water extraction regime for each water system. National guidelines developed over the past decade have outlined a general approach to using scientific processes and techniques to determine this regime which minimises the risks to the resource and users that depend on it.

This chapter analyses the evolution of the use of 'sustainable yield' in Australian groundwater management and presents how four themes have come to shape a shared conceptual framework for groundwater management. These are: appreciating how the timing and location of extraction impacts on recharge and discharge processes; accepting that setting sustainability limits necessarily requires value judgements; employing a risk-based management approach that includes socio-economic considerations and greater stakeholder engagement; and using resource condition limits together with volumetric allocations to set optimal management rules.

Keywords Acceptable extraction limits · Resource condition limits · Risk-based management · Sustainable yield

D. Pierce (✉)
Department for Environment and Water, Adelaide, SA, Australia
e-mail: daniel.pierce@sa.gov.au

P. Cook
Flinders University, Adelaide, SA, Australia
e-mail: peter.cook@flinders.edu.au

J.-D. Rinaudo et al. (eds.), *Sustainable Groundwater Management*,
Global Issues in Water Policy 24, https://doi.org/10.1007/978-3-030-32766-8_14

14.1 Introduction

The establishment of a limit for extraction is a fundamental requirement for the long-term sustainable development of groundwater resources. As in many countries, the context and methodology for determining these limits in Australia has evolved over time. The instigation of the National Water Initiative (NWI) in 2004 was a major milestone in Australia which enabled the development of a nationally consistent framework for water management. A key component of this major reform process has been a commitment across the States and Territories to the concept of establishing a sustainable water extraction regime for each water system. National guidelines developed over the past decade have outlined a general approach to using scientific processes and techniques to determine this regime which minimises the risks to the resource and users that depend on it.

This chapter analyses the evolution of the use of 'sustainable yield' in Australian groundwater management and presents how four themes have come to shape a shared conceptual framework for groundwater management. These are: appreciating how the timing and location of extraction has an impact on recharge and discharge processes; accepting that setting sustainability limits necessarily requires value judgements; employing a risk-based management approach that includes socio-economic considerations and greater stakeholder engagement; and using resource condition limits together with volumetric allocations to set optimal management rules.

14.2 The Evolution of Sustainable Yield Estimation in Australia

For the first 200 years of European settlement in Australia, water resource policies were focused on development to promote economic and demographic growth (Tisdell, Ward, & Grudzinksi, 2002). The role of water authorities was to develop dams and water supply systems to capture water and to encourage water use. This often resulted in over-allocation of water at below-cost, and a lack of incentives for water conservation (Mulligan & Pigram, 1989). Although bore construction was regulated in most Australian States and Territories from the 1960s (Clark & Myers, 1969), and mechanisms were put in place to control groundwater use, a focus on development remained. This development focus suited the concept of 'safe yield', in which sustainability was equated with maintenance of the resource and extracting less than the annual rate of recharge (Meinzer, 1920). The role of groundwater in sustaining ecosystems was generally not considered.

As discussed earlier in Chap. 6 of this book, groundwater development in Australia rapidly expanded during the 1960s and 1970s. However, increases in rates of water use were accompanied by increased concern about long-term impacts. In the late 1970s, there was a realisation that licensed surface water volumes within the

Murray-Darling Basin exceeded the available supply, and that further development would reduce the security of supply to existing users (Turral & Fullagar, 2007). In the 1980s, water management therefore began to consider broader issues, including efficiency of delivery services and environmental degradation (Tisdell et al., 2002). In 1981, the first national survey of water use in Australia took place (Turral & Fullagar, 2007), followed in 1983–84 by what was at that time, the most reliable and comprehensive assessment of Australian water resource availability (DPIE, 1987). Changes in water management coincided with the international focus on sustainability (culminating in the Brundtland Commission in 1987).

14.3 Navigating the Limitations of 'Sustainable Yield'

The concept of 'safe yield', where groundwater extraction from a resource is viable as long as it stays below the long-term recharge rate, has now been discredited (Bredehoeft, 1997; Bredehoeft, 2002; Sophocleous, 2000). Fundamentally, it ignores groundwater discharge processes and so can have serious environmental consequences. To avoid this misstep, it is helpful to remember that any groundwater extraction upsets the pristine equilibrium of a groundwater system – i.e. any amount of sustained extraction will produce a long-term impact. Initially the effect is to produce a loss in storage and a corresponding decline in water levels; but in time a new equilibrium in the water balance will be reached (assuming stable climatic conditions), with a portion of the former natural discharge now being extracted (Ponce, 2007). Ecosystems that depend on natural groundwater discharge will necessarily be impacted. In certain cases, such as where there are losing streams or where the natural recharge is limited by the aquifers being at full storage capacity, extraction can be, in part, accommodated by what is called 'captured' or 'induced' recharge (Ponce, 2007; Theis, 1940). This phenomenon is thought to be rare in the Australian landscape (Cook, Herczeg, & Harrington, 2001), although it is given due consideration in parts of the Murray-Darling system (DPI, 2015; MDBA, 2012).

The concept of sustainability should therefore apply to ecosystems that are dependent on groundwater, as well as to the resource itself. All groundwater pumping will have an environmental impact, and there is a need to assess what the impact of different levels of pumping will be once the system reaches a new equilibrium. The 'sustainable yield' of the system can therefore be thought of as a combination of the amount of discharge that can be acceptably diverted and the captured recharge, if any occurs. The 'sustainable yield' therefore depends on the value that is assigned to the discharge – for example, the impact on baseflow that can be tolerated by downstream users, or the acceptable impact on groundwater-dependent ecosystems (GDEs). Depletion of an aquifer eventually occurs if the extraction rate exceeds the natural and captured recharge. In this situation, a new equilibrium does not occur and water levels will continue to decline (i.e., mining the resource). Groundwater mining is a legitimate management strategy (DLRM, 2016), but it should be specifically acknowledged, and not confused with sustainability.

Recognising that the environment is a 'user' of groundwater, it is tempting to determine an 'environmental allocation' and hence to define the 'sustainable yield' to be some fraction of the long-term recharge rate. However, there are several problems with this approach. The first of course, is in defining the volume of water that is required by the environment to maintain its function. The second, is that if groundwater extraction lowers the watertable, the environment may no longer have access to the water resource (for instance if the watertable drops belong the root zone of groundwater-dependent vegetation). Managing groundwater using volumetric limits alone will therefore usually be insufficient to protect ecosystems that depend on groundwater, and other management tools will be required. These typically include either (i) monitoring water levels near high priority GDEs, and reducing groundwater extraction when water levels drop below specified limits; and/or excluding groundwater extraction from within defined distances of these GDEs (Noorduijn, Cook, Simmons, & Richardson, 2019). These tools are likely to be more effective than volumetric limits alone, although defining maximum watertable depths and exclusion zone distances can be problematic.

Another problem in using the long-term recharge rate to determine 'sustainable yield' is that recharge cannot be measured directly, it varies from year to year depending on the climate and also varies spatially across the landscape due to changes in soil type and land use. Estimates of recharge typically have large error bands, ranging up to +/−50%.

This cursory review alludes to some of the difficulties involved in 'getting it right': not under-estimating the yield of a system and therefore limiting its development, and not over-estimating what can be sustainably drawn without unwanted environmental impacts, some which may be irreversible, and to the challenge of having to navigate the political and social constraints to reducing allocations. There is the potential to misinterpret water level declines and to misjudge the impact on discharges; aquifers are not necessarily being over-exploited where there is a protracted decline in levels as it may be that a new equilibrium is still to be reached. In unconfined aquifers, such a decline may be predominantly driven by below average rainfall and reduced recharge, rather than groundwater extractions. In certain situations, pumping even a small fraction of the recharge can have dramatic and adverse impacts on baseflow into a stream if it is conducted in proximity to the stream, while in others, groundwater extraction well in excess of the mean recharge rate can safely proceed for hundreds of years before environmental effects become evident due to long residence times and large distances between the extraction and discharge areas (Cook et al., 2001). The environmental consequences of groundwater extraction will depend on the characteristics of the groundwater system and the particular extraction scheme being evaluated, not simply on the volume of water to be taken.

The sustainable yield concept is nevertheless a primary means of achieving good stewardship of a resource. Some of its main limitations can be accommodated by applying certain principles. The first is to acknowledge the trade-off between groundwater use and environmental impact; we should think of 'acceptable extraction limits', and this involves value judgements. This approach manages the inherent expectation – one that is misleading and unfounded – that a certain volume of annual extraction equates to the tipping point where a resource moves from being used sustainably to being overused. Another is to approach management as an ongoing endeavour and not as an engineering design problem that can be solved once-and-for-all. It requires investing adequate resources on an ongoing basis to build up an increasingly sound understanding of the hydrogeology of a system and quantify the inherent uncertainties, including through the use of more sophisticated numerical models and adequate monitoring data. However while adaptive management is essential, it should not replace appropriate planning, as reducing allocations can be difficult and environmental impacts may not be reversible. As explored in the next sections, determining the acceptable whole-of-system extraction rate needs a risk-based management approach to work out where most attention is given when seeking to understand and manage a resource. It accepts the reality that extraction limits are one tool in a water resource manager's toolbox, and that the risk of more localized impacts can be effectively managed using other management rules.

14.4 The Impacts of Water Management Reforms on Sustainable Groundwater Management

Towards the end of the twentieth century, Australia's surface water resources had become fully allocated and the demand for groundwater was increasing. Groundwater managers were concerned about ensuring the future viability of the resource. They were also increasingly expressing concern about the protection of natural environments that were dependent upon the groundwater (e.g., Hatton & Evans, 1998).

To address these issues, a national approach to water reform started in 1994 through the landmark COAG water reform framework and then continued through subsequent initiatives such as the National Water Initiative (2004), the Water Act 2007 and the Murray–Darling Basin Plan, which came into effect in November 2012.

The National Water Initiative (NWI) consolidated previous efforts and produced a significant advance by developing a nationally consistent framework for water management. A key component of this major reform process has been a commitment across the governments that are party to the NWI to the concept of establishing a *sustainable water extraction regime* for each system.

The position taken by the national agency guiding the reform process, the National Water Commission (NWC) centred on the concept that there is a level of water extraction from a particular system "which, if exceeded, would compromise key environmental assets or ecosystem functions and the productive base of the resource".

This concept was already in application by various States and Territories in the management of groundwater using the terminology of "sustainable yield". However, the NWC soon concluded that while the intent of the sustainable use definition was clear, there was a large degree of ambiguity in its detailed application. This meant that by 2010, it was still not possible to develop a consistent picture across the country of the level and distribution of over-allocation and over-use of water resources (NWC, 2010a).

National guidelines were subsequently developed outlining a general approach to using scientific processes and techniques to determine this regime (NWC, 2010b) for both surface water and groundwater resources. In terms of groundwater, this involved assessing the impacts of abstraction, including on groundwater dependent ecosystems, setting resource objectives, and then determining the extraction regimes that minimise the risks to the resource and users that depend on it. The level of detail was to be relative to the nature of the water resource, the level of risk and the type of plan to be prepared. In terms of the interaction between surface water and groundwater, and important policy change was the assumption that surface water and groundwater systems were fully connected, and that "water planning and management of the resource should be conjunctive" (NWC, 2009). The assumption of connection unless and until it could be proven otherwise was the reverse of the policy position in place until that time.

The general approach that was set nationally, reflects approaches that had by this time been largely adopted by various state and territory jurisdictions in the form of guidelines to the development of water resource management plans (for example, NSW: Bish, Williams, Gates, & Gill, 2006; WA: DFW, 2011a, 2011b; SA: Department of Environment, Water and Natural Resources, 2012a, 2012b). The national guidelines provide a common terminology that assists in building up a level of consistency across the country and facilitating the exchange of best practice. The guidelines have been accompanied by toolboxes describing technical approaches that can be used for implementing the frameworks (e.g., Richardson, Evans, & Harrington, 2011)

Thus within the early part of the twenty-first century, a two-tiered process for determining the 'acceptable yield' of an aquifer system had generally been adopted:

- For a low-risk resource, where the current use is a fraction of the estimated long-term annual recharge, the acceptable extraction limit is often initially set using the most straightforward method, which is as a percentage of rainfall recharge estimates. There are a number of 'rule of thumb' approaches or decision trees that have been developed for different types of aquifer to set the recharge percentage. The Basin Plan uses the 'recharge risk assessment method' (CSIRO & SKM, 2010); New South Wales uses a similar process in its "macro water sharing plans" for "less highly connected systems" with set infiltration factors for aquifer types (DPI, 2015). This approach also has been largely adopted by Western Australia (DFW, 2011a).
- For resources where the overall level of extraction from a system is a significant proportion of the recharge (and hence discharge), then the risk of sustaining extraction over the long term and on impacting on GDEs and other water resources is likely to be higher. In these situations, numerical modelling becomes a useful tool since groundwater systems have dynamic responses to time-varying and spatially distributed stresses. Groundwater models are important for under-

standing the impact of long term pumping, and can also be used to: (i) assist in setting volumetric extraction limits and developing other management tools; or (ii) assess the impact on the environment and other groundwater users of individual groundwater licence applications (Noorduijn et al., 2019). The Australian Groundwater Modelling Guidelines (Barnett et al., 2012; MDBC, 2001) have rapidly been adopted throughout the groundwater industry as a benchmark for best industry practice, and are almost universally referenced by environmental regulators and model developers. There are numerous examples of groundwater models used to inform the process of setting acceptable extraction limits and other management rules for groundwater management areas. Reviews of best practice in Australia have been commissioned by the Murray Darling Basin Authority (MDBA) and the NWC (Anderson, Cauchi, Hamstead, et al., 2014).

14.5 The Murray Darling Basin Plan

Efforts to systematize the definition of sustainable water use have been further advanced as a result of the Murray Darling Basin Plan (Basin Plan) which affects the states of South Australia, Victoria, New South Wales and Queensland. The Basin Plan was envisaged by the Commonwealth *Water Act 2007* and commenced in November 2012. It sets limits on the quantities of surface water and groundwater that can be taken from the Murray-Darling Basin within water resource plan areas. Water resource plans for these areas must be consistent with the requirements set out in the Basin Plan in order to be accredited or adopted under the Act. The Basin Plan is discussed in more detail in Chap. 8 of this book.

In effect, the Basin Plan specifies a two-stage process for managing groundwater use in a sustainable manner. Firstly, it requires whole-of-system extraction limits for each groundwater resource unit to be set. These are known as the long-term average sustainable diversion limits (SDLs) and can be revised during a review of the Basin Plan itself. Within the limits set by the SDL, localised impacts will be managed through water management arrangements in water resource plans which will be developed and implemented by the Basin states and accredited by the Authority. It then requires a set of rules to support the SDLs: rules which address residual risks due to spatial and temporal variations or uncertainties that are not able to be managed solely with the SDL.

The resource extraction limits in the Basin Plan are determined using either numerical modelling or a recharge estimation method (Anderson, Cauchi, Hamstead, et al., 2014; MDBA, 2012). Numerical models have the ability to directly relate extraction scenarios to impacts on the environmentally sustainable level of take for a water resource by the use of Resource Condition Limits (RCLs). RCLs define an acceptable upper limit to the impact on the groundwater resource and are typically described as water levels at key monitoring sites, which might include sites in stressed parts of the aquifer or sites near key environmental assets. A series of sites that are considered to be representative of the aquifer and for which historical records are available are established in each management area. The sites can be considered as barometers of aquifer condition and are used to determine SDLs that do not compromise the sustainability criteria in the Plan (which are described below in the discussion of the second stage of the management regime). For areas not covered by models, extraction limits are linked directly to recharge estimates. The 'sustainable yield' is determined by multiplying the estimates by a sustainability factor derived from a risk assessment of four areas: key environmental assets; key ecosystem functions; productive base; and key environmental outcomes.

To manage localized impacts of extraction, rules are developed that consider potential spatial and temporal impacts, as outlined in sections 10.18–10.21 of the Basin Plan. In summary these provisions state that a water resource plan must be prepared 'having regard to whether it is necessary for it to include rules which ensure that':

- For priority environmental assets and priority ecosystem functions that depend on groundwater, the operation of the plan does not compromise the meeting of environmental watering requirements.
- For groundwater that has a significant hydrological connection to surface water, the operation of the plan does not compromise the meeting of environmental watering requirements (for example, base flows).
- There is no structural damage to an aquifer (whether within or outside the water resource plan area).
- Hydraulic relationships and properties between groundwater and surface water systems, and within groundwater systems are maintained.
- Elevated levels of salinity and other types of water quality degradation are prevented.

The Murray-Darling Basin Authority has placed particular emphasis on the localized management through appropriate rule development. It has prepared the *Handbook for Practitioners - Water resource plan requirements* (MDBA, 2013) that includes the kinds of information and approaches that could contribute to demonstrating that the jurisdiction has 'had regard to' these matters. It has also made other information available to assist in identifying the types of rules that could be included in water resource plans to manage local impacts of groundwater extraction and has provided a suggested framework on how an assessment of the need for rules could be undertaken (Anderson, Cauchi, Hamstead, et al., 2014). The framework sets out each step of the assessment in detail, providing guidance on how it might be done and suggests tools and sources of information that could be used. The result of applying the framework to a particular groundwater Sustainable Diversion Limit (SDL) resource unit would be an identification of where and how priority environmental assets, groundwater and surface water connections, the aquifer productive base and groundwater quality are at risk from spatial variations in extraction or temporal variations in extraction and recharge.

Resource Condition Limits are central to both developing SDLs and addressing specific risks in the Water Resource Plans. These are defined in the Basin Plan as the limits beyond which the taking of groundwater will affect specified sustainability requirements (MDBA, 2013). As explained in (CSIRO and SKM, 2010):

> RCLs are effectively an upper limit to the impact on the groundwater resource – in that they define when an impact moves beyond being acceptable – and they will inform further work in the area. An RCL could be indicative of acceptable impact to other users, the groundwater resource itself, groundwater dependent ecosystems (GDEs) and to surface water resources. Implicitly, this assumes that there is a metric (i.e. a resource condition indicator (RCI)) which reflects these key constraints. RCIs are defined in terms of piezometric levels (or drawdown), since these are easily monitored and of direct relevance to most impacts e.g. GDEs, surface water, other users, entrainment of surface water, subsidence. However, other RCIs could include water quality (usually salinity) and flow volumes.

Resource condition limits can either be used in setting the overall extraction limit, or in the design of rules (e.g. bore distance rules are developed based on protecting a maximum level of drawdown at a groundwater dependent ecosystem), or they can be expressed in the rule itself (e.g. contingent rules that are activated when a RCL is approached or exceeded).

The development of MDBA-accredited management plans for 23 groundwater areas by four state jurisdictions has built on existing experience with modelling, management rules and using RCLs. It has provided a consistent management arrangement at a national level that provides a baseline for future improvements.

14.6 A New Way of Thinking

The development and use of guidelines at the national, state and territory levels over the last decade and a half has led to considerable progress in addressing some of the fundamental challenges in undertaking sustainable groundwater management in the country. Australia has succeeded in articulating a suitable conceptual framework and has demonstrated how this can be refined over time through collaboration between policy makers at state and national levels, and through rigorous scientific endeavour and the engagement of various community groups who have a stake in the use of groundwater resources.

There appear to be four key interconnected developments that have moved the management paradigm beyond the simplistic and often misleading approach of setting a single volumetric extraction limit beyond which a groundwater management area or aquifer system is 'over-used', towards a way of thinking that sees sustainable use as a complex problem that requires nuanced management based on ongoing monitoring, applied science and the definition of unacceptable aquifer conditions. These developments are: appreciating the way the timing and location of extraction impacts on recharge and discharge processes; accepting that setting sustainability limits necessarily requires value judgements; employing a risk-based management approach that includes socio-economic considerations and greater stakeholder engagement; and using resource condition limits to set optimal management rules.

As pressures on groundwater resources continue to increase around Australia, due to both increased demand and the uncertainties of climate change, each of these developments will be advanced further and integrated together more skillfully. The following sections provide a review of the concepts, approaches and methods currently being used, highlighting best practice across the country and outlining the opportunities for further improvement.

14.7 Risk-Based Approach to Groundwater Management

The approach now in place across all Australian jurisdictions is to adopt the right level of allocation management to suit each management area, and to give more planning effort and resources into the areas where pressures and risks are higher (eg. Bish et al., 2006; DEWNR, 2012b; DFW, 2011a; NWC, 2010b). Risk management and risk-based approaches are therefore central to achieving improvements in the efficiency and effectiveness of water planning and management activities, especially in times of declining financial resources.

There is a general commitment to following the provisions of the Australian and New Zealand Standard for risk management (AS/NZS ISO 31000: 2009) which provides overarching principles and a clear process for carrying out risk management.

- The initial phase involves identifying the environmental, social and economic objectives for managing the groundwater resource.
- When the objectives for water resource use have been set, the next phase involves identifying, assessing and managing factors that might threaten the ability to meet those objectives.
- The level of risk posed by each of the threats is assessed or examined by considering the likelihood and consequences of the threatening events occurring.
- The risks are evaluated and prioritised, and options for managing (or treating) the risks are identified.
- The risk treatments or management options are implemented.
- The success (or otherwise) of the management strategies is monitored and reviewed. The entire process is repeated periodically or as needed.

This approach ensures that more time and effort is directed to monitor, mitigate or respond to the threats that may pose the highest overall risks, and to ensure that management is targeted at the appropriate part of the water system.

The types of risks posed by groundwater extraction generally follow the following categories used by the MDBA (2012):

- Maintaining the ability of aquifers to continue to be productive over time.
- Sustaining groundwater dependent ecosystems.
- Sustaining surface water resources that are fed from groundwater.

- Preserving groundwater quality (salinity).

The value of the risk-based approach is that it facilitates greater stakeholder engagement, which is likely to lead to greater ownership of water management plans and compliance with their provisions (see examples in Richardson, Irvine, et al., 2011). Most jurisdictions have used a risk management framework to embed a significant amount of stakeholder consultation in the process of developing or revising water sharing plans (eg. SA DEWNR, 2012a, 2012b).

14.8 Resource Condition Limits

With increasing pressures on groundwater resources, groundwater managers around Australia began to increasingly encounter localised areas of declining groundwater conditions, even where the overall volumetric extraction for a management area was within the estimated sustainable yield (NGC, 2003). 'Water Level Response Management' was put forward as a useful tool for groundwater resources nearing full allocation. This refers to the regulation of groundwater take in a local area based on the response of the aquifer to pumping. Early on it was noted that a locally agreed "band-width" of water levels could be negotiated with the local community and other stakeholders (NGC, 2003). The approach naturally followed a risk-based framework, where targets could be pre-determined based on the level of impact on priority ecosystems and on water users that was agreed to be unacceptable.

Following trials in parts of NSW with long-term declining water levels, the use of water level management has become increasingly adopted around Australia. It has been employed both in setting the overall 'sustainable yield' for a system, and in the management of 'hot-spot' areas. Some detailed comparisons with the use of 'flux-based' management (mainly in the form of proportion of recharge estimates) have been undertaken (e.g., Werner et al., 2011), but are relatively rare. Nevertheless, over time, the terminology of resource condition limits has increasingly being adopted, recognising that there are several types of indicators of resource condition, beyond water levels alone. The possibilities and the benefits of community engagement in setting condition limits is increasingly acknowledged, as is the need to develop tools that explicitly link changes in the condition of groundwater to indicators of social and economic value (McIntyre & Wood, 2011; Richardson, Irvine, et al., 2011). However, this approach requires a greater level of monitoring than volumetric approaches, and also requires that governments are prepared to promptly respond to water level declines by reducing allocations (Noorduijn et al., 2019). This has not always been the case. A country-wide review of localized management, including the use of RCLs was conducted by the MDBA in 2014 (Anderson, Cauchi, Hamstead, et al. 2014; Anderson, Cauchi, Mozina, Smyth, 2014). The review noted a wide range in application of the RCL concept, and was able to provide useful guidelines based on best practice around the country.

Perhaps the greatest advantage of approaching the quandary of sustainable groundwater management by setting resource condition limits is its ability to integrate the social, economic and environmental needs. The sustainability question becomes a discussion of the level of risk and how the resource is valued: volumetric extraction limits become seen as 'acceptable extraction limits'.

14.9 What Next?

The previous sections have demonstrated how two decades of water planning reform in Australia has developed a strong conceptual framework and community of practice to guide the development and use of volumetric extraction limits in sustainable groundwater management. What is there left to do?

Firstly, there remain a number of technical challenges, which impede implementation of policy developments. Identifying ecosystems dependent on groundwater usually requires detailed field studies, and significant science gaps exist in determining their water requirements, and predicting impacts of changed water regimes (Eamus, Zolfaghar, Villalobos-Vega, Cleverly, & Huete, 2015). Development of appropriate approaches for monitoring impacts on GDEs is also in its infancy.

At a policy level, while the NWI and the development and implementation of the Murray-Darling Basin Plan have been important catalysts for progress, there is a clear need for the next generation of improvements in how volumetric limits are used in sustainable groundwater management. This comes from having to address the uncertainties of climate change, to incorporate tools that explicitly link changes in groundwater condition to indicators of social and economic value, to channel efforts to where they are most required from an economic, social and environmental perspective, to address uncertainty about impacts, to manage community expectations and engage various stakeholders meaningfully, including indigenous Australians.

What is required to meet this need is a deliberate, ongoing and systematic process of improvement that links the States and Territories together, and that allows sharing of information and provides access to the latest scientific developments. It is difficult to envisage this occurring without the leadership of a body similar to the National Water Commission and sufficient funding at the federal level to support the States and Territories in tackling the next generation of water resource management plans and in refining the state-level risk-based management frameworks that are currently guiding efforts. The Australian Government's recent assessment of the progress with national water reform (Productivity Commission, 2017) may provide some impetus in this direction – it identified certain priorities that should be incorporated into a renewed NWI by 2020, including the need to respond to the challenges of climate change. At the least, however, it is in the interest of the state jurisdictions to commit to funding the monitoring and modelling efforts that underpin the sound evidence-based management of the country's groundwater resources.

References

Anderson, T., Cauchi, T., Hamstead, M., Merrick, N. P., Mozina, M., & Phillipson, K. (2014, February). *Approaches to achieve sustainable use and management of groundwater resources in the Murray–Darling Basin using rules and resource condition limits*. Prepared for Murray–Darling Basin Authority.

Anderson, T., Cauchi, T., Mozina, M., & Smyth, B. (2014, February). *Approaches to achieve sustainable use and management of groundwater resources in the Murray–Darling basin using rules and resource condition limits – Literature review reference report*. Prepared for Murray–Darling Basin Authority.

Barnett, B., Townley, L. R., Post, V., Evans, R. E., Hunt, R. J., Peeters, L., Richardson, S., Werner, A. D., Knapton, A., & Boronkay, A. (2012). *Australian groundwater modelling guidelines* (Waterlines report). Canberra, Australia: National Water Commission.

Bish, S., Williams, R. M., Gates, G., & Gill, J. (2006). Assessment of sustainable limits for groundwater sources in NSW, MDBA conference Canberra pp. 107 – 115, NSW Department of Natural Resource.

Bredehoeft, J. (1997). Safe yield and the water budget myth. *Editorial, Ground Water, 35*(6), 929–929.

Bredehoeft, J. (2002). The water budget myth revisited: Why hydrogeologists model. *Ground Water, 40*, 340–345.

Clark, S. D., & Myers, A. J. (1969). Vesting and divesting: The Victorian groundwater act 1969. *Melbourne University Law Review, 7*, 237–257.

Cook, P. G., Herczeg, A. L., & Harrington, G. A. (2001, April 1–5). *Is groundwater sustainability possible?* 2001 A Water Odyssey, Australian Water Alliance 19th Federal Convention, Canberra, Australia.

CSIRO and SKM. (2010, December). *Sustainable extraction limits derived from the recharge risk assessment method*. Report to Murray-Darling Basin Authority.

DEWNR (Department of Environment, Water and Natural Resources, South Australia). (2012a, December). *Risk management framework for water planning and management*.

DEWNR (Department of Environment, Water and Natural Resources, South Australia). (2012b, December). *Risk management policy and guidelines for water allocation plans*.

DFW (Department for Water, Western Australia). (2011a, November). *Water allocation planning in Western Australia: A guide to our process*. Water resource allocation planning series.

DFW (Department for Water, Western Australia). (2011b, January). *Groundwater risk-based allocation planning process*. Water resource allocation and planning series. Report no. 45.

DLRM. (2016). *Alice springs water allocation plan 2016–2026*. Northern Territory government, Department of Land Resource Management. Report 01/2016A.

DPI (Department of Primary Industries, NSW). (2015, November). *Macro water sharing plans – The approach for groundwater*.

DPIE (Department of Primary Industries and Energy, Commonwealth of Australia). (1987). *1985 review of Australia's water resources and water use*. (Vol. 2). Department of Primary Industries and Energy and The Australian Water Resources Council, Canberra.

Eamus, D., Zolfaghar, S., Villalobos-Vega, R., Cleverly, J., & Huete, A. (2015). Groundwater-dependent ecosystems: Recent insights from satellite and field-based studies. *Hydrology and Earth System Sciences, 19*, 4229–4256.

Hatton, T., & Evans, R. (1998). *Dependence of ecosystems on groundwater and its significance to Australia*. Occasional Paper No. 12/98. Land and Water Resources Research and Development Corporation, CSIRO Australia.

McIntyre, N., & Wood, C. (2011). *Developing adaptive groundwater management arrangements in the South East of South Australia: A pilot study*. Government of South Australia, through Department for Water, Adelaide.

MDBA (Murray-Darling Basin Authority). (2012). *The proposed groundwater baseline and sustainable diversion limits: Methods report, MDBA publication no: 16/12, Murray-Darling Basin Authority, Canberra.*

MDBA (Murray-Darling Basin Authority). (2013, October). *Handbook for practitioners: Water resource plan requirements.*

MDBC (Murray–Darling Basin Commission). (2001, January). *Groundwater flow modelling guideline.* Report prepared by Aquaterra.

Meinzer, O. E. (1920). Quantitative methods of estimating ground-water supplies. *Bulletin Geological Society of America, 31*, 329–338.

Mulligan, H. K., & Pigram, J. J. (1989). *Water administration in Australia: agenda for change.* Centre for Water Policy Research. Occasional Paper No. 4, University of New England, Armidale.

NGC (National Groundwater Committee). (2003, March 14). *Water level response management as a micro-management tool.* Issue Paper 4. Australian Department of the Environment and Heritage.

Noorduijn, S., Cook, P., Simmons, C., & Richardson, S. (2019). Protecting groundwater levels and ecosystems with simple management approaches. *Hydrogeology Journal, 27*, 225–237.

NWC (National Water Commission). (2009). *Australian water reform 2009.* Second biennial assessment of progress in implementation of the National Water Initiative.

NWC (National Water Commission). (2010a, May). *Sustainable levels of extraction: Position statement.* Australian Government.

NWC (National Water Commission). (2010b). *NWI policy guidelines for water planning and management.*

Ponce, V. M. (2007). *Sustainable yield of groundwater.* http://groundwater.sdsu.edu

Productivity Commission. (2017). *National water reform (*Report no. 87). Canberra. Released on 31 May 2018.

Richardson, S., Evans, R., & Harrington, G. (2011). Connecting science and engagement: Setting groundwater extraction limits using a stakeholder-led decision-making process. In D. Connell & R. Q. Grafton (Eds.), *Basin futures: Water reform in the Murray-Darling basin.* Canberra, Australia: ANU Press.

Richardson, S., Irvine, E., Froend, R., Boon, P., Barber, S., & Bonneville, B. (2011). *Australian groundwater-dependent ecosystems toolbox part 2: Assessment tools* (Waterline report series no. 70). Canberra, Australia: National Water Commission.

Sophocleous, M. (2000). From safe yield to sustainable development of water resources – The Kansas experience. *Journal of Hydrology, 235*(1–2), 27–43.

Theis, C. V. (1940). The source of water derived from wells: Essential factors controlling the response of an aquifer to development. *Civil Engineering, 10*(5), 277–280.

Tisdell, J., Ward, J., & Grudzinksi, T. (2002, May) *The development of water reform in Australia* (Technical report 02/5). Melbourne, Australia: Cooperative Research Centre for Catchment Hydrology.

Turral, H., & Fullagar, I. (2007). Institutional directions in groundwater management in Australia. In M. Giordano & K. G. Villholth (Eds.), *The agricultural groundwater revolution: Opportunities and threats to development* (pp. 320–361). Wallingford, UK: CAB International.

Werner, A. D., Alcoe, D. W., Ordens, C. M., Hutson, J. L., Ward, J. D., & Simmons, C. T. (2011). Current practice and future challenges in coastal aquifer management: Flux-based and trigger-level approaches with application to an Australian case study. *Water Resources Management.* https://doi.org/10.1007/s11269-011-9777-2

Daniel Pierce is a senior hydrogeologist with the Water Science Unit of the South Australian Department for Environment and Water (DEW). He is part of a group of technical staff who provide technical advice and collaborate with policy and decision makers within DEW. Drawing on many years of private sector experience across Australia and the Pacific Islands, including 5 years working in French Polynesia, Daniel has worked on projects to support the periodic revision of Water Allocation Plans for groundwater resources around South Australia and the management of groundwater issues affecting the River Murray.

Peter Cook is a Professorial Research Fellow with Flinders University, and is internationally recognized for his work on groundwater recharge and surface water – groundwater interactions. He has co-written books on environmental tracers and ecohydrology, and has published more than 100 research papers. He frequently advises the Australian government on water management, and was a member of the Victorian Government Technical Audit Panel for Water Resources from 2002 – 2008, and the National Groundwater Committee from 2002 – 2007. Between 2009 and 2014 he was Deputy Director of the National Centre for Groundwater Research and Training (NCGRT).

Chapter 15
Case Study: An Integrated Approach to Determining Sustainable Abstraction Limits in Perth's North West Urban Growth Corridor

Mal McGivern and Clive Hampton

Abstract The North West urban growth corridor (NWGC) is an important area accommodating Perth's expanding population which is expected to increase by 50% to 3.5 million by 2050. Groundwater from a shallow Quaternary aquifer is the preferred source for the reticulated water supply provided by the Water Corporation (WC). Due to rising demands, the Department of Water and Environmental Regulation (DWER) reviewed groundwater allocations in the area to provide increased security for environmental values related to wetlands and to manage the saltwater interface. Because of inadequacies in the regional numerical model in the NWGC area, DWER developed a spread sheet analytical model to calculate groundwater balances and discharge to the ocean. This assessment proposed lower allocations for users than derived by the model. After consultations with DWER, WC developed an analytical assessment using additional data and an alternative method. Instead of individual aquifer test results, the regional impacts of significant extraction from a wellfield over 15 years were used to develop a relationship between the aquifer's response to pumping and transmissivity. This resulted in a larger estimate of throughflow which was more consistent with the original modelled estimates. These results were also confirmed by recently acquired drilling and aquifer testing data in the area.

Keywords Analytical model · Environmental values · Groundwater allocations · Groundwater model · Throughflow · Transmissivity

M. McGivern (✉) · C. Hampton
Water Corporation of Western Australia, Perth, WA, Australia
e-mail: Mal.McGivern@bhp.com; clive.hampton@watercorporation.com.au

J.-D. Rinaudo et al. (eds.), *Sustainable Groundwater Management*,
Global Issues in Water Policy 24, https://doi.org/10.1007/978-3-030-32766-8_15

15.1 Introduction

Perth is considered to be the first major city in Australia to feel the impact of climate change on water resources. With declining winter rainfalls, and an increase in average temperatures in the south western corner of Western Australia, the access to sustainable water resources for a fast growing population is becoming more problematic and costly for both water regulators and water utilities.

Perth's North West urban growth corridor (NWGC) represents some of the most important areas of future urban growth in the Perth Metropolitan Area to accommodate the region's expanding population (Fig. 15.1). The provision of locally sourced groundwater is considered the preferred solution to provide public water supplies for this area.

This case study examines how the sustainable yield of the aquifer designated to supply the NWGC was determined, and demonstrates that groundwater flow models are not the only tools that can be used – simple spread sheet analytical models using representative hydraulic parameters can also play a role. This study also highlights how co-operation between water providers and regulators and flexibility in the management approach are important ingredients for successful outcomes.

15.2 Background

15.2.1 Impacts of Climate Change

Perth experiences a Mediterranean type climate with hot, dry summers and mild, wet winters. Rainfall occurs mainly between May and September. However an estimated 10–15% decrease in annual rainfall has occurred since 1975 (CSIRO, 2009), and according to the Australian Bureau of Meteorology (BOM), the annual mean temperature has increased by 1 °C in southwest Western Australia (WA) during the past 40 years. The combination of these factors has meant that soil profiles have become subject to a drying trend, meaning that more rainfall is lost to evapotranspiration rather than running off into rivers and storage reservoirs, or recharging aquifers. Inflows into Perth dams have fallen from approximately 280 billion litres a year, to less than 50 billion in recent years.

Due to these declining inflows into surface storages, innovation has been required to ensure a growing water demand can be met. Since the late 1970s, Perth has increasingly used groundwater rather than surface water. Seawater desalination which commenced in 2005, has also grown to nearly 50% of the total water supply for the Perth metropolitan area. Over the past decade, Perth has been trialing a groundwater replenishment scheme to recharge deep confined aquifers with treated wastewater. In 2017, approval was granted by the Department of Health (DOH) and the Department of Water and Environmental Regulation (DWER) to fully imple-

Fig. 15.1 Locality plan

ment the scheme to inject 28 GL of treated wastewater per year (almost 10% of Perth's annual water requirements), into the deep confined aquifers beneath Perth. Collaboration between government agencies and water providers has been critical in implementing these projects, as well as gaining the acceptance of the general public.

15.2.2 Hydrogeology

Perth is situated mostly on the Perth Basin, a graben structure infilled with around 10 km of sediments ranging from Permian to Cretaceous in age, with Quaternary coastal plain sediments up to 80 m thick at the surface. The Superficial aquifer, which consists of a variety of Quaternary sands, sandy limestone and minor clay interbeds (Fig. 15.2) is the main aquifer of interest in the area. It has a saturated thickness ranging from 20 to 40 m, and groundwater flows from the crest of the Gnangara Mound (in the east of 15.2) in a south westerly direction and discharges over a saltwater interface at the coast. In the east, the sediments are mainly sand, whereas along the coastal strip, the sediments are sand and sandy limestone which is karstic in places. There is a hydraulic discontinuity between the sand and limestone (15.2), with a very low hydraulic gradient in the aquifer along the coast, reflecting the high permeability due to karstic conditions.

The Superficial aquifer is underlain by the Cretaceous Leederville aquifer and the deeper Jurassic Yarragadee aquifer. The Leederville and Yarragadee aquifers are deep confined aquifers (between 80 and 2000 m below ground level), and are predominately used for public water supply.

Approximately 210 GL per year is abstracted from these three aquifers to the north of Perth, servicing agriculture, industry, and the irrigation of parks and gardens. An estimated 110 GL of this water is provided for public water supply. About 65% of the abstracted water comes from the Superficial aquifer, with just over 15% abstracted from both the Leederville and Yarragadee aquifers. (DOW, 2009). Due to

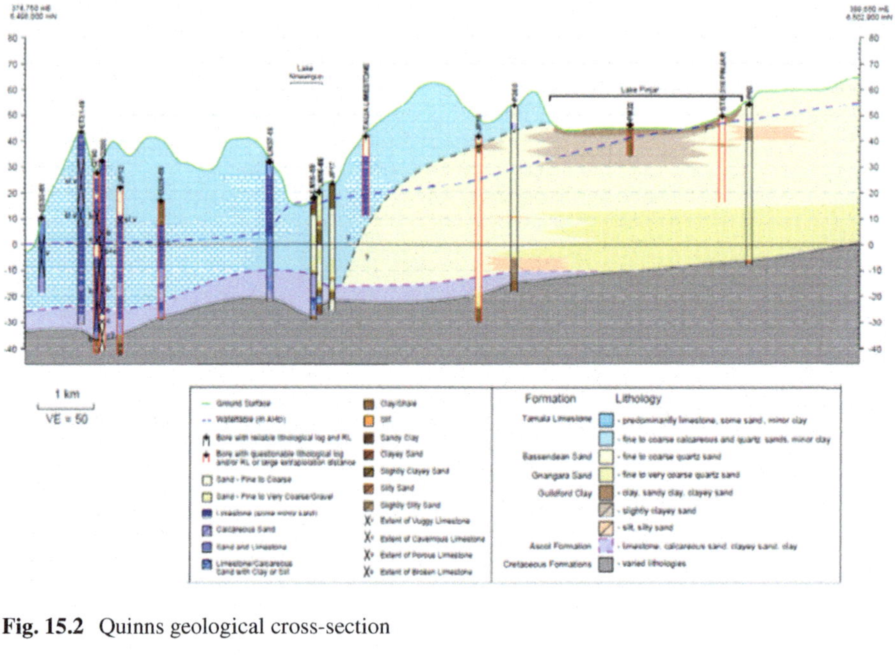

Fig. 15.2 Quinns geological cross-section

allocation constraints and demands from agriculture and industry, the confined aquifers do not provide significant opportunities for additional abstraction outside the current licensing/allocation arrangements.

Environmental considerations such as the prevention of sea water intrusion and maintenance of groundwater dependent ecosystems, together with social considerations relating to lake and wetland health are inextricably linked to groundwater abstraction, land use and climate change, and need to be taken into account in assessing the feasibility of groundwater abstraction.

15.2.3 Management Framework

The Western Australian Department of Water and Environmental Regulation (DWER) plans and manages all water resources throughout Western Australia, and manages water abstraction by issuing water extraction licences under the *Rights in Water and Irrigation Act 1914*.

The Water Corporation (WC) is the government owned utility and is the primary provider of water supply, wastewater treatment and drainage infrastructure across the state of Western Australia (including Perth), with services, projects and activities covering over 2.6 million km^2.

As part of a formal commitment between the WC and the DWER to collaborate on significant groundwater management issues, a working group was instigated to reconcile any differing technical interpretations that may occur with groundwater resources across the entire state of WA. This working group was known as the Water Resource Management Committee (WRMC).

Increasing urbanization and rapid population growth has placed significant stress on both surface and groundwater resources in the Perth region. This has seen the need for DWER to review groundwater allocations across the Perth region, to ensure that the previously issued groundwater allocations are sustainable, and will meet the climatic challenges Perth is currently facing and will likely continue to face into the future.

15.3 Case Study

Perth's North West urban growth corridor (NWGC) represents some of the most important areas of future urban growth in the Perth Metropolitan Area to accommodate the region's expanding population. The state government has approved over 9000 ha of new land development from north of Two Rocks to Alkimos (Fig. 15.1), to facilitate this growth and service Perth's growing population. This development will require water for household supply (scheme water) and for the irrigation of public open spaces. In addition, the Water Corporation's *Water Forever* (2009) identifies the North West Coastal Scheme as a new water source option that could potentially supply up to 25 GL per year of water from the Superficial aquifer, to the Integrated Water Supply Scheme to meet growing demand.

Due to the project's importance and complexity, DWER and the WC jointly agreed that all groundwater issues related to the urban growth corridor should be discussed and managed within the structure of the WRMC.

Northwest coastal groundwater is included in the Integrated Water Supply Scheme (IWSS) water sources identified through the Water Forever (Water Corporation, 2009) planning process. The provision of locally sourced groundwater is considered the preferred solution to meet the social, environmental and economic needs to resource the North West Corridor with scheme water.

In 2012, DWER reviewed groundwater allocations in the NWGC area to provide increased security for environmental values related to wetlands and the salt water interface. DWER reviewed water allocation limits for the superficial aquifer in the NWGC in response to:

- demand for water for planned urbanisation
- the Water Corporation's interest in the area for a new Integrated Water Supply Scheme (IWSS) source
- potential saltwater intrusion at the coast
- declining water levels in wetlands and groundwater-dependent ecosystems within the area.

Groundwater also supports important wetlands to the east of the NWGC. These are already under pressure from declining groundwater levels associated with regional and local abstraction and reduced rainfall recharge. The DWER review of allocation limits considered these factors, including the responsibilities under the *Rights in Water and Irrigation Act 1914* and *Environmental Protection Act 1986*.

The Perth Regional Aquifer Modelling System (PRAMS), a numerical groundwater model designed and constructed with input from both the WC and the DWER, was perceived by the DWER as not having an adequate calibration in this area to derive allocation volumes from modelled water balances. This is partly due to the problem of modelling the hydraulic discontinuity at the sand/limestone boundary, and the problem of determining a suitable transmissivity in the karstic zone of low hydraulic gradient. Consequently, DWER developed a spread sheet analytical model to calculate groundwater balances and the discharge to the ocean for the Quinns, Yanchep and Eglington Groundwater Sub-areas, based on the hydraulic conductivity values determined from a literature review.

A summary of the new proposed allocation limits and the projected demand in 2030 estimated by WC is provided in Table 15.1.

The DWER resource assessment calculated a reduction of throughflow of almost 10,000 ML/yr across the three sub-areas. The significantly lowered groundwater allocations shown in Table 15.1 would not meet future demand in two of the sub-areas. The technical assessment was reviewed by the WC who expressed concerns that the modelling work was not suitably comprehensive, and did not include all current and available data.

Table 15.1 DWER current and proposed allocation limits and projected demand (ML/yr)

Sub – area	Current allocation limit	Proposed allocation limit	Estimated demand
Yanchep	10,870	12,056	2456
Eglinton	15,450	3522	9000
Quinns	24,650	9208	14,000
Total	**50,970**	**24,786**	**25,456**

Table 15.2 Previous and revised throughflow estimates (ML/year)

Sub – area	Previous DWER (2012)	Revised WC (2014)
Yanchep	40,428	46,909
Eglinton	10,634	13,102
Quinns	37,703	36,833
Total	**88,765**	**96,844**

The WC decided to develop their own analytical assessment and water balance using additional data and an alternative method for estimating transmissivity. The standard approach is to estimate transmissivity (and hence hydraulic conductivity) from pumping test data, typically from a 24 hr constant rate and step test. The theory of pumping test analysis can be problematic in aquifers dominated by preferential flow and dual porosity environments and consequently, analysing and interpreting pumping test data from the karstic and highly heterogeneous limestone environment encountered near the coast proved to be problematic, particularly due to the absence of monitoring bore test data.

The Quinns borefield has been in operation since 1999, and represented a unique opportunity to examine the effects of significant groundwater abstraction on groundwater levels in the Superficial aquifer environment and develop a relationship between the aquifer's response to pumping and the aquifer transmissivity without any reliance on pumping test data. The WC was then able to input the new transmissivity estimates into a flow net and water balance model.

This resulted in larger estimates of throughflow (an increase of 8000 ML/yr) shown in Table 15.2 which are more consistent with the PRAMS modelled water balance estimates.

These results were also confirmed by recently acquired data from drilling and aquifer testing programs in the area. The WC successfully negotiated with DWER to carry out a re-assessment of the region's allocations based on the new throughflow estimates. The final outcome was that the allocation limits developed for the forthcoming Gnangara Allocation Plan due for release later in 2018 will be the same as the Current Allocation Limits listed in Table 15.1 and hence sufficient to meet the anticipated demand in the NWGC area.

15.4 Lessons Learned

The methodology, parameters and results contained in this program represent a collaborative approach between the Department of Water and Environmental Regulation and the Water Corporation to develop an agreed understanding of the hydrogeology encountered at the North West urban growth corridor (NWGC). A simple spread sheet analytical model was able to provide throughflow estimates that prevented a potential water supply shortfall.

The utilization of the WRMC to present and discuss key components of the assessment for the NWGC, was instrumental in a resolution being sought for this project. The ability for technical and water governance professionals from both DWER and the WC to discuss, challenge and collaborate on this project ensured a thorough interrogation of all issues regarding the availability of water resources within the NWGC. It also ensured the exchange of information relating to water planning and compliance related matters, and continued to foster working relationships, as well as share understanding of each agency's strategic and water related policy positions.

References

CSIRO. (2009). *Surface water yields in the south-west Western Australia* (A report to the Australian Government from the CSIRO South-West Sustainable Yields Project). CSIRO Water for a Healthy Country Flagship, Australia.

Department of Water. (2009). *Gnangara groundwater areas allocation plan.*

Water Corporation. (2009). *Water forever – Whatever the weather.*

Mal McGivern began groundwater career as a hydrogeological field technician with Rio Tinto in Paraburdoo North West Western Australia in 2005. After several years working in consulting, I began working for the Water Corporation in WA in 2010, where my work primarily focuses on both borefield development and optimization, with a strong focus on allocation and regulator compliance in various schemes around the state.

Clive Hampton is principal hydrogeologist at the Water Corporation, Perth, Western Australia.

Chapter 16
Using Resource Condition Limits to Define Groundwater Management Objectives in the Barossa Valley, South Australia

Daniel Pierce, Roger Cranswick, and Megan Hancock Lane

Abstract Groundwater resources are of vital importance to the iconic Barossa Valley wine producing region in South Australia. This case study outlines a new approach to determining water entitlements for a revision of the Water Allocation Plan that regulates the use of groundwater. Resource managers have been engaging stakeholders in a discussion about resource condition limits in newly defined management areas. This approach is becoming more widely adopted and provides greater transparency in linking rates of groundwater extraction to unacceptable impacts on users, including the environment. In this instance, a numerical flow model has been used to estimate the likelihood that a certain level of pumping will result the condition of the system declining in coming decades beyond a certain state as measured by resource condition indicators such as water levels, groundwater discharge to streams and the ingress of higher salinity groundwater. In management areas where the resource is more vulnerable to short term changes in condition, a more responsive management regime is being developed where allocations can vary on an annual basis. This case study presents a useful prototype for more responsive and participatory management for other regions which face the uncertainties of climate change and increased demand due to economic pressures.

Keywords Groundwater entitlements · Groundwater management · Groundwater model · Management plan · Resource condition · Stakeholder consultation

---

D. Pierce (✉) · R. Cranswick · M. Hancock Lane
Department for Environment and Water, Adelaide, SA, Australia
e-mail: daniel.pierce@sa.gov.au; cranswickrh@cdmsmith.com; megan.hancock@sa.gov.au

J.-D. Rinaudo et al. (eds.), *Sustainable Groundwater Management*,
Global Issues in Water Policy 24, https://doi.org/10.1007/978-3-030-32766-8_16

16.1 Introduction

Groundwater resources are of vital importance to the economy of the iconic Barossa Valley wine region in South Australia, and have been managed using a system of licenced water entitlements since 1989. The regional natural resources management body and the state government are currently undertaking a second revision of the Water Allocation Plan (WAP) that regulates use of groundwater and have opted for a new approach to determine the capacity of the resource to meet various demands placed on it.

The approach draws on recent scientific studies that recommend new management areas defined primarily by hydrogeological considerations, and that provide information that allows resource managers to engage various stakeholders in a discussion about establishing resource condition limits in each management area.

The use of condition limits is becoming more widely adopted in South Australia, as it provides greater transparency in linking rates of groundwater extraction to the impacts on users, including the environment.

This chapter describes how numerical groundwater modelling scenarios (applying different levels of pumping under projected climatic conditions) have been used to estimate the likelihood of specific and measurable resource condition limits being exceeded (e.g. water levels, groundwater discharge to streams and the inferred ingress of higher salinity groundwater).

As a step in the process of community engagement around the revision of the WAP, resource managers have worked with a representative community group to investigate how to use this information in a risk assessment and community engagement process. The process of determining the unacceptable level of risk to the users that are dependent on the resource is currently underway, with the aim of setting a limit on how much water is available for consumptive use.

16.2 Background

Located approximately 60 km north-east of Adelaide (Fig. 16.1), the Barossa Prescribed Water Resources Area (PWRA) covers an area of approximately 528 km^2. It incorporates the Barossa Valley and the surrounding highland areas of the Mount Lofty Ranges. It is characterised by a Mediterranean climate, with an average annual rainfall ranging between approximately 500 and 750 mm. The Barossa PWRA has been extensively cleared of native vegetation and supports around 7000 ha of irrigated crops, with vineyards predominating.

Surface drainage occurs predominantly via the North Para River, which originates in the Ranges to the east and flows in a south-westly direction along the

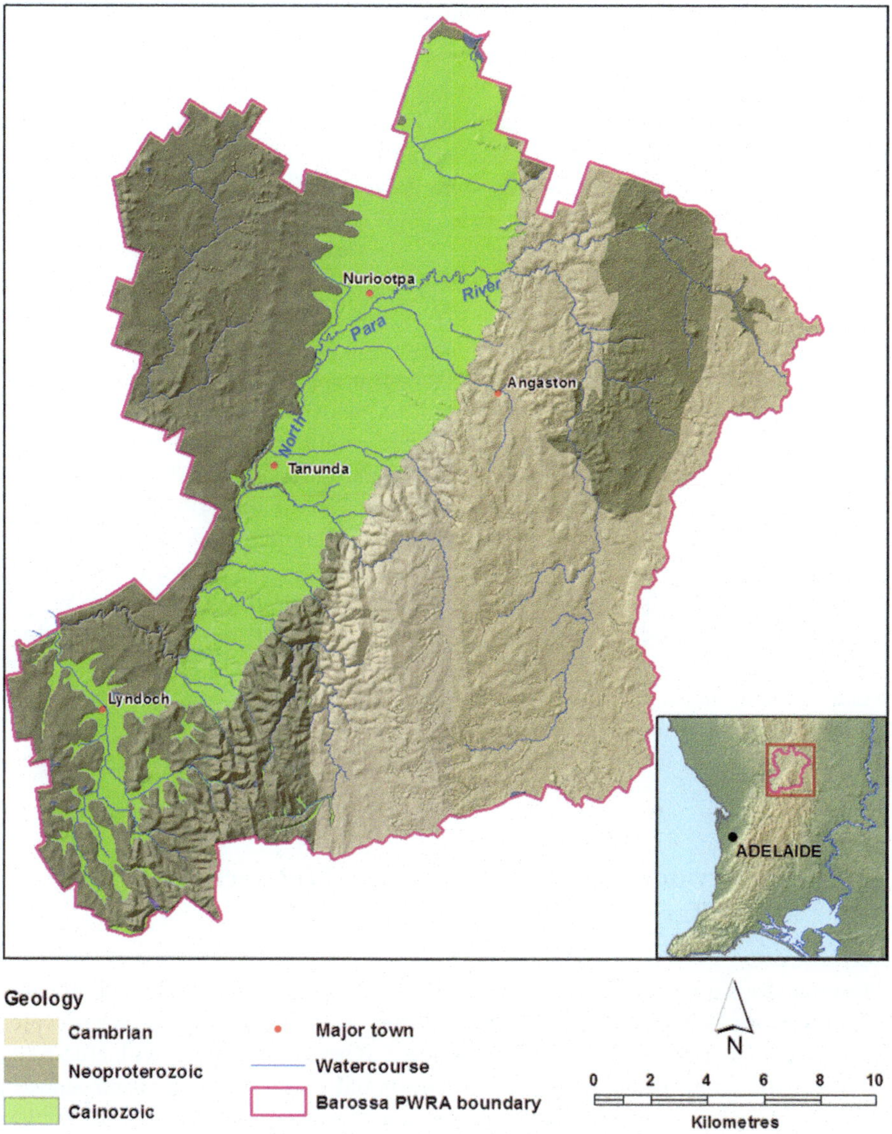

Fig. 16.1 Location and simplified geology of the Barossa prescribed water resources area

Barossa Valley. Streamflow in this ephemeral river and its tributaries is heavily impacted by infrastructure such as farm dams or in-stream weirs. Within the watercourses, numerous permanent pools are maintained by groundwater discharge (Hancock, Stewart, & Green, 2014).

16.3 Hydrogeology of the Barossa PWRA

The Barossa Valley is approximately 25 km long, 6 to 8 km wide, and up to 140 m deep. It has been in-filled with Tertiary fluvio-lacustrine sediments comprising sand, silt, clay and lignite, which is overlain by up to 30 m of Quaternary clays (Cobb, 1986). The valley sediments were deposited from the north and east in a series of overlapping alluvial fan deposits, resulting in a complex system of aquifers and confining beds (Fig. 16.1).

Surrounding the valley are faulted and folded metasediments of late-Proterozoic and Cambrian age which consist mainly of metamorphosed dolomitic siltstone, shales and marble and form the regional Fractured Rock Aquifer (FRA).

The complex sedimentation has led to the grouping of the basin sediments into a Lower Aquifer (LA) and an Upper Aquifer (UA) which are separated by a thin layer of carbonaceous deposits (Brown, 2002). The weathering of the basement rocks beneath the valley sediments is highly variable and influences the degree of connection between the LA and the underlying confined FRA (Cobb, 1986).

The lateral connection between the sedimentary aquifers and the adjacent FRA is also not well understood but a general flow from the ranges westward into the basin is indicated in Fig. 16.2 by the salinity distribution (Cranswick, Pierce, Wright, & Videka, 2015). The characteristics of the FRA vary according to the host geological formation. Most studies have highlighted the complexity of the overall groundwater system, noting variable hydraulic connections between the three major aquifers depending on the location (Cranswick, Harrington, & Harrington, 2016; Cranswick, Pierce, & Green, 2016).

16.4 History of Groundwater Use and Management in the Barossa PWRA

The groundwater resources were protected in 1989 by the declaration of a Prescribed Water Resources Area (to manage groundwater, as well as surface water in watercourses and farm dam storages), which granted existing water users with area-based entitlements, controlled the granting of additional entitlements, and introduced a program of metering of groundwater extraction. Estimated groundwater use for irrigation in the 1990s ranged from 3 to 5 GL/year, with extractions mainly concentrated in areas of lower salinity groundwater within the Valley.

Building on earlier hydrogeological studies, a water balance approach was used to estimate the overall sustainable yield of the sedimentary aquifers. Acknowledging considerable uncertainty in estimating the water balance components (especially recharge), this limit was defined as 5 GL/yr in the first Water Allocation Plan (WAP) which was prepared in 2000. This plan also included controls such as a moratorium on increased entitlements in 'stressed areas' (where historically salinity had increased and/or water levels had declined by an unacceptable amount), the

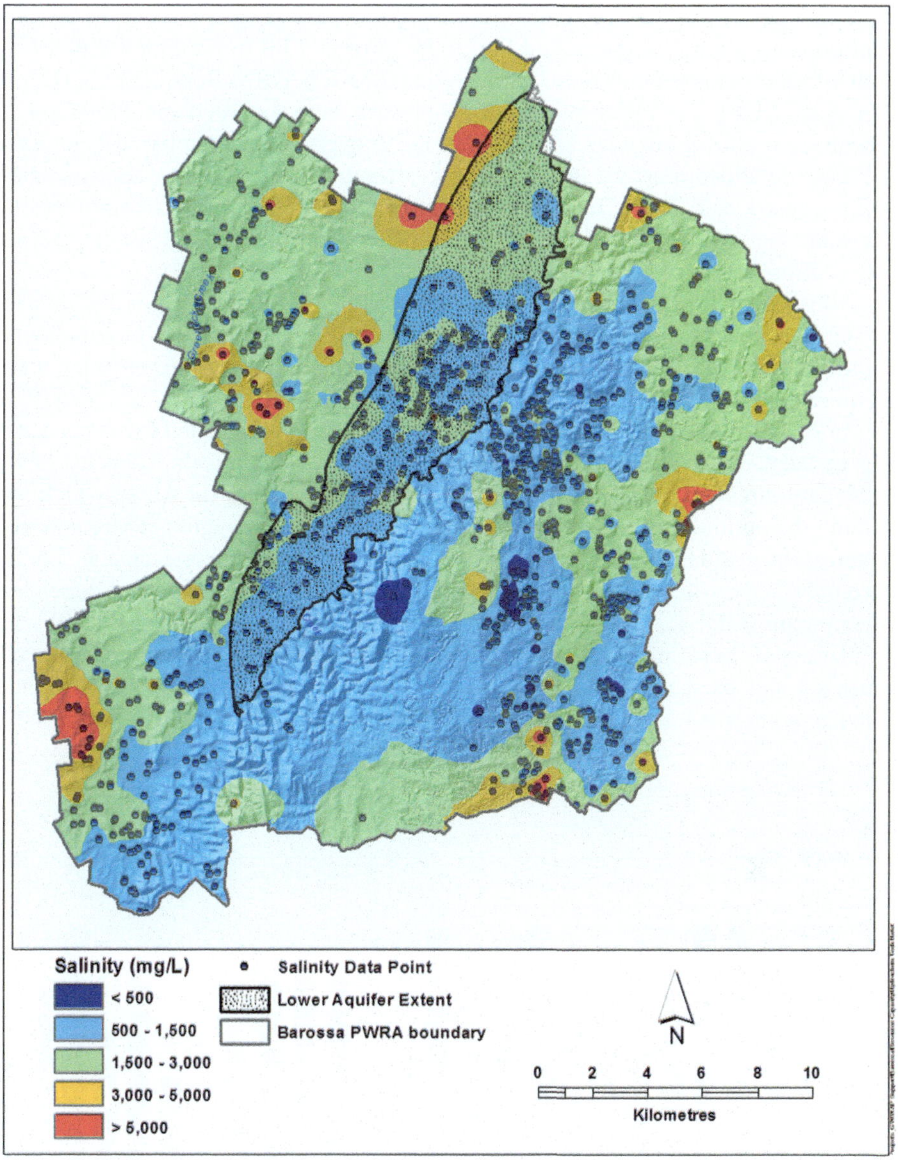

Fig. 16.2 Salinity levels in the Lower Aquifer and surrounding fractured rock aquifers

provision for managed aquifer recharge and for permanent and temporary trade of entitlements.

As part of the process to review the WAP in 2009, groundwater level and salinity trends were analysed in detail to detect any declines in the condition of the resource under the current management regime and climatic conditions. In light of mostly stable trends, the management approach was largely maintained albeit with the introduction of more adaptive determination of 'stressed areas' where increases in entitlements should be prevented (i.e. to be informed by future monitoring of the rate of salinity increase or water level decline) (AMLR NRM Board, 2009). Despite the considerable expansion of irrigated vineyards between 2000 and 2009, and the increase in allocations to 7.1 GL due to the conversion of area based licences to volumetric allocations, the use of the groundwater resources by the approximately 450 licensed users had in fact, slightly declined to 3–4 GL/yr due to improvements in water-use efficiency and the uptake of better quality imported water from the River Murray (Fig. 16.3).

Subsequent to the implementation of the 2009 WAP, further studies into the water-dependent ecosystems of the Barossa were instigated in order to improve the ability of the next WAP revision to better accommodate environmental water requirements.

In 2014, the scheduled five-yearly review of the WAP concluded that it needed to be updated, with a main consideration being a re-evaluation the capacity of the water resources to meet the considerable growth of demand for water, which was mainly being met through water imported from outside the region. There was also interest in evaluating the potential impact of projected climate change on the groundwater resources, and incorporating the demands of groundwater dependent ecosystems (GDEs).

To assist these investigations, a numerical groundwater flow model was constructed and used to model the impacts of both extraction and various climate change scenarios for the revision of the WAP (Li & Cranswick, 2016).

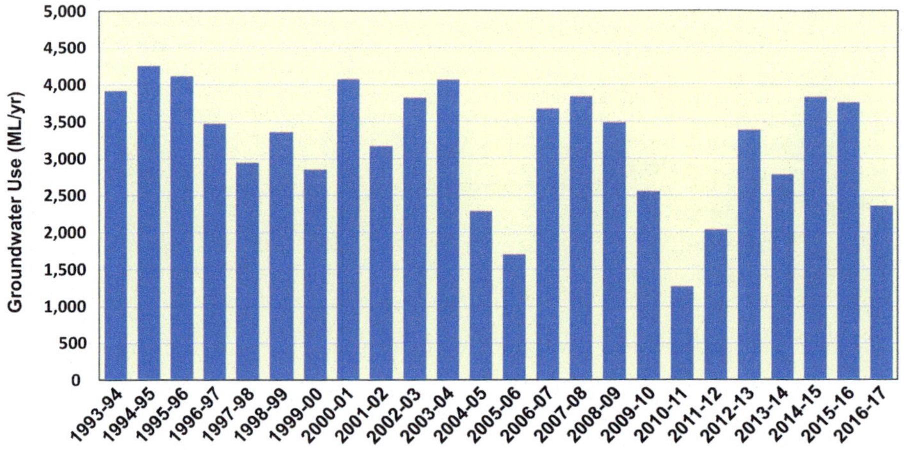

Fig. 16.3 Estimated and metered groundwater use. (After Cranswick et al., 2015)

16.5 A New Approach to Groundwater Management

The groundwater modelling scenario that simulated the extraction of full entitlement volumes predicted the impact on the resource would be considerable, and in some areas could not be sustained for more than a few years without declines in groundwater level (and probable increases in salinity) well beyond the conditions experienced historically. It is likely that this would not be acceptable from an environmental or consumptive water user perspective.

A new management approach was therefore proposed for consideration in the development of the revised WAP. This approach, which is becoming more widely adopted in South Australia, provides greater transparency in linking rates of groundwater extraction to impacts on users, including the environment. It involves using 'resource condition limits' to set overall volumetric limits, and the use of further risk mitigation measures, such as varying the volume which can be extracted from specific parts of the groundwater system in the short term (for example, annually) based on the local resource condition.

In the Barossa PWRA, a number of management zones have been proposed (Fig. 16.4). These were defined by the most important hydrogeological processes occurring in each proposed zone, the critical stresses, and the groundwater responses to those stresses. The cumulative impact of pumping in each proposed management zone can result in clear impacts on the condition of the resource, as measured by lowered water levels, reduced groundwater discharge to streams, and deterioration in water quality caused by the ingress of higher salinity groundwater. These resource condition indicators are also impacted in many parts of the system by climate variability.

The question to all those with a stake in the management and condition of the resource is: How much water can be taken before the risk of these impacts occurring becomes unacceptable? This determination requires a series of steps.

Firstly, estimating the likelihood of impacts occurring at various levels of pumping, which can be achieved by using predictive modelling, is required. Secondly, the level of consequence needs to be estimated, which requires consideration of the social, economic and environmental impacts; and thirdly, a determination is needed of the how tolerable the risks are, which requires a process of balancing competing interests, such as the requirements of GDEs versus consumptive water users.

The groundwater modelling predictions provide some of the inputs to the risk assessment process. The model has been used to estimate the likelihood that a certain level of pumping under projected climatic conditions will result in the resource condition declining beyond a certain pre-defined baseline in the coming decades. These pre-defined baselines are termed *potential* resource condition limits, and are set for each proposed management zone depending on which of the potential impacts are relevant to that zone.

Figure 16.5 introduces the two types of impacts for which resource condition limits can be set for the unconfined aquifer:

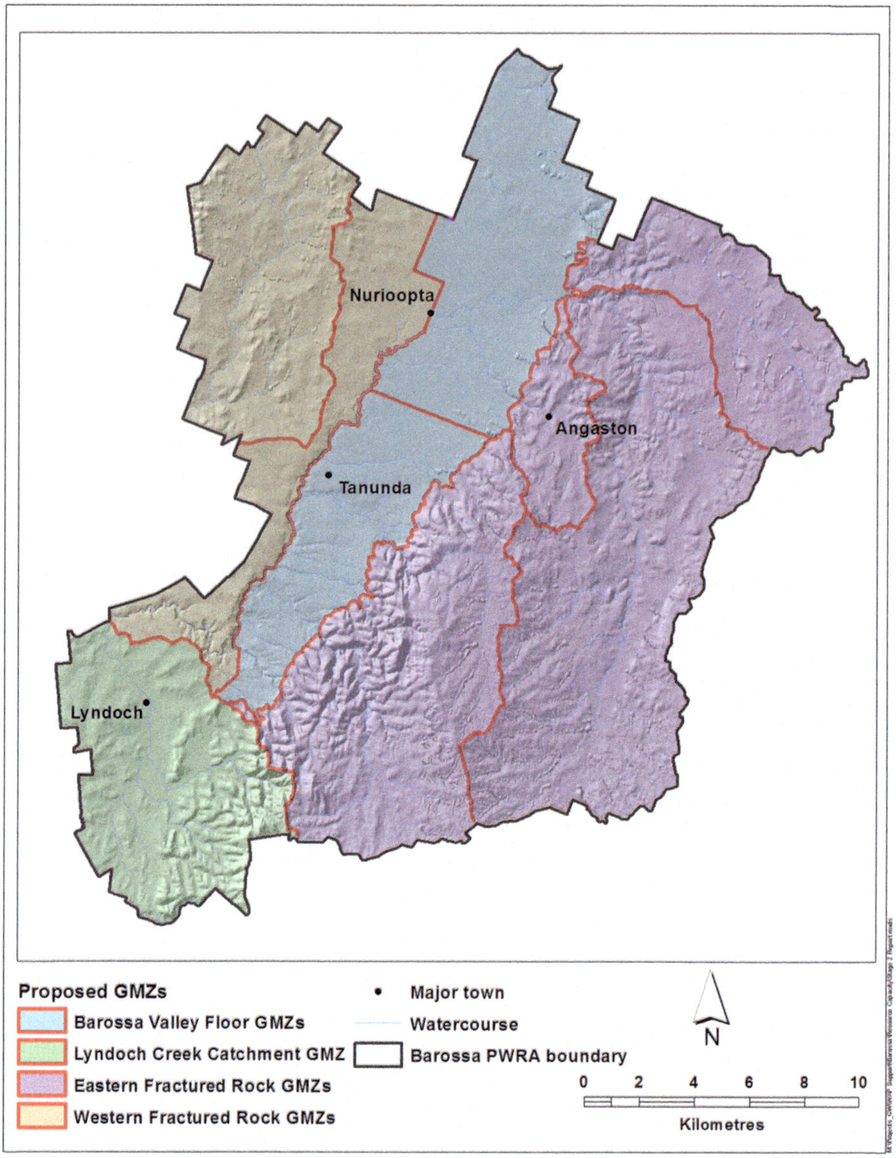

Fig. 16.4 Proposed groundwater management zones for the revised WAP

(a) Where declines in water levels in an unconfined aquifer will see declines in baseflow to streams, and even a reversal of flow gradients resulting in losing streams.
(b) Where declines in water levels in an unconfined aquifer will see a decline in the productivity of production wells.

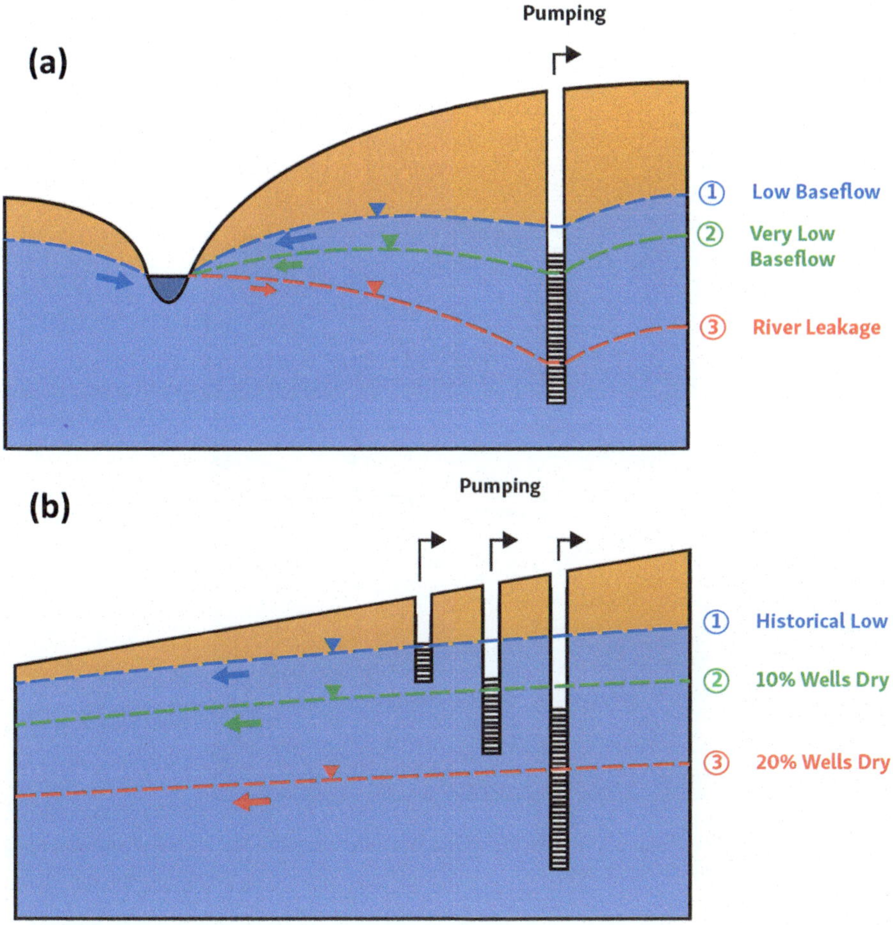

Fig. 16.5 Resource condition limits for the unconfined aquifer (**a**) declines in baseflow and (**b**) declines in storage

Similarly, Fig. 16.6 presents the two types of impacts for which resource condition limits can be set for the confined aquifer:

(a) Where declines in pressure levels in a confined aquifer may cause pumping costs to become uneconomical or present a risk to the structural integrity of the aquitard.
(b) Where declines in pressure levels in a confined aquifer may see induced leakage of more saline water from adjoining formations, leading to rising salinity levels in production wells.

Following initial consultations with a representative community stakeholder group, potential resource condition limits were identified as being a level of impact that has a severe consequence on either consumptive use or on GDEs. Secondary

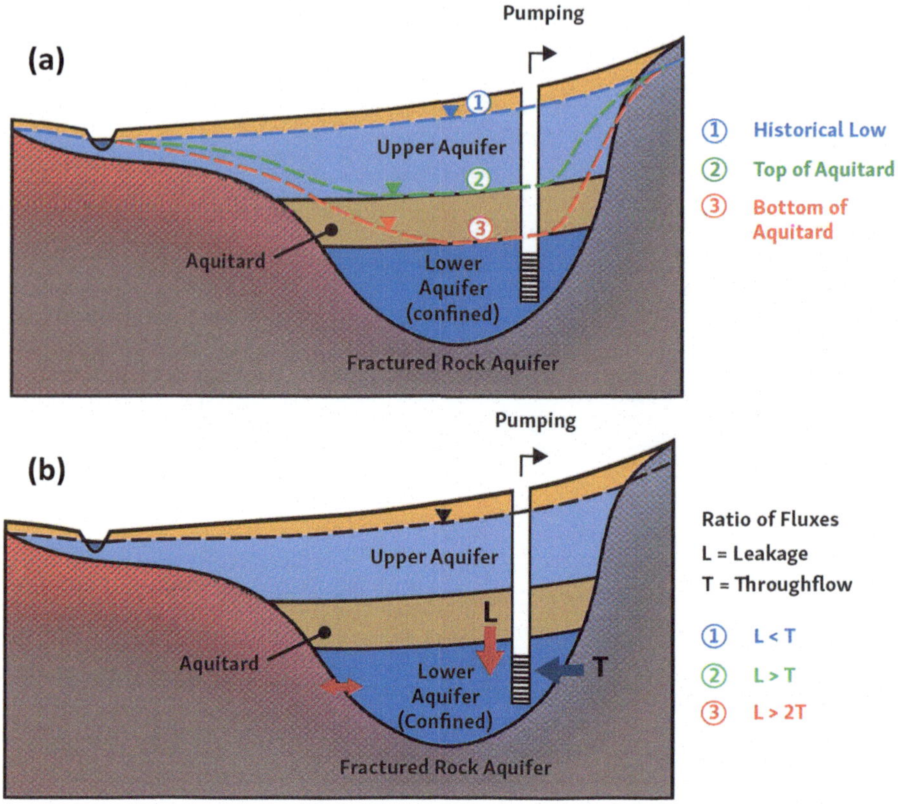

Fig. 16.6 Resource condition limits for the confined aquifer (**a**) pumping costs and aquitard integrity and (**b**) risk of salinity increase

impacts, such on impacts on the visual amenity of permanent pools in riverbeds, were also considered. Some examples are:

- Water storage in Upper Aquifer: groundwater levels are above the lowest historical level recorded in winter for at least 80% of the time (i.e. 4 out of 5 years).
- Water storage in Lower Aquifer: the pressure drawdown in summer remains above the top of the aquitard for at least 66% of the time (i.e. 2 out of 3 years).
- Salinity of modelled groundwater inflows into the Lower Aquifer: the volume of high salinity inflows from the overlying Upper Aquifer is less than half the volume of low salinity inflows from the FRA for at least 66% of the time (i.e. 2 out of 3 years).
- Streamflow conditions: baseflow is maintained such that the low flows experienced in a low flow year do not occur more frequently than in the past.

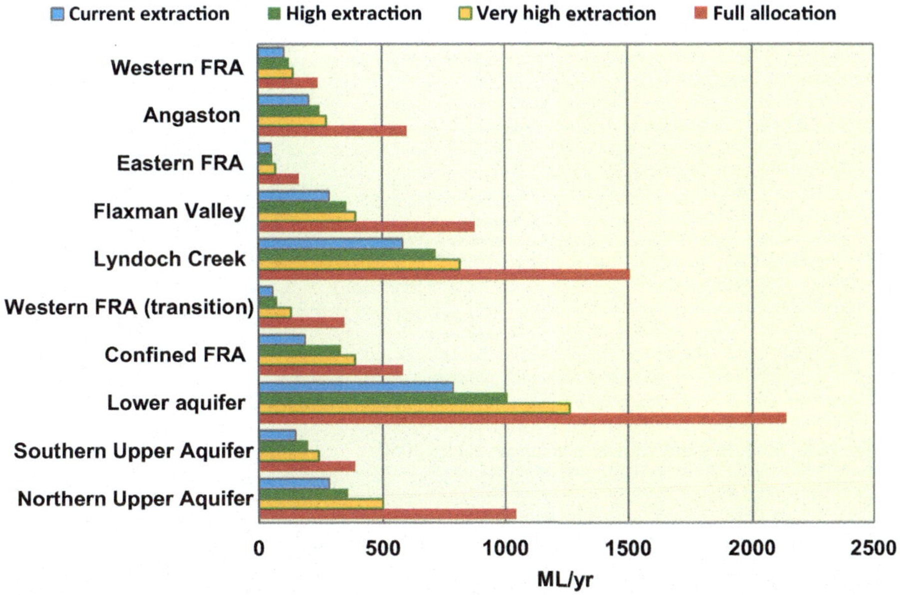

Fig. 16.7 Extraction rates in each proposed management zone under various predictive scenarios

A number of predictive modelling scenarios were then run to predict the likelihood of these impacts occurring. Eight scenarios were tested for each proposed management zone – four water use scenarios (Fig. 16.7) and two different climate scenarios.

The water use scenarios were as follows:

1. Current use – average groundwater use over the last 10 years
2. High use – average of the five highest water use years over the last 10 years
3. Very high use – average of the five highest water use years over the last 10 years, with use at full allocation during dry periods (15 out of 40 years)
4. Full allocation – water use at full allocation occurs every year

The two climate scenarios were:

1. Historical observed climate (rainfall & potential evapo-transpiration)
2. Climate change projections – historical rainfall was reduced into the future based on the regional summary of average projected change in rainfall (Charles & Fu, 2014) and a recharge elasticity factor of 3 was applied)

The outputs from the predictive modelling scenarios have been used to produce groundwater drawdown maps. Figure 16.8 presents the drawdown in water levels by 2050 compared to the current water levels for (a) extractions at current levels and (b) extractions at full entitlement volumes.

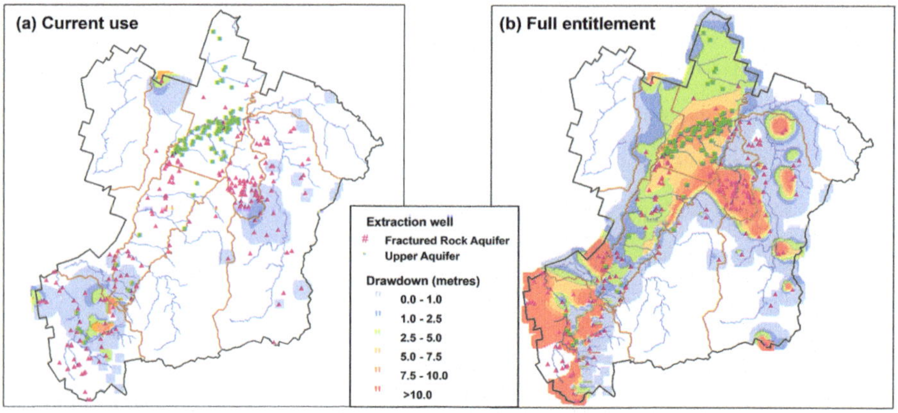

Fig. 16.8 Modelled drawdown in water levels by 2050

16.6 Setting Volumetric Extraction Limits

The outputs from the technical investigations – the different levels of decline in the resource condition, and the likelihood that these declines would occur under a range of pumping rates – can then be combined in a risk assessment process that engages the users of the resource in setting extraction limits.

Determination of the consequence has a socio-economic dimension – for example, how low can groundwater levels go before a certain number of irrigation wells run dry resulting in significant impacts on the livelihoods of irrigators. Determining the overall level of risk that is considered unacceptable also requires inputs of the community. The outcome from this decision-making process is the creation of suitable extraction regimes that do not cause the agreed resource condition limits to be exceeded.

The use of resource condition limits is therefore an elegant way of setting extraction limits in each proposed management zone, in a manner that is transparent to the users of the resource and various parties who have a stake in the condition of the resource. Table 16.1 presents the recommended extraction limit for each management zone derived from this risk management approach and a comparison with the existing use. These numbers provide a starting point for wider community discussions during the revision of the Water Allocation Plan and may be refined if further studies are carried out.

Table 16.1 Recommended extraction limits based on proposed resource condition limits in each proposed management zone

Management zone	Current extraction (ML/yr)	Recommended extraction limit (ML/yr)
Northern upper aquifer	281	357–501
Southern upper aquifer	147	147
Lower aquifer	786	1007–1264
Confined fractured rock	186	329–389
Western fractured rock transition zone	56	56
Lyndoch Creek	581	581
Flaxman Valley	284	284
Eastern fractured rock	52	52
Angaston	202	202
Western fractured rock	104	104
Total Barossa PWRA	**2679**	**3119–3580**

16.6.1 Lyndoch Creek Catchment Test Case

The process of setting extraction limits can be explored using the Lyndoch Creek Catchment as a test case. This catchment is located at the southern end of the PWRA (Fig. 16.4). A north-south trending valley is surrounded and underlain by the Neoproterozoic FRA and contains Upper Aquifer sediments up to 50 m in thickness.

- In the Lyndoch Creek Catchment, groundwater in the Upper Aquifer and FRA is recharged by rainfall. Groundwater level (as an indicator of aquifer storage) is therefore the critical resource condition indicator. Reduced recharge under a changing climate may result in declines in groundwater storage in addition to those caused by extraction.
  - In all extraction scenarios, groundwater storage is likely to be lower in the future than historical conditions.
  - Under the high, very high and maximum extraction scenarios, groundwater storage would be permanently lower than historical conditions by 2035.
  - Under the very high and maximum extraction scenarios, it is likely that wells will run dry for 20% of licensed users by 2035, and 30% of users by 2055.
- In this management zone, wells running dry for 10% of licensed users by 2030 at the current usage rates, is the potential resource condition limit to use in the risk assessment process engagement with various stakeholders.
- The adoption of this resource condition limit results in a recommended extraction limit of no more than 581 ML/yr. Further work is required to define the extraction limit more precisely.

16.7 Further Refinements in Quantitative Management

The use of extraction limits is a longer-term strategy in that they are reviewed every 5–10 years to evaluate their effectiveness in achieving management objectives. To deal with variations in climate which can affect water availability from year to year in thinner "less robust" unconfined aquifers, additional safeguards can be implemented.

The water available for use in certain management areas can be dynamically linked to the condition of the resource (i.e. by making a percentage of total entitlements available each year in response to changes to resource condition). Having worked with the community to determine the resource condition limit, a participatory approach can also be used to set appropriate triggers that determine when a management response is initiated. This helps groundwater users with short-term planning, as they can then see at the beginning of each irrigation season the likelihood of receiving lower water entitlements. A similar approach has been implemented in several areas of France (Chaps. 5, 13, and 18). Some of the advanced irrigators in the region have shown interest in this approach, where water availability becomes simply another factor in their business planning. Adopting this adaptive management approach would be advantageous for the Barossa PWRA as it would engage stakeholders at an early stage in dealing with the uncertainties of a drying climate before the impacts become significant.

Conjunctive use with surface water resources is restricted by the current policies and presents another potential area of refinement that is being considered in the current revision of the WAP.

16.8 Conclusions

As South Australia's water resources face the uncertainties of climate change and increased demand due to economic pressures, the Barossa PWRA approach presents a potential prototype for more responsive and participatory management.

This approach introduces the merits of the resource condition limit approach for relatively small groundwater management areas where groundwater extraction can be linked to specific impacts. In the areas where the resource is particularly vulnerable to short-term changes of condition, it can also be used to develop a more responsive management regime where entitlements can be varied annually or bi-annually.

While the resource condition limit approach can be adopted with only a basic understanding of a groundwater system, a high-value and relatively complex groundwater system similar to that in the Barossa PWRA, may require substantial investment over a number of years for a program of scientific investigation coupled with substantial community engagement. This example illustrates the potential for conducting hydrogeological and socio-economic studies based on a substantial monitoring network of resource condition indicators, and of undertaking numerical modelling in a manner that can provide inputs to be used in a risk assessment process and the development of appropriate management responses.

References

Adelaide and Mount Lofty Ranges (AMLR) Natural Resources Management (NRM) Board. (2009). *Water allocation plan – Barossa prescribed water resources area.*

Brown, K. G. (2002). *The hydrogeology of the Barossa Basin, South Australia.* Report DWLBC 2002/18, Government of South Australia, through Department of Water, Land and Biodiversity Conservation, Adelaide.

Charles, S. P., & Fu, G. (2014). *Statistically Downscaled Projections for South Australia – Task 3 CSIRO Final Report.* Goyder Institute for Water Research Technical Report Series No. 15/1, Adelaide, South Australia.

Cobb, M. A. (1986). *Groundwater resources of the Barossa Valley: Report of investigations 55.* Department of Mines and Energy, Geological Survey of South Australia.

Cranswick, R. H., Harrington, N., & Harrington, G. (2016). *Apparent groundwater age and hydrochemistry of the Barossa PWRA aquifers.* DEWNR Technical note 2016/07, Government of South Australia, through the Department of Environment, Water and Natural Resources, Adelaide, Australia.

Cranswick, R. H., Pierce, D., & Green, G. (2016). *Barossa PWRA groundwater resource capacity: Report 3 – Projected changes to resource condition indicators due to groundwater extraction and climate change.* DEWNR Technical report 2016/06, Government of South Australia, through Department of Environment, Water and Natural Resources, Adelaide, Australia

Cranswick, R. H., Pierce, D., Wright, S., & Videka, H. (2015). *Barossa PWRA groundwater resource capacity: Report 1 – Hydrogeological assessment and conceptual model development.* DEWNR Technical report 2015/17, Government of South Australia, through Department of Environment, Water and Natural Resources, Adelaide, Australia.

Hancock, M., Stewart, S., & Green, G. (2014). *Interactions between groundwater and surface water systems, Barossa prescribed water resources area.* DEWNR Technical Report 2014/09, Government of South Australia, through Department of Environment, Water and Natural Resources, Adelaide, Australia.

Li, C., & Cranswick, R. H. (2016). *Barossa PWRA groundwater resource capacity: Report 2 – Numerical groundwater flow modelling.* DEWNR Technical report 2016/05, Government of South Australia, through Department of Environment, Water and Natural Resources, Adelaide, Australia.

Daniel Pierce is a senior hydrogeologist with the Water Science Unit of the South Australian Department for Environment and Water (DEW). He is part of a group of technical staff who provide technical advice and collaborate with policy and decision makers within DEW. Drawing on many years of private sector experience across Australia and the Pacific Islands, including 5 years working in French Polynesia, Daniel has worked on projects to support the periodic revision of Water Allocation Plans for groundwater resources around South Australia and the management of groundwater issues affecting the River Murray.

Roger Cranswick is a senior hydrogeologist at CDM Smith Australia. He has over 13 years of experience including roles in environmental consulting, groundwater research and most recently in applied science. His primary areas of technical expertise include: river–aquifer exchange processes; environmental isotopes and hydrochemistry; water resource management; developing groundwater modelling tools; and science communication.

Megan Hancock Lane is a hydrogeologist & environmental scientist at the South Australian Department for Environment and Water (DEW). She conducts research to support water management. Her areas of expertise include: advanced hydrochemistry and isotopes; surface and downhole geophysics; aquifer and pumping tests; surface water/groundwater interaction; drilling; aquifer characterisation including geometry/recharge/discharge; and extended and remote field work.

Chapter 17
Reducing Entitlements When Groundwater Has Been Over-Allocated: Policy Issues and Options

Stefanie Schulte and Gabriela Cuadrado Quesada

Abstract Reducing entitlements when groundwater is over-allocated in Australia has evidenced both challenges and successes. This chapter examines policy pathways for reducing entitlements when groundwater has been over-allocated. It explores the definitional challenges that initially hampered progress within Australia's federated structure, before examining attempts to reduce over-allocation and over-use across Australia's numerous groundwater allocation plans and catchments. The chapter highlights the challenges that led to slower than expected progress in addressing over-allocation and over-use, as well as highlighting some of the policy pathways that have been pursued to attain sustainable levels of groundwater extraction.

Keywords Groundwater management · Reform · Over-allocation · Over-use · Australia

17.1 Introduction

This chapter identifies and examines policy pathways for reducing entitlements when groundwater has been over-allocated. The focus of this chapter is Australia, where issues of over-allocation and the related challenge of over-use have been, and continue to be, a significant and ongoing water management and sustainability concern.

S. Schulte (✉)
NSW Irrigators' Council, Sydney, Australia

G. Cuadrado Quesada
Department of Integrated Water Systems and Governance, IHE-Delft Institute for Water Education, Delft, The Netherlands
e-mail: g.quesada@un-ihe.org

J.-D. Rinaudo et al. (eds.), *Sustainable Groundwater Management*,
Global Issues in Water Policy 24, https://doi.org/10.1007/978-3-030-32766-8_17

The first section of the chapter outlines Australia's approach to the issue of over-allocation. It defines the concept of "over-allocation", and the associated term of "over-use", which lies at the heart of Australia's national water reforms. The chapter notes that a lack of consistent definitions and methods for assessing over-allocation and over-use has been a continuing problem for progressing regulatory reforms and policies to address over-allocation in groundwater systems in Australia. The second section then briefly reflects on Australia's approach to groundwater allocation under various water allocation planning processes, before outlining some of the core policy tools to move over-allocated and over-used groundwater systems to environmentally sustainable levels of extraction. The chapter also highlights some of the core challenges and draws lessons from the various policy pathways being pursued to reduce groundwater over-allocation and over-use across Australia's eastern states.

In a chapter of this size, it is not possible to fully explore every policy pathway to addressing groundwater over-allocation. Thus, the chapter focuses on four case studies to discuss a range of possible mechanisms and policy tools, namely phasing in allocation reductions and carry-over provisions over time; compulsory reductions of allocations with compensation; moratoriums; conjunctive forms of management through collective action, including donations of groundwater rights in return for surface-water rights; water licence/entitlement purchases by governments in the water market, as well as government-funded infrastructure for water systems that improve efficiency and provide more water to meet environmental needs. Through the case studies, the chapter examines the strengths and weaknesses of these mechanisms and their economic, social and environmental impacts. Finally, the chapter reflects on future groundwater issues in Australia, including the long timeframes involved in achieving major changes through allocation reduction efforts.

17.2 Overview of Australia's Approach to Addressing Over-Allocation of Groundwater

Australia has undergone several water reform processes to address water scarcity issues and deal with the competing demands of different water users (see Chaps. 7 and 8 of this book). The reforms also confronted a historical legacy of past decisions by State and Territory governments to distribute more water access entitlements/licences than could be delivered at their full allocations in any given year (Holley & Sinclair, 2013; Connell & Grafton, 2011). Some of the earliest attempts to set caps and achieve a sustainable use of water resources also did not include groundwater (e.g. Murray-Darling Basin Cap on surface water diversions in the mid-1990s, Murray-Darling Basin Authority (MDBA) 2018). Over time, the 1994 water reforms and the National Water Initiative (NWI) (2004) saw Australian governments acknowledge the importance of groundwater and a total water cycle approach (paras 23(x), 56; see also Council of Australian Governments (COAG), 1995) and commit to returning currently over-allocated or over-used systems to environmentally sustainable levels of extraction (NWI, 2004, para 23(iv), see also COAG, 1995). Environmentally sustainable levels of extraction were defined as "the level of water extraction from a particular system which if exceeded, would compromise key environmental assets, or ecosystem functions and the productive base of the resource" (NWI, 2004, Schedule B(i)). In order to achieve this, States and Territories had to determine the precise pathway by which any of those systems found to be over-allocated and/or over-used would be adjusted to meet environmental and other public benefit outcomes. (NWI, 2004, para 43).

As explained in Chap. 7, under the NWI, water entitlements/licences give a right to a share of the water made available for extraction at a particular time, and a responsibility to use this water in accordance with usage conditions set by the government (NWI, 2004, para 2). Thus, over-allocation referred to "situations where with the full development of water entitlements/licences in a particular system, the total volume of water able to be extracted by entitlement/licence holders at a given time exceeds the environmentally sustainable level of extraction for that system" (NWI, 2004, Schedule B(i)). Over-use, in turn, referred to "situations where the total volume of water actually extracted for consumptive use in a system at a given time exceeds the environmentally sustainable level of extraction for that system". The NWI acknowledged that over-use could arise in systems that were over-allocated, or where the planned allocation was exceeded due to inadequate monitoring and accounting" (NWI, 2004, Schedule B(i)).

The COAG and NWI commitments to address over-allocation and over-use in each system was an important and ambitious objective. However, it faced difficulties in quantifying and implementing changes. After slower than expected progress on 1994 National Competition Council commitments, the National Water Commission's (NWC) updated 2007 Biennial assessment identified over-allocation as a "central national challenge [that was] not being managed as envisaged" (NWC, 2008, p. 19, 21). This was followed by a second biennial assessment (NWC, 2009, pp. viii-ix) which found that progress in dealing with the over-allocated and over-used systems was unlikely to be met by the NWI suggested timeline of 2010.

One of the problems that slowed progress in delivering the NWI's goals was the wide discretion available to State and Territories in developing water plans. Certainly, such discretion was necessary, given the many uncertainties associated with changing hydrological and hydrogeological systems, the unique local characteristics of groundwater and surface water systems, and the diversity in rural water users and uses which arguably require decisions and rules to be tailored to local conditions (as opposed to uniform "one size fits all" rules) (Holley, 2016). This local tailoring (NWI, 2004 para [38]), combined with stakeholder consultation during plan development (NWI, 2004, Schedule E) aspired to work through the social, economic and environmental trade-offs and identify appropriate responses to address the impacts of water entitlement/licence reductions in the face of over-allocation and over-use (Holley, 2016; NWI, 2004 paras 36 and 97; Tan, Bowmer, & Mackenzie, 2012). However, the flexibility that was provided gave rise to variations in the determinations of groundwater system's permissible average annual extraction figures and prolonged the implementation of approaches to address groundwater over-allocation and over-use.

In addition, complaints were also levelled at State and Territory governments for their use of different criteria and interpretations of the terms "over-allocation" and "over-use", and a tendency to put in place "short term" responses (NWC, 2008, 2012, p. iv; 2014, p. 8). There was also a reported "reluctance to publicly acknowledge over-allocated and over-used water systems", as well as disagreements over whether the NWI outcomes traded off environmental values to support consumptive requirements within water plans (Bunn, 2017, p. 102). Indeed, the process of setting environmentally sustainable levels of extraction and identifying over-used systems through water planning has at times been highly contentious, as stakeholders have debated the economic and social trade-offs associated with reallocating water to the environment (Hamstead, 2009; New South Wales Irrigator's Council (NSWIC), n.d.-a; Productivity Commission (PC), 2018, p. 355). In particular, there have been concerns, albeit contested ones (Carmody, 2018) about the adverse effects of reducing allocations for irrigators and the flow on effects on regional communities (PC, 2018, p.310).

Such tensions were particularly evident at the introduction of the Water Act, 2007 (section 22) and the Basin Plan, 2012 which introduced new sustainable diversion limits (SDLs) in the Murray-Darling Basin (MDB). The SDLs represent new maximum long-term annual average extraction levels that can be sustainably taken from the 80 SDL water resource areas (Water Act, 2007, s 22; for further see Chap. 8). As Danielle (Daniell, 2011, p. 416) explained, the introduction of these reforms was controversial because of "the lack of transparency of the planning process in clarifying underlying assumptions on which synthesis and planning decisions are made, and the common lack of openness to engage in discussions about these assumptions—so that community members and other land and water managers at different administrative levels can understand them". Indeed, debates over the determination of long-term average annual extraction limits, adequate annual allocations and the methodology to determine appropriate allowable average extraction figures remain a live issue for some stakeholders, who argue over whether the allocations among various entitlement/licence holders are appropriate (NSWIC, n.d.-a).

Although disagreements and the diversity in defining over-allocation and over-use have historically made it difficult to identify and evaluate whether and what steps were taken to deal with over-allocation and over-use (NWC, 2011, p. 101), some progress towards addressing it has been made over time. Following criticism on the 2009 and 2011 National Water Commission (NWC) assessments, the Commission's study on water stress in Australia (2012) noted that there were some 35% (~100 out of 293) of Australia's groundwater management units that fell into the most or highly water-stressed categories, with some statutory plans expressly recognising they were over-used (32 out of 101 GMUs) or over-allocated (30 out of 101 GMUs). Over the next 6 years, improvements were identified, partly due to efforts under the Basin Plan (NWC, 2014). By 2018, the PC (2018 pp. 57, 59) noted that "In over-allocated systems, pathways to achieving a more sustainable balance between consumptive and environmental use have been established —although there is more work to do before they are completed". In addition, the PC (2018, p.72) noted the partial achievement of bringing over-allocated and over-used systems back to sustainable levels of extractions, noting "there are still a number of systems identified as over-allocated and/or over-used".

Given this context, this chapter seeks to examine how progress has been made (albeit perhaps slower than expected) in different catchments and groundwater systems across Australia. There are, of course, numerous options for responding to issues of over-allocation and over-use (NWC, 2013; Harrington & Cook, 2014, p. 14–15). For example, the then chairman and CEO of the NWC in 2008 (Matthews, 2008) noted there were at least 13 possible responses, namely:

- Allocate less per entitlement holder.
- Invest to improve irrigation system efficiency.
- Invest to improve the efficiency of environment watering (e.g. improve wetland infrastructure).
- Extract more environmental benefits from consumptive water (return flows from irrigation to the environment).
- Buyback entitlements through the water market (and re-direct to the environment).
- Revise water plans as they expire and then "re-set" entitlements.
- Compulsorily acquire certain entitlements e.g. high salinity or low-efficiency irrigation areas.
- Retire less viable (e.g. less efficient) irrigation districts.
- Compulsorily acquire a percentage of entitlements across the board.
- Reduce target levels of reliability (security).
- Suspend water plans and arbitrarily revise entitlements.
- Regulate water use to reduce consumption.
- Lower environmental aspirations.

In a chapter of this size and given the numerous groundwater management areas that (arguably) face the challenge of over-allocation and over-use, there is not the space to catalogue the experiences of all possible options.[1] Instead, a selection of policy responses is presented that illustrate the challenges and successes of addressing groundwater over-allocation and over-use in Australia.

[1] Nor can we cover related attempts at recovering water for the environment such as in the Great Artesian Basin Sustainability Initiative, which involved public and private investment in capping and piping free flowing bores, elimination of open-bore drains and installation of piping to deliver water to stock, Bunn, 2017, p104; GABCC, 2000.

17.3 Example 1: NSW – Achieving Sustainable Groundwater Entitlement Program

The NSW government commenced addressing the issue of groundwater over-allocation in 2005 when it introduced the *Achieving Sustainable Groundwater Entitlement* (ASGE) program (NSW Department of Natural Resources (NSWDNR), 2005). The ASGE program was proposed in recognition of previous state policies that had resulted in an over-allocation of groundwater licences[2] whose combined permissible extraction volume in each of the state's six major alluvium groundwater systems exceeded the volume that could be sustainably extracted (NSW Department of Industry (DI), n.d.).

The ASGE program aligned with the NSW government's broader water reform process which occurred following the signing of the NWI in 2004. A part of the water reform required an assessment and determination of the long-term sustainable yield of each inland groundwater system, which in NSW was defined as the proportion of the recharge that could be extracted without compromising the integrity of the water sources, ecosystems or communities (NSWDNR, 2005). Following the determination that all six major alluvium groundwater systems in NSW had permissible extraction levels that exceeded the long-term sustainable yield, the NSW government decided to gradually reduce the licenced groundwater volumes in line with the sustainable yield calculations, whilst taking into consideration "other social, economic and environmental factors" (NSWDNR, 2005).

While several options to reduce the licenced volume were considered (e.g. an across the board cut, history of use or other approaches, see Holley & Sinclair, 2013), it was decided that the reduction should be based on *History of Extraction* which factored in past uses of, and dependencies on, the groundwater resource. This approach relied on confirmed metered extraction over a specified time period[3] and calculated the required reductions in licenced volume on the basis of the active and inactive component of a licence holder's entitlement holdings (NSWDNR, 2005). If a licence holder was able to prove greater dependence on groundwater (e.g. greater extraction volumes), the reductions were lower than for an equivalent licence holder who had a lower level of dependence (NSWDNR, 2005). *History of Extraction* was preferred by many (although not all) licence holders because it utilised a weighted reduction approach which took into account past usage and limited the possibility of stranded assets (Kuehne & Bjornlund, 2006).

The implementation of the licensed volume reductions was scheduled to be phased in over the course of the NSW water sharing plans (WSP) that were made under the *Water Management Act 2000* (NSW). As documented in detail elsewhere (Holley & Sinclair, 2013; Tan et al., 2012; see also the case study on the Lower Murrumbidgee Chap. 21), the decision was followed by a series of court cases, lob-

---

[2] These licences were temporary groundwater licences (five-year) that had specified maximum annual extraction volume and access condition pursuant to the NSW Water Act, 1912.

[3] The time period under consideration varied between groundwater sources.

bying and other processes. In response, the NSW Minister put numerous WSPs on hold as a review of the draft WSPs was undertaken and a Groundwater Adjustment Advisory Committee (including representatives of the NSWIC, Catchment Management Authorities and the NSW and Australian governments) was convened to consider the adjustment methodology and financial assistance (See Chap. 21; Holley & Sinclair, 2013).

With the subsequent implementation of the WSPs[4], the licensed groundwater volumes in the six major alluvium groundwater systems were adjusted in line with the sustainable yield assessment. The management process around the reduction and the future allocation of groundwater was specified in each individual WSP and tailored to the specific water sources and with consideration of the needs of different water users, local communities and the environment. For example, state and federal government authorities acknowledged that all previous groundwater licences carried a certain monetary value and groundwater users had invested in groundwater dependent infrastructure to utilise the allocations made against these licences (NSWDIN, n.d). Given the financial impact of the reductions in licenced volumes through the ASGE program, licence holders received ex-gratia payments to help with the adjustment process (NSWDNR, 2005).

As the WSPs allowed for the issue of replacement groundwater licences, works and use approvals (NSWDNR, 2005), new groundwater licences were issued which carried specific rules around water use, trade and carry-over (NSWDI, 2017) (Table 17.1).

If a previous licence holder could establish that their *History of Extraction* exceeded the new licensed volume, they were able to obtain a second supplementary licence. This supplementary licence was issued for an equivalent volume to the difference between the licence holder's new licenced volume and their history of extraction volume (NSWDNR, 2005). These supplementary licences were issued for a period of 10 years and were not tradable and could not carry-over unused water allocations. In addition, the volume of water available under the supplementary licences was reduced annually to allow for a gradual adjustment to the lower

Table 17.1 Licenced volume reduction for each groundwater area under the ASGE program (NSWDI, 2017)

Valley	Pre-WSP licensed volume (ML)	New licensed volume (ML)	Reduction (ML)	Reduction (%)
Lower Gwydir	65,885	28,719	37,166	56
Lower Lachlan	206,455	105,654	100,801	49
Lower Murray (deep)	267,440	83,580	183,860	69
Lower Murrumbidgee	512,409	267,500	244,909	48
Lower Macquarie	133,730	65,524	68,206	51
Upper and Lower Namoi	438,475	167,102	271,373	62

[4] For the six major alluvial groundwater systems, these WSPs commenced in either 2006 or 2008.

licenced volumes. By the end of the 10-year period, these supplementary licences received no further allocations and were ultimately cancelled by 1 July 2018 (*Water Sharing Plan for the Lachlan Unregulated and Alluvial Water Sources 2012*).

While the ASGE program provided a policy pathway to bring groundwater use in line with the NSW sustainable yields assessment, it also gave certainty to licence holders in the form of new perpetual groundwater licences that were tradable and had a market value. However, the implementation of the ASGE program has been heavily criticised by some groundwater users for being inequitable in the allocation of new licence and in the provision of compensation (NSWIC, n.d.-b). Furthermore, criticism emerged around the administration of a Community Development Fund that was established as part of the program (NSWIC, n.d.-b). As the ASGE program was a joint State and Federal government funding initiative, friction occurred when a change in government at the Federal level altered the eligibility criteria for some previously accepted community development projects. Also, recent assessments have shown that some groundwater systems continue to experience stepwise drawdowns in groundwater levels, suggesting that the previous efforts under the ASGE program may not have been sufficient to address over-use in NSW (NSWDI, 2018a).

More than a decade since the ASGE program was first initiated, NSW is preparing a second iteration of the WSPs. With the introduction of the Basin Plan in 2012, NSW is required to develop new groundwater management plans (e.g. Water Resource Plans (WRPs)) which incorporate and build on the previous WSPs. In order to receive accreditation for these new plans, the NSW government has to prepare a risks assessment for its inland groundwater system. Several preliminary risk assessments have been released and show ongoing risks to groundwater sources across NSW, particularly in respect to the environmental watering requirements of priority environmental assets and ecosystem functions that depend on groundwater (NSWDI, 2018b). Also, the risk assessments point to potential risks to groundwater dependent Aboriginal cultural asset (NSWDI, 2018b). If NSW classifies any of these risks as "medium" or "high", section 10.41(2)(a) of the Basin Plan, 2012 requires the state government to include new rules into the WRPs to address the identified risk.

In addition, the progressive implementation of the Basin Plan, 2012 has raised further questions about the potential impact on groundwater resources of the previous infrastructure investment by the Australian government in the MDB. In particular, there are concerns that funding for upgrades to on-farm and off-farm infrastructure assets have decreased seepage and hence impacted on groundwater recharge in the MDB (Wang, Walker, & Horne, 2018).

17.4 Example 2: Queensland – Moreton Moratorium

Another approach to address over-use of groundwater resources was applied by the Queensland (Qld) government in 2005 when it issued a moratorium on granting new licences to take water in the Moreton catchment of south-eastern Qld and halt applications for the construction of surface water and groundwater works that could take water from the artesian and sub-artesian basin (Robertson, 2005). The moratorium was designed to halt further depletion of the region's groundwater resources and ensure the ongoing water supply for the Greater Brisbane and Ipswich areas, the Lockyer Valley and parts of the Sunshine Coast as well as other adjacent inland areas.

Within the Moreton catchment, the Lockyer Valley has experienced significant over-use of its groundwater resource. Groundwater is the main irrigation water supply for vegetable and Lucerne production with annual extractions ranging between 8000 and 18,000 ML/year compared to a calculated sustainable yield of 15,000 ML/year (Harrington & Cook, 2014, p. 14). The continuous over-use of groundwater has led to a progressive depletion of groundwater reserves and caused access issues for resident groundwater users. In addition, groundwater over-use and increasing salinity levels have impacted on water quality in the Lockyer Valley.

Prior to the moratorium, the Qld government attempted to address the over-use of groundwater resources by using surface water to recharge groundwater sources and installing specific infrastructure that was designed to assist the recharge process (e.g. weirs) and monitoring groundwater levels (e.g. meters). In addition, the Qld government tried to manage the progressive groundwater depletion through the introduction of allocation limits. However, none of these policy and regulatory approaches was successful to prevent further over-use of groundwater resources in the valley.

Thus, the Qld government implemented two further measures. Firstly, it issued a moratorium on new water take licences and applications for the construction of surface water and groundwater works - with the exception of works for stock and domestic use. The moratorium also specified that any replacement of existing infrastructure must be preceded by an application for a development permit under the *Sustainable Planning Act 2009* (Qld) (Australian government, n.d, p. 19). Secondly, the Qld government divided the entire Lockyer Valley into different groundwater management areas and progressively implemented management plans that targeted better monitoring of groundwater extractions and determined recharge levels in order to identify, establish and implement appropriate sustainable levels of groundwater extractions (Australian government, n.d, p. 19). The Qld government developed and implemented two water plans which applied to the Lockyer Valley (e.g. the Great Artesian Basin Water Resource Plan in 2006 and the Moreton Water Resource Plan in 2007) (Australian government, n.d, p. 19; Harrington & Cook, 2014, p.14). Under the new water plans, the Qld government issued new groundwater licences and implemented compulsory metering requirements for groundwater take. Since these measures were introduced, over 400 monitoring bores have been installed which are assessed quarterly (Australian government, n.d.).

Despite these regulatory and policy approaches, the Australian government's bioregional assessments have shown that the Lockyer Valley groundwater resource remains under stress (Australian government, n.d.). Thus, the Qld government has proposed to convert 315 surface and groundwater entitlements into tradable water allocations in 2016 (Gunders, 2015). These allocations are designed to limit the volume of groundwater that can be pumped each water year based on an assessment of groundwater levels. Whilst irrigators in the valley have expressed concerns about the potential economic impact of these changes, the Qld government has argued that by quantifying the volume of water that can be used within a water year, water users would have greater certainty in their annual groundwater availability which would lead to more sustainable and efficient agricultural development in the central Lockyer Valley (e.g. through the annual allocations and the ability to trade) (Qld Government, 2018).

Although similar processes have been undertaken across the state, areas within in the Moreton Catchment remain under interim arrangements which do not allow for water trading. In November 2018, the Qld government stated that further consultation with landholders will be conducted to determine the amount of groundwater individual landholders would receive (Qld Government, 2018). As these policy changes remain ongoing, it is difficult to evaluate whether the previous initiatives have been successful to address over-use of groundwater resources in the catchment. Further monitoring of groundwater use and assessment of future groundwater trading will be required to evaluate the effectiveness of previous policy and regulatory amendments.

17.5 Example 3: Conjunctive Management and Collective Action in South Australia – Angas Bremer

The Angas Bremer district is located 60 km south-east of Adelaide, capital of South Australia (SA), near the town of Strathalbyn, beside Lake Alexandrina. The district lies in the rain shadow of the Mt. Lofty Ranges and has a relatively low annual rainfall of 400 mm. Rainfall is winter dominant and occurs mainly between April and October (Zulfic & Barnett, 2007, p. 8). In this area, groundwater resources were over-allocated and over-used to irrigate different crops for many decades. The area has undergone dramatic changes in land, surface and groundwater use (Zulfic & Barnett, 2007). High groundwater extractions in the 1970s and 1980s induced downward leakage of saline groundwater (Zulfic & Barnett, 2007). By 1981, the use of groundwater for irrigated agriculture had reached 26,600 ML/year, four times the estimated annual recharge of 6000 ML (Thomson, 2008).

This unsustainable use caused a decrease in groundwater levels by up to 10 m and a decline in water quality (Muller, 2006). The increases in groundwater salinity made many realise that they were facing serious water problems. As a result, the community formed the Angas Bremer Water Management Committee (ABWMC) and started work with the government to develop and implement innovative mechanisms to counter over-exploitation and rising salinity.

The initiatives implemented in the Angas Bremer district to address over-allocation and over-exploitation included a reduction in groundwater allocations; aquifer storage and recovery systems; and annual monitoring and reporting. These are discussed briefly below.

First, echoing notions of conjunctive management through collective action (Holley, Sinclair, Lopez-Gunn, & Schlager, 2016), the Angas Bremer irrigators (along with the South Australian government) reduced their groundwater allocations in order to reduce groundwater use. This mechanism consisted in exchanging groundwater licences for River Murray surface water licences which allowed pumping from nearby Lake Alexandrina. This was the first step taken in order to tackle groundwater over-allocation. The idea behind that action was to use alternative water sources such as rivers and lakes that could satisfy the district's water needs. Aquifer storage and recovery systems were also established which recharged the aquifer with surface water during the wet winter months, and improved groundwater levels and quality (Thomson, 2004). Moreover, monitoring and reporting of groundwater use have been implemented, which has helped to address groundwater over-use because it encouraged groundwater use recording by irrigators. The monitoring and reporting are legal requirements of their groundwater licences whereby each irrigator collects and records data including their annual groundwater meter readings and the area of crops under irrigation.

All the mechanisms discussed above have contributed to addressing groundwater over-use during the last 30 years, with the ABWMC being fundamental to the success. Not only did the ABWMC commit to addressing the problems of over-allocation and over-exploitation, but it promoted the knowledge of all growers about groundwater systems for improving the long-term use and management of groundwater in their district.

Although the ABWMC initially received strong technical and funding support from SA government agencies, such support has decreased significantly in recent years as new programs and priorities across the state have arisen. Although the overall activities and role of the ABWMC are now more limited, it has provided useful insights on how the local community and water users can provide a valuable contribution to address over-allocation and promotion of sustainable groundwater use when they are informed and aware of the adverse impacts resulting from over-exploitation.

17.6 Example 4: MDB – Sustainable Rural Water Use and Infrastructure Program

A final approach to address the over-use of groundwater resources involved the use of market mechanisms (as opposed to the uncompensated attenuation of water rights) to purchase water licences/entitlements from consumptive water users (PC, 2018, p. 18–19).

Following prolonged drought conditions in the 1990s and early 2000s, the then Australian Prime Minister John Howard introduced a $10 billion National Plan for Water Security in 2007 that aimed at improving water use efficiency in the MDB and address over-use of surface and groundwater resource (Howard, 2007).

Following, the Australian government passed the *Water Act 2007 (Cth)* and the *Basin Plan 2012 (Cth)*. The Act allowed the Australian government to assume significant planning and management responsibilities over water resources in the MDB and established the MDBA and the Commonwealth Environmental Water Holders (CEWH) as two independent statutory authorities which are critical to addressing over-use of water resources in the MDB (*Water Act 2007* (Cth) s.104 and s.171).

Drafting the Basin Plan, 2012 required the determination of long-term sustainable diversion limits in each surface water and groundwater area in the MDB.[5] The determination of the SDLs effectively quantified the required reductions in surface water and groundwater use by consumptive water users. After extensive consultation, the basin-wide groundwater SDL was set at 3334 GL[6] (LTAAY) (MDBA, 2017). In its determination, the MDBA stated that it had considered the effect of groundwater use on groundwater dependent ecosystems, hydrological connectivity, long-term productivity of the groundwater resource and water quality (including salinity). Further, the MDBA's assessment found that compared to the existing long-term average basin-wide groundwater use of 1375 GL, groundwater extractions in most systems were below the SDL, except for the Qld Upper Condamine Alluvium which required a long-term average reduction in groundwater use by 40.4 GL.

After considering various approaches to achieve reductions in surface water and groundwater use, the Australian government allocated $10 billion to recover water licences/entitlements from willing sellers (e.g. either via direct purchases of water licences/entitlements or via infrastructure funding arrangements). To date, the Australian government has spent over $2 billion on direct water licence/entitlement purchases and over $4 billion on infrastructure projects (House of Representatives Standing Committee on Regional Australia, 2011, Chap. 5; Murray-Darling Basin Ministerial Council, 2017). The recovery of water licences/entitlement by the Australian government is conducted by the Department of Agriculture and Water Resources in accordance with the Water Recovery Strategy which was released in June 2014 (Australian Government, 2014). Under the strategy, the Australian government has committed to prioritising infrastructure projects over direct water licence/entitlement purchases and has legislated a cap on direct surface water licence/entitlement purchases in 2015 (Hunt, 2015).

For groundwater resources, the Australian government decided that it would run several open tender processes to purchase groundwater licences/entitlements from willing sellers in the Upper Condamine Alluvium (Australian Government, 2018b). As of January 2019, the Australian government had recovered 3.097 GL of groundwater in the Condamine Alluvium, however another 31 GL was contracted (Australian Government, 2018c). As of 31 March 2019, the overall recovery targets in the Upper Condamine Alluvium was approximately 7.7 GL (Australian Government, 2018c). Given previous challenges with the tender process (e.g. there

[5] The determination of the Sustainable Diversion Limits had to consider social, environment and economic impacts.

[6] The 2017 Basin Plan Amendments would increase the SDL for groundwater to 3494 GL.

was no successful tender in 2014), risks remain whether the Australian government will be able to meet its groundwater recovery target for the Qld Upper Condamine.

In addition to the Qld Upper Condamine Alluvium, the Australian government also holds 5.077 GL of groundwater entitlements in the NSW Murrumbidgee Valley and 1.522 GL in the NSW Murray Valley which were recovered early in the water recovery program (Australian government, 2018a). Similar to the groundwater licences/entitlements that have been recovered in the Upper Condamine, the groundwater licences in the NSW Murray and Murrumbidgee Valleys have not been used by the CEWH for any direct environmental watering activities, partially due to the high energy costs associated with groundwater extraction (CottonInfo, n.d.).

In addition to the recovery of groundwater licences/entitlements, the Basin Plan, 2012 also addresses any future "over-use" of groundwater resources through a newly established monitoring and compliance framework. Under the Basin Plan, 2012, the quantity of groundwater extracted in a groundwater WRP area must not exceed the determined SDL. To ensure this is the case, the State governments must monitor groundwater extraction and compare the sum of the annual amounts of groundwater permitted to be taken from the WRP area with the sum of the annual amounts of water actually taken each year. The two volumes and the cumulative balance between the two will be compared and if the cumulative balance is in debit by an amount that is equal to or greater than 20% of the respective groundwater SDL then a breach of the SDL has occurred unless there is a reasonable excuse (Basin Plan, 2012 (Cth), s.6.12; see also proposed amendments to groundwater SDL compliance rules, Basin Plan Amendment Instrument 2017 (No 1) s. 6.12C). Although two slightly different compliance methodologies are proposed pre-2028 and post-2028, the post-2028 SDL compliance approach could be problematic for groundwater users because the State governments currently do not have climatically adjusted groundwater models (i.e. which would adjust the figures for annual permitted take of groundwater) (Climate Council of Australia, 2015; MDBA, n.d.; Neave, McLeod, Raisin, & Swirepik, 2015). In the absence of such a model, states will have to rely on the groundwater SDLs as the "permitted" take figures, meaning that any over-use (compared to the SDL) will need to be offset in a future period.

Depending on the compliance approach that will be adopted by individual state governments, there is a risk that the new method could have a distributional impact on different groundwater licence/entitlement holders. For example, there were discussions to apply either proportional reductions to all groundwater licence/entitlement holders' allocations in an area where a breach has occurred. Alternatively, suggestions have been made to apply a weighted allocation reduction approach that depends on how active a groundwater licence/entitlement holder has been in the past (e.g. similar to the History of Use approach taken under the ASGE program earlier). States are still in discussions with local stakeholders about which approach to take should a breach of the groundwater SDL occur post 1 July 2019. However, it is also possible that the states will remain non-committal on a particular compliance approach at the commencement of the WRPs and will assess future SDL non-compliance on a case by case basis, thereby creating further potential delays in addressing over-use.

The evaluation and review of the Basin Plan in 2026 and future NWI assessments by the PC will highlight whether these recent water reforms and policy initiatives have been effective in addressing over-use of groundwater sources in the MDB.

17.7 Conclusion and Future Challenges

This chapter has outlined a range of policy tools and mechanisms to address over-allocation and over-use of groundwater resources in Australia. These tools and mechanisms included phasing in allocation reductions through compulsory reductions in licenced volumes with compensation (AGSE, NSW); moratoriums and trading (Moreton, Qld); conjunctive forms of management through collective action, including donations of groundwater rights in return for surface-water rights (Angas Bremer, SA); and allocations purchased by governments on the water market from willing sellers (MDB). Based on the case studies, what can be said about conditions for successful pathways?

While moratoriums can work in areas where significant problems are prevalent, more successful programs have often included redress for reductions - be it ex-gratia payments (AGSE, NSW), alternative sources of water (Angas Bremer, SA) or monetary compensation and other funding (MDB and to a lesser extent Moreton, Qld). Such redress can win support from affected groundwater users, however, the slow progress, individualistic approaches and questions raised over Australia's policy mechanisms to address over-allocation and over-use suggest that delivering change is a complex and difficult process that takes time and ongoing investment in monitoring and evaluation. Also, slow implementation of these regulatory and policy changes was often attributable to ongoing resistance by stakeholders to the proposed reductions and compensation payments. Further delays in the implementation of these programs have occurred as a result of staffing changes within departments and uncertainty around future funding commitments from governments for ongoing program implementation, monitoring and evaluation.

As part of the development of the WRPs, many of the previous issues (e.g. previous licenced volume reductions in 2006) have re-emerged. Whilst preliminary feedback from the State governments suggest that these issues will not be revisited in the WRP development (e.g. due to concerns about changes in asset values and interference with existing groundwater markets and partnership agreements), ongoing calls for amendments will most likely remain. Professor Brad Karkkainen nicely captures the challenges here when he noted: "Both environmentalists and irrigators frequently complain that their side is getting short shrift, and this has led to policy reversals seeking to rebalance the equation in favour of one set of interests or the other" (Karkkainen, 2018).

Recent controversies around water theft and maladministration of water resources in NSW has renewed the water reform efforts by all governments (e.g. under the NSW Water Reform Action Plan (NSW Government, 2017) and the Compliance Compact (Murray-Darling Basin Ministerial Council, 2018)), although groundwater resources continue to remain an afterthought in the debate. This is regrettably, particular as ongoing drought conditions in Australia have again illustrated the variability of Australia surface water supply and the risks it poses to the environment, communities and industries (NSW Government, 2018; Vincent, 2018). In times of water scarcity, available groundwater resources are often sourced to supplement insufficient surface water supplies, suggesting that further stress on groundwater resources are likely to continue. In the context of climate change, demands on groundwater resources are likely to increase causing additional concerns for those groundwater sources that are slower to respond to the changes (Cuthbert, M.O. Cuthbert et al., 2019).

However, opportunities exist under the renewed water reform processes to improve the current measurement and monitoring of groundwater resources (e.g. through recent regulatory changes requiring comprehensive metering and measurement of all water take (Murray-Darling Basin Ministerial Council, 2018)), estimate future water needs of different water users (e.g. including water utilities in rural areas and other inception activities like mining or forestry) and evaluate the impacts of previous regulatory and policy reforms (e.g. the impact of the Basin Plan and infrastructure investments). Whether these opportunities will be seized and acted upon remain to be seen, but further regulatory and policy changes are likely unavoidable to ensure Australia's groundwater resources are used sustainably into the future.

References

Australian Government. (2014). Department of the Environment, Water Recovery Strategy for the Murray-Darling Basin.

Australian Government. (2018a). Department of the Environment and Energy, Environmental water holdings. http://www.environment.gov.au/water/cewo/about/water-holdings

Australian Government. (2018b). Queensland Upper Condamine Alluvium Groundwater Purchase Tender. http://www.agriculture.gov.au/water/markets/commonwealth-water-mdb/groundwater-purchase-tender

Australian Government. (2018c). Queensland Upper Condamine Alluvium Groundwater Purchase Tender Outcomes. http://www.agriculture.gov.au/water/markets/commonwealth-water-mdb/groundwater-purchase-tender/groundwater-uca-tender-evaluations

Australian Government. (n.d.). *Groundwater flow*. https://www.bioregionalassessments.gov.au/assessments/11-context-statement-clarence-moreton-bioregion/1143-groundwater-flow

Basin Plan 2012 (Cth).

Bunn, S. E. (2017). Environmental water reform. In B. Hart & J. Doolan (Eds.), *Decision making in water resources policy and management: An Australian perspective* (pp. 97–110). London: Elsevier.

Carmody, E. (2018). The unwinding of water reform in the Murray-Darling Basin: A cautionary tale for transboundary river systems. In C. Holley & D. Sinclair (Eds.), *Reforming water law and governance* (pp. 35–55). Singapore: Springer.

Climate Council of Australia. (2015). *W Steffan: Thirsty country: Climate change and drought in Australia*. https://www.climatecouncil.org.au/droughtreport2015.

Connell, D., & Grafton, Q. (Eds.). (2011). *Basin futures: Water reform in the Murray-Darling Basin*. Canberra, Australia: ANU Press.

CottonInfo. (n.d.). *Fundamentals of energy use in water pumping*. https://www.cottoninfo.com.au/sites/default/files/documents/Fundamentals%20EnergyFS_A_3a.pdf

Council of Australian Governments (COAG). (1995). Agreement to implement the national competition policy and related reforms, Australian Government.

Cuthbert, M. O., Gleeson, T., Moosdorf, N., Bfus, K. M., Schneider, A., Hartnmann, J., et al. (2019). Global patterns and dynamics of climate – groundwater interactions. *Nature Climate Change, 9*, 137–141.

Daniell, K. A. (2011). Enhancing collaborative management in the basin. In D. Connell & R. Q. Grafton (Eds.), *Basin futures: Water reform in the Murray-Darling Basin* (pp. 413–438). Canberra, Australia: ANU E-Press.

GABCC. (2000). *Great Artesian Basin strategic management plan*. Canberra, Australia: Great Artesian Basin Consultative Council.

Gunders, J. (2015). Lockyer Valley irrigators concerned about planned water reform, *ABC News*. https://www.abc.net.au/news/rural/2015-12-04/lockyer-valley-irrigators-concerned-water-reform/7002294

Hamstead, M. (2009). *Improving environmental sustainability in water planning. Waterlines report*. Canberra, Australia: National Water Commission.

Harrington, N., & Cook, P. (2014). *Groundwater in Australia*. Adelaide, Australia: National Centre for Groundwater Research and Training.

Holley, C. (2016). *National water reforms and pragmatism* (Learning from Discretion, The relevance of Professor Charles Sabel's experimental approach to Australian Public Administration, Canberra 18–19 February 2016).

Holley, C., & Sinclair, D. (2013). Deliberative participation, environmental law and collaborative governance: Insights from surface and groundwater studies. *Environmental and Planning Law Journal, 30*, 32–55.

Holley, C., Sinclair, D., Lopez-Gunn, E., & Schlager, E. (2016). Conjunctive management through collective action. In A. J. Jakeman, O. Barreteau, R. J. Hunt, J. D. Rinaudo, & A. Ross (Eds.), *Integrated groundwater management* (pp. 229–252). Cham, Switzerland: Springer.

House of Representatives Standing Committee on Regional Australia. (2011). Of drought and flooding rains: Inquiry into the impact of the Guide to the Murray-Darling Basin Plan. https://www.aph.gov.au/Parliamentary_Business/Committees/House_of_Representatives_Committees?url=ra/murraydarling/report/chapter5.htm

Howard, J. (2007). Transcript of the Prime Minister, the Hon John Howard MP address to the National Press Club (25 January 2007). http://parlinfo.aph.gov.au/parlInfo/download/media/pressrel/K81M6/upload_binary/k81m68.pdf;fileType=application%2Fpdf#search=%22media/pressrel/K81M6%22

Hunt, The Hon. Greg. (2015). The Hon Bob Baldwin MP, Coalition delivers election commitment with 1500GL water buyback cap. http://www.environment.gov.au/minister/hunt/2015/mr20150914.html

Karkkainen, B. (2018). Multi-jurisdictional water governance in Australia: Muddle or model? In C. Holley & D. Sinclair (Eds.), *Reforming Water Law and Governance* (pp. 56–78). Singapore: Springer.

Kuehne, G., Bjornlund, H. (2006). 'Frustration, confusion and uncertainty – Qualitative responses from Namoi Valley irrigators' *Water*, 33(3) 78–82 May 2006.

Matthews, K. (2008). Addressing over-allocation of water entitlements CEDA water series, Thursday 21 February 2008.

Muller, K. L. (2006, November 25–26). *A partnership approach to environmental stewardship in Langhorne Creek, South Australia*. Second National Wine Industry Environment Conference and Exhibition, South Australia. www.angasbremerwater.org.au/documents/WIEC%20Full%20paper.pdf. Accessed 28 Nov 2015.

Murray-Darling Basin Authority. (2018). *Groundwater*. https://www.mdba.gov.au/basin-plan-roll-out/groundwater

Murray-Darling Basin Authority. (2017). *Basin plan amendments – Proposed changes to groundwater*. https://www.mdba.gov.au/sites/default/files/pubs/773-BP-amendments-groundwater-snapshot-24%20Feb.pdf

Murray-Darling Basin Authority. (n.d.). *Climate variability and change*. https://www.mdba.gov.au/basin-plan-roll-out/climate-change

Murray-Darling Basin Ministerial Council. (2017). *Implementing the basin plan*. https://www.mdba.gov.au/sites/default/files/pubs/Report-by-Minco-implementing-the-Basin-Plan.pdf

Murray-Darling Basin Ministerial Council. (2018). *Murray-Darlin Basin compliance compact*. https://www.mdba.gov.au/sites/default/files/Basin-Compliance-Compact-180702-D18-31184.pdf

National Water Commission (NWC). (2014). Australia's water blueprint: National reform assessment 2014, Canberra, Australia : National Water Commission (Australian Government)

National Water Initiative (NWI). (2004). *Intergovernmental agreement on a National Water Initiative*. Canberra, Australia: Council of Australian Governments.

Neave, I., McLeod, A., Raisin, G., & Swirepik, J. (2015). Managing water in the Murray-Darling Basin under a variable and changing climate. *AWA Water Journal, 42*(2), 102–107.

NSW Department of Industry – Lands and Water. (2017). Groundwater WRP Discussion Paper – Groundwater Shares (31 October 2017).

NSW Department of Industry – Lands and Water. (2018a). NSW Groundwater Baseline Project, at https://www.industry.nsw.gov.au/about/our-business/statutory-officers-and-independent-appointments/land-and-water-commissioner/nsw-groundwater-baseline-project

NSW Department of Industry – Lands and Water. (2018b). *Water resource plans currently on public exhibition*. https://www.industry.nsw.gov.au/water/plans-programs/water-resource-plans/drafts

NSW Department of Industry – Lands and Water (NSWDI). (n.d.). *Achieving sustainable groundwater entitlements program*. http://www.water.nsw.gov.au/water-management/water-sharing/plans_commenced/achieving-sustainable-groundwater-entitlements-program

NSW Department of Natural Resources (NSWDNR). (2005). *Achieving sustainable groundwater entitlement program*. http://www.water.nsw.gov.au/__data/assets/pdf_file/0006/548565/changes_draft_131205.pdf

NSW Government. (2017). Securing our water – NSW Government water reform action plan. https://www.industry.nsw.gov.au/__data/assets/pdf_file/0016/136204/nsw-government-water-reform-action-plan.pdf

NSW Government. (2018). NSW Department of Primary Industry, Water access licences. http://www.water.nsw.gov.au/water-licensing/about-licences/new-access-licences

NSW Irrigators' Council (NSWIC). (n.d.-a). *Over-allocation or over-use?* http://www.nswic.org.au/pdf/fact_sheets/Over-Allocation%20or%20Over%20Use.pdf

NSW Irrigators' Council (NSWIC). (n.d.-b). Achieving Sustainable Groundwater Entitlements (ASGE). http://www.nswic.org.au/pdf/fact_sheets/ASGE%20Fact%20Sheet.pdf

NWC (National Water Commission). (2008). Update of progress in water reform. Appendix 1. In *WGCCW (Working Group on Climate Change and Water), 2008* (Report to Council of Australian Governments, pp. 16–44), March 2008: Water.

NWC (National Water Commission). (2009, September). Australian water reform 2009: Second biennial assessment of progress in implementation of the National Water Initiative, Canberra, Australia, ACT: National Water Commission

NWC (National Water Commission). (2011). *The national water initiative – Securing Australia's water future: 2011 assessment*. Canberra, Australia: ACT: National Water Commission.

NWC (National Water Commission). (2012, June). *Assessing water stress in Australian catchments and aquifers*. National Water Commission.

NWC (National Water Commission). (2013). *Water management and pathways to sustainable levels of extraction: Issues paper*. NWC.

Productivity Commission (PC). (2018). *National water reform* (Inquiry report).

Qld Government. (2018). 'Greater certainty and security for Lockyer Valley water users. http://statements.qld.gov.au/Statement/2018/11/5/greater-certainty-and-security-for-lockyer-valley-water-users

Robertson, Hon. S. (2005). *Moreton catchment water licence moratorium*, media release. http://statements.qld.gov.au/Statement/Id/40041

Tan, P. L., Bowmer, K., & Mackenzie, J. (2012). Deliberative tools for meeting the challenges of water planning in Australia. *Journal of Hydrology, 474*, 2–10.

Thomson, T. (2004, May 12). Learning together at Angas Bremer. *Irrigation Association of Australia Conference*, Adelaide. http://www.angasbremerwater.org.au/documents/Learning%20Together.pdf. Accessed 28 Nov 2017.

Thomson, T. (2008). More water management innovations in the Angas Bremer district of south Australia. In M. Lambert et al. (Eds.), *Proceedings of the water down under*. (Causal productions 2008) 1090–1101.

Vincent, M. (2018). Walgett has two rivers but no water left to drink, *ABC News*. https://www.abc.net.au/news/2018-12-19/walgett-has-two-rivers-but-no-water-left-to-drink/10558428

Wang, J. Q., Walker, G., & Horne A. (2018). *Potential impacts of groundwater sustainable diversion limits and irrigation efficiency projects on river flow volume under the Murray-Darling Basin Plan*. Melbourne School of Engineering Water, Agriculture and Environment Program.

Water Act 1912 (NSW).

Water Act 2007 (Cth).

Water Management Act 2000 (NSW).

Water Sharing Plan for the Lachlan Unregulated and Alluvial Water Sources 2012.

Zulfic, D., & Barnett, S. R. (2007). *Angas Bremer PWA – Groundwater status report* (Report DWLBC 2007/27). Adelaide, Australia: Department of Water, Land and Biodiversity Conservation, Government of South Australia.

Stefanie Schulte has over 10 years of water resource management and policy experience in the private and public sector. She has worked extensively across Australia's Murray-Darling Basin and California on issues concerning water resources, energy and land management. She has extensive experience in Australia's regional water resource management and policy and previously worked as a policy manager at NSW Irrigators' Council.

Gabriela Cuadrado Quesada joined IHE-Delft in April 2017. Gabriela completed her PhD at the Faculty of Law at the University of New South Wales, Sydney, Australia in 2017. Her doctoral research focuses on groundwater governance looking in particular at the design and implementation of legal and policy frameworks in Australia and Costa Rica. Currently she is expanding her research to new jurisdictions such as the Netherlands and India. Gabriela has previously completed a Masters in Water and Coastal Management and Environmental and Infrastructure Planning at the University of Oldenburg (Germany) and at the University of Groningen (The Netherlands). She also completed a Masters in Human Rights and Education for Peace at the National University in Costa Rica. She obtained her Law Degree at the University of Costa Rica. Gabriela also worked in the NGO sector in Costa Rica as an environmental lawyer for many years. Her current research interests spring from a dual interest in the socio-legal implications of water crisis on the one hand, and on the other hand, practices that mix elements of community organisation and government responses. She explores the ways in which these practices engage with (socio)-legal and governance frameworks.

Chapter 18
Developing Substitution Resources as Compensation for Reduced Groundwater Entitlements: The Case of the Poitou Marshes (France)

Olivier Douez, Jean Eudes du Peuty, Daniel Lepercq, and Marielle Montginoul

Abstract This chapter describes the groundwater management policy implemented in the Poitou marshes, a 100,000 ha wetland located on the Atlantic coast in Western France. Similarly to other French basins, irrigated agriculture has rapidly developed since the 1980s, mainly based on groundwater exploitation. Clear signs of groundwater overexploitation appeared in 1992–1995, with the intrusion of brackish water in the aquifer. Because of the overexploitation, ecosystems were severely affected and the French Government was sued by the European Commission for noncompliance with the Bird Directive (1999). The chapter describes the progressive implementation of a groundwater management policy aiming at ensuring the long-term sustainability of an emblematic groundwater dependent wetland. To do so, the State imposed a very significant reduction in historical water entitlements. This case study illustrates the difficulties encountered in implementing this reduction, in a context of extreme competition between economic uses (agriculture, urban and touristic) and environmental objectives. The case study also reports on the complexity of developing an integrated management plan in basins where groundwater,

O. Douez (✉)
BRGM (French Geological Survey), Bordeaux, France
e-mail: O.douez@brgm.fr

J. E. du Peuty
EPTB Marais Poitevin, Luçon, France
e-mail: jeaneudes.dupeuty@epmp-marais-poitevin.fr

D. Lepercq
CACG, Tarbes, France
e-mail: d.lepercq@cacg.fr

M. Montginoul
INRAE – UMR G-Eau, Montpellier University, Montpellier, France
e-mail: marielle.montginoul@inrae.fr

J.-D. Rinaudo et al. (eds.), *Sustainable Groundwater Management*,
Global Issues in Water Policy 24, https://doi.org/10.1007/978-3-030-32766-8_18

rivers, wetlands and canals are highly interdependent. It highlights the importance of a (shared) knowledge on water resource and uses, of involving stakeholders in the different steps, and of trying to share scarcity in an equitable way.

Keywords Groundwater overexploitation · Groundwater management · Irrigation · Wetland · Hydrogeological model · Substitution resources

18.1 Introduction

This chapter describes the groundwater management policy implemented for the Poitevin Marshes, a 100,000 ha wetland located on the Atlantic coast in Western France. Just like other basins in central and western France (see Chaps. 5 and 13), irrigated agriculture has rapidly developed in this area since the 1980s, mainly based on groundwater extraction. Clear signs of groundwater overexploitation first appeared in 1992, with the intrusion of brackish water into the aquifer directly impacting some farmers as groundwater quality became unsuitable for irrigation. Ecosystems were also severely affected (with impacts on migratory birds) and the French government was sued by the European Commission for non-compliance with the Bird Directive (1999).

The chapter describes the progressive implementation of a groundwater management policy aiming at ensuring the long-term sustainability of an iconic groundwater-dependent wetland. To reach this environmental objective, the State has imposed a very significant reduction in historical water entitlements. This case study illustrates the difficulties encountered in implementing this reduction in a context of extreme competition between economic uses (agriculture, urban uses, and tourism) and environmental objectives. The case study also reports on the complexity of developing an integrated management plan in basins where groundwater, rivers, wetlands, and canals and highly interdependent.

The chapter is organized as follows: The first section presents the case study area, water resources and their uses. Section 18.2 describes the historical evolution of the quantitative management strategy progressively implemented in the Poitevin Marshes. It ends by pointing out the process that took place to define the maximum volume to be abstracted. Section 18.3 depicts the groundwater model developed to assess sustainable pumping limits and define operational rules for refilling reservoirs. Section 18.4 focuses on the new established governance, looking in particular at the coordination between the State, the local water management board (EPMP – "Établissement Public du Marais Poitevin": "Poitou Marshes Public Establishment"), and the users' associations. The last section concludes by analyzing the lessons learned from this experience in terms of the conditions for success and the limits of such a process for establishing quantitative groundwater management in agricultural areas.

18.2 The Case Study Area, Water Resources, and Their Uses

The Poitevin Marshes is the largest wetland on the Atlantic coast and is located halfway between the Loire and Gironde estuaries. It is the second largest in size in France behind the Camargue, and covers an area of 100,000 ha spanning three departments (Fig. 18.1): Deux-Sèvres, Charente-Maritime, and Vendée. Its watershed basin extends over 640,000 ha.

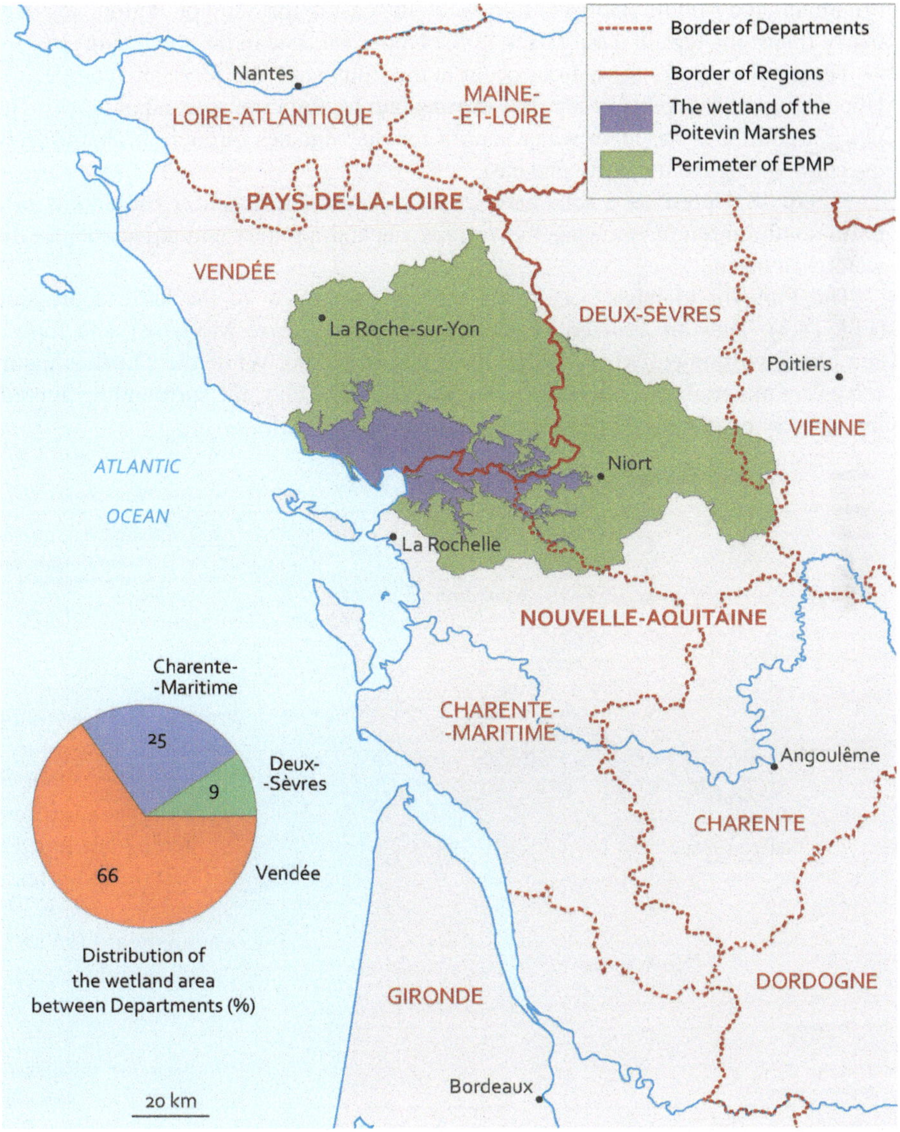

Fig. 18.1 Location of the Poitevin Marshes

18.2.1 A Unique Geological and Hydrogeological Context

Located on the boundary between the Aquitaine Basin to the south and the Armorican Massif to the north, the area used to be a gulf that was progressively filled in by fluvio-marine clay called "Bri" during the so-called "Flandrian" transgression estimated to date back approximately 7000 years (Anongba, 2007). The watershed basins that supply the Marshes with surface water spread over the Hercynian basement in the north, over the sedimentary terrain dating mainly back to the Lower (Toarcian) and Middle (Dogger) Jurassic in the east to the Seuil du Poitou, and over marly limestone terrain dating back to the Upper Jurassic in the south (Fig. 18.2).

Three main aquifer formations occur in the study area: the Toarcian, Dogger, and Upper Jurassic, separated by low-permeable to impermeable aquicludes. The Dogger aquifer is the main water supply for the Marshes through discharge from overflow springs to the north and east.

Figure 18.3 presents a schematic cross-section of the aquifer formations in a north-south direction, showing the various stacked aquifers and aquicludes in the watershed basin.

The Poitevin Marshes occupy the entire lower zone of the large depression (Fig. 18.3). They are crossed by rivers (notably the Sèvre Niortaise), and form a complex environment where water plays a central role. While the Marshes themselves are underlain by a layer of rather impermeable clay, the surrounding altered, fractured or even karstic limestone units form good aquifers.

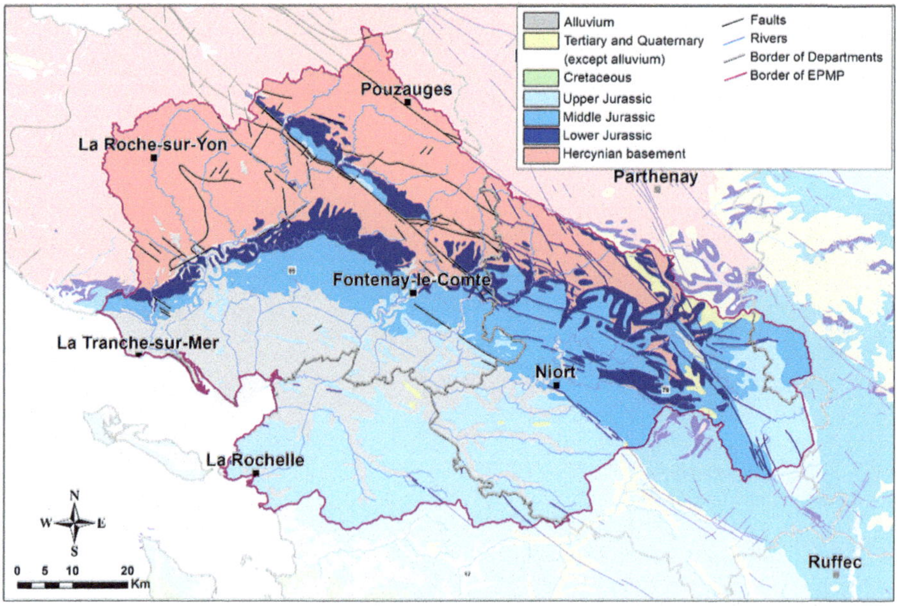

Fig. 18.2 Geological outcrop formations on the Marsh watershed basin

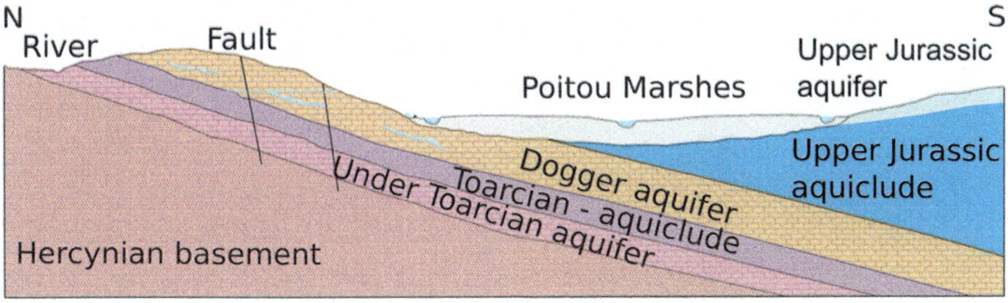

Fig. 18.3 Schema illustrating the hydrogeologic context of the Marshes – north-south cross-section

18.2.2 The Development History of the Poitevin Marshes

The Poitevin Marshes as they are known today have experienced strong human intervention mainly through many developments aiming to exploit the land, notably for farming. However, these developments made it necessary to manage flooding and the risks related to summer droughts.

Monks in the eighth century began draining the zone so that the wet soil could be cultivated. They built dikes and dug canals to grow crops on land that had until then been flooded. In the seventeenth century, Dutch investors developed most of the land that is currently cultivated, with the last polder taken from the sea in 1960. Later, lateral dikes along watercourses were built to prevent flooding during high flows. Indeed, crops in the Poitevin Marshes have long depended on flooding. The first crops were market garden crops; then grains were grown in areas drained and protected by dikes and canals.

These developments led to the creation of two types of zones: wet marshes and dried marshes (Fig. 18.4). Pastures are dominant in the wet marshes which serve as overflow basins during flooding, thus protecting the dried marshes which are used for larger crops.

Later, work was carried out to prevent sea water from encroaching during high tide, and then finally, sea dikes were built to protect the land from ocean storms.

18.2.3 Water Use

Currently, water from the Poitevin Marshes watershed basin is used for three main purposes:

- Irrigation, now the main use for groundwater: in 2010, the volume of water taken was 77.9 million m^3, with 67% coming from groundwater (Morardet & Boulfrad, 2013). In 2017, water abstraction authorizations were approximately 50 million m^3 in the summer and 40 million m^3 in the winter (inter-Prefecture decree of July

Fig. 18.4 The Poitevin Marshes wetlands

12, 2016) making it possible to irrigate approximately 25,000 ha from 1500 abstraction points throughout the watershed basin. While crop diversification is now considerable, irrigation of corn crops began in the 1980s. The climate is favorable for this crop, but the shallow and permeable soils require the regular addition of large amounts of water.

- Drinking water supply: the average annual abstraction for this purpose over the 2008–2011 period was approximately 50 million m^3 of which more than 10 million m^3 came from groundwater (information taken from the Loire Water Agency's database). In the Vendée Department, 90% comes from surface storages created for this purpose in response to the very sharp increase in demand during the summer tourist season along the coast.
- Aquaculture: the Bay of Aiguillon, the third largest mussel-growing area in France, is where fresh groundwater discharge (from overflow when groundwater levels are higher than the Marshes) mixes with sea water and provides essential nutrients for the growth of marine plants (notably plankton) on which mussels feed. This is one of the key requirements for shellfish development and therefore maintaining freshwater discharge during the summer low-water periods is important.

18.2.4 The Southern Vendée, an Emblematic Sector

Half of the surface area of the watershed basin for the Poitevin Marshes is located in the Vendée Department (Fig. 18.1). Because of its abundant water resources, it is the section of the study area that supports the largest users, with abstractions for irrigation authorized up to 25 million m^3, or half of the total volume for the basin.

Since it is also the main area for supplying the Marshes, the environmental concerns are significant and led to the very early investment in water management. It was in this context that an original collective management system experiment was set up for irrigation extractions in one sector. This experiment is described in this next chapter.

18.3 Implementation of Collective Water Management

18.3.1 The Irrigation Expansion Period and the First Problems

In the 1980s, irrigation developed throughout France as it did in the southern Vendée, thanks to sizeable financial grants for drilling wells and installing pumps for individual use. This led to a growth in corn cropping that made it possible to stabilize and then expand an efficient agricultural economy.

But this greater demand for irrigation water lowered the groundwater levels in the southern Vendée. Regular monitoring of levels was thus initiated in 1987, with the progressive installation of additional reference piezometers managed by the Vendée Department.

The drought in 1990–1991 and the soaring demand for water that it caused highlighted the vulnerable nature of the resource. Agricultural abstractions caused groundwater levels to drop below sea level; the decline in the north of the Poitevin Marshes caused a localized rise of connate salt water from the geological formations underlying the marsh (this salt is derived from sea water stored in the sediments that were deposited during the Flandrian transgression).

18.3.2 1992: First Steps in Collective Management

Following those two difficult years, water stakeholders decided to set up collective management of water resources. The first summer abstraction management plan for agricultural use in the southern Vendée groundwaters was signed in 1992. It defined piezometric warning thresholds that triggered restrictions on abstractions and then their cessation, if they were exceeded. The objective was to prevent the overly rapid decline in groundwater levels through scheduled restrictions. In the event of threshold alerts, irrigation was first prohibited on Sundays, then Saturdays, and sometimes totally banned if the situation became too critical.[1]

[1]The alert level corresponded to the lowest levels reached during the 1991 drought. In reality, the protocol allowed this threshold to be exceeded under certain conditions without triggering the total cessation of abstractions.

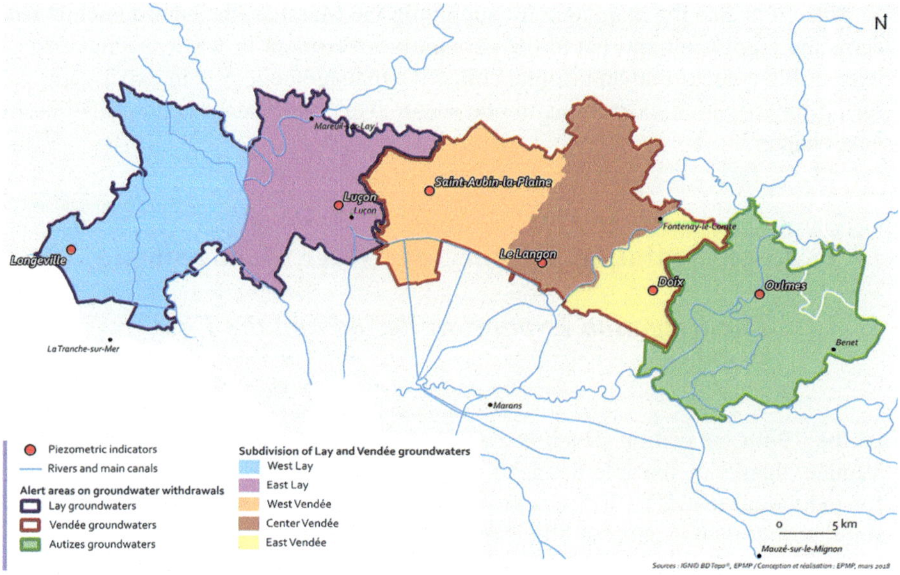

Fig. 18.5 Southern Vendée groundwaters with management zones and monitoring indicators

The first reference piezometer used to define the thresholds was located at Oulmes in the Autizes sector (Fig. 18.5). Indeed, this sector has always been the most fragile and most sensitive, being very close to the ecological and tourist heart of the Marshes known as the "Green Venice." Very rapidly, additional reference piezometers were selected and used either individually or by the averaging of readings from several of them. However, the minimum levels in the 1992 management plan were still below sea level and not sufficient to prevent salt water intrusion.

Although this protocol was an important first step because it ended a situation in which abstractions were completely unlimited, the restriction schedules had a very limited and short-term effectiveness. Irrigation users became used to no longer irrigating on weekends and bought additional and more powerful pumps in order to be able to irrigate the same crop area in less time. The increased pumping infrastructure quickly cancelled out the effects of the restrictions, which as a result, became more frequent. A vicious circle thus began, leading to an overhaul of the management approach.

18.3.3 1995: The Start of Collective Management Fully Involving Farmers

The Water Law of 1992 set obligations that profoundly changed water management in the area (see Chap. 3). It became compulsory for irrigation users to measure their abstraction, which was a revolution that would make it possible to establish a collective management regime based on individual accountability.

In response to the difficulties being encountered, the farmers took control of their future in 1995 by grouping together (based on their type of water use) in irrigation users' associations to defend their interests.

The establishment of these associations made it possible to build dialogue between the association leaders, the State, and the departmental water management stakeholders, and was the start of collective management approach.

The State then initiated a new management model that sought to make each farmer individually accountable: The administration gave each user an individual authorization specifying the maximum volume that could be abstracted for the summer season. This volume was defined in a variable manner based on the demand and negotiations with each irrigation users' association:

- on the basis that one season corresponded to 1000 h of irrigation (about 42 days), the volume assigned was 1000 × pump capacity; or
- the volume depended on the requirement for the crops in question; or
- the volume was calculated based on a combination of the two previous methods, starting with pump capacity as the base and adjusting the results with a coefficient based on the crops in question.

This volume, which became a historic baseline, was itself based on equipment history. Even though a few spoke out to express their disagreement with these rules which some deemed unfair, they have not been challenged since.

The volumetric management regime covered the summer period only, with abstraction limits applied after June 1. The goal was to encourage farmers to consume water earlier in spring to avoid overly large declines in groundwater levels during summer. Spring consumption was therefore initially not counted in these volumes.[2] In the event of non-compliance with the warning levels in the month of June, the summer allocation was reduced. All consumption was declared by the irrigation users themselves at the end of the crop year; administrative verifications were rare and only targeted inconsistencies or doubts regarding declarations for past years.

In the event that summer consumption was higher than the overall authorized volume, the authorized volumes for the following year were reduced. Over-usage estimates were not done on individual farm level, but rather collectively, with the application of a reduction coefficient to the sum of authorized volumes (i.e. a collective penalty). This system was a first step toward volumetric management. However, it made those extracting water only moderately accountable, as they did not know the results of their actions until after the end of the irrigation season when all abstractions were known. This system was then overhauled to progressively include individual accountability.

---

[2] Starting in 2005, spring consumption was also limited because the management regime had led to an improvement in summer conditions by shifting the over-extraction problem to the spring.

18.3.4 Recent Evolution: Toward a Suppression of the Structural Deficit

Extreme weather events then drove changes in the management approach. The 2003 drought led to the realization that such crisis situations need to be anticipated and planned for, and made it possible to generalize framework agreements which define responsive measures for these exceptional situations. Henceforth every year before the start of the irrigation season, a Prefectoral Decree defines the crisis response measures which include restricting use and protecting priority uses.

Progressively, it was noted that the effort was to move from crisis management to management of a structural shortfall in the water balance. The aim was to secure drinking water extractions, and meet the needs of environmental and economic uses (including agriculture) in 8 out of 10 years and reach the 'right water status' by 2021. Thus, the water resource must be the subject of balanced quantitative management and crisis management modes were only to be mobilized during exceptional climate episodes.

All this was given concrete form in planning documents (the *SDAGE* on the regional level, and *SAGE*s on the level of the watershed basin). These documents defined groundwater level targets that were sufficiently high enough to ensure good water supply to the Marshes in all seasons. The threshold for the start of the low-water period was therefore set to allow considerable supply in the spring at a time when the risk for biodiversity is at its maximum; the threshold for the end of the low-water period sought to guarantee a minimal level higher than that of the marsh to avoid any saltwater intrusion. The principle was to define ambitious target levels with a lengthy delay for their application (some indicators were thus set for compliance as early as 2021, and others only after 2021).

After 2000, this principle sparked extensive debate with numerous studies carried out that were for or against, the ambitious target levels. Ultimately, BRGM was called upon to develop a groundwater flow model that would provide better understanding of how the aquifer system worked.

To prevent significant reductions in authorized volumes, irrigation users' associations mobilized to generate substitution reservoir projects. The substitution principle consists of filling surface reservoirs by pumping groundwater in winter when piezometric levels are high. The stored water is then used later in the summer, making it possible to reduce groundwater abstractions during this period.

In the southern part of the Vendée Department, the first reservoirs were built under the management of an inter-communal joint association (SMVSA) and commenced operation in 2007 (Table 18.1). SMVSA entrusted reservoir management to a manager (the CACG – Compagnie d'Aménagement des Coteaux de Gascogne).

In this context of a highly variable climate and rather dry periods in 2016 or very dry periods in 2017 (with little or no groundwater recharge), the building of the first reservoirs made it possible to relieve water stress on the environment. A 3 m rise in groundwater levels was noted in Oulmes to the east, and a 1 m rise in Saint Aubin in the center and Luçon further to the east. These levels stayed above or near the target piezometric levels.

Table 18.1 Number of reserves and storage levels produced as at January, 2018

Sector	Number of reservoirs (created/planned)	Storage in million m^3 (created/planned)
Autizes	10/10	3.2/3.2
Vendée	8/10	4.4/5.4
Lay	4/5	1.75/2.4

18.3.5 The Effectiveness of Management Measures

The establishment of quantitative management made it possible to progressively reduce the magnitude of the summer decline in groundwater levels. Figure 18.6 shows the evolution of measured levels over several years. In 1990 before abstractions were limited, the level dropped rapidly between April and August, with warning thresholds being passed on July 15 and the end-of-season level being 0.5 m below sea level. The situation improved in 1995 and 2005, clearly showing the temporary effect of scheduled restrictions with a small rise in groundwaters on days when abstractions were banned (resulting in a saw-tooth curve). However, the end-of-season levels remained very low. Eventually, the level recovered above the warning threshold in 2010 and 2015.

18.4 The Groundwater Model Developed to Improve Knowledge, Assist Management, and Guide Investments

Setting up the previously described management required improving knowledge of how aquifers function, and their interactions with surface water. This knowledge acquisition process was initiated at the end of the 1980s with the installation of the first piezometers. A modeling tool was then developed in 2007 at the request of the State to assist in the calculation of the volumes authorized for abstraction. As a result, the Jurassic hydrogeologic model (Putot & Bichot, 2007) was adopted on this sector (Douez, 2010; Douez et al. 2010) and was subsequently updated following various investigations (Douez, 2015a, 2015b; Douez, Bichot, & Petit, 2011), in partnership with stakeholders in the field, and notably with the Etablissement Public du Marais Poitevin (EPMP, Poitevin Marshes Public Establishment).

This model is a response to the need to have a water resource management tool integrating all watershed basins and in particular those supplying the Marshes. It is part of the set of tools developed in the west-southwest of France to help manage groundwater on a regional scale (Douez et al., 2016; Wuilleumier, Saltel, Douez, & Cabaret, 2016), with the aim to:

- better understand the operation of all aquifer formations and, for some, analyze groundwater/river relationships; and
- help answer various questions relating to water resource issues such as availability, management, impact of global warming, etc.

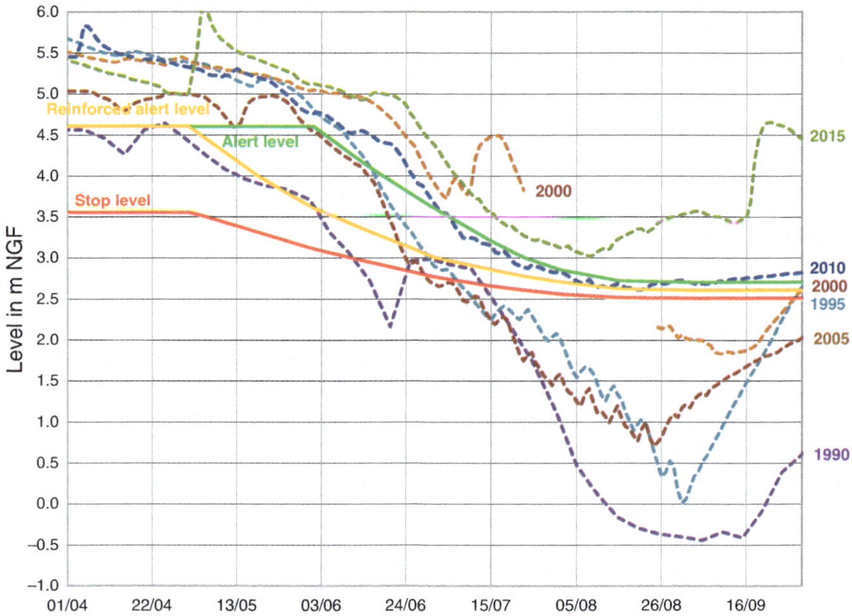

Fig. 18.6 Historical groundwater levels

18.4.1 Presentation of the Hydrogeologic Model

The modelling code used in this model is MARTHE[3] developed by the BRGM (Thiéry, 2015). The Jurassic regional hydrogeologic model is calibrated over the 2000–2011 period on a monthly and weekly time scale (June to August). It covers a surface area of 20,195 km^2 (Douez, 2015a, 2015b). The model has a 1 km grid size, except in the northern periphery of the Poitevin Marshes where the grid size is 333 m in order to better represent the interaction with waterways and abstraction zoning. It contains eight layers corresponding to the various aquifers and aquitard layers in the area.

The groundwater recharge calculations for the entire surface area of the model were based on hydro-climatic balance sheets (a breakdown of recharge and runoff) drawn up from spatialized evapotranspiration data and rainfall.

Since groundwater/river exchanges play an important role in regional water dynamics, the main waterways were included in the model. Abstractions and discharge of water, either for groundwater or surface water, were integrated for the 2000–2011 period (abstractions for drinking water supply, agriculture and industry,

[3] MARTHE: *Modélisation d'Aquifères par un maillage Rectangulaire en régime Transitoire pour le calcul Hydrodynamique des Ecoulements* (modeling of aquifers in rectangular grids in the transitional regime for the hydrodynamic calculation of groundwater flows).

as well as discharge from wastewater treatment plants). Seven dams (low flow support, drinking water supply, etc.), located along waterways, were also considered.[4]

18.4.2 Uses of the Model to Manage the Marshes

18.4.2.1 2007–2010. First Use of the Model to Manage Groundwater in the Poitevin Marsh Sector

The model (assuming conditions experienced during the 2000–2007 period) made it possible to test various agricultural abstraction reduction scenarios (Douez et al., 2010). This reduction was first simulated by applying single reduction coefficients to the total annual volumes extracted, and then by differentiating the spring and summer reduction. This made it possible to calculate an extractable volume based on targets for groundwater levels and waterway flows (the volume extracted that would enable the targets to be met in 8 years out of 10).

18.4.2.2 2011–2016. Simulations of Establishing Substitution Reservoirs

Following this work, and based on the extractions volume limits set in 2010, the model was used to test different scenarios for the establishment of reservoirs to substitute groundwater abstractions in the Lay (Douez, 2011), Vendée (Douez, 2012), and Sèvre-Niortaise (Abasq, 2016) sectors.

The analysis of the simulation results indicated that setting up substitution reservoirs would greatly improve the summer piezometric levels as well as the flow in waterways throughout the study zone. The simulations also predicted it would be possible to comply with most of the piezometric targets set in the SDAGE. The negative impact during the winter reservoir filling period was low compared to a very significant positive impact in summer for both groundwater and surface water supplies in the Marshes when the supply from reservoirs replace extraction from boreholes.

18.5 The Reduction of Entitlements

The model simulations performed also made it possible to determine the volumes that could be extracted in order to attain the piezometric level targets for the management zone. These volumes were generally much lower than current abstractions.

[4] It should be stated that the model for the Poitevin Marshes does not aim to simulate the hydraulic behavior of the "marsh" zone (i.e. flows in canals) where the manipulated hydraulic operation are complex.

18.5.1 Reduction Process

The authorized volume over the entire territory for the spring-summer period needed to fall from 49.6 to 32 million m^3. This structural reduction was obtained in part, by a reduction in volume (by a minimum of 20%) without compensation, and in part by the water source substitution programs which had the greatest impact on the environment.

Once the overall limit had been set, the maximum annual authorized volumes were divided among irrigation users by sector, by abstraction installation, and by period. The volume assigned annually per installation depended on demand, past consumption and the impact of the abstraction. Volumes freed up by an irrigation user ceasing irrigation were allocated as a priority to new irrigation users, and then to increasing volumes for existing users.

The distribution plan was drawn up with collaboration from irrigation user representatives, owners, and the managers of substitution infrastructures. This distribution plan was adjusted annually to take into account any new reservoirs built and abstraction reduction targets in zones where the extractable limits were lower than actual abstractions. In this way, it was possible to accompany the abstraction reductions in these challenged zones with the redistribution of available water to new irrigation users.

18.5.2 Temporal and Operational Management

After abstraction reductions and reservoir capacities were considered, the extractable volume limits that were set aimed to reduce the structural deficit between actual extractions and the authorized volume limit over the entire territory. Compliance with them did not however, guarantee that the ‘right ecological state’ of water resources and the environment would systematically be achieved every year. To account for climatic variations, a series of operational management rules was established and set up for each management sector.

These operational management rules covered zones with collective and pooled management in which all irrigation users contributed to lowering the level of abstractions in summer low-water periods.

At the end of May, each irrigation user indicated the projected distribution of their authorized volume based on their projected needs for their crops. This distribution would define the management rules to which each user would be subject during the irrigation season.

During the season, the volume could be reduced based on changes in groundwater levels with reference to a management curve called the warning threshold, which considered target piezometric levels for the start of the low-water period as well as the natural discharge of groundwaters.

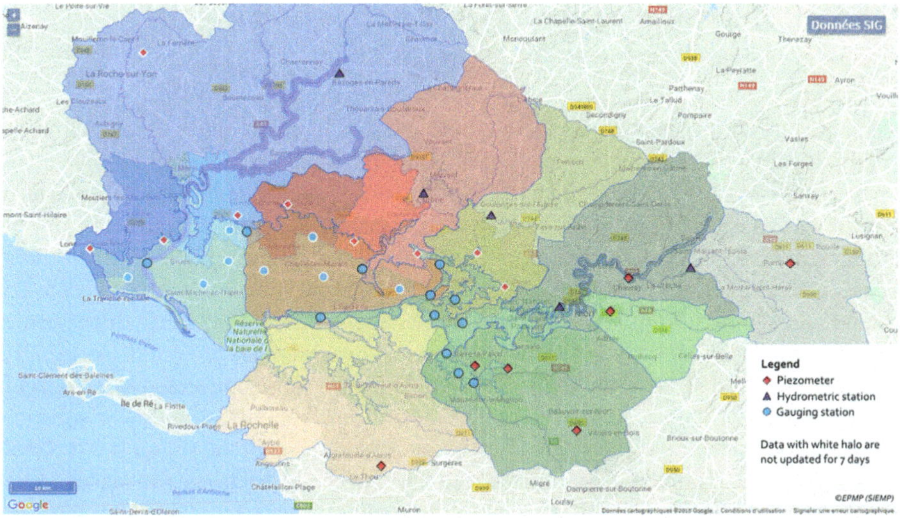

Fig. 18.7 Location of reference piezometers to monitor water extraction

- If the groundwater level was above the warning curve, consumption was not limited.
- If the groundwater level was located between the warning curve and the heightened warning level, the volume for the fortnight could be reduced by up to 50%.
- If the groundwater level fell below the heightened warning threshold, the volume was reduced by at least 50%.
- Finally, as soon as the stoppage level was crossed, irrigation was banned.[5]

Each management zone was monitored daily by at least one reference piezometer (Fig. 18.7). A public application was set up by the EPMP to allow monitoring: the *Système d'Information de l'Eau sur le Marais Poitevin* (SIEMP, or Poitevin Marsh Water Information System).

As soon as stress on the resource was detected during the irrigation period, a management committee would meet every 15 days. Chaired by the EPMP and assisted by its manager, it brought together representatives of the administration, the agricultural industry and the irrigation users' association. After consultation, it decided on the appropriate limits required to maintain the target groundwater levels by using predictions from the groundwater model which used knowledge of future crop needs, upcoming weather, observed groundwater levels and the monitored consumption in real time as inputs.

[5]The OUGC was in charge of system management so long as the groundwater levels were above the heightened warning threshold; below that level, the State intervened (Fig. 18.6).

18.5.3 Water Management Based on Unity Among All Irrigation Users on Sector Scale

The novelty of the new management approach was to create unity among all water extractors in a given management sector. The same management rules applied to all, whether or not they were connected to a substitution reservoir. This took form in two ways: firstly, a single water tariff was established, whereby investment costs for the reservoirs (part of which was not subsidized by the State) and all operational costs were thus shared among all users. Secondly, all water users were subject to the same management rules. This volumetric management was applied to three sectors in the southern Vendée, starting in 2006. If the groundwater level fell too rapidly, volumetric restrictions were applied to all users, even those connected to a reservoir storage. This principle of sharing both the costs and benefits of reservoir substitution was a key factor in the success of the operation through the social ties it developed. In order to prevent opportunistic breaches of the restrictions, the management system included sanction and verification measures:

- Individual over-usage of the authorized volume was penalised by a minimum of an equivalent reduction the following year. Over the entire South Vendée sector, a financial penalty for excess use on a fortnightly basis as well as the total volume over the irrigation season, resulted in a considerable reduction in individual over-usage.
- The CACG set up a random enhanced verification system to ensure the proper reading of all meters every fortnight. Later, abstraction points were progressively equipped with smart meters (that can be read remotely by telemetry) that made it possible to obtain data on a daily basis automatically.

Figure 18.8 shows the effectiveness of the management approach by comparing the average water levels during the summer low-water period before the implementation of the governance system in 2012, to those measured after 2012.

18.5.4 Governance Report by a Non-agricultural OUGC

There are two main contributing factors to the farmers' acceptance of the management measures that were implemented. Firstly, various hydrogeological experts and the groundwater model helped them understand how groundwater system worked, in particular how abstractions downstream were influenced by those carried out upstream. In addition, although the planning documents (SDAGE) forced a reduction in extracted volumes to the extractable volume limit, the establishment of substitution reservoirs allowed them to maintain their irrigation potential.

The choice of substitution zones was not made by the irrigation users but was determined through use of the groundwater model. The quality of the modeling was acknowledged by agricultural bodies, reservoir project implementers, nature fore-

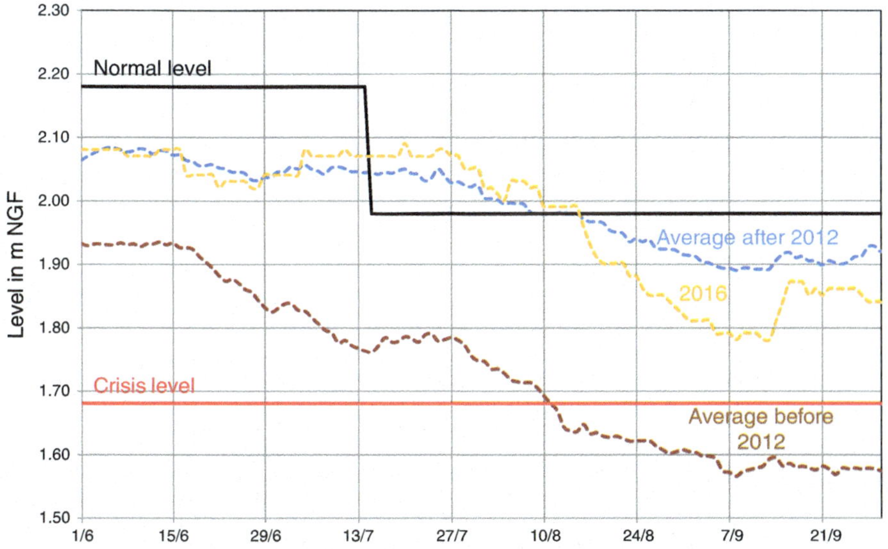

Fig. 18.8 Comparison of water levels at Saint Arnault before and after implementation of the new water governance system

casting associations, and the EPMP. This helped make irrigation users confident in their decision to accept the substitution reservoirs as a replacement for their individual boreholes. The reservoir project implementation was driven by local territorial governments and not by agricultural bodies. This provided an additional guarantee that the substitution was not done to favor any given group of water users. The neutrality of the expertise of the public service delegate (the CACG) also helped with project implementation.

For its part, the EPMP obtained management autonomy from the State provided that a protocol for managing and monitoring consumption was established. Such a protocol would be only accepted if:

- The limits had a visible and predictable effect on monitoring indicators. Analysis of changes in groundwater in relation to changes in consumption was therefore vital.
- The information was widely shared. The EPMP met every 2 weeks with all stakeholders to analyze the information and propose appropriate management actions. These decisions were made collectively. During this process, the importance of irrigation users' representatives on this management committee must be emphasized as they played a vital role in relaying information to all irrigation users in the field.
- The volumes abstracted were metered and verified. Since the accuracy of the analysis of the effect of abstractions on groundwater levels was crucial, reliable monitoring and verification of abstractions needed to be established. This was done by installing water meters, by obliging all irrigation users to report their

own consumption every 15 days, and finally by installing smart meters. Indeed, the verifications showed some discrepancies between the declarations and the actual volumes extracted; smart meter readings introduced transparency and equal treatment for all, even though it felt intrusive to some.

- Conflict was reduced. Established in 2012, the EPMP was able to establish respect and authority with stakeholders by assuring all users that the collective, shared management regime in place sought to respect extractable volume limits and if implemented well, would allow everyone to irrigate with a minimum of restrictions. One group could not gain precedence over another; and that in the event of management failure, the State would intervene and there would be a risk of a sharp reduction in authorized abstractions.

The creation of substitution reservoirs did not happen smoothly. The wider community was opposed to it, and environmental associations filed legal action against the first authorization decree. Here, it should be noted that this system for southern Vendée, which was based on setting up clear management rules and monitoring and verification of abstractions, allowed two other projects (Vendée and Lay) to be accepted without contest later. The relevance of the analysis and information sharing allowed increasingly smooth governance in this territory and the reconciliation of the quantitative water balance target for biodiversity and the preservation of effective agricultural activity.

18.6 Lessons Learnt

The Poitevin Marshes experiences very unique interactions between surface water and groundwater resources that contribute to the development of an ecosystem with a rich diversity of flora and fauna. The marshes can be subdivided into two broad areas; the older marsh supply upstream zone that has been drained and where water extraction occurs for irrigation and drinking water supplies, and the marsh itself which is a living environment downstream for flora and fauna, pastures and leisure. The marshes are therefore divided into two zones which have different water uses and stakeholders. This situation complicated the establishment of the management regime that relied on both groups of water users being aware of connectivity between their water resources and required those located in the supply zone to accept restriction measures that did not provide them any benefits.

The success of the novel water management approach described in this case study can be contributed to many factors, the most important of which are summarized below:

- Unity among irrigation users, irrespective of the source of the water they use (waterways, groundwater or reservoirs). This unity is seen in the willingness to pay for infrastructure even among those who do not benefit from it directly, and acceptance of restrictions on abstractions even when their resource is not directly affected by overuse.

- A shared effort to attain the allowed agricultural abstraction volume limit, through demand reduction (water saving) measures and the creation of substitution reservoirs that have minimal impact on the resource condition because they are filled outside of the high demand period.
- Spatial management of restrictions based on the impact of abstractions. These restrictions were also scaled to individual circumstances previously declared by each farmer in order to impact them as "fairly" as possible, by considering their actual needs.
- Management carried by a non-agricultural body with the aim of seeking cooperation among the various stakeholders. This body was also entrusted with allocating the maximum volume able to be extracted for agriculture amongst farmers. But it entrusted enforcement of measures to a manager used to sharing water among farmers (the CACG).
- Excellent knowledge of the resource and the use of the water. This made it possible to manage the resource for the best outcomes, gain acceptance for the measures taken, and optimize structural investments (for example, the size and location of substitution reservoirs). This knowledge was widely shared via a website. This allowed all stakeholders to be informed of management decisions, and potentially be able to provide feedback on the proposals.
- An information system on agricultural abstraction volumes that made it possible to guarantee and verify enforcement of restriction measures.
- A double sanction system. Both financial and volumetric penalties encourage compliance with restriction measures.
- Reactive joint management. Bimonthly meetings of a management committee are held as soon as adverse resource trends are detected which allow stakeholders to decide collectively on the adoption of appropriate restriction measures and to inform users.

This success can also be explained by more practical elements: management decisions relayed to water users in the field by the farmers' representatives, a non-agricultural management body in which all stakeholders placed their trust, and an early awareness of potential crisis situations.

However, certain tensions remain as the wider community did not easily accept the creation of substitution reservoirs. The balance that has been achieved is fragile and could be threatened if there are considerable modification of the targets for the groundwater and extraction levels in future planning documents.

References

Abasq, L. (2016). Simulation de l'impact des projets de réserves de substitution du bassin de la Sèvre Niortaise/Marais poitevin avec le modèle hydrodynamique du Jurassique - Phase 2. Brgm, Rapport final BRGM/RC-65669-FR, 159 p.

Anongba, B. (2007). *Identification du système hydrogéologique des formations quaternaires et callovo-oxfordiennes du Marais poitevin par approche couplée minéralogique, hydrodynamique et géochimique* (379 p.). PhD Thesis. http://www.theses.fr/2007POIT2320

Douez, O. (2010, November 25–27). Apports des modèles hydrodynamiques régionaux pour contribuer à la gestion des ressources en eaux souterraines: Exemple du modèle Jurassique du Poitou-Charentes pour le Marais poitevin. In *17èmes Journées Techniques du Comité Français de l'Association Internationale des Hydrogéologues - La DCE 10 ans après : une dynamique pour la connaissance et la gestion des eaux souterraines*. Toulouse.

Douez, O. (2011). Simulation de l'impact des projets de réserves de substitution du secteur du Lay avec le modèle hydrodynamique du Jurassique. Brgm, BRGM/RP-60122-FR, 85 p.

Douez, O. (2012). Simulation de l'impact des projets de réserves de substitution du secteur Vendée avec le modèle hydrodynamique du Jurassique - simulations complémentaires. Brgm, Rapport final BRGM/RP-60956-FR, 139 p.

Douez, O. (2015a). Actualisation 2008-2011 du modèle maillé des aquifères du Jurassique. Brgm, Rapport final BRGM/RP-64816-FR, 209 p.

Douez, O. (2015b). Mise au point d'un outil pour la gestion des autorisations de prélèvements annuelles de l'EPMP. Brgm, BRGM/RP-65052-FR, 108 p.

Douez, O., Bichot, F., Dequidt, D., Dugrillon, D., Putot, E., & Petit, L. (2010). Contribution à la gestion des prélèvements à la périphérie du Marais poitevin par modélisation hydrodynamique. BRGM/RP-58297-FR, 239 p.

Douez, O., Bichot, F., & Petit, L. (2011). Contribution à la gestion quantitative des ressources en eau à l'aide du modèle Jurassique de Poitou-Charentes. Brgm, BRGM/RP-59288-FR, 411 p.

Douez, O., Cabaret, O., Saltel, M., Abasq, L., Wuilleumier A., & Pedron N. (2016). *Groundwater management of large aquifers in southwestern France by regional hydrodynamic models*. 43rd IAH CONGRESS - Groundwater & Society - 60 Years of IAH, Montpellier, France, 25–29 September 2016.

Morardet, S., & Boulfrad, Y. (2013). Etat des lieux de l'agriculture irriguée dans le Marais Poitevin. Irstea, 117 p.

Putot, E., & Bichot, F. (2007). CPER 2000-2006 Phase 4 – Modèle Infra-Toarcien Dogger: calage du modèle hydrodynamique en régime transitoire. Brgm, BRGM/RP- 55742-FR, 94 p.

Thiéry, D. (2015). Code de calcul MARTHE – Modélisation 3D des écoulements dans les hydrosystèmes – Notice d'utilisation de la version 7.5. Brgm, BRGM/RP-64554-FR, 308 p.

Wuilleumier, A., Saltel, M., Douez, O., Cabaret, O., Abasq L., & Pedron N. (2016). *A platform to harmonize the regional hydrodynamic models in the southwest of France*. 43rd IAH CONGRESS - Groundwater & Society - 60 Years of IAH, Montpellier, France, 25–29 September 2016.

Olivier Douez is senior a hydrogeologist at Brgm (French Geological Survey). PhD in hydrogeology, he coordinates and develops the "water" activity within the Bordeaux BRGM (team of 10 people). He is also the program manager on "water body relationships": groundwater/rivers, groundwater/coastal waters, groundwater/wetlands. Olivier has over 15 years of experience in modeling large regional aquifer systems, and in particular on coupled models with hydrographic network. The developed models are used for public service issues (quantification of volumes that can be pumped, analysis according to piezometric objectives and/or flow river objectives, management plan tests of water resources, etc.) and research (climate change, rise of sea-level, quantification of groundwater/river flow exchanges, etc.). He has also worked on many topics of hydrogeology, particularly on sedimentary environments, whether on quantity or quality of water resources.

Jean Eudes du Peuty is deputee director of Établissement Public du Marais Poitevin (EPMP). He is in charge of quantitative water resources management. He supervised the establishment of Water Users' Associations (OUGC) and the development of water management (including allocation rules), in tight collaboration with Agricultural Chambers. Prior to joining EPMP, Jean-Eudes worked for several government departments in the Vienne and Charentes counties, where in was in charge of forest and water management.

Daniel Lepercq is an expert in water management. He has been working at the Compagnie d'Aménagement des Coteaux de Gascogne (CACG) for more than 30 years. His expertise includes water allocation, designing water tariffs for cost recovery as well as the design of operational water management rules. During the last ten years, he was in charge of developing new reservoir in Poitou Marshes and related water allocation rule. In addition, he is also in charge of conducting studies on the social, environmental and economic impact of water pricing policies at CACG.

Dr. Marielle Montginoul is senior researcher in Economics at the National Research Institute for Agriculture, Food and Environment (INRAE – previously IRSTEA) and she is a member of the Joint Research Unit G-Eau. Her work focuses on understanding and modeling farmers and households' water consumption behaviors. She more specifically studies instruments that can be used to reveal these behaviors when information is incomplete. Her research also focuses on economic tools to manage water withdrawals, with a focus on water pricing. She mobilizes a wide range of methodologies including surveys, experimental economics, and scenarios workshops. Marielle is member of the scientific council of the Rhône Méditerranée and Corsica Water agency. She coordinates a Master in Social sciences applied to water management in Montpellier University.

Chapter 19
New Approaches for Allocation Reductions and Groundwater Salinity Management in South Australia

Steve Barnett and David Williamson

Abstract The Tintinara area is located in the Upper South East of SA and lies 200 km southeast of the capital city of Adelaide. An unconfined Quaternary Limestone aquifer lies at a shallow depth (< 5 m), provides high well yields (< 200 L/sec) and is extensively used for the irrigation of lucerne (alfalfa). In 2003, the first management plan introduced licences, a volumetric entitlement and required metering of extractions. The size of each entitlement was determined by the theoretical crop irrigation requirements (TCR) for the existing area of the range of crops irrigated by a variety of systems (flood, sprinkler, drip). A review the management plan in 2008 found that metered extractions had been reasonably consistent at about 15,000 ML/yr which is only about half of the volume that had been allocated to irrigate the same area of land. This large gap between usage and allocation will make future management responses difficult and ineffective. After extensive consultation with the affected irrigators, an allocation reduction program was initiated that minimised impacts on existing users and would be staged over several years so that irrigators had time to adjust their operations. This program included a review of the TCRs using more recent information.

Keywords Groundwater entitlements · Irrigation requirements · Stakeholder consultation · Salinity management · Irrigation recycling · Management plan

---

S. Barnett (✉)
Adelaide, SA, Australia
e-mail: Steve.Barnett@sa.gov.au

D. Williamson
Department for Environment and Water, Mount Gambier, SA, Australia
e-mail: david.williamson@sa.gov.au

J.-D. Rinaudo et al. (eds.), *Sustainable Groundwater Management*,
Global Issues in Water Policy 24, https://doi.org/10.1007/978-3-030-32766-8_19

19.1 Introduction

This chapter presents a case study of a successful exercise to reduce irrigation entitlements in a groundwater management area in South Australia (SA). Although the main driver for the reduction was not a direct and imminent threat to the sustainability of the groundwater resource, but a longer-term risk to effective management of the resource, the 'recipe for success' is considered to have widespread applicability.

19.2 Hydrogeology

The Tintinara area is located in the Upper South East of SA and lies approximately 200 km southeast of the capital city of Adelaide. It overlies part of the Murray Basin which is a large sedimentary groundwater basin covering 300,000 km^2 of southeastern Australia. Managers of the groundwater resources in this PWA are dealing with a number of complex issues.

Groundwater flows through two major aquifer systems: a regional unconfined Quaternary Limestone aquifer and an underlying Tertiary Buccleuch Group confined aquifer comprising sand and limestone layers as shown in Fig. 19.1 (Barnett, 2002). The upper, unconfined limestone aquifer is the most extensively used of the two aquifers because in the area of interest for this case study, it lies at a shallow depth (less than 5 m) and is also high yielding (up to 200 L/s). The main use of the water is for the irrigation of lucerne (alfalfa).

As a result of the rapid expansion of irrigation activity during the 1980s and 1990s, concerns were raised in the community that detrimental impacts on the groundwater resource could occur in the form of declining groundwater levels.

19.3 Management Intervention

In response to the community's concerns, the area was prescribed in November, 2000 (this process as it is undertaken in SA, is described in more detail in Chap. 8). Subsequently, after an assessment of the groundwater resource and its capacity to meet demands placed upon it, and extensive community consultation on management approaches, a Water Allocation Plan (WAP) for the Tintinara-Coonalpyn Prescribed Wells Area (TCPWA) was prepared by the South East Catchment Water Management Board and released by the Minister of Environment and Conservation in May 2003 (SECWMB, 2003). This WAP required irrigators to have a licence with a volumetric limit and a meter to measure extractions.

The 2003 Plan also limited extractions in the case study area (the Tintinara Management Area (MA)) to the current levels at that time although due to a lack

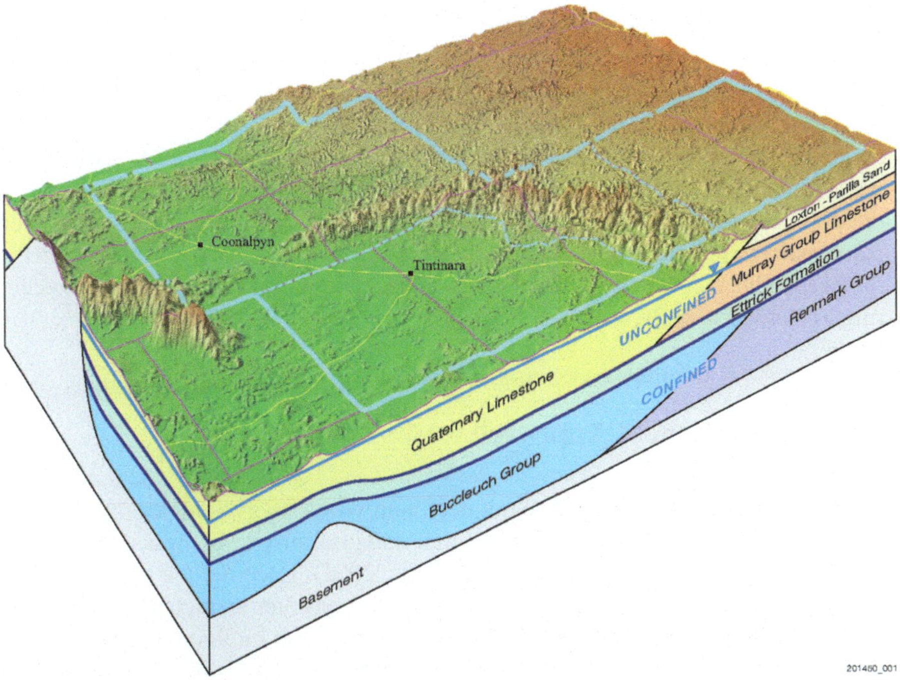

Fig. 19.1 Hydrogeology of the Tintinara area

of meter data, the exact volume of those extractions was not known. Theoretical crop irrigation requirements (TCR) were used to determine the volumetric allocations given to irrigators throughout the TCPWA (also described in Chap. 7). An area limitation was also imposed so that the area of irrigation could not be increased.

19.3.1 Determination of Volumetric Allocations

A study was undertaken to determine appropriate allocations for irrigators by calculating the irrigation requirements of a range of crops irrigated by a variety of irrigation systems. The study utilised data from a desktop study of irrigation requirements, information from irrigator workshops and field validation of data (ICMS, 2002). The water allocations issued to irrigators consisted of a base allocation (the volume of irrigation water required by the crop for maximum growth and assumes that water is available at all times, based on Allen, Pereira, Raes, & Smith, 1998), and delivery component (the volume of water required to deliver the net irrigation requirement to the cropped area and also includes water for the leaching of salts, frost control and the establishment of cover crops).

19.3.2 Community Consultation

During the formulation of the WAP, there was extensive consultation with irrigators and other community stakeholders. The management agency hydrogeologist (the author) attended several public meetings and numerous meetings with the local water planning committee to explain the hydrogeology of the area and the causes of the observed monitoring trends for water level and salinity. Numerous visits were made to irrigation properties to collect water samples for salinity testing. The irrigators were advised of their salinity results so they could understand the trends occurring in their own wells.

19.4 Management Approach

Ironically, despite the concerns about declining water levels, long term monitoring and the groundwater assessment found that rising salinity levels was a greater sustainability issue in the Tintinara MA (Barnett, 2002). Lucerne, the major irrigated crop grown in the area, has a high water use. Groundwater extracted from the shallow Quaternary Limestone aquifer is applied to the crop by flood and pivot irrigation, and as water is drawn up through the root system, most of the dissolved salt is not taken up by the plant and accumulates in the root zone (Fig. 19.2). This salt is then flushed back down into the shallow aquifer during subsequent irrigation applications or the infiltration of rainfall, resulting in increasing groundwater salinity due to the recycling of the irrigation drainage water. Over 50% of irrigation wells sampled showed a salinity increase due to this process.

Typically, groundwater degradation problems in aquifers are managed by reducing extractions to sustainable levels. However in this situation, it was recognised

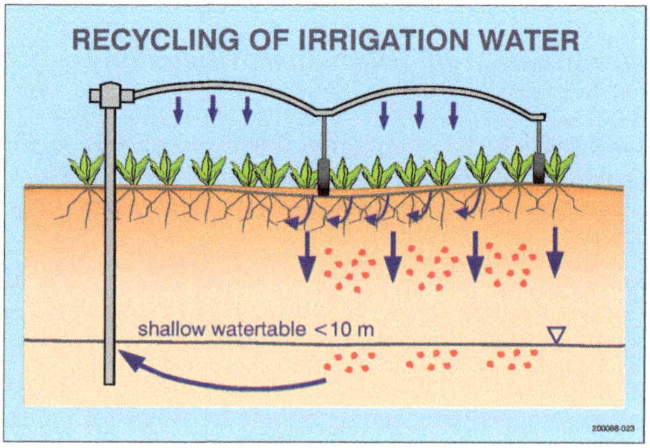

Fig. 19.2 The process of irrigation recycling

that the degradation due to recycling is caused by application of water to the crop rather than the physical removal of water from the aquifer. If the extracted water was piped elsewhere for public water supply or industrial purposes, no salinity increases would have occurred.

A new management approach was therefore required. A buffer zone method was instigated in the WAP whereby no new irrigation wells could be drilled within a 2 km buffer applied around each existing irrigation well. This method prevents concentrations of irrigation (and the resultant salt accessions) in any given area and allows natural dispersion to occur and dilution of the salt added to the aquifer by rainfall recharge. This method will also prevent excessive drawdowns in water levels caused by pumping which may prevent lateral groundwater flow through the aquifer which removes salt from the region. Fortunately, there are no other groundwater users downgradient of the Tintinara MA due to naturally high salinity levels.

19.5 Review of the 2003 Water Allocation Plan

Under the legislation, each WAP must be reviewed every 5 years. This review should include items such as the current condition of the groundwater resources, whether the management approach adopted for the sustainable use of the resources is appropriate, the current levels of use and allocation in each of the management areas within the TCPWA and whether any detrimental impacts have occurred as a result of that use.

A hydrogeological review of the WAP for the TCPWA (Barnett, 2008) found that in the Tintinara MA case study area, watertable levels had steadily declined by 1.5 m in response to below average rainfall and irrigation extraction. This represented about 5% of the unconfined aquifer storage volume. Very little, if any recharge occurred during the 2006 drought, leading to the lowest groundwater levels on record. Average winter rains during 2007 resulted in a strong recovery of levels close to pre-drought levels. Previously observed salinity rises had stabilised over several years prior to 2008, possibly in response to increased irrigation efficiency and ongoing below average rainfall.

The hydrogeological review also found that since metering of extraction began in 2001, extractions in the Tintinara MA have been reasonably consistent at about 15,000 ML/yr which is only about half of the volume that had been allocated to irrigate the same area of land (Fig. 19.3). A significant decrease to just over 10,000 ML occurred in 2005–2006 due to the very wet spring delaying commencement of irrigation. The 2006 drought resulted in a significant increase to 26,100 ML in the 2006–2007 season compared to previous years.

This obvious gap between usage and allocation to irrigate the same area of land indicates that theoretical irrigation crop requirements adopted were too generous. The imposition of an area limitation proved crucial in the gathering of evidence that proved over-allocation because irrigators could not increase their area of irrigation to take advantage of their unused allocations. A large gap between usage and

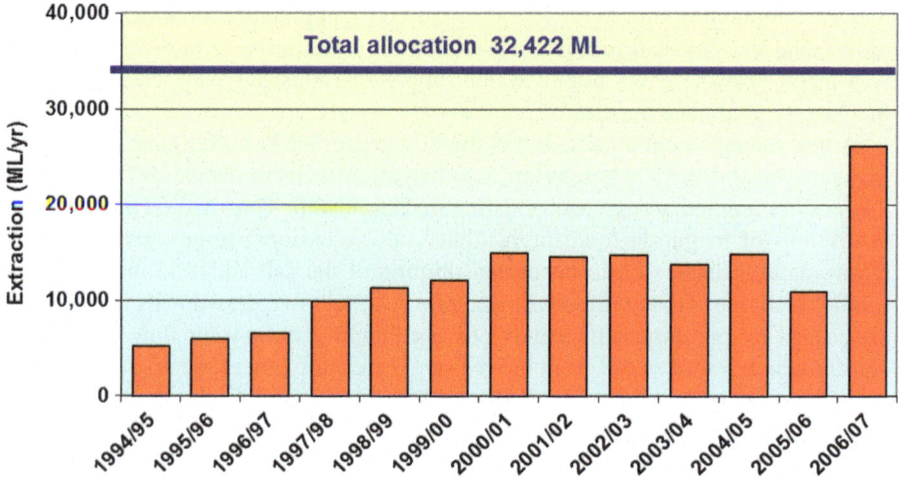

Fig. 19.3 The volumes of extraction and entitlement in the Tintinara MA

allocation will make future management responses more difficult and ineffective. For example, a reduction in allocations to alleviate significant declining groundwater levels and salinity impacts due to climate change, will have little or no impact on actual levels of extraction.

The review (Barnett, 2008) consequently recommended that a reduction in the total allocation of the order of 12–14,000 ML/year be made over time in the Tintinara MA, based on the usage trends before the 2006 drought. When this reduction had been made, removal of the area limitation on licences would occur. This adjustment will ensure that if further reductions in allocations are required in the future, actual reductions in extractions will occur.

This evidence also prompted a review of the process to determine the theoretical irrigation crop requirements which was carried out by the South East Natural Resources Management Board (SENRMB) which superseded the SECWMB. This review recommended some adjustments to the delivery component for some irrigated crops (SENRMB, 2009) based on observed extractions and new industry standards.

19.6 Allocation Reduction Process

As the process for reducing allocations was required to be included in the revised WAP, the SENRMB conducted extensive consultation with the affected irrigators. Because the reductions were not driven by adverse impacts caused by extraction, a methodology that minimised impacts on existing irrigation operations was required. Also in recognition of the fact that a considerable component of flood irrigation actually returns to the shallow aquifer through infiltration, the target for reduction

Table 19.1 Reductions in allocations for lucerne irrigation

	Centre pivot	Flood
Original allocation (ML/ha/yr)	7.69	9.29
Original delivery component	27%	54%
New delivery component	18%	118%
New allocation (ML/ha/yr)	7.13	13.19
Percentage change	−7%	42%
Reduction after 5 years	10%	10%
New allocation (ML/ha/yr)	6.44	11.92
Overall change	−16%	28%

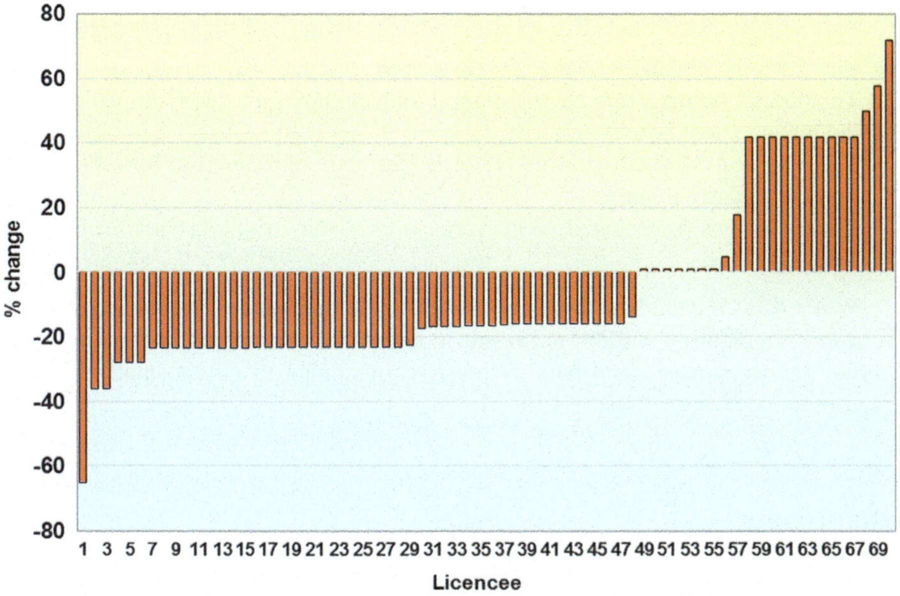

Fig. 19.4 Changes in allocation volumes in theTintinara MA

was revised to 26,000 ML/year. The reductions were to be achieved over a period of 5 years to allow irrigators to adjust their operations and increase their efficiency.

For lucerne irrigation, the following table lists the original allocations defined in the 2003 WAP (comprising the base allocation + delivery component), the new delivery components and new allocations, and the additional reductions to be achieved over time to reach the target (Table 19.1).

Figure 19.4 shows the difference between the original allocation and new allocation for individual irrigators. Those with the positive change are almost all flood irrigators, while those who experience reductions are those using centre pivot irrigation.

It should be noted that for 75% of the irrigators, these changes in allocation resulted in no change in usage.

19.7 Revised Water Allocation Plan

After community consultation, the allocation reductions and other management issues raised during the review, were incorporated in a revised WAP which was adopted by the Minister for Environment in 2012 (SENRMB, 2012). This process was considered successful because when the public consultation had concluded, there was not one submission from an irrigator against the Plan. There are several factors that contributed to this successful outcome.

- The metered extraction data clearly showed evidence of over-allocation, even during the very dry years. This was made possible only because of the area limiting condition that was imposed.
- The management agency hydrogeologist had built up a good relationship and trust with the irrigators over a 10 year period
- This led to the irrigators having a good understanding of the hydrogeology and the management issues
- The irrigators worked with the NRM Board and contributed to how the reductions were carried out
- The reductions were staged over several years so that irrigators had time to adjust their operations.

While this case study refers to allocation reductions that were not triggered by resource degradation and did not result in significant reductions in extraction, the above factors should contribute to positive outcomes in circumstances that do require reduced extractions.

References

Allen, R. G., Pereira, L. S., Raes, D., & Smith, M. (1998). *Crop evapotranspiration; guidelines for computing crop water requirements* (FAO irrigation and drainage paper, 56). Rome: Food and Agriculture Organization of the United Nations.

Barnett, S. R. (2002). *Water resource assessment of the Tintinara – Coonalpyn prescribed wells area* (Report, DWLBC 2002/20). Adelaide, Australia: Department of Water, Land and Biodiversity Conservation.

Barnett, S. R. (2008). *Hydrogeological review of the Tintinara–Coonalpyn PWA water allocation plan* (DWLBC Report 2008/09). Adelaide, Australia: Department of Water, Land and Biodiversity Conservation.

Irrigated Crop Management Service. (2002). *Working document on irrigation requirements for the Tintinara Coonalpyn prescribed wells area.* Rural Solutions, Government of South Australia, through Department of Primary Industries.

South East Catchment Water Management Board. (2003). *Tintinara – Coonalpyn prescribed wells area water allocation plan.*

South East Natural Resources Management Board. (2009). *Review of irrigation water use and delivery components in the Tintinara Coonalpyn prescribed wells area.*

South East Natural Resources Management Board. (2012). *Tintinara – Coonalpyn prescribed wells area water allocation plan.*

Steve Barnett is a Principal Hydrogeologist at the Water Science and Monitoring Branch of the Department for Environment and Water in South Australia. He has been involved in the investigation, monitoring and management of groundwater resources in SA for over 40 years and has contributed technical and policy input into ten groundwater management plans, which incorporate a variety of different aquifers and management issues. He is a past president of the Australian Chapter of the International Association of Hydrogeologists.

David Williamson is Team Leader for Water Licencing in the South East Region for the Department for Environment and Water. He has over 10 years experience working in water policy development and water licencing and has led development of five Water Allocation Plans in the region, some of which have included a reduction of water entitlements for groundwater users. One of these Plans won the United Nations Association of Australia 2016 Environmental Day Award for Excellence in Water Management.

Chapter 20
Reducing Groundwater Entitlements in the Lower Murrumbidgee Groundwater Management Area

Ken Schuster, Amanda Kennedy, and Cameron Holley

Abstract This chapter explores the case study of the Lower Murrumbidgee Groundwater Management Area in New South Wales, Australia. In particular, it illustrates the contours of two policy approaches for water entitlement reduction: one was a failure (unilateral reductions imposed uniformly on all water users); and one was a success (financial compensation for cutbacks in entitlements, negotiated in the shadow of court action). The long-standing problem of over-allocation in the Lower Murrumbidgee was addressed initially through a process of entitlement reduction, driven by the government and involving a heated and contested policy approach. The primary method of reduction was an approximate 50% cut to all entitlements (regardless of capital commitments). This was challenged by a group of groundwater irrigators in the Land and Environment Court, who preferred to regulate pumping by managing the water level within a sustainable bandwidth. Although the case was unsuccessful, the judge raised concerns about the fairness of the new arrangements, and the irrigators planned an appeal. The litigation and threat of an appeal proved a catalyst for cooperation amongst groundwater users across the state, producing a policy shift that saw the government pursue a program known as Achieving Sustainable Groundwater Entitlements. This program recognised historical extraction in calculating entitlement reduction, and provided financial assistance to licence holders. Overall, this case study illustrates important lessons for policy approaches for reducing entitlements, not least the need to account for local knowledge and concerns, as well as providing adjustment mechanisms (e.g.

K. Schuster (✉)
Farmer, Founding Member of the Murrumbidgee Groundwater Preservation Association, Darlington Point, NSW, Australia

A. Kennedy (✉)
Faculty of Law, Queensland University of Technology, Brisbane, QLD, Australia
e-mail: amanda.kennedy@qut.edu.au

C. Holley
Faculty of Law, University of New South Wales Sydney, Sydney, NSW, Australia
e-mail: c.holley@unsw.edu.au

J.-D. Rinaudo et al. (eds.), *Sustainable Groundwater Management*,
Global Issues in Water Policy 24, https://doi.org/10.1007/978-3-030-32766-8_20

economic compensation) to ensure the long-term sustainable management of groundwater.

Keywords Water planning · Entitlement reduction · Court challenge · Consultation · Adjustment package

20.1 Introduction

This chapter presents a case study of the Lower Murrumbidgee Groundwater Management Area in New South Wales, Australia. Specifically, it illustrates the contours of two policy approaches for groundwater entitlement reduction to reduce over-allocation and their success in obtaining buy-in from affected water users. Good policy development typically requires consideration of a policy's impacts, the engagement of affected parties, and efforts to minimise negative effects (Holley & Sinclair, 2016; Syme & Nancarrow, 2008). In the context of reducing entitlements to restore the groundwater balance in the Lower Murrumbidgee, one reduction policy was a failure (involving unilateral reductions imposed uniformly on all water users), and one was a success (involving financial compensation for cutbacks in entitlements, negotiated in the shadow of court action).

The Lower Murrumbidgee management area is 3.3 million hectares and contains a shallow, middle and deep level aquifer consisting of a sequence of semi-consolidated and unconsolidated alluvial fan deposits (Hope & Wright, 2003; Kumar, 2010). Estimated storage is over two hundred million mega-litres, with an estimated one hundred mega-litres of water with a salt content of less than 500mg/L. During the late 1800s to the early 1900s, groundwater use was for stock and domestic supply needs, and measurement of the aquifer and its sustainable yield was a fairly inexact science. Demand was modest, and groundwater was used sparingly. However, over the next 60 years, purpose built dams on the upper river catchment allowed the Lower Murrumbidgee area to become increasingly irrigated by surface water. Bores were put in place to monitor accessions to groundwater as a result of the surface irrigation, which revealed that the aquifer was absorbing considerable amounts of river water.

Concerned about the impact of rising water tables, government authorities actively encouraged groundwater users to pump more water, declaring that current licence holders could pump 150% of their allocation, while simultaneously issuing new licenses. As pumping increased, bore irrigators began to lower the water table until pumps also had to be lowered. New, deeper bores were drilled at considerable cost, increasing pressure on the aquifer and raising concerns that the groundwater demand was unsustainable. By the end of the twentieth century, temporary moratoriums and an embargo on licences were declared.

Confronting this over-allocation problem, the ensuing process of entitlement reduction involved a heated and contested policy approach. The primary method of reduction was an approximate 50% cut to all entitlements (regardless of the capital commitments) via a statutory water sharing plan (see Chap. 17 in this book). This was challenged by a group of groundwater irrigators – the Murrumbidgee Groundwater Preservation Association (MGPA) – in the Land and Environment Court of New South Wales. Although the MGPA's case was unsuccessful, the judge raised concerns about the fairness of the new arrangements, and the MGPA planned an appeal. This case, and the threat of an appeal, proved a catalyst for cooperation amongst groundwater users across the state. In the shadow of this case (and other threats of litigation), 'behind the scenes' negotiations with Ministers at state and national levels took place. This helped to catalyse a shift in policy that saw the government pursue a program known as 'Achieving Sustainable Groundwater Entitlements'. This program recognised historical extraction in calculating entitlement reduction, and provided financial assistance to licence holders. Under this program, and as a result of more localised negotiations between the MGPA and state ministers, a new plan was put forward that recognised historical rates of extraction, with most irrigators relatively satisfied with the new plan despite having their entitlements reduced (cf *Harvey v Minister Administering the Water Management Act 2000* (2008) 160 LGERA 50).

Drawing on the lived history and experience of Ken Schuster, who was involved with the Murrumbidgee Groundwater Preservation Association during the above events, this case study chapter illustrates important lessons for policy approaches that seek to reduce groundwater entitlements, not least of all the need to account for local knowledge and concerns, as well as providing adjustment mechanisms (e.g. economic compensation) to ensure the long term sustainable management of groundwater.

The chapter starts by providing a history of the Murrumbidgee Groundwater resource, before discussing one of the primary tools for reducing water use entitlements, namely water sharing plans. Although Australia's national water reforms provided the framework for driving the use of water sharing plans in New South Wales, we have not discussed these national policies in detail given they are discussed elsewhere in the book (see Chap. 7). The chapter maps out the entitlement reduction approach of the water sharing plan as initially introduced, namely proportionate reductions imposed uniformly on all water users. As will be seen, this approach (along other weaknesses in the planning process) produced disquiet and subsequent court challenges. The chapter then examines the subsequent emergence of the so called Achieving Sustainable Groundwater Entitlements (ASGE) program, which offered a more palatable and ultimately more successful policy of financial compensation for cutbacks in entitlements (for further analysis of ASGE, see Chap. 17). It concludes with a discussion of the broader implications from the case study.

20.2 A Brief History – The Murrumbidgee Groundwater Resource

The Murrumbidgee Groundwater resource is located in the Murrumbidgee Irrigation Area in the south west of New South Wales, and underlies an area that extends the full length of the Murrumbidgee River from Narrandera in the east to Balranald in the west (Fig. 20.1).

Although three separate aquifers are described for management purposes, there is movement of water vertically between the aquifer levels. The upper aquifer is known as the Shepparton, which is near the surface to a depth of 50–70 m. The Calival formation is the most productive layer and extends to a depth of 100–140 m. The thickness of this layer is generally between 50 and 70 m. The proportion of sand at this level is between 50% and 70% (Hope & Wright, 2003; Kumar, 2010). The Renmark group contains the oldest deposits, and extend to the bedrock at a maximum depth of about 280 m in the Murrumbidgee low salinity area.

It is believed that there was a continuous depositing of sands and gravels by prior streams. These prior streams have been gradually silted over and new streams have formed new channels eventually covering the area with braided deposits at different levels. The thickness of these sediments varies from 170 m at the eastern end to 400 m at the western end. The porosity of the aquifer is estimated to be 25% (Hope & Wright, 2003; Kumar, 2010). The estimated storage of this aquifer is calculated to be two hundred million mega-litres with an estimated one hundred million mega-litres of water with a salt content of less than 500mg/l.

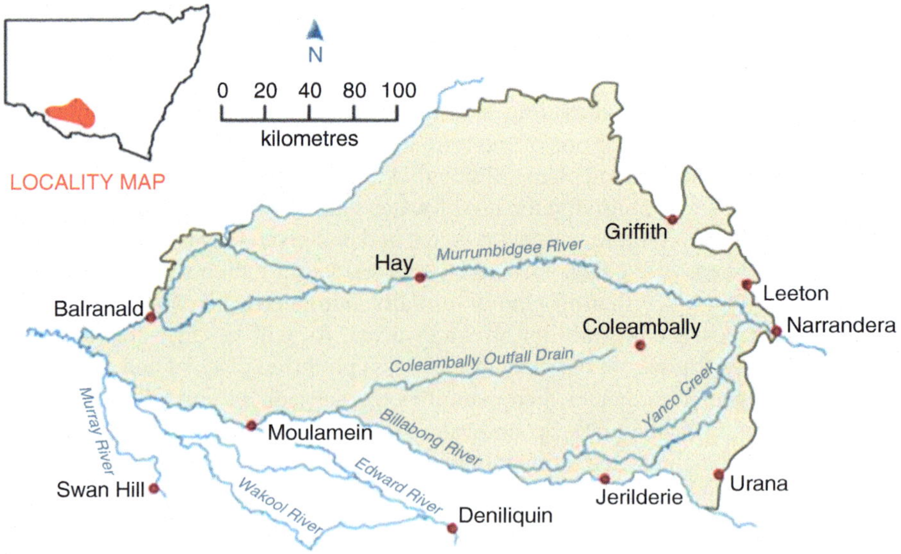

Fig. 20.1 Map of Lower Murrumbidgee Groundwater. (Source NSW Government, 2018)

The highest yielding area in the aquifer with the highest quality of water is at the eastern end, nearest the Murrumbidgee River. Proceeding west, the water quality becomes more saline. Also, further west away from the source of the feeder streams the sand deposits become much finer, restricting water flow. While this makes it more challenging to establish a high yielding bore, it is still possible to find an ancient prior stream and procure a useful yield. The assessed high risk to this aquifer is the fact that the unconsolidated medium containing the water is hydraulically suspended. Pumping to lower the water table below the Calival could accordingly result in consolidation of the medium and lead to subsidence at the surface.

As Gross (2014) details, the most recent phase of water law and policy in Australia has attempted to respond to the problems associated with overallocation and overuse that characterised earlier periods of water infrastructure development. The situation in the Murrumbidgee Basin is no exception. Historically, small-scale irrigation along the Murrumbidgee was undertaken from the mid-1800s (Barwick, 1979), then in the early 1900s the Murrumbidgee Irrigation Area Scheme was established (Lewis, 2012). The Murrumbidgee Irrigation Area (MIA) is north of the Murrumbidgee River, while the Coleambally Irrigation Area (CIA) is situated south of the Murrumbidgee River, and is considered part of the overall irrigation scheme. Upon its establishment, the purpose of the Scheme was not entirely clear – whether it was to facilitate intensive cultivation, or for drought relief – but reports from the opening ceremony nonetheless capture the aspirations held for the Murrumbidgee Irrigation Area Scheme:

> Not only was it a memorable ceremony, but it was a joyous occasion full of hope for the future, for this was a scheme for the poor man - the 'little' man - to enable thousands of new settlers to make homes for their families on the land (Lewis, 2012 citing Chessbrough, 1982).

During the late 1960s, the first irrigation bores were installed. Initially, water bores required a licence, but these were issued in perpetuity with no area or volumetric restrictions (Kumar, 2010). Despite the expansion of irrigation, groundwater extraction was still seen as a 'second-best' option, due to the availability of surface water and the high cost and unknown quantities of groundwater. However, during the mid-1970s – when a moratorium on the issuing of new Murrumbidgee River irrigation licences was instated – there was a marked increase of interest in groundwater for irrigation (Wilkinson, 1997). Licences were issued on a 5-year basis, covering an authorised area of 162 ha (400 acres) (Kumar, 2010). A period of rapid expansion followed, with large tracts of land to the north and north-west of the Coleambally being subdivided into 'bore' blocks. As Udoye has noted in the context of other Irrigation Schemes (e.g. the Gwydir), the interpretation of the 162 hectare rule and joint irrigation schemes enabled companies and family partnerships to secure multiple licences on a speculative basis (Udoye, 1984). As a point of reference, at June 1983 the total entitlement issued in the Murrumbidgee Basin was 127,000 ML, and extraction for the season 1982–1983 was 40,000 ML.

Under the *Water (Amendment) Act 1980*, volumetric allocations were subsequently introduced throughout NSW. In 1984, the then Water Resources Commission implemented a volumetric system of supply in the Murrumbidgee Basin (Lewis, 2012). Various formulae were trialled on a ML per hectare basis, until a formula with a sliding scale depending on farm size, and a maximum allocation of 4000ML, was adopted. This formula was also linked to a "One Resource Policy" whereby pumpers who had access to surface water were required to subtract that allocation from their groundwater allocation. In a typical case, the net result of these policy changes was a reduction of allocation from 4000 ML to 2056 ML. This was from an original allocation of around of 7000 ML prior to the introduction of the 4000 ML maximum clause. The introduction of these measures effectively halted the development of bores, and also, in conjunction with the *Water (Amendment) Act 1986* (which introduced the ability to transfer water allocations), provided the basis for trading water entitlements (Wilkinson, 1997).

It was around this time that bore hydrographs (originally furnished by the planners of the Murrumbidgee Irrigation Scheme to monitor groundwater levels) indicated rising water tables, mainly in the Shepparton Aquifer. During the late 1980s, it was discovered that pumping from the aquifer was beneficial in controlling the rise of groundwater tables within the Coleambally Irrigation Area (CIA), and consequently addressing salinity. Through a process of 'controlled groundwater depletion', farmers were encouraged to remove water from deep groundwater sources, which would in turn lower the water table and reduce the salt content of the water at or near the surface. The then Department of Land and Water Conservation (DLWC) decided to increase allocations to 150% in order to explore the limits of the aquifer and to observe the effects of this on the Shepparton water table. This was to be reviewed on an annual basis. Licences were issued by the state government department responsible for water (the Department) upon application for a fee of $150.

However, an impediment to the uptake of this opportunity was the capital cost of developing an irrigation project. Irrigators intending to invest in a bore irrigation project needed to conduct due diligence before proceeding. This was especially the case as it was hard to convince bank managers that such projects would be secure and sustainable. A typical investment to establish a viable working scheme (in the mid 1990's) would cost in the order of $750,000, comprised of the bore (approximately $250,000), pump and installation (approximately $250,000), and land forming and channels including provision for recycling of run off (approximately $250,000). There would also be the cost of energy for pumping and fairly high maintenance costs to factor into a business plan.

With development at a standstill and very few irrigators accessing the 150% allocation, the DLWC called a meeting of groundwater pumpers to ascertain what would be required to increase the amount of water pumped, such was the concern over the rising water table. Those present at the meeting determined that a profitable crop, as well as certainty about the long-term security of allocations, would be required to prompt greater water use. At that time, the most profitable crop to grow was rice. However, in order to grow rice, a licence was needed from the Rice Board. Licences were strictly controlled, and not issued to bore irrigators. The only other viable option was corn (maize), which required investment in specialised machinery, and carried a higher marketing risk being a competitive unregulated market.

It was around this time that the Murrumbidgee Bore Pumpers Association Inc (later renamed as the Murrumbidgee Groundwater Preservation Association (MGPA)) was formed, a group of pioneer investors in bore irrigation in the area. They took a very keen interest in the Departmental reviews of groundwater use, and the effect of pumping on groundwater levels, holding regular meetings and concerning themselves with the status of groundwater levels. Shortly thereafter, the Government announced the deregulation of the rice industry, paving the way for rice to be grown under bore irrigation. Partly in light of these factors, a further rapid expansion took place with a surge in applications for groundwater licences throughout the 1990s. Owing to buoyant returns within the rice industry, licences were sought in areas that were outside the previously accepted locations of high yielding aquifers, into areas of unknown recharge. During the mid-1990s, the DLWC monitoring bores were showing a significant rise in the Shepparton aquifer. This was largely due to accessions to the water table from the surface irrigation, sourced from the dams on the Murrumbidgee, i.e. the Burrenjuck and Blowering dams (Kingsford, 2003) As a result of continuing concern with regard to the rising water table in the Shepparton aquifer, the DLWC continued to vigorously promote the greater use of groundwater, with little regard for the impacts of extraction.

Water use increased modestly in 1994/95 by about 50,000 mega litres. However, during the same period the DLWC issued approximately 100,000 mega litres of new licences. The Bore Pumpers Association members realised that there was now the potential to more than double the levels of usage, and over time it became clear that water levels were declining. The Bore Pumpers Association later resolved to urge the Department to place a moratorium on issuing new groundwater licences until usage caught up with the licences that had already been issued. The moratorium was declared in 1997, 3 months after the written request was sent from the Bore Pumpers Association. The moratorium also coincided with 1997 *NSW State Groundwater Policy Framework Document* (DLWC 1997), which advocated groundwater management plans be developed across the state (and was followed by the prioritisation of aquifers most at risk and the establishment of Groundwater Management Committees, see DLWC 1998; *Arnold v Minister Administering the Water Management Act 2000* (No 6) [2013] NSWLEC 73, Biscoe J, [20] – [24]).

Notwithstanding the moratorium, a surge of licence applications had already been received by the Department, prompted by the threat of licence restriction as well as drought conditions. Water use had reached a peak in the 1997-8 season, an increase of some 80,000 mega litres, when irrigators experienced a sudden drop in drawdown levels mid-season. By the end of 1998, it was clear that the Lower Murrumbidgee Groundwater Source was facing serious trouble, as a result of over-allocation. At a meeting with the DLWC at the end of the 1997/8 season, it was decided that the 150% allocation would be terminated. An embargo was finally declared in 1998/9; however, the DLWC continued to process applications that were in hand until 2000/01, by which time the issued licences had been finalised at 520,000 mega litres – more than double the usage at that time, which had plateaued at 220,000 mega litres (see Fig. 20.2).

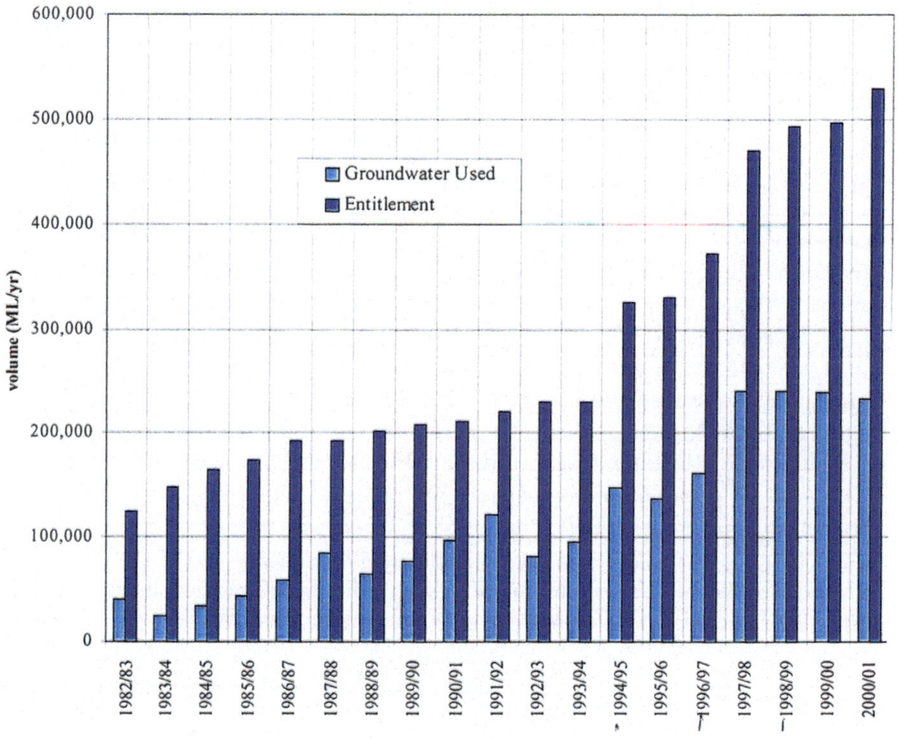

Fig. 20.2 Groundwater entitlements and use. (Kumar, 2002)

As hydro graph levels in high use areas were indicating serious depletion (see below), many irrigators were forced to lower their pumps in order to maintain their cropping programs. The DLWC then announced in 2002 that the recharge to the aquifer was estimated to be 335,370 mega litres, with an estimated sustainable yield of 270,000 mega litres. This prompted the government to announce that there would be a cut to all entitlements in the Murrumbidgee Groundwater Area of 52%. This decision also applied to the five other Catchment Management Authorities in NSW - Border Rivers/Gwydir, Namoi, Central West, Lachlan and Murray. The percentage of the cuts varied in each case, according to local conditions (Fig. 20.3).

20.3 The Lower Murrumbidgee Water Sharing Plan

At the turn of the 21st Century, the Lower Murrumbidgee groundwater resource was identified as a high-risk system as a result of over-allocation, salinity and water quality (Bowmer, 2003; DLWC, 1998). On 1 January 2001, the *Water Management Act 2000* came into force, which established rules and procedures for water sharing. This included the development of water management plans, which could be made either by collaborative committees (*Water Management Act 2000*

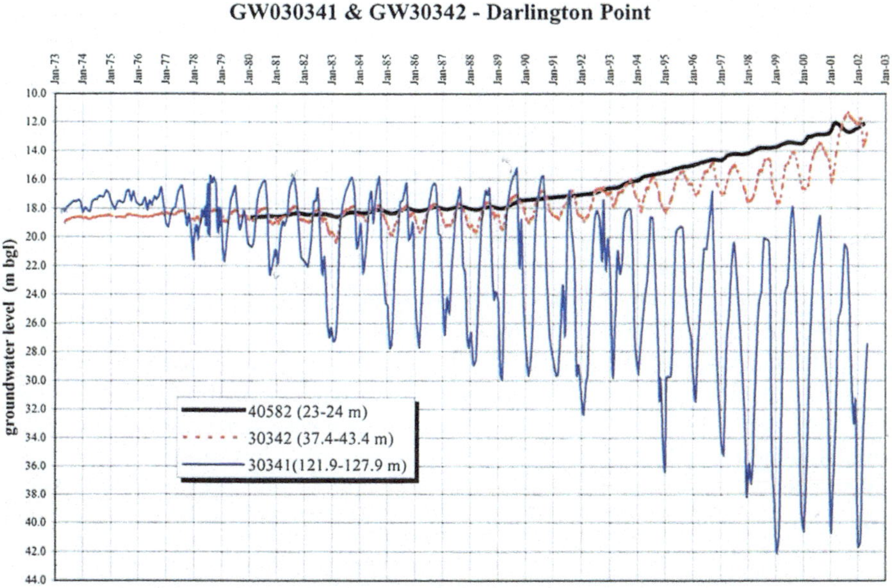

Fig. 20.3 Example hydrograph from groundwater monitoring sites. (Kumar, 2002)

(NSW), ss12, 13, 14(a), 15) or by Ministerial decree (*Water Management Act 2000* (NSW), s50). In most cases, water sharing plans were made as "Minister's Plans" rather than under a collaborative committee (Gardner & Bowmer, 2007; Holley & Sinclair, 2013). While advisory committees were typically consulted by the Minister in preparation of these plans, they were not seen as equivalent to the more formal collaborative committee structure envisaged under the *Water Management Act* (Gardner & Bowmer, 2007; Holley & Sinclair, 2013). In the present case, the Minister established the Murrumbidgee Groundwater Management Committee in 1998 to provide a means of community consultation (Tan, 2008). However, as discussed below, the draft Water Sharing Plan for the Lower Murrumbidgee Groundwater Sources was ultimately made as a Minister's Plan pursuant to s50 of the *Water Management Act*, which ostensibly had the same requirements as plans made by collaborative committee (e.g. requirements to develop a management plan with a vision, objectives, strategies and performance indicators; and rules for sharing water between environmental and other consumptive uses (*Water Management Act 2000* (NSW), ss20, 35)).[1] However, later amendments to the Act saw this requirement amended such that the Minister only needed to deal with such matters "in general terms" (Millar, 2005).[2] The lack of clarity as to the nature of the roles of the Minister and committees – not only in the case of the Lower Murrumbidgee,

[1] For more discussion of this point see Millar, I. (2005). Testing the waters: Legal challenges to Water Sharing Plans in NSW. Presented at the *Water Law in Western Australia Conference, July*.

[2] As Millar notes, later amendments again in 2004 went so far as to exempt Minister's plans from certain public consultation requirements altogether.

but for several other Water Sharing Plans throughout the state – was to become a particular source of criticism.

Amidst this confusion, the *Water Sharing Plan for the Lower Murrumbidgee Groundwater Sources* was finalised and gazetted in 2002, to become effective from July 2003. Consistent with a broader state-wide policy that proposed to reduce groundwater licences in a proportional manner, the primary method of groundwater reduction preferred by the Minister was an approximate 50% cut to all theoretical entitlements, regardless of capital commitments or whether the licence holder had previously used the water, along with the creation of a market for access licences (Tan, 2008). For the Murrumbidgee Bore Pumpers Association, a long-term sense of frustration gave way to outrage. Members believed that it was possible to work collaboratively with the government through its Departmental officers to establish a plan and rules for the sustainable extraction of groundwater, and felt a sense of betrayal. This feeling of betrayal was shared by many similarly placed committees across the state, who also felt misled as to the nature of their role in the decision-making process (Millar, 2005; Tan, 2008; *Murrumbidgee Groundwater Preservation Association v Minister for Natural Resources* [2004] NSWLEC 122).

These sentiments led to a well-attended meeting of more than 95% of licence holders to decide almost unanimously to challenge the validity of the Water Sharing Plan in the NSW Land and Environment Court. In order to test the resolve of the membership, the then-President announced that a pad would be circulated through the hall and that members should sign up at a suggested rate of at least 50 cents per mega litre of entitlement in order to finance the court challenge. The President also said that unless a minimum of $250,000 was pledged, the challenge could not proceed. It was over-subscribed! At the same time, the Association changed its name to the Murrumbidgee Groundwater Preservation Association (Incorporated) (MGPA). In order to have a sense of unity of purpose, it was deemed necessary to determine a policy position that would appeal to all licence holders, regardless of whether they had fully developed irrigation farms or simply held a paper license with no development. The policy preferred by the MGPA, with unanimous approval, was that there should be no cuts to entitlements, but that sustainable use of the aquifer be achieved by regulating use within a 'bandwidth', so that the annual extraction limit would be declared according to the performance of the aquifer.

The MGPA proceeded to engage a lawyer and construct a case to present to the Land and Environment Court. The lawyer's advice was that the MGPA could proceed by arguing that the plan was invalid and should be set aside. The arguments presented on behalf of the MGPA canvassed issues regarding fairness and irrationality, and that the action of cutting entitlements by such a significant degree could not be justified by the evidence of aquifer performance.

20.3.1 *Murrumbidgee Groundwater Preservation Association v Minister for Natural Resources [2004] NSWLEC 122*

The MGPA was not the only group seeking to challenge a Water Sharing Plan made pursuant to s50 of the *Water Management Act*, and due to the large number of cases on similar issues the MGPA case was put forward as a 'test case' on the matters arising (Holley & Sinclair, 2013; Millar, 2005). The MGPA argued that the Lower Murrumbidgee Water Sharing Plan significantly reduced both theoretical entitlements as well as the actual amount of water available to users, which, given the significant capital expenditure in anticipation of receiving water entitlements, would produce extreme inequities. It was also put forward that the uniform regulation and control of the entire area by an across-the-board reduction was inappropriate, given limited interconnectivity within the aquifers, the history of extraction, and differing site specific hydrogeological impacts. The MGPA also took issue with the exercise of Ministerial power under s50 to make the Water Sharing Plan, arguing that a failure to make the plan according to Part 3 of the Act (which sets out procedures for plans made by management committees) was in breach of statutory requirements for procedural fairness and thus an abuse of Ministerial power. Accordingly, the MGPA submitted that the plan should be deemed invalid.

On the facts presented, McClellan CJ was not persuaded that the MGPA had met the threshold for judicial review of administrative proceedings, and set aside the application. It was noted that it was "for the Minister, and not the Court to balance the desired environmental outcome, and the chosen method of achieving it, with the beneficial and adverse social and economic consequences" (*Murrumbidgee Groundwater Preservation Association Inc. v Minister for Natural Resources* [2004] NSWLEC 122 at 184 per McLellan CJ.), and that a court must be cautious when reviewing the validity of decisions not to stray towards evaluating the merits of the case.

20.3.2 *Murrumbidgee Groundwater Preservation Association Inc v Minister for Natural Resources (2005) 138 LGERA 11*

The MGPA appealed the decision of the Land and Environment Court to the NSW Court of Appeal. It was again argued, *inter alia*, that the Minister's power to make the plan had been exercised for an extraneous purpose – that is, to avoid the statutory consultation procedures that would have taken place for a plan developed by a formal management committee. Further, it was once more submitted that the limited interconnectivity in the aquifers rendered an across-the-board pro-rata reduction to pre-existing entitlements perverse, and not attuned to sustainable usage limiting extraction. It was argued that this caused the plan to be 'illogical to the point of irrationality', and involved a substantial degree of unfairness.

Once more, the decision went against the MGPA, and the appeal was dismissed with costs. While the *Water Management Act* provided for a management committee to develop a draft plan, it was found that such bodies could also be consulted in an advisory capacity, and the failure to hand over decision making power to a committee did not constitute a breach of the statutory procedures. It was noted by the Court of Appeal that the Minister's power to establish a management committee was discretionary, and that public consultation could be gleaned through a variety of mechanisms. Moreover, it was determined that there was nothing in the *Water Management Act* to suggest that a Ministerial plan was a secondary or subordinate method of making a plan. Ultimately, while there was clear evidence that the Water Sharing Plan would result in differential and unfair impacts for particular members of the irrigation community, it was held by Spigelman CJ (with whom Beazley JA and Tobias JA agreed) that:

> "Inevitably, when significant changes are made to an established regulatory regimes, there will be winners and losers. Considerations of equity are quintessentially matters for political decision-making. I am not satisfied that anything in the nature, scope and purpose of the Act prevents the Minister from implementing a scheme which operates to the detriment of some persons and to the advantage of others, in a manner not determined by availability of water but by broader considerations of what the Minister regards as equitable" (*Murrumbidgee Groundwater Preservation Association Inc. v Minister for Natural Resources* (2005) 138 LGERA 11 at 144 per Spigelman CJ).

The MGPA members who attended the hearing believed that there was a good chance that they would be successful in having the plan set aside, enabling a new plan to be formulated. There was great disappointment when the appeal was dismissed. Meanwhile, discontent over Water Sharing Plans continued to play out in the judicial system, with eleven other cases from other groundwater areas also before the courts at the same time.[3] Matters of procedural fairness remained largely unresolved, as the courts focused more on procedural issues; this did little to quell the widely-held belief that participatory processes had been effectively trumped by a strategic use of Ministerial Plans under the Act (Gardner, Bartlett, & Gray, 2009; Holley & Sinclair, 2013).

While the MGPA's case was unsuccessful, both the initial decision and the appeal proved a catalyst for cooperation amongst groundwater users across the state. Their discontent was escalated to meetings of the collective interests of the groundwater users' forum, the NSW Irrigators Council, which Ministers sometimes attended. In addition, there was dialogue between the Upper Namoi Water Users Association and the MGPA. The Upper Namoi Group had the advantage of having the then-

[3] See, for example, *Upper Namoi Water Users Association Inc. v Minister for Natural Resources* [2003] NSWLEC 175; and *Nature Conservation Council of NSW v Minister administering the Water Management Act 2000* [2005] NSWCA 9.

Deputy Prime Minister as their local Federal member in the House of Representatives. This direct contact was to be a significant factor in gaining Federal government finance to facilitate structural adjustment. The NSW State government at this time was going through a somewhat unstable period, with not only leadership changes, but also a change of Minister in the Department's portfolio. During this period, the Department was served by five different Ministers.

One of these changes, to a Minister with a property valuation background, proved to be an important development toward a solution in the Murrumbidgee. Among other things, the Minister was shown a fully developed bore irrigation property contrasted with an undeveloped bore covered with an empty two-hundred-litre drum, and he immediately recognised the significance of developmental costs and the history of use/history of extraction (HOE). The inevitable result of 'across the board' cuts to entitlements would leave stranded assets and unviable businesses. HOE utilises a weighted reduction process which takes into account past usage, enabling economic activity to be maintained and limiting the impact of stranded assets (Kuehne & Bjolmund, 2006). Despite the fact that the MGPA was not successful initially, the new Minister made the decision to put the Water Sharing Plan on hold while a review of the draft plan was undertaken and the 2004 Groundwater Adjustment Advisory Committee (including representatives of the NSW Irrigators' Council, Catchment Management Authorities and the NSW and Australian governments) convened to consider revised entitlement methodology and assistance (NSW Government, n.d.).

20.4 Achieving Sustainable Groundwater Entitlements (ASGE)

The MGPA was invited to participate in the formation of the revised Water Sharing Plan, and became involved in consultations with the Murrumbidgee Catchment Management Authority (CMA), with whom the government consulted in developing the new plan. The two court cases had exhaustively examined every aspect of the reform needed, and the new plan would take account of the issues involved. Around this time, dialogue between the State and Federal governments resulted in a significant amount of money to be provided to facilitate implementation of proposed entitlement reductions under Water Sharing Plans. This program, known as 'Achieving Sustainable Groundwater Entitlements', opened up a new perspective: namely, that there was a possibility of some form of 'buy back' of unused entitlement, or compensation for loss of entitlement. The program had four main components – reducing entitlements based on HOE (History of Extraction) a $125 million financial assistance (for the whole state) package for licence holders, a $9 million Community Development Fund, and up to $1 million for implementation of the program (NSW Government, n.d.; Parliament of New South Wales. (2005)). Of particular importance was its recognition of historical extraction in calculating entitlement reductions, and financial assistance for licence holders to transition to the new arrangements. Under the auspices of this program, and as a result of ongoing discussions between the MGPA, the CMA and State Ministers, a revised Water Sharing Plan was put forward.

The ASGE had addressed all the salient issues within this difficult policy space; most notably, it took into account the HOE, it acknowledged that undeveloped licences had a value, and that groundwater entitlements should have a mortgageable legal title. This was all predicated on the declaration of a sustainable yield of 270,000 mega litres. The entitlement reductions, previously in the vicinity of 50% across-the-board, would still be achieved by reducing entitlements according to HOE by between 15% and 85%. An *ex gratia* payment would be made for the loss of 'active' water, i.e. water that had been used but would be lost. The total amount of money available for structural adjustment in the Murrumbidgee was $5.313 million. The loss of 'active' water would be reduced by means of a supplementary licence reducing by 10% each year. This was the 'in principle' proposal of the Government. In the end, modifications were made to enable the funding available to cover the plan. In the final settlement of the ASGE, the ratio of entitlement reduction was determined at 82/18. That is, high HOE licence holders would lose 18% of their licence, and licence holders with no history of extraction would lose 82% of their licence.

Figure 20.4 is an extract from an actual settlement statement for licence of a 100% user. The licence was reduced by 229 ML and there was an *ex gratia* payment of $34,034. The reduction of the 229ML was scheduled to reduce by 10% per annum. There would have been a reasonable prospect that the licensee could be able to achieve layout efficiencies to compensate for the loss of the water. Alternatively, it may have been possible to purchase the water back (subject to zone restrictions) at a cost of approximately $275,000. The financial consequences are quite significant in either case. Figure 20.5 below shows the outcome for a 1000 ML licence with 0% history of extraction:

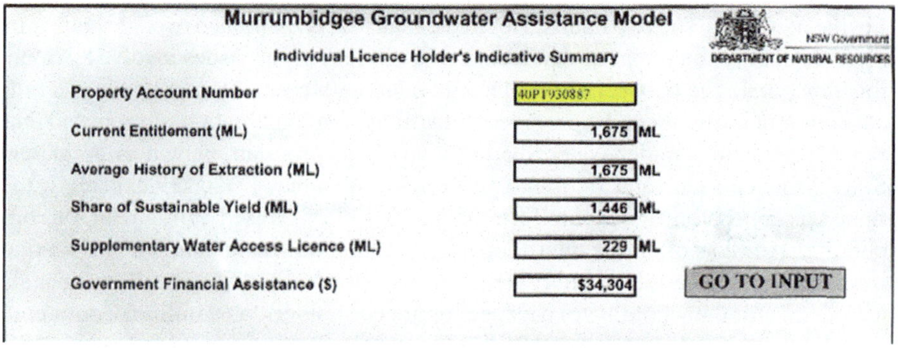
Murrumbidgee Groundwater Assistance Model

Individual Licence Holder's Indicativo Summary

NSW Government
DEPARTMENT OF NATURAL RESOURCES

Property Account Number	40PT930887
Current Entitlement (ML)	1,675 ML
Average History of Extraction (ML)	1,675 ML
Share of Sustainable Yield (ML)	1,446 ML
Supplementary Water Access Licence (ML)	229 ML
Government Financial Assistance ($)	$34,304

GO TO INPUT

Fig. 20.4 Example of NSW Department of Natural Resources Murrumbidgee Groundwater Assistance Model Licence Holder's Indicative Summary ASGE statement provided by Ken Schuster

Fig. 20.5 Example of NSW Department of Natural Resources Murrumbidgee Groundwater Assistance Model Licence Holder's Indicative Summary ASGE statement provided by Ken Schuster

	HOE Default Model
New entitlement:	**FNE: 186 ML** **Supp water: 0 ML**
Determining financial assistance:	*Asset component:* Old ent. 1000ML*$65/ML = $65,000 New ent. 186ML*$350/ML = $65,100 *Lost active water:* 0ML
Financial assistance:	**$0**

It can be seen from this chart that there was no theoretical loss of value to the licence holder. Subsequent to the final settlement of the new plan, the value of groundwater increased quite dramatically, exceeding $1200 per ML. In the case of the theoretical licence above, the value of the licence would amount to $223,000. Given the fee for the issue of the licence was initially $150, it could be seen ostensibly as a windfall. However, it was not often the case that such a windfall was a reality. If a property had devised a business model that included the exploitation of the 1000ML to underpin the viability of the enterprise, then the remaining 186ML would be inadequate to create a viable irrigation scheme. Alternatively, to try to purchase the 814 ML lost entitlement would cost in the order of $976,000. Clearly, the business model would be in great jeopardy, and would need major revision. The issue of equity against financial viability would be very challenging indeed.

The $5.3million fund allocated to the Murrumbidgee reform has, in hindsight, been seen as inadequate to fully compensate licence holders. However the figure was arrived at, and the adjustment scheme was manipulated to fit within, the money available and was not designed to fully compensate every licence holder. Hence the payments were termed "Ex Gratia", as an alternative to "Structural Adjustment", which would have implied that each individual business was assessed on its merits. Although the Government intended the ASGE scheme to be a fair settlement, in reality all licence holders experienced a loss to a greater or lesser extent according to their particular circumstances.

There were no winners in this exercise, or were there? The facts are that after the implementation of the new plan, the adjusted licences were secured by a bankable title and a conservatively estimated sustainable yield. So, in reality, the sustainability and security of the resource was assured. The final plan also declared three management areas (see Fig. 20.6) with restrictions between these zones on transfers, based on the principle of managing water level within a 'bandwidth', which reflected the MGPA's desires for the plan.

From the point of view of highly developed licence holders, the plan was tolerable. However, there were many licence holders with low to zero HOE who were

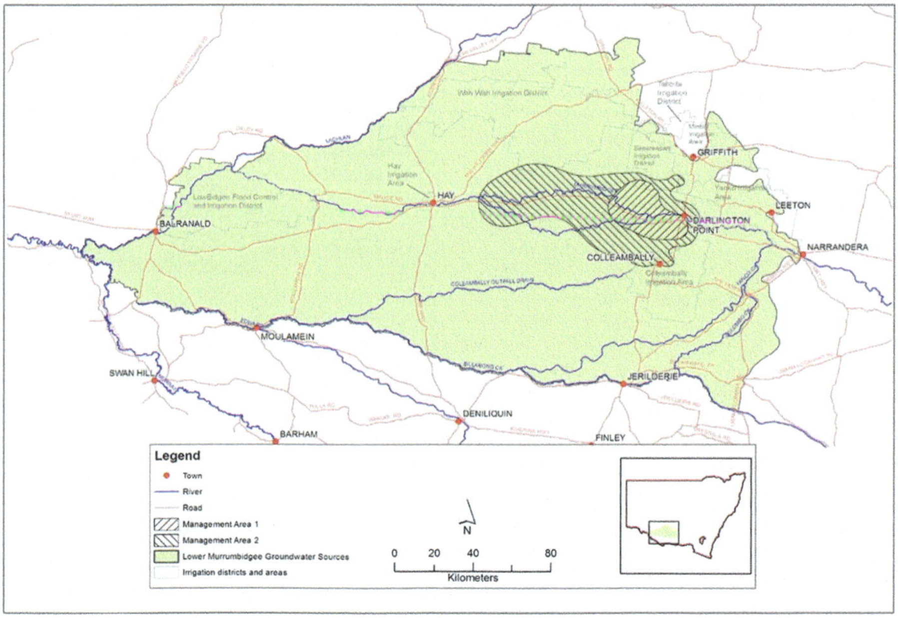

Fig. 20.6 Local impact management areas (Kumar 2002, 2010)

aggrieved; in fact, at least half of the licence holders threatened to mount a further court case. In the end, too many licences had been issued without a clear idea of the capacity of the resource to sustain the volume. This was arguably misleading to those who applied for the licence in good faith, only to be disappointed (see also *Harvey v Minister Administering the Water Management Act 2000* [2008] NSWLEC 165).

Putting to one side the water resource planning developments spurred by the Basin Plan 2012 (Cth) (see Chap. 8), more than 10 years on from the implementation of the AGSE, the aquifer is now more sustainably managed. The security of tenure and sustainability of the aquifer has created investor confidence in an increasingly scarce resource. There has been a surge in development in the cotton industry, with four new cotton gins being established in the Murrumbidgee valley (APG Workforce, 2017). Quite recently, significant property and water licence transactions have been undertaken by large corporate investors to increase permanent plantings of nut crops.

20.5 Discussion and Recommendations

As with other analyses of the New South Wales experience with Water Sharing Plans (Holley & Sinclair, 2013; Kuehne & Bjolmund, 2006), this chapter provides further insight into the ideal conditions for reducing entitlements and achieving more sustainable groundwater management, particularly in an environment of over-extraction and other concerns. In particular, at a time where there was a growing awareness of the benefits of community engagement in water law and governance, the case study illustrates two stark approaches to reducing entitlements in difficult and contested policy spaces. While the *Water Management Act* had aspirations of collaborative decision-making for groundwater governance, initially the reality of the development of the Lower Murrumbidgee Water Sharing Plan was far removed, producing uncertainty, perceptions of unfairness, and anger (Holley & Sinclair, 2013; Kuehne & Bjolmund, 2006).

First and foremost, this case study has illustrated that universal policy approaches that seek to unilaterally reduce entitlements to water or other scarce natural resources will rarely garner support from those who are to be governed. While this is somewhat axiomatic, time and again governments attempt such policy approaches and then remain perplexed when widespread controversy and disputes ensue. Failure to appreciate perceptions of fairness and justice in the allocation of resources will invariably court conflict (Gross, 2014; Kennedy, 2017). In this case, it was not until the HOE was recognised, and attempts were made to reasonably account for it in calculating and compensating for reductions, that atleast a substantial proportion of the discontent eased.

Beyond recognising historical extraction, and adjustment mechanisms for previously held entitlements, this case study also demonstrates the influence of authentic as opposed to more 'tokenistic' forms of engagement on situations of conflict. The mere existence of participatory mechanisms will often not guarantee effective, multi-directional and multi-phase consultation (Gross, 2014; Kennedy, 2017). And the courts do not seem willing to find a duty to provide procedural fairness either (eg. *Harvey v Minister Administering the Water Management Act 2000* [2008] NSWLEC 165). As others have explained, asymmetry in participatory capacity – whether technical knowledge, financial or otherwise – can also create further burdens to effective engagement (Gross, 2014; Kennedy, 2017).

The course of this conflict shifted path with the threat of litigation. A growing awareness of the need for more effective and meaningful methods of engagement (Holley, 2010; Syme and Nancarrow, 2008; Tan, 2008), coupled with a serendipitous change of Minister to an individual more attuned to the issues at hand ultimately proved pivotal to resolving a heavily contested policy issue. It is, of course, easy to imagine a very different outcome had those factors not aligned.

20.6 Conclusion

Australia is arguably the driest continent on earth (Australian Government, n.d.). While its water resources are limited, careful development of semi-desert inland areas can create highly productive agricultural enterprises of huge value to the nation and an increasingly hungry world. The sustainability of the nation's water resources is thus imperative. This chapter has been a story of what it takes to achieve a sustainable policy that seeks to reduce groundwater entitlements. Without doubt, the policy will still need constant monitoring and fine tuning over time, not least because of the ongoing implementation of the Basin Plan, as well as the growing demand for energy sources (e.g. coal seam gas) that are likely to bring new pressures to bear on groundwater (Hayter, 2014; NSW Government, 2014).

References

APG Workforce. (2017). *Local round up – cotton growing in the Riverina*. https://www.apglobal.com.au/latest-news/entry/local-round-up-cotton-growing-in-the-riverina. Accessed 17 Apr 2018.

Australian Government. (n.d.). *The Australian continent*. https://www.australia.gov.au/about-australia/our-country/the-australian-continent. Accessed 20 Apr 2018.

Barwick, W. (1979). *The Murrumbidgee irrigation areas: A History of irrigation development in New South Wales from 1884, with special emphasis on the Murrumbidgee irrigation area from 1906–916* (B. Litt. thesis). University of New England, Australia.

Bowmer, K. H. (2003). Learning from existing practice. Reflections on developing a water sharing plan. In B. P. Wilson & A. Curtis (Eds.), *Agriculture for the Australian environment*. Paper presented at the 2002 Fenner conference on the environment, Charles Sturt University, Albury, 30 July–1 August (pp. 201–222). Deakin West: Conference Solutions.

Chessbrough, S. (1982). *A spread of green: Griffith – Its people and past events*. Sydney, Australia: Kalora Publishers.

Department of Land & Water Conservation (DLWC). (1997). *The NSW State groundwater policy framework document*. Sydney, Australia: NSW Government.

Department of Land & Water Conservation (DLWC). (1998). *Aquifer risk assessment report*. Sydney, Australia: NSW Government.

Gardner, A., & Bowmer, K. H. (2007). Environmental water allocations and their governance. In K. Hussey & S. Dovers (Eds.), *Managing water in Australia: The Social and institutional challenges* (pp. 43–47). Melbourne, Australia: CSIRO Publishing.

Gardner, A., Bartlett, R., & Gray, J. (2009). *Water resources law*. Chatswood, Australia: LexisNexis Butterworths.

Gross, C. (2014). *Water under the bridge: Fairness and justice in environmental decision-making*. London: Routledge.

Hayter, M. (2014, March 6). Murrumbidgee irrigation area CSG exclusion zone bid fails in state parliament's Upper House. *ABC News*. http://www.abc.net.au/news/2014-03-06/irrigation-csg/5302004

Holley, C. (2010). Public participation, environmental law and new governance: Lessons for designing inclusive and representative participatory processes. *Environmental and Planning Law Journal, 27*, 360–391.

Holley, C., & Sinclair, D. (2013). Deliberative governance: Insights from surface and groundwater studies. *Environmental and Planning Law Journal, 30*(1), 32–55.

Holley, C., & Sinclair, D. (2016). Rethinking water law and governance: Successes, challenges and future directions. *Environmental and Planning Law Journal, 33*(4), 275–283.

Hope, M., & Wright, M. (2003) *Murrumbidgee catchment irrigation profile*. Water use efficiency advisory unit, Dubbo.

Kennedy, A. (2017). *Environmental justice and land use conflict: The governance of mineral and gas resource development*. London: Routledge.

Kingsford, R. T. (2003). Ecological impacts and institutional economic drivers for water resource – A case study of the Murrumbidgee River, Australia. *Aquatic Ecosystem Health Management, 6*(1), 69–79.

Kuehne, G., & Bjorlund, H. (2006). Frustration, confusion and uncertainty: Qualitative responses from Namoi River irrigators. *Water, 33*(3), 78–82.

Kumar, P. B. (2002). *Review of groundwater use and groundwater level behaviour in the lower Murrumbidgee groundwater management area 002* (Groundwater status report no. 6). NSW Department of Land and Water Conservation, Murrumbidgee Region.

Kumar, P. B. (2010). *Lower Murrumbidgee groundwater sources: Groundwater management area 002 groundwater status report – 2009*. Sydney, Australia: Office of Water.

Lewis, G. (2012). *The growers' paddy: Land, water and co-operation in the Australian rice industry to the 1990s*. Ricegrowers' Association of Australia. http://sydney.edu.au/business/__data/assets/pdf_file/0008/155969/Lewis-GrowersPaddy-2012.pdf. Accessed 20 Apr 2018.

Millar, I. (2005). Testing the waters: Legal challenges to water sharing plans in NSW. Presented at the *water law in Western Australia conference*.

Murrumbidgee Groundwater Preservation Association Inc. v Minister for Natural Resources. (2004). NSWLEC 122.

Murrumbidgee Groundwater Preservation Association Inc. v Minister for Natural Resources. (2005). 138 LGERA 11.

Nature Conservation Council of NSW v Minister administering the Water Management Act 2000. (2005). NSWCA 9.

NSW Government. (n.d.) *Achieving sustainable groundwater entitlements program*. https://www.water.nsw.gov.au/water-management/water-sharing/plans-commenced/achieving-sustainable-groundwater-entitlements-program. Accessed 30 Mar 2018.

NSW Government. (2014). *NSW Gas Plan*. Sydney, Australia: NSW Government.

NSW Government. (2018). *Lower Murrumbidgee groundwater sources*. https://www.water.nsw.gov.au/water-management/water-sharing/plans-commenced/water-source/lm-gw-sources. Accessed 30 Mar 2018.

Parliament of New South Wales. (2005, November 17). *Water management amendment bill (Legislative Assembly Hansard)*. https://www.parliament.nsw.gov.au/Hansard/Pages/HansardResult.aspx#/docid/HANSARD-1323879322-82768/. Accessed 30 Mar 2018.

Syme, G. J., & Nancarrow, B. E. (2008). The social and cultural aspects of water use. In L. Crase (Ed.), *Water policy in Australia: The impact of change and uncertainty* (pp. 230–247). Washington, DC: Resources for the Future.

Tan, P.-L. (2008). *Collaborative water planning: Legal and policy analysis. Report to the tropical rivers and coastal knowledge (TRaCK) program*. Canberra, Australia: Land and Water Australia.

Udoye, C. (1984). *An economic study of the Gwydir irrigation scheme* (M.Agr.Sci. Thesis). University of Sydney.

Upper Namoi Water Users Association Inc. v Minister for Natural Resources. (2003). NSWLEC 175; and *Water Management Act 2000* (NSW).

Wilkinson, J. (1997). *Water for rural production in NSW: Grand designs and changing realities*. Sydney, Australia: NSW Parliamentary Library Research Service.

Ken Schuster, retired farmer, contractor and irrigator has served in many capacities in the public interest, starting with Rural Youth. Ken was a pioneer irrigator in the South East of South Australia and was secretary of the South Australia Seed Producers Association, and a founding director of SEEDCO (S.A.Seed Producers Cooperative). Ken served as Board member and Chairman of The Advisory Board of Agriculture in South Australia. Director of Harvest Technology Pty.Ltd. In 1995 Ken transferred farming and contracting operations to Darlington Point in New South Wales and was elected President of The Murrumbidgee Groundwater Pumpers Association (MGPA) which is now known as Murrumbidgee Groundwater Preservation Association.

Amanda Kennedy joined the Queensland University of Technology as a Professor of Law in 2019. She was previously the Deputy Director of the Australian Centre for Agriculture and Law at the University of New England (2007–2018). Amanda's research interests fall predominantly within the area of agri-environmental law, and include: environmental justice, land use law and conflict, effective systems for natural resource governance and food systems governance. She has conducted funded research on these issues under Australian Research Council and other industry grants, and has published her work in both domestic and international journals and books, including a recent monograph with Routledge, Environmental Justice and Land Use Conflict: The Governance of Mineral and Gas Resource Development.

Dr. Cameron Holley is a Professor at University of New South Wales Law and is a member of the Global Water Institute and Connected Waters Initiative, University of New South Wales Sydney. Cameron has worked closely with Australian and international government and non-government organizations on a range of water and natural resource management research projects. He currently holds ARC Discovery Grants (DP170100281 DP190101584) on Non-Urban Water Regulation and Integrating the Governance of Water and Coal Seam Gas that supported this chapter. He is an Editorial Board member on the Environmental and Planning Law Journal, and in 2016 was the guest editor of a Special Issue (EPLJ Vol 33 Part 4), entitled Rethinking Water Law and Governance.

Chapter 21
Development of Groundwater Markets in Australia: Insights from Victoria in the Murray Darling Basin

Julia De Luca and Darren Sinclair

Abstract Markets are designed to be an efficient policy mechanism to a deal with water scarcity, by enabling market participants to adjust their consumption in accordance with a flexible price signal. However, groundwater presents some challenges to the use of markets to achieve sustainable water use in that there are physical and policy constraints that may determine where markets operate. This chapter examines how the legal rights to use groundwater are managed throughout Australia through application of markets, the success or otherwise of this policy approach, and its capacity to adapt to future pressures on water availability as a consequence of climate change. We begin by outlining the principles underpinning groundwater markets across Australia. This includes key statistics, data and trends in relation to the history of groundwater trade. We then evaluate the experience of groundwater markets in practice, using Victoria as a case study in the Murray Darling Basin – this outlines how trade is administered by local authorities, possible influences on groundwater trade and markets, together with issues relating to physical connectivity between systems that can enable or stymie trade. We conclude by considering the future possibilities of markets as a tool for groundwater management.

Keywords Groundwater · Markets · Murray Darling Basin · Sustainable management

J. De Luca (✉)
Department of Environment Land Water and Planning (DELWP), Melbourne, VIC, Australia
e-mail: Julia.DeLuca@delwp.vic.gov.au

D. Sinclair (✉)
Institute for Governance and Policy Analysis (IGPA), University of Canberra, Canberra, Australia
e-mail: Darren.Sinclair@canberra.edu.au

J.-D. Rinaudo et al. (eds.), *Sustainable Groundwater Management*,
Global Issues in Water Policy 24, https://doi.org/10.1007/978-3-030-32766-8_21

21.1 Water Market Reform in Australia

Australia's water management underwent a transition during the late 1800s and early 1900s where common law water rights were replaced by state legislative regimes forming a system of water licensing (Gardner, Bartlett, & Gray, 2009). In 1994, intergovernmental action was taken to arrest widespread water resource degradation, taking into account broad economic, environmental and social considerations. A strategic framework was introduced to guide state government implementation of new market-based reforms to achieve efficient and sustainable water use (Council of Australian Governments COAG, 1994).

Under Australia's water reforms, there has been an ongoing commitment to improving trade in water allocations and water entitlements (Wheeler, Loch, Zuo, & Bjornlund, 2014). The National Water Initiative (NWI) was agreed in 2004 to provide a framework for how each state and territory would improve trade in both surface and groundwater markets. The states and territories were required to progressively remove barriers to water trading on an open market, develop water accounting to meet the needs of different water systems for planning, monitoring and trading, and recognise the connectivity between surface and groundwater resources (Crase, 2008). An independent statutory body, the National Water Commission (NWC), conducted biennial assessments to track state progress. Collectively, these reforms led to the introduction of groundwater trading in Australia (NWC, 2014).

21.2 The Rationale for Market Trading

By international standards, Australia has limited fresh water resources, with highly variable rainfall and intermittent river flows (MDBC, 2007). Historically, governments across Australia have aimed to overcome this by encouraging water infrastructure investment and use to increase agricultural activity, especially through irrigation, and thereby foster economic development and prosperity (NWC, 2011a, 2011b). Governments played an active role in determining how, when and where water should be used. As water availability became more limited, through competing uses and successive droughts, and the negative environmental impacts emerged, governments recognised the need to make best use of existing resources, and determined that water markets and trading were the preferred policy means to achieve this (National Competition Policy, 2013).

The rationale for development of water markets rests on a number of key principles (Productivity Commission, 2010). Prominent amongst these are that: water users are best placed to decide how to use water to meet their needs; being able to buy and sell entitlement to access water provides consumers with financial incentives to not waste water; markets are flexible, and allow users to adjust to different water restrictions, whilst still producing goods and services, as licences will move

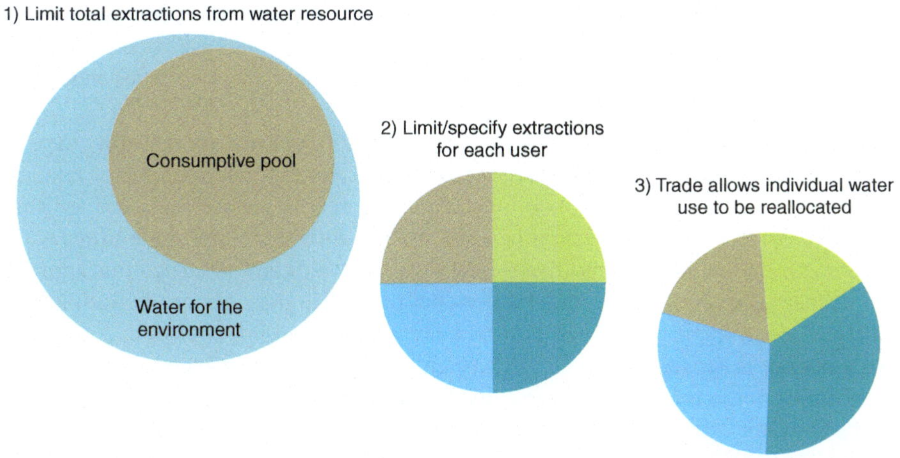

Fig. 21.1 Water markets aim to facilitate the economically efficient allocation of water while ensuring environmental sustainability. (Source: NWC, 2011a, 2011b)

to where water is needed most; and governments should not dictate how commercial water is used, for example, what type of crop should be watered and when, as compared to another type of water use.

Reducing government interference in determining how, where and when water is used for consumptive purposes allows market participants to make decisions for themselves about how much water they need and which type of entitlement product suits their circumstances (Grafton & Horne, 2014). In summary, markets aim to support regional economies by allowing participants to choose where and when they need water, whilst not exceeding overall caps (MDBA, 2015). The presence of caps is a crucial feature that requires substantial intervention by government in terms of their establishment and enforcement, but they are not inevitable features – for example, many water markets in the United States do not have caps.

A cap and trade approach to management, then, indirectly supports the sustainable consumption of water by reducing unacceptable impacts on third parties and the environment (MDBA, 2017). Ideally, market trading offers both flexibility and certainty as a management tool to accommodate Australia's droughts and floods whilst supporting equal, fair access to available water (Bjornlund, 2003) (Fig. 21.1).

21.3 How Water Markets Work

In order to facilitate water trading it was necessary to separate water rights from land ownership, so that land transfers and water transfers can occur independently of one another, although this can mean different things in different jurisdictions (Productivity Commission, 2010). A works licence is intrinsically linked to the point of extraction; how much groundwater can be sustainably taken from the infrastructure at its specific location and what licence conditions must be followed. However, the separation from land titles involved a ‘take and use’ licence – access to groundwater held by the Crown – to be traded; bought and sold independently of land sales.

Some jurisdictions have examined the concept of licensing reforms that may follow the example of regulated surface water markets; for ‘unbundling’ of take, use and works licences. In such examples, rights to access groundwater are split into three separate licences that can mix and match to allow for faster trade times within shared groundwater systems. To date, New South Wales has advanced this approach to licensing beyond that of its jurisdictional peers (Hughes, Gupta, & Rathakumar, 2016). Generally, in the case of groundwater, market trading requires that an extraction licence is separate from an infrastructure (bore) works licence.

Limits, in the form of the physically available water or through the imposition of caps on water extraction, are also essential to the operation of functional water market (NWC, 2011a, 2011b). It is one thing to have separate water and land title; it is another thing to have active trading. In this respect, it is the cap that provides the incentive for water users to trade. Once a cap is reached, no new entitlement may be granted, and access to water can only occur through trading with existing entitlement holders. The cap sets the overall limit on the amount of water that can be extracted and used within a given jurisdiction. As the volume of overall entitlements approaches a given cap, the only way for individual water users to meet their additional consumption requirements is through purchasing entitlements from other water users. The demand for those existing water entitlements, once a cap has been reached, determines the market price of water trades. The ideal policy outcome is that those trades will go to the water users with the highest productive and economic value for a given unit of water consumption, thus achieving the dual objectives of sustainability and economic efficiency.

It is important to recognise water markets are not a ‘free lunch’. They require government support to function properly, and even then there are some inherent limitations (Holley & Sinclair, 2016a, 2016b). Markets are based on the price of water entitlement being determined by market demand, with those accessing water in a given system being on equal footing in how they access water trades. Markets also require transparency and a mechanism for conducting and recording trades. Beyond this, there are two further requirements, compliance and metering, that are fundamental to the successful operations markets.

Unauthorised take above the volume stated on a licence is not permitted (NWC, 2011a, 2011b). It is also illegal to exceed a revised volume of permitted take, as reduced by a qualification of rights, which is commonly a ‘water restriction’, which may occur in seasons where rainfall has significantly impacted recharge and groundwater levels. An example of the legislative powers of a Water Minister to qualify rights to water is section 33AAA of Water Act 1989 (Vic) which allows temporary qualification of rights to water (such as seasonal water restrictions in a drought with low rainfall and recharge into groundwater aquifers). Section 33AAB of the Water Act 1989 provides for permanent qualification of rights to water. Given the amount of permitted take is subject to change within seasons and years, particularly in times of drought, resource monitoring and measurement of extraction is central to establish the physical availability of shared water resources.

A crucial component of water trading, therefore, is an effective compliance and enforcement regime (see Chap. 22; Holley & Sinclair, 2013; Matthews, 2017). As Shearing (1993) has noted, even in the most unfettered of circumstances, markets require institutional and regulatory underpinnings. At a minimum, this entails basic licensing procedures and a regulatory, legal and judicial system to enforce the conditions of those licences and address disputes if and when they arise. In the case of water, this means that the volumes on water licences within a trading market cap are adhered to. This provides confidence in the market system that the value of trades will not be undermined by illegal take. This was a major problem for early versions of the Australian water market, where trades were more expensive than fines for illegal water use (see Young et al., 2000).

When entitlement holders are confident that their supply of groundwater is secure (that is, not over exploited or illegally tapped), they can make financial investments to improve production rather than spend extra money holding licensed entitlement that is surplus to their needs. This means that when entitlement holders have confidence in resource management it may help to facilitate confidence in trading.

Compliance and enforcement governing a shared resource like groundwater reduces conflict between users by preventing over extraction and interference with environmental assets dependent on groundwater, and is beneficial to broader socio-economic community outcomes. The challenge, then, is to ensure that compliance and enforcement of entitlements is effective and comprehensive enough to support a functional water market (Holley & Sinclair, 2016a, 2016b).

Accurate measurement of water extractions through effective metering is also vital to maintaining a market cap (Holley & Sinclair, 2015). Again, it is difficult for individual water traders to have confidence in a market price if there is uncertainty about the actual size of water takes in a given trading zone. Why would water users contemplate purchasing additional water rights if they believe that other water users are extracting more than their fair share through inadequate metering? The temptation is that they will similarly undermine the trading rules. Further, without accurate metering, regulators are unable to enforce entitlements within a given cap. So reliable and accurate metering is essential to market trading, and to the compliance and enforcement regime that underpins that trading.

Beyond these essential elements of water markets, there is considerable scope for policy variation and refinement (MBDA, 2015). Within Australia's national water reforms, these include: allocations based on seasonal conditions, with allowable water takes adjusted accordingly as a percentage of overall entitlement (within a financial year); allowing for temporary trades within a single season, with the volume based on the groundwater allocated to an entitlement that can be traded, while permanent trades can occur for longer term entitlements (although these may have a maximum period, for example, 10 years in Victoria); entitlements may be traded in full or in part between buyers and sellers; trades are accounted for in annual reporting; and water registers provide collective accounting mechanisms for water trading and, together with metering and monitoring, record the amount of water used and traded in locations at multiple scales (for example, within properties, local neighbourhoods, aquifers, groundwater catchments and sub basins of state jurisdictions).

In addition, governments maintain regulatory oversight of water markets to ensure broader policy objectives are adhered to, in particular, to avoid negative impacts on other parties and the environment (NWC, 2011a, 2011b). Government approval is required for individual trades to take place. In this regard, market rules are documented and publicly accessible to inform existing and potential market participants. Such rules for groundwater are most commonly published within management plans that are developed in consultation with affected users, environmental agencies and local councils. When water availability is limited, governments have the power to intervene in the market to reduce the overall water take. Water restrictions may lead to increased trade of temporary entitlement (for example, a licensed volume which is transferred for 1 year can allow those who sow temporary crops to downsize production to save water by selling the excess to those with permanent crops). Other factors that may impact on trade approvals is when groundwater is being transferred between locations (for example, Section 40 of the *Water Act 1989* (Vic) requires an assessment to inform such decisions), and when there is interaction between surface and groundwater in particular locations such as highlands and unconfined sedimentary plains.

21.4 Progress in Water Markets

During the 1960s to 1980s the states introduced volume-based water licences to replace area-based water rights. Then, in the 1990s, caps on new water diversions started to be introduced to prevent further deterioration in the environment (NWC, 2011a, 2011b). This applied to surface water first, with caps introduced in 1994 for the Murray Darling Basin (MDB). In contrast, the MDB Plan did not apply caps to groundwater until 2012. To date, such caps include Sustainable Diversion Limits (SDLs) in the MDB and Permissible Consumptive Volumes (PCVs) in Victoria, Long-term Average Annual Extraction Limits (LAAELs) in New South Wales, Annual Volumetric Limits in Queensland and Water Take Limits (WTL) in South Australia.[1]

As part of the national water reform process, from the 1990s onwards, states and territories have significant discretion in determining how best to implement water trading. A diversity of approaches, when combined with distinct basin and catchment caps, have produced a composite of many separate markets in Australia, each

[1] See relevant legislation as follows: SDLs in MDBA Basin Plan 2012; PCVs in Victorian Water Act 1989; LAAELs e.g. Water Sharing Plan for the NSW Border Rivers Unregulated and Alluvial Water Sources 2012 – Regulation 28; AVLs in QLD e.g. Water Plan (Mitchell) 2007: Natural Resources Management Act 2004 (SA).

defined by water system boundaries and administrative arrangements (NWC, 2013). Further, within those markets, there are segments for different water products, such as access entitlements and allocations (often of varying levels of security), and different trading transactions (NWC, 2013).

While the process of separating water rights and land title has been slow in some states and territories, in particular Western Australia and the Northern Territory, in other jurisdictions, particularly those within the MDB (New South Wales, Victoria, Queensland and South Australia), much more progress has been made. Indeed, at present, the majority (approximately 100) of the 172 completed Water Sharing Plans across Australia have separate water and property rights.

To date, the most connected markets occur in the surface water context, largely because of the hydrological connectivity between systems within the MDB (Grafton & Horne, 2014). In contrast, groundwater markets are less developed and tend to be much smaller, reflecting the scale of how they are managed. As expected, in most instances, it is surface water trading that has led the way, with groundwater trading following some time later.

Successive reforms in the MDB have sought to increase confidence in water markets by improving trading rules, developing central online access points for comparing water products and rules, and introducing a MDB compliance strategy (see also MDBA, 2014). While public access to jurisdictional registers and 'searchability' could be improved, reductions in transaction costs and online access to many state trade registers have occurred since 2004 (Grafton & Horne, 2014). These improvements were confirmed in surveys of irrigators asking whether trading had become easier in the last 5 years. Most agreed that trade had become easier (~40%), with strongest agreement found in Southern MDB (although over 50% disagreed outside the MDB) (NWC, 2014). There is also evidence of growing confidence in markets and trading in terms of the level of trade and prices. It is important to note that information on groundwater trade is housed in various documents and need to be sourced to get a picture on where someone may decide to propose to trade a licence. The availability of quality data on market price and volume is 'not optimal' for groundwater and climate variability and regulatory changes have affected market prices and trading (see NWC, n 78, 41 and ABARES, 2016).

In other aspects of the trading regime, such as compliance, enforcement, monitoring and metering, progress has also been mixed. Historically, environmental regulation in Australia has favoured 'soft' regulatory approaches such as information, education, self-regulation and voluntarism (Gunningham, Grabosky, & Sinclair, 1998; Holley & Sinclair, 2013). As such, there has been a reticence to impose robust compliance and enforcement (Gunningham & Sinclair, 2004). Further, there are inherent complexities confronting water compliance and enforcement. These include constrained regulatory resources (see for example Holley & Sinclair, 2012 where regulators in New South Wales had low numbers of compliance officers), multiple diffuse points of extraction, large geographic areas, numerous variable and dynamic surface and groundwater systems, and water users who are often resistant to government intervention compared to other sectors of the economy (The NCCARF Water Governance Research Initiative, Undated; Head,

2010). This is relative, as in the United States irrigators may be even more fundamentally opposed to government intervention (see Garrick & O'Donnell, 2016). Enforcing groundwater extractions presents additional difficulties in that, by its very nature, it is less visible than surface water, does not attract a similar level of scrutiny by the public, and is susceptible to extraction from unlicensed bores and/or or bores that have deficient metering (see Nelson, 2018).

As noted earlier, a lack of compliance and enforcement capability has the capacity to undermine the integrity of water markets. While the Australian Competition and Consumer Commission has observed that compliance in both water market and water charge rules has also improved, and costs in some areas have declined (Grafton & Horne, 2014), there are concerns that compliance and enforcement of individual water takes is inadequate, and that existing approaches have weaknesses (see Chap. 22; WG, 2017; Holley & Sinclair, 2015; Boizard et al., 2016; Brown, 2017).

Licensing and resource management functions, together with compliance and enforcement, rest with state and territory government regulators in Australia (NWI, 2004). Despite significant federal and state investment to improve water metering, compliance and enforcement (e.g. DSEWPC, 2012), recent Commonwealth government reports suggest that regulation has largely been ineffective (Matthews, 2017; MDBA, 2017; NWC, 2014; PC, 2017). Although there is some uncertainty as to what constitutes water theft, and the overall levels of non-compliance (Brown, 2017; Holley & Sinclair, 2015; Matthews, 2017), there is growing community and government concern about the systemic failure of compliance and enforcement.

As evidence of this, the Productivity Commission, (2017, pp. 51, 401) notes the need 'for improvement specifically relating to implementation of national frameworks for non-urban water metering, and compliance and enforcement systems for water'. The Matthews Inquiry (Matthews, 2017, 7) reported that 'water related compliance and enforcement arrangements in NSW have been ineffectual and require significant and urgent improvement' (see also NSW Ombudsman, 2017) and that, in the MDB, state budgets and reporting have significantly reduced in recent years (Matthews, 2017; MDBA, 2017); and Queensland, New South Wales, and to a lesser extent Victoria, have demonstrated a 'notable lack of transparency' and remain 'bedevilled by patchy metering … the lack of real-time, accurate water accounts … [and] a low level of compliance resourcing' (MDBA, 2017, 14; WG, 2017). 'Considerable frustration' has also been levelled at the role of the MDB Authority and its response to alleged serious breaches (MDBA, 2017, 14). It is critically important to have compliance and enforcement approaches that discourage users who might illegally take water that is already allocated to other users and the environment.

In regard to the adequacy of water metering, although various metering technologies have been implemented in non-urban water contexts, their application has been patchy and uneven (Grafton & Peterson, 2007; Holley & Sinclair, 2013; Lavau, 2013). While surface water use has often been metered, the monitoring of groundwater extraction remains weak (or completely absent in some jurisdictions) (Holley

& Sinclair, 2013). The accuracy of many current water meters (for both surface and groundwater) is also said to be 'not high due to their age, lack of maintenance and improper installation' (Commonwealth of Australia, 2009; Department of the Environment Water Heritage and the Arts, 2009; Hamblin, 2009; Holley & Sinclair, 2013).

Some recent government reforms have attempted to address these deficiencies (Holley & Sinclair, 2013; NSW Office of Water, 2015). However, the net effect of inadequate metering is that the volume of water diverted may exceed entitlement volumes, and undermine overarching goals of fair and efficient water use (Commonwealth of Australia, 2009; Holley & Sinclair, 2013). The lack of accurate meters is a particular impediment to the operation of water markets and their ability to guide water to the highest value uses (Raft & Hillis, 2010). Nevertheless, there have been some positive, developments in groundwater metering. For example, Grampians Wimmera Mallee Water in Victoria has installed 'smart meters' across its jurisdiction to record data for groundwater use in real time, which can be read remotely. This has allowed the licensing authority to monitor overall resource consumption more efficiently, reduced travel times for staff and provided users with a better understanding of their consumption habits.

21.5 Social Barriers to Trade

There are potential barriers to trade from lack of knowledge and information and some market participants may prefer to retain their allocations even when it appears not to be in their economic interests to do so (see Pérez-Blanco et al., 2016; Loch et al., 2018). An emerging research field is seeking to investigate why groundwater trade markets are underdeveloped. Participant survey responses from Gill et al., 2017 have shown there is a complex mix of social, economic, institutional and technical reasons. Barriers to trade are influenced by the circumstances of each groundwater user, administrative process and resource management rules. Water brokers deal with few trades at low margins and noted unrealistic selling prices and administrative difficulties. Irrigators who have successfully traded identify that there are few participants in trading, technical appraisals are expensive and administrative requirements and fees are burdensome, when compared to surface water trading. Opportunities to facilitate trade include groundwater management plan refinement and improved information provision. Simplifying transaction processes and costs, demonstrating good resource stewardship and preventing third party impacts from trade could address some concerns raised by market participants. The study concluded there are, however, numerous circumstances that may inhibit an individual from groundwater trading. In time understanding of social barriers to trade is likely to evolve, particularly as institutional and administrative settings become clearer.

21.6 Physical Barriers to Trade

In addition to issues that may undermine the overall integrity of water markets, there are a series of potential barriers to trade between individual water users, as well as within physical regions. In terms of the former, this includes transaction costs as a result of regulatory requirements, and there may also be hydrogeological and/or infrastructure constraints (Brooks & Harris, 2014).

In this respect it important to recognise that trade is a form of substitution, that is, a transfer of the right to take and use groundwater, as opposed to a physical transfer of groundwater that is subject to a licence. As such, it is impossible to use the same physical litre as was approved for someone else at another location to take and use. Consequently, rural water corporations treat each application for a licence transfer as the equivalent of a new licence application in applying s40 of the Victorian Water Act 1989. Approval of a trade offsets the equivalent volume from being taken and used as licenced entitlement from a bore that may be located elsewhere, in a unit, aquifer or groundwater catchment.

Clear management boundaries for trading zones are necessary to avoid confusion of where users can trade. Ideally, boundaries, which set the confines of a market, are based on current resource knowledge. However, it can take years or decades and considerable effort for local authorities to review markets boundaries' spatial units. The topic of boundary issues, including the potential for gaps and overlaps, and the complexities of managing a three dimensional resource in a constant state of flux is a challenge for resource managers. The review process never ceases when there are multiple forms of boundaries to be managed, new approaches and GIS mapping techniques.

Such barriers have meant trading tends to be concentrated in surface water (as opposed to groundwater) (Burdack, Biewald, & Lotze-Campen, 2014; NWC, 2012–2013)[2] and mostly within the MDB. For example, in 2012/13, ~80% of entitlement trade and 98% of allocation trade occurred in the MDB (NWC, 2012–2013). Collectively, the markets outside the MDB accounted for only 300 GL of entitlement trade and 126 GL of water allocation trade in 2012–13 (or 22% and 2% of the total Australian markets for those products, respectively) (NWC, 2012–2013).

[2] As discussed below, water access entitlement trading until now primarily involves surface water rights rather than groundwater rights (which amount to around 21% by volume of entitlements on issue).

21.7 Interaction of Commonwealth Legislation and State Management of Trade

Commonwealth legislation may be perceived to create new barriers to groundwater trade, under ss12.24–12.26 of the Basin Plan 2012, through the potential for misinterpretation of the term 'sufficient hydraulic connectivity', which may be debated from legal, management, policy and hydrogeological science perspectives. Since groundwater flow is not static, this term may be challenged on the basis of scientific uncertainties as to who connects to a shared resource, as aquifers are not engineered. In effect, all of the MDB is connected, including groundwater and surface water (a key reason the Basin Plan was introduced), with the exception of confined aquifers, e.g. the West Wimmera. It is possible for states to define what is connected based on evolving understanding and management of groundwater.

It is worth noting, the NWI is still recognised as a binding commitment between Commonwealth and signatory states.[3] It is possible the NWI objectives and those of the Basin Plan combine to have a positive influence on states in that trade could be facilitated across whole groundwater systems, subject to caps and licensing decisions. Reaching full allocation of entitlement need not result in bans on trade, so long as volumes traded are substituted between buyers and sellers if caps are not exceeded, trading rules are transparent, publicly accessible and treat users equally. It is also desirable to recognise that the development of groundwater markets has beneficial socio-economic outcomes in supporting communities and rural economies by allowing access to water, whilst operating within overall caps on take.

21.8 Case Study: Groundwater Markets in Victoria

In Victoria, the rights to take and use groundwater are distributed to commercial users as licences under Section 51 of the *Water Act 1989*. Rural water corporations are the delegated authorities who, on behalf of the Minister for Water, make licensing decisions, including whether a trade of a Section 51 licence is made. Any individual, subject to a judicial review by an independent body, the Victorian Civil Administrative Tribunal, can contest these decisions. Commercial users of groundwater must hold a Section 51 licence that states the amount of groundwater they can legally extract over the course of a financial year, and the licence conditions they are to meet.

[3]As the NWI was cited in Water Act 2007 Murray Darling Basin Plan 2012 Implementation Agreement 7 August 2013, clause 1.3.

If there is a shortage of available groundwater from the overall consumptive pool, then licence holders in the affected management unit or zone are placed on a restriction, which is a percentage (allocation) of the groundwater volume (entitlement) they can extract in a year. Restrictions can also be applied to manage other resource concerns such as reducing the risk of seawater intrusion in coastal aquifers. At the end of the millennium drought, due to lack of rainfall and its subsequent impact on recharge, several units and zones across the state were placed on water restrictions, with entitlements ranging from 0% to 100% depending on the location and type of aquifer that groundwater users tapped.

In 2012, the Victorian Government announced a Groundwater Management Framework (GMF) in order to define and manage groundwater across the state. This was developed from a process of community engagement that aligned with the physical hydrogeology of groundwater systems. The Groundwater Securing Allocation Future Entitlement project held 14 workshops, with over 350 participants, across regional Victoria. The workshops included groundwater users (irrigators and businesses), domestic and stock users (which do not participate on the water market but are concerned with security of supply), catchment management authorities, local government and urban water corporations. There was debate about the scale at which groundwater resources should be managed, however, the need for vertical integration between different scales (from small to large groundwater systems) and the desire for sustainable supplies into the future emerged as common themes. The completed GMF was a combination of the positive improvements that stakeholders requested together with detailed geo-spatial mapping. Technical reports were supplemented by cost benefit analysis of the intended outcomes, to socio-economic development in rural Victoria, from the broadening of access to groundwater markets.

In terms of scale and extent, the GMF comprises five major groundwater basins, supplemented by 20 smaller groundwater catchments, whilst retaining units and zones at local scales (that are subject to regular updates). The GMF was a politically bipartisan policy outcome that spanned across different government terms. A key to the success of its development was the comprehensive engagement process that allowed stakeholders a high degree of ownership of the policy process (Fig. 21.2).

Since the inception of the GMF, water authorities are gradually improving groundwater management. Goulburn-Murray Water, a large rural water corporation responsible for managing over 50% of Victoria's allocated groundwater, has almost completed installing groundwater management units and zones at local scales across all of its jurisdiction. They have also sought to integrate the local scales with the management of the overarching Goulburn-Murray Basin (which is one of Victoria's five major groundwater basins in the framework).

A key outcome of the framework in northern Victoria (which encompasses the Goulburn-Murray Basin) has been to facilitate groundwater trade. Reviewing local units and zones takes time and resources, in particular, for water authorities to consult with communities on groundwater management and undertake reviews of local trading rules. Nevertheless, recent data trends indicate that the framework has

Fig. 21.2 State-wide groundwater management framework in Victoria, comprised of groundwater basins and catchments, introduced in 2012. (Source: Aither, 2017)

provided the necessary policy guidance and management rules for broadening access to groundwater trading opportunities.

An evaluation of trading in Victoria was undertaken in 2016–17. Supply and demand for a capped resource within an individual market is a determinant of trade. With these conditions assumed as precursors to the development of a water market, the evaluation covered different groundwater zones and complex rules, with the number and volumes of entitlement traded in each market identified and analysed. Given the number of trading zones and markets in Northern Victoria, several hypothesises were examined (De Luca & Wiltshire, 2016; DELWP and Cardno, 2017).

One of these was to identify whether the number of water users (i.e. potential buyers and sellers) in a market (i.e. the potential market size) was a determinant of its trade activity. A key finding was that the Goulburn-Murray Basin, with 245 GL (billion litres) of groundwater distributed to commercial users via 2895 licences across the basin, had the most trading activity. This was significantly more trade activity than comparable basins with similar amounts of entitlement volume, thus supporting the hypothesis. Another contributor to the trade activity in northern Victoria was the presence of rules that allow entitlement to be traded *outside* of local zones and units, across broader resources that are connected inside the basin – this had expanded the number of participants who can trade with each other.

By mid-2016, there were 174 groundwater zones across Victoria, with each zone further aggregated into larger groundwater units (See Fig. 21.3, De Luca & Wiltshire, 2016). Another key finding of the evaluation was that several zones combine to define the bounds in which trade can occur, in doing so, create a single and larger

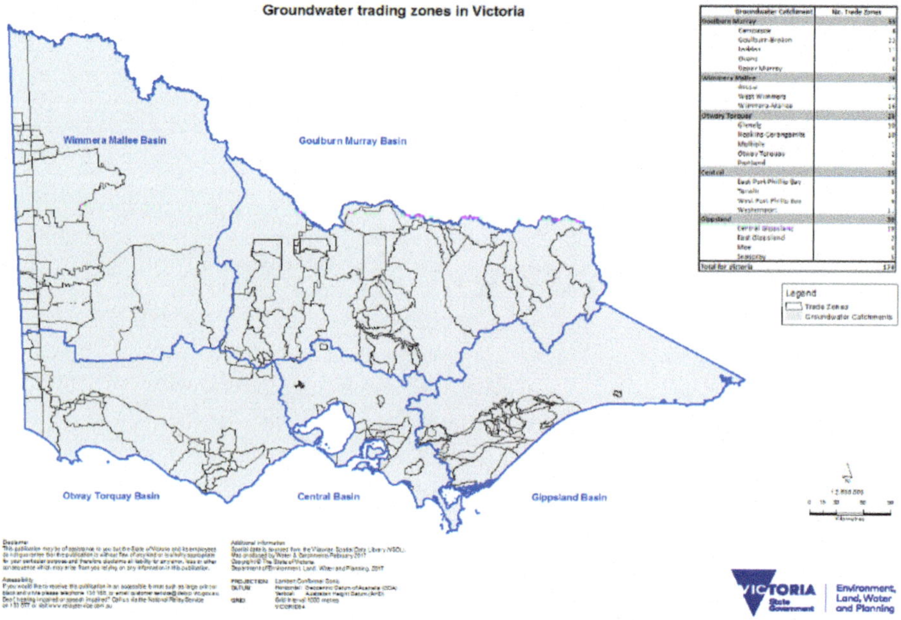

Fig. 21.3 State-wide map showing 174 trading zones in Victoria 2017. (Source: De Luca & Wiltshire, 2016)

groundwater market. The number of zones involved in this amalgamation of trading rules was surprisingly complex, reflecting the complicated history of groundwater management. Most zones are found inside a unit, and there are over 50 units across Victoria. Traditionally, local zone boundaries followed rural development spatial patterns and *not* the underlying groundwater systems. This was due to a lack of hydrogeological information at the time of regional development.

Groundwater extraction now occurs from every aquifer in the state with low levels of salinity. Even hyper-saline geothermal groundwater is becoming popular, with a recent licensing dispute on the Mornington Peninsula (Peninsula Hot Springs Pty Ltd. v Southern Rural Water [2017] VCAT 2103 (19 December 2017) and St Andrews Beach Country Golf Course Pty Ltd. v Southern Rural Water [2017] VCAT 919 (23 June 2017)). Further, development of the resource has expanded from groups of local irrigators, across the sedimentary and volcanic aquifers near the surface, into lower quality resources at basin margins, resources in fractured bedrock and fresh confined resources at considerable depth that require significant investment in drilling of bores to reach.

In order to detail the extent of market formation, the study examined how many zones and units there are, and how many markets have been created at local scales. A market was defined by the spatial bounds in which a person can trade entitlement (volume on their licence) from one location to another. It was found that by mid-2016, the groundwater trade in Victoria comprised 95 markets for temporary trade and 93 markets for permanent trade (Fig. 21.4).

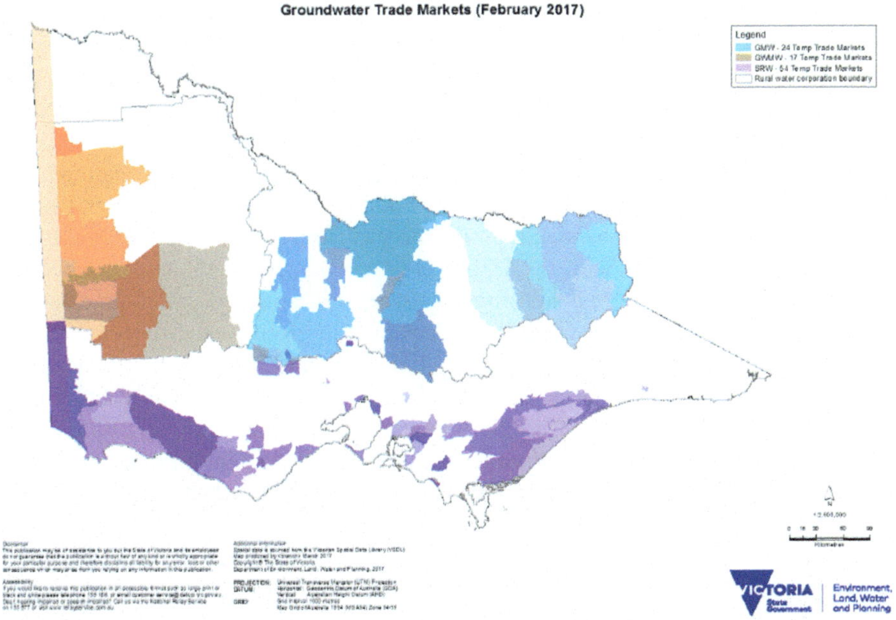

Fig. 21.4 State-wide markets in Victoria. Map of temporary groundwater trade markets – based on areas where trade can occur without restriction between zones. (Source: De Luca & Wiltshire, 2016)

The net result of this 'bottom-up' approach to spatial groundwater management in Victoria has been a complex mosaic of trading opportunities. Further analysis revealed that within this overall mosaic, five types of trading rules exist within individual zones. These range from 'no trade allowed', 'trade allowed between users within your zone only', 'trade allowed between users within your zone and into your zone only', 'trade allowed between users within your zone and out of your zone only', to 'trade allowed between users within your zone and into or out of another zone'. In other words, from less to more open trading regimes.

The sheer complexity of the different trading rules has created an administrative challenge for resource managers. Indeed, local users within any one of the Victoria's 174 zones are more likely to be familiar with their particular trade opportunities than water managers. Consequently, water authorities are on a lengthy journey to consolidate and streamline local groundwater management rules to provide greater consistency across Victoria, and in doing so increase the equal access to trade.

Having catalogued the complexity of the administrative settings at localised scales, water authorities are in a better position to plot a path forward to expand groundwater market trading in shared resources. In particular, there are opportunities to consolidate and streamline administrative settings to provide more timely licensing and management services. This can make groundwater markets more understandable for participants and resource managers alike. The evolution of markets presents an opportunity to trade in connected systems at different scales of management. Future management will likely consolidate different zones together into shared markets and involve more stakeholders in regionally broader markets within connected resources, based on catchment management approaches, inside the overall groundwater GMF (Figs. 21.5 and 21.6).

The Upper Ovens river valley in northern Victoria is a unique example of a shared market that allows entitlement to be traded between groundwater and surface water resources. This innovative regime emerged from the treatment of highly connected alluvial aquifers with the Upper Ovens River, with a shared management plan. In this instance trading rules treat both resources as one, interconnected system, whereby surface water trade rules also apply to groundwater. This approach is the first trading regime of its kind in the MDB and, indeed, Australia as a whole.

In conclusion, trade is best supported by a transparent rules based framework, that is documented in legal instruments to inform licensing decisions, with simplified rules that are contained in consistent management plan formats, which are reviewed and updated on a regular basis, as new resource knowledge emerges.

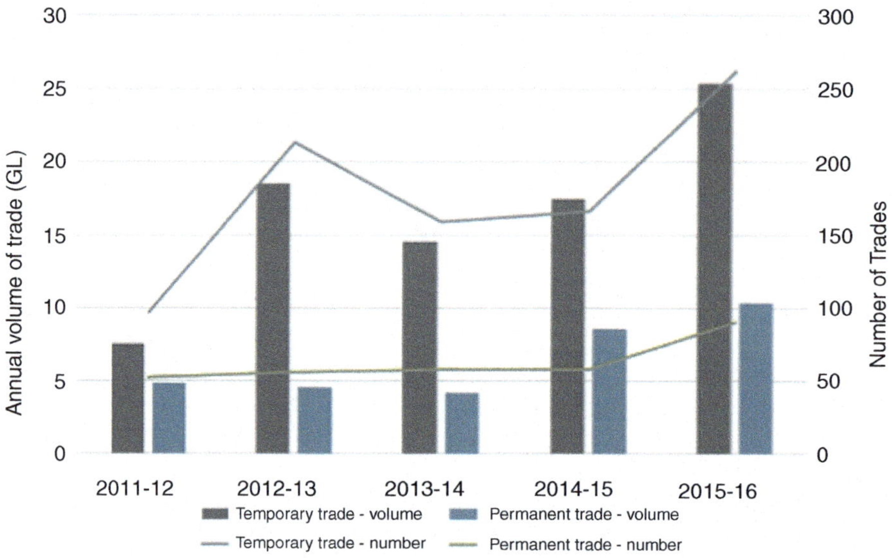

Fig. 21.5 Trade activity. Total volume of groundwater traded, and number of trades in Victoria, 2011–12 to 2015–16. Note: Trades involving changes in land ownership are not included. Total annual permanent Victorian groundwater trade increased from 4799 ML in 2011–12 to 10,290 ML in 2015–16. Over the same period, total annual temporary groundwater trade fluctuated significantly, peaking at 25,363 ML in 2015–16. Overall, there has been a substantial increase in trade (volume and number) for both permanent and temporary groundwater trade over the period considered. (Source: Aither, 2017)

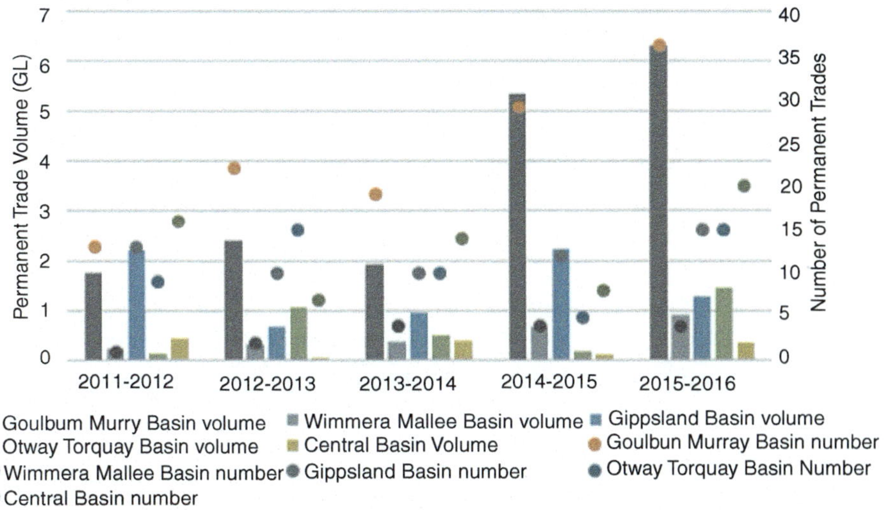

Fig. 21.6 Basin-scale trade. Groundwater trade has been most extensive in the Goulburn-Murray Basin, for both permanent and temporary trade. The Gippsland and Wimmera Mallee Basins are the next most intensive in most years. Most of the increase in trade over the 5-year period is accounted for by permanent trading in the Goulburn-Murray Basin in the 2014–15 and 2015–16 water seasons. Similarly, in 2015–16, trades in the Goulburn-Murray Basin accounted for 63% (by volume) of all temporary groundwater trades in Victoria that year. (Source: Aither, 2017)

21.9 Lessons from Victoria and the Future of Groundwater Trade with Climate Change

As with other Australian jurisdictions, and countries and regions around the world, Victoria confronts challenges in managing its water resources into the future as the impact of climate change becomes more pronounced. Climate change is adding uncertainty to variable weather patterns, particularly in relation to rainfall. Research by the CSIRO for example, indicates that cold fronts from the Southern Ocean have retracted southwards towards the Antarctic pole (CSIRO, 2010).[4] This correlates with reduced rainfall for Victoria, particularly during the cooler seasons (*Our Water Our Future: The Next Stage of the Government's Water Plan*, 2007)[5]. Consequently, it is projected that climate change will increase both the severity and frequency of droughts (Fig. 21.7).

[4] It is worth noting that the rainfall in south-west Western Australia has continued to be below its long- term mean for more than 30 years, and that there is some evidence of links between the climates of that region and of south-eastern Australia with both regions being reliant on similar mid-latitude weather systems for their rainfall (Hope, Timbal, & Fawcett, 2009) which have been affected by surface pressure increases (Timbal & Hope, 2008).

[5] This responded to the Millennium Drought's record low inflows across the state.

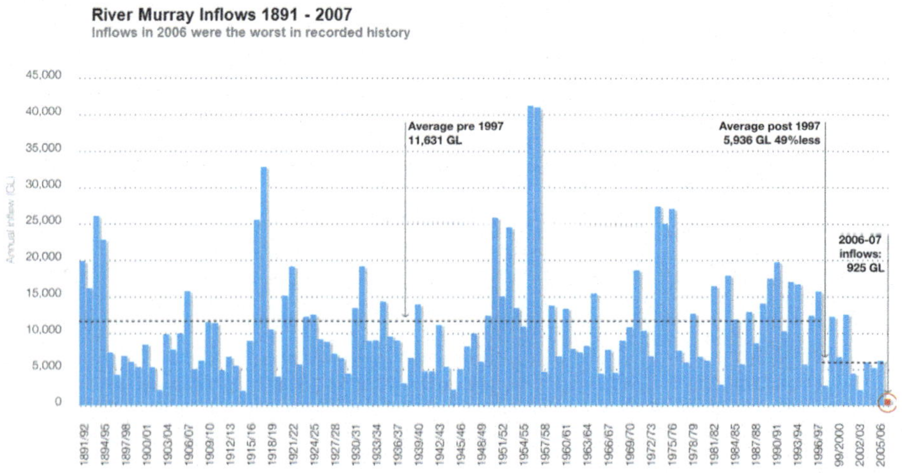

Fig. 21.7 Millennium drought impacts on streamflow. (Source: Murray Darling Basin Commission, in Our Water Our Future: The Next Stage of the Government's Water Plan, Victoria 2007)

Water trading increases in importance during times of drought. The greater the scarcity, the greater the value of available water resources. Those with higher value products and/or susceptibility to water shortages have a greater incentive to purchase water. It is vital then, that effective and efficient trading is available to facilitate the distribution of water to where it is needed most. And in this respect, groundwater is crucial in that it is far less susceptible to short to medium seasonal rainfall fluctuations, and, indeed, may be the only secure water supply in a prolonged drought. Of course, groundwater recharge rates are susceptible to the impact of climate change over the longer term, but highly variable large flood events may bank future water supplies when water is more available (CSIRO, 2010).[6]

It is imperative for sustainable management of water, and the environments, industries and communities that depend on steady water supplies, that water governance is able to adapt to and address the challenges that climate change is projected to bring. Groundwater markets work best when they are monitored and adjusted regularly to ensure participants are treated equally, and where transaction costs (both financial and temporal) are minimised. Publicly accessible information, with transparent documentation of each trading rule is also vital to supporting market development and compliance with requirements of the ACCC (Australian Competition and Consumer Commission) and *Basin Plan 2012*.

Although the State Government governs the legal framework that determines groundwater trade in Victoria, increasingly the Commonwealth Government of

[6] The record rainfalls and flooding experienced across much of the region, and throughout Australia, in the 2010/11 and 2011/12 summers highlighted the importance of the status of three oceans – Pacific, Indian and Southern – in influencing seasonal and inter-annual rainfall variability.

Australia is centrally influencing water management. Commonwealth legislation defines the Commonwealth role as requiring States to report on activities they undertake on trade rather than directly making decisions. While states and territories will continue to oversee water trade and licensing decisions, the Commonwealth may have substantial influence on operational matters in the future via prescriptions and interpretations as to how the MDB Plan is to be implemented.

There are numerous issues arising from complexities in the current legal requirements, including: conflicting messages from the Commonwealth on whether it promotes trade (NWC, 2011a, 2011b; Productivity Commission, 2017); creation of dual systems in water governance from mixing state and federal laws, with several Acts and related legislation; contention over Commonwealth jurisdiction on state water management (Gardner 2012); and problems caused by splitting Victoria into two states of water management, given Commonwealth investment in the north could detract the from the south, presenting a challenge for state-wide policy approaches.

Nevertheless, Commonwealth policy does not prevent Victoria seeking to improve groundwater trade at a broad level. Indeed, the NWI supports the development of the groundwater trading market, and even allows for groundwater trade between MDB state and territory jurisdictions (NWI, 2004, Action 60). State policies cover the legal framework of licensing, policy, and management that determines how and where people may trade a licence to take and use groundwater. In this respect, the Victorian experience highlighted above sheds light on the important contribution of water trading, and groundwater, in particular, as key planks of the policy response, and ways in which a trading-based policy approach might be enhanced. Key insights are as follows.

First, that trading needs to be able to adapt to seasonal variations, which are likely to more pronounced with climate change. In particular, Victoria's approach of adjusting allocations of a consumptive pool within a trading market allows policy-makers to respond to seasonal variations whilst maintaining the flexibility that trading brings.

Second is the need to align groundwater-trading zones with the physical hydrology of groundwater systems, and to integrate this across multiple scales, from small to very large groundwater systems. Victoria has pursued this approach through the creation of a state-wide framework that aligns management with groundwater catchments and basins. A critical component of this process is comprehensive geospatial mapping, together with regular reviews and updates of local zones and units as more information becomes available.

Third, that extensive consultation and engagement is essential to obtaining community and political support necessary to generate broad ownership and policy longevity. In this respect, it is important to communicate with stakeholders from the start of the planning process. A key feature of the Victorian approach was detailing the socio-economic benefits that could be derived from an expanded groundwater trading market.

Fourth, increasing the number of participants in a trading market leads to more trading, and by extension, better economic outcomes for trading regions. Trading in the Goulburn-Murray Basin, with the largest number of licence holders, was greater than in smaller regions. Further, expanding the trading of entitlements *between* local zones and units also increased the amount of trading. A key component of this was aligning trading zones with the geospatial mapping of aquifers in groundwater systems.

Fifth, reducing administrative complexity of water trading and divergent trading rules between different zones and units is essential to facilitating greater trade from both within and between trading markets. In this respect, Victoria has some considerable way to go, but nevertheless demonstrates the desirability of doing so, with a commitment to several policies for realising the potential of markets (*Water for Victoria* 2016).

Sixth, trading provides the flexibility to tailor water consumption to different conditions and different types of groundwater. For example, Victoria has several climate types and associated groundwater resources, including in the north-west where there are pockets of underlying low salt groundwater near the border with South Australia. Here, water markets have developed in order to access confined and secure supplies of groundwater, with management plans setting out rules that promote longevity. Permanent trade can promote and support regional development in these arid regions. Victoria and South Australia also have a legislative agreement on management of these shared resources (BGARC) that may eventually lead to interstate groundwater trade.

And seventh, allowing integrated trading between groundwater and surface water catchments can create more trading opportunities. Although Victoria has only recently begun to go down this path, early results are encouraging, in particular, with the Upper Ovens River and its groundwater aquifers. The alpine region of Victoria has interconnected groundwater-surface water resources that are responsive to rainfall and snowmelt along the highlands. Here, groundwater trading provides users with the ability to adjust to variable rainfall, especially through temporary trades.

21.10 Concluding Remarks

Beyond these particular insights from the Victorian experience, there are some broader issues that need to be addressed in order to maximise the benefits of a trading-based approach to water management. Central amongst these is the fundamental necessity of a comprehensive and effective compliance and enforcement regime. Without effective compliance and enforcement, not only is there considerable risk that water caps are not achieved, but also, there is the potential to undermine the integrity of the market and the incentive to engage in individual trades. This is an area that deserves policy attention, implementation resources and support to effectively manage Australia's established and emerging water markets.

Priority developments for improving compliance and enforcement are increased inspection of resource users, a clear and consistent compliance and enforcement policy based on the principles of responsive regulation, including a regulatory 'peak' of prosecution, the avoidance of regulatory capture and an institutional culture that emphasises risk-based compliance and enforcement, and effective monitoring and measurement of groundwater extractions and levels.

Technology, in the form of metering and telemetry offers the opportunity to substantially enhance compliance and enforcement, and also offers benefits to water users in the form of better on-farm water management (Holley & Sinclair, 2016a, 2016b). The paradigm of monitoring is changing, with computing systems able to transmit data and information in real time to provide regulated entities with the ability to self-regulate and monitor their activities (Snyder et al., 2013; see also Markell & Glicksman, 2016). In this respect, the Victorian experiment with the use of smart meters by Grampians Wimmera Mallee Water is instructive. As climate change increases the frequency and size of droughts, the temptation for water users to engage in illegal extractions increases. Into the future, then, metering and telemetry will be essential to ensure the integrity of market trade caps.

The water market in Victoria is worth billions of dollars annually, across both surface water and groundwater (see for example, trade reports on the Victorian Water Register website). Increasingly, water users are viewing groundwater as an opportunity to access a secure resource, particularly when its management is supported by market mechanisms to enable them to adjust use to their needs into the future. Groundwater is also likely to grow in demand as it is buffered from drought compared to surface water, which in turn will become vital for adaptation as climate change continues to impact Australia and as the population grows. To the extent that reforms facilitate trade within the connected groundwater systems of Victoria, more water users will be able to access these benefits, at the same time as helping to secure sustainable water supplies.

Future opportunities to access trade are increasingly likely to be centred on how local management relates to surrounding basins, and the broadening of groundwater markets. When a choice is provided to water licence holders to alter the volume of their water take, with the flexibility to make temporary or permanent adjustments, they can decide how and when they use groundwater to match their needs. In short, the market provides options for increasing or decreasing the use of groundwater in a timely and efficient manner. In the absence of a capped market, there is as strong incentive to increase water usage (Loch et al., 2018). Provided it is embedded within a strong regulatory framework, the market provides an economic incentive to not waste groundwater, and encourages water going to the highest value uses.

References

ABARES. (2016). Australian water markets report 2014-2015. Commonwealth of Australia.

Aither. (2017). *Early trends in groundwater trade in Victoria.* http://waterregister.vic.gov.au/water-trading/trade-reports

Bjornlund, H. (2003). Efficient water market mechanisms to cope with water scarcity. *International Journal Water Resource Development, 19*, 553–567.

Boizard, P. C. et al. (2016). *Audit of the implementation of water and biodiversity policies in the department of Lot-et-Garonne* MEES/MAA.

Brooks, R., & Harris, E. (2014). Price leadership and information transmission in Australian water allocation markets. *Agricultural Water Management, 145*, 83.

Brown, V. F. (2017). *Water theft and water smuggling*. Washington, DC: Brookings.

Burdack, D., Biewald, A., & Lotze-Campen, H. (2014). Cap-and-trade of water rights: A sustainable way out of Australia's rural water problems? *GAIA Ecological Perspectives for Science and Society, 23*, 318.

COAG. (1994). *Communiqué, attachment A, water reform*. Australian Government.

Commonwealth of Australia. (2009). *National Framework for non-urban water metering policy paper*. Australian Government.

Crase, L. (2008). An introduction to Australian water policy. In L. Crase (Ed.), *Water policy in Australia: the impact of change and uncertainty*. Washington, DC: Resources for the Future.

CSIRO. (2010). *Climate variability and change in south-eastern Australia: A synthesis of findings from Phase 1 of the South Eastern Australian Climate Initiative (SEACI).*

De Luca, J. (DELWP), & Wiltshire, E. (CDM Smith). (2016). *Groundwater Trade in Victoria Baseline Report*, unpublished, summarised in Aither (2016) *Early Trends in Groundwater Trade in Victoria*.

DELWP (Department of Environment, Land, Water and Planning) and Cardno. (2017). Audit of Water Entitlement Purchases – GMW Connections Project. *Report*, 3606–3644.

Department of the Environment, Water Heritage and the Arts. (2009). *National Framework for Non- urban Water Metering Regulatory Impact Statement* 3, Commonwealth of Australia.

DSEWPC. (2012). *National framework for compliance and enforcement systems for water resource management*. Australian Government.

Gardner, A., Bartlett, R., & Gray, J. (2009). *Water resources law*. Chatswood, Australia: LexisNexis.

Gardner, A. (2012). Water Reform and the Federal System. In I. P. Kildea, A. Lynch, & G. Williams (Eds.), *Tomorrow's Federation Reforming Australian Government*. Sydney, Australia: Federation Press.

Garrick, D., & O'Donnell, E. (2016). Exploring private roles in environmental watering in Australia and the US. In E. Bennett (Ed.), *Protecting the environment privately* (pp. 203–232). Singapore, Singapore: World Scientific Publishing.

Gill, B., Webb, J., Stott, K., Cheng, X., Wilkinson, R., & Cossens, B. (2017). Economic, social and resource management factors influencing groundwater trade: Evidence from Victoria, Australia. *Journal of Hydrology, 550*, 253–267.

Grafton, Q., & Horne, J. (2014). Water markets in the Murray Darling basin. In Q. Grafton, P. Wyrwoll, C. White, & D. Allendes (Eds.), *Global water: Issues and insights*. Canberra, Australia: ANU Press.

Grafton, R., & Peterson, D. (2007). Water trading and pricing. In K. Hussey & S. Dovers (Eds.), *Managing water for Australia: The social and institutional challenges* (pp. 73–84). Melbourne, Australia: CSIRO Publishing.

Gunningham, N., Grabosky, P., & Sinclair, D. (1998). *Smart regulation: Designing environmental policy*. Oxford, UK: Oxford University Press.

Gunningham, N., & Sinclair, D. (2004). Non-point pollution, voluntarism and policy failure: Lessons for the swan-canning. *Environmental and Planning Law Journal, 21*(2), 93–104.

Hamblin, A. (2009). Policy directions for agricultural land use in Australia and other post-industrial economies. *Land Use Policy, 26*, 1195–1198.

Head, B. (2010). Water policy: Evidence, learning and the governance of uncertainty. *Policy and Society, 29*, 171.

Holley, C., & Sinclair, D. (2012). Compliance and enforcement of water licences in NSW. *Australian Journal of Natural Resources Law and Policy, 15*, 149–189.

Holley, C., & Sinclair, D. (2013). Non-urban water metering policy: Water users' views on metering and metering upgrades in NSW. *Australian Journal of Natural Resources Law and Policy, 16*(2), 101.

Holley, C., & Sinclair, D. (2015). Metering, measurement and water policy reform. *Australian Environment Review, 30*(6–7), 131–135.

Holley, C., & Sinclair, D. (2016a). Governing water markets: Achievements, limitations and the need for regulatory reform. *Environmental and Planning Law Journal, 33*(4), 275–283.

Holley, C., & Sinclair, D. (2016b). Regulation, technology and water: 'Buy-in' as a precondition for effective real-time advanced monitoring, compliance and enforcement. *George Washington Journal of Environment and Energy, 7*(1), 52–66.

Hope, P., Timbal, B., & Fawcett, R. (2009). Associations between rainfall variability in the southwest and southeast of Australia and their evolution through time. *International Journal of Climatology, 30*, 1360–1371.

Hughes, N., Gupta, M., & Rathakumar, K. (2016). *Lessons from the water market: The southern Murray-Darling Basin water allocation market 2000–01 to 2015–16 (Research Report 16)*. Australian Bureau of Agricultural and Resource Economics and Sciences, Department of Agriculture and Water Resources.

Lavau, S. (2013). Going with the flow: Sustainable water management as ontological cleaving. *Environment and Planning D: Society and Space, 31*, 416–420. 426.

Loch, A., Perez-Blanco, C. D., Rey, D., O'Donnell, E., & Adamson, D. (2018). Regulatory and economic instruments: A useful partnership to achieve collective objectives? In C. Holley & D. Sinclair (Eds.), *Reforming water law and governance: From stagnation to innovation in Australia* (pp. 123–139). Singapore, Singapore: Springer.

Markell, D., & Glicksman, R. (2016). Next generation compliance. *Natural Resources and Environment* (Winter), 1–17.

Matthews, K. (2017). *Independent investigations into NSW water management and compliance*. NSW Government.

MDBA (Murray Darling Basin Authority). (2017). Water Markets in the Murray Darling Basin. In *Murray Darling Basing Authority*.

Murray Darling Basin Authority (MDBA). (2014). *Murray-Darling basin authority annual report 2013–2014* (MDBA No. 34/14), Australian Government.

Murray Darling Basin Authority (MDBA). (2015). *Water markets in the Murray Darling Basin*. Australian Government.

Murray Darling Basin Authority (MDBA). (2017). *The MDB water compliance review*. Australian Government.

Murray Darling Basin Commission (MDBC). (2007). *A brief history of the Murray-Darling basin agreement*. Canberra, Australia: Murray Darling Basin Commission.

National Water Commission (NWC). (2011a). *Strengthening Australia's water markets*. Canberra, Australia: National Water Commission.

National Water Commission (NWC). (2011b). *Water markets in Australia: A short history*. Canberra, Australia: National Water Commission.

National Water Commission (NWC). (2012). *Impacts of water trading in the southern Murray-Darling basin between 2006–07 and 2010–11*. Canberra, Australia: National Water Commission.

National Water Commission (NWC). (2013). *Australian water markets: Trends and drivers 2007–08 to 2011–12*. Canberra, Australia: National Water Commission.

National Water Commission (NWC). (2014). *Australia's water blueprint: National reform assessment 2014*. Australian Government.

National Water Initiative (NWI). (2004). *Intergovernmental agreement on a national water initiative*. Canberra, Australia.

Nelson, R. (2018). Regulating cumulative impacts in groundwater systems: Global lessons from the Australian experience. In C. Holley & D. Sinclair (Eds.), *Reforming water law and*

governance: From stagnation to innovation in Australia (pp. 237–256). Singapore, Singapore: Springer.
NSW Office of Water. (2015). *NSW water take measurement strategy – Water take measurement in NSW: A way forward' discussion paper*. NSW Department of Primary Industries.
NSW Ombudsman. (2017). *Investigation into water compliance and enforcement 2007–17: A special report to Parliament under section 31 of the Ombudsman Act 1974*.
Our Water Our Future: The Next Stage of the Government's Water Plan. (2007). *Department of sustainability and environment*. Melbourne, Australia: Victorian Government.
Pérez-Blanco., C.D. Gomez, C.M., Loch, A., Adamson, D. (2016). Why do economic instruments fail? The role of water trading and pricing at a River Basin Scale. In *AGU Fall Meeting Abstracts*.
Productivity Commission. (2010). *Market mechanisms for recovering water in the Murray-Darling basin*. Final Report, Canberra.
Productivity Commission (PC). (2017). *National water reform draft report*. Australian Government.
Raft, S., & Hillis, G. (2010). *NSW sustaining the basin program: NSW metering project business case*. NSW Office of Water, vi.
Shearing, C. (1993). A constitutive conception of regulation. In P. Grabosky & J. Braithwaite (Eds.), *Business regulation and Australia's future* (pp. 67–80). Canberra, Australia: Australian Institute of Criminology.
Snyder, E., Watkins, T., Solomon, P., Thoma, E., Williams, R., Hagler, G., et al. (2013). The changing paradigm of air pollution monitoring. *Environmental Science & Technology, 47*(20), 11369–11377.
The NCCARF Water Governance Research Initiative. (Undated). *Strengthening Water Governance in Australia* (Water Policy Briefing No 1). NCCARF.
Timbal, B., & Hope, P. (2008). Observed early winter mean sea level pressure changes over southern Australia: A comparison of existing datasets. *CAWCR Research Letters, 1*, 1–7.
Wentworth Group (WG). (2017). *Review of water reform in the MDB*. Wentworth Group.
Wheeler, S., Loch, A., Zuo, A., & Bjornlund, H. (2014). Reviewing the adoption and impact of water markets in the Murray-Darling basin, Australia. *Journal of Hydrology, 518*, 28–41.
Young, M., Hatton McDonald, D., Stringer, R., & Bjornlund, H. (2000). *Interstate water trading: A two year review*. Canberra, Australia: CSIRO Publishing.

Julia De Luca is a Senior Policy Officer in Licensing, Groundwater and Unregulated Systems, Water Entitlements and Markets Brach, Water Resources, Department of Environment Land Water and Planning. Graduating from Monash University with a BSC (hons) in Environmental Hydrogeology, Julia worked in the private sector, urban stormwater and recycling programs and managed review of legislation; State Environment Protection Policy (Waters of Victoria and Groundwaters of Victoria) for the Environment Protection Authority. Julia Chairs the IPAA Victoria's Sustainability Community of Practice and has broad experience in Water Resources in monitoring and reporting, resource assessment and planning teams. Julia managed delivery of new ministerial guidelines for protection of high value groundwater dependent ecosystems, departmental resource share guidance and represented the department to review a management plan in Murrayville, Victoria. Julia oversaw innovative research on groundwater trade, that quantified the number of trading zones and markets for the first time in Victoria's history of water management. Julia went on secondment to Victoria's Energy Emergency Management team, in the first summer since Hazelwood Power Station closed, to develop and deliver summer preparedness plans with stakeholder engagement, before the electricity grid transitioned to new renewable supplies. Since returning to water policy, Julia is currently managing several projects for unregulated systems, resource planning policy, reasonable domestic and stock use and overseeing a new online platform for drillers to send Bore Construction Reports.

Dr. Darren Sinclair is an Associate Professor at the Centre for Change Governance, University of Canberra. He has a BSc (Hons I) from the University of Sydney, and a Masters and PhD from the Australian National University. He represented Australia at the United Nations International Framework Convention on Climate Change (Geneva 1993 and 1994). He co-designed the self-regulatory and trading scheme to phase out the use of ozone-depleting gases under Australia's commitment to the Montreal Protocol. He currently holds a DECRA Fellowship and an ARC Discovery Grant (DP190101584) that supported this chapter. Prior to this, he was a Research Fellow at RegNet, the Australian Centre for Environmental Law (Faculty of Law) and the Fenner School of Environment and Society, all at the Australian National University. He has published widely on environmental governance, regulation and policy, and mining occupational health and safety regulation and policy. He has also worked as a consultant to government and industry.

Chapter 22
Groundwater Regulation, Compliance and Enforcement: Insights on Regulators, Regulated Actors and Frameworks in New South Wales, Australia

Cameron Holley, Tariro Mutongwizo, Susan Pucci, Juan Castilla-Rho, and Darren Sinclair

Abstract Compliance and enforcement is a major issue for groundwater management. Yet it remains untheorised and underexamined. This chapter drills down into Australian compliance and enforcement efforts, which have been on a significant reform journey over the last two decades, oscillating between being an under resourced, low priority water reform task, to taking primacy within national and state water reform frameworks. The chapter begins by developing an analytical framework for studying groundwater compliance and enforcement. Using a case study of the state of New South Wales, the chapter examines the experiences of a government regulator and the compliance and enforcement experiences of water users. It concludes with a summary of challenges and policy implications for groundwater compliance and enforcement regimes.

C. Holley (✉)
Faculty of Law, University of New South Wales Sydney, Sydney, NSW, Australia
e-mail: c.holley@unsw.edu.au

T. Mutongwizo
University of New South Wales, Sydney, NSW, Australia
e-mail: t.mutongwizo@unsw.edu.au

S. Pucci
NSW Department of Planning, Industry and Environment – Water, Sydney, NSW, Australia
e-mail: susan.pucci@dpi.nsw.gov.au

J. Castilla-Rho
UTS (University of Technology Sydney), Sydney, NSW, Australia
e-mail: Juan.Castilla@uts.edu.au

D. Sinclair
Institute for Governance and Policy Analysis (IGPA), University of Canberra, Canberra, Australia
e-mail: Darren.Sinclair@canberra.edu.au

J.-D. Rinaudo et al. (eds.), *Sustainable Groundwater Management*,
Global Issues in Water Policy 24, https://doi.org/10.1007/978-3-030-32766-8_22

Keywords Water compliance · Water enforcement · Water crime · Water regulation

22.1 Introduction

Australia and many other nations have implemented a range of groundwater policies, including setting sustainable abstraction limits, allocating water rights and reducing entitlements. Despite these substantial developments managing groundwater, academics and policymakers increasingly agree on the need to improve approaches to groundwater (and surface water) regulation, compliance by users and enforcement of breaches – together known in the literature as 'compliance and enforcement' (Brown, 2017; PC, 2017; Rinaudo, Montginoul, Varanda, & Bento, 2012; WC, 2017).

Although groundwater and surface water both have confronted (and continue to confront) compliance challenges, groundwater presents a particularly diabolical compliance and enforcement problem. For instance, compared to surface water, groundwater's subterranean location makes it far less visible. This concealed nature means that individual and collective extraction impacts are often not immediately evident (compared to more visible and easily monitorable drops in rivers and other surface waters). This can mean illegal extraction of groundwater becomes comparatively more clandestine than in surface water contexts, and thus limits the effectiveness of peer or other forms of monitoring and persuasion that can foster social norms of compliance (Castilla-Rho, Rojas, Mariethoz, Andersen, & Holley, 2017). Augmenting this problem is the historic under-investment in monitoring of groundwater, which in turn increases uncertainty over the extent and condition of aquifers and creates significant challenges in measuring and modelling consumption levels and the impacts of extraction (Allan, 2010; Jakeman, 2010). More generally, ensuring groundwater users comply with extraction limits and other rules is a truly complex challenge (Rittel & Webber, 1973). This is because groundwater use often involves multiple points of extraction (e.g. numerous bores) and sometimes multiple uses for a single bores (e.g. irrigation and stock and domestic, which can have different regulatory obligations). Groundwater also tends to be used sporadically (representing a small proportion of overall water use, but relied on heavily during dry periods) (MDBA, 2017; WEF, 2016, p. 33; CC, 2015). Bores are also dispersed over large geographic areas (often being the only form of water available in some remote/rural areas) and can have impacts on numerous locally variable and dynamic systems (often with long time lags between extraction and response) (Head, 2010; MDBA, 2017; Rubenstein, Wallis, Ison, & Godden, 2009; WEF, 2016, p. 33).

Despite these unique challenges, groundwater compliance and enforcement in Australia has largely been subject to the same compliance and enforcement policies and regulatory regimes as surface water.[1] This global approach to regulating water sources means it is often difficult to discretely separate the compliance and enforcement trends and experiences of groundwater from surface water. Even so, it is clear that compliance and enforcement remains a major issue for both types of water in Australia and internationally (INTERPOL, 2016; Matthews, 2017; MDBA, 2017; NWC, 2014; PC, 2017; WC, 2017; WG, 2017). Indeed, no matter how many novel groundwater governance tools are designed, they will all be "insufficient if enforcement is absent or inadequate" (Brown, 2017, p. 8). If sustainable use is exceeded due to illegal water extraction, if the various licences, approvals or tradable rights are not observed, and if stakeholders lack confidence there is an even-handed sharing of water resources, then the entire edifice of sustainable groundwater management is destabilised (Holley & Sinclair, 2012).

To date, compliance and enforcement studies in groundwater have been limited to specific contexts and are still nascent (e.g. Blomquist, 1992; Holley & Sinclair, 2012; Ostrom, 1990; Rinaudo et al., 2012). To add to this literature, the chapter drills down into compliance and enforcement efforts in Australia to draw insights from practice and the literature. There are various aspects to compliance in Australian groundwater governance, including states complying with their obligations under the Murray Darling Basin Plan 2012 (Cth), compliance with water trade rules, and even the groundwater impacts of the mining and resources sector (see cl 34 NWI, 2004). This chapter is concerned with highlighting both the successes and challenges of ensuring non-urban water users (e.g. farmers) abide by their individual conditions and extraction limits imposed for using water for irrigation or other purposes. This accounts for the vast majority of non-urban water consumption in Australia (50–60%, see ABS, 2017) and is the core activity devolved to Australia's system of state-based water regulators. Under Australia's national water reforms (COAG, 1994; NWI, 2004), states were required to determine how best to undertake these regulatory activities. The approach has differed between states, some devolving responsibility to regional water authorities (Victoria), while others, as discussed below, approached compliance through centralised regulatory agencies (e.g. New South Wales). Regardless of the approach taken, groundwater compliance and enforcement has been on a significant reform journey over the last two decades, oscillating between being an under resourced, low priority water reform task, to taking primacy within national and state water reform frameworks (DSEWP, 2012).

Given the emerging nature of the groundwater compliance and enforcement literature, Sect. 22.2 commences by examining wider regulatory scholarship on compliance and enforcement to propose an analytical framework for studying groundwater compliance and enforcement. It connects this framework to the aspirations of Australia's national water reform efforts. Section 22.3 then analyses

[1] Note that groundwater and surface water can involve different compliance elements e.g. groundwater may involve compliance with an approval to construct the bore and a bore licence, while surface water may involve compliance with an approval for a pump and an access licence. However, in terms of compliance and enforcement frameworks (e.g. education strategies or prosecution approaches) the two sources tend to be subject to the same regulatory framework.

Australia's experience in pursuing compliance and enforcement aspirations, using a case study of the state of New South Wales (NSW) as its primary focus. While we reflect on other states and trends, NSW was chosen as the primary source of data because it has been a particularly rich site of compliance challenges and reforms, as well as being home to the "lion's share" of water use in the Murray Darling Basin. Moreover, although the chapter's analysis applies to the regulatory responsibilities and practices for groundwater and surface water use, we place particular emphasis on the regulation of groundwater (Holley & Sinclair, 2012, p 153). Our analysis of compliance and enforcement in NSW proceeds in two parts, examining the experiences of the government regulator and then examining the compliance and enforcement experiences of water users. Section 22.4 then concludes the chapter.

22.2 Compliance Framework for Understanding and Analysing Groundwater Compliance and Enforcement in Australia

Studies of compliance and enforcement in quantitative groundwater contexts are few and far between (Blomquist, 1992; Holley and Sinclair 2012; Ostrom, 1990). Some studies have begun to explore regulated actors in the related areas of fisheries, agriculture and rural crimes (Anderson & McCusker, 2005; Barclay & Bartel, 2015; Bartel & Barclay, 2011; Fisher, 2011; Honneland, 1999; Jagers, Berlin, & Jentoft, 2012; Martin & Gunningham, 2011) and increasingly the role of technology in compliance activities (Holley & Sinclair, 2013; Paddock & Wentz, 2014; Purdy, 2011). However, much of this research in the water context (and groundwater context in particular) is preliminary, raising more questions than answers.

In contrast, there is a related and more substantial regulatory literature that has examined what are sometimes termed "first wave" environmental challenges, like point source water pollution produced by large factories, that are regulated by standalone environmental protection agencies (Gunningham, 2011; Zaelke, Stilwell, & Young, 2005). This broader compliance and enforcement literature has made significant progress in understanding more general compliance behaviour and motivations of regulated actors (Parker & Nielsen, 2011). Drawing on these and related studies, below we identify four core pillars that can be used to constitute a conceptual framework for analysing compliance and enforcement in groundwater contexts. These pillars arise from the objectivist tradition in compliance research, which suit this chapter's focus on sharing applied and normative policy design and implementation lessons from Australia (Parker & Nielsen, 2011, p. 3). This tradition aims to build and test theories identifying characteristics (motives, capacities, resources) and external factors (the nature of the regulatory policy area, enforcement strategy and style, third party actors and stakeholders) that are associated with compliance, non-compliance and effective enforcement (Parker & Nielsen, 2011 p. 3). The 'pillars' include motivations, characteristics, enforcement strategies and pluralism.

22.2.1 *Motivations*

Motivations are the factors that inspire individuals and businesses to comply with regulation. Consistent with the sizeable research arguing for a plural and interactive account of motives (e.g. Ayres & Braithwaite, 1992; Gunningham, Kagan, & Thornton, 2003), motivations are broadly recognised to include social (a commitment to comply in order to earn the approval and respect of significant people with whom an actor interacts, e.g. perceived social norms, or informal sanctions inflicted by local communities and others, see Gunningham et al., 2003; Tyran & Feld, 2006); economic (a commitment to comply in order to maximise one's economic or material utility, e.g. the threat of fines, Braithwaite, 2009); and normative factors (a commitment to obey the regulation for its own sake – e.g. a belief in the legitimacy of regulation, Tyler, 2006) (Parker & Nielsen, 2011, p. 10). Australia's National Framework for Compliance and Enforcement Systems (the National Framework) (discussed further below) implicitly recognised at least some of these plural motives by targeting most compliance action at encouraging and assisting groundwater users to voluntarily comply with the rules (DSEWP, 2012, p. 1). Identifying and understanding specific motivations of groundwater users, and considering their interaction and consequences for enforcement is an important aspect that we explore in NSW below.

22.2.2 *Characteristics*

Recognising that motivations to comply can be of secondary importance to the capacity and ability of waters users to achieve groundwater use reductions, this pillar focuses on the characteristics of water users that may explain compliance behaviours. These are provisionally conceived as including economic resources, knowledge, cognitive capabilities (e.g. education), technologies (e.g. monitoring) and management capacity (Borck & Coglianese, 2011; Parker & Nielsen, 2011 pp. 14–15). Australia's National Framework recognises the importance of these characteristics to compliance, including addressing issues such as the use of meters (see also National Framework on Non-urban Water Metering 2010) and the provision of "information to educate the public and the stakeholders on the importance of compliance and enforcement" (DSEWP, 2012, p. 7). Below we begin to paint a picture of groundwater users in NSW and their water management practices so as to shed light on when and how such characteristics explain compliance behaviour.

22.2.3 Regulation and Enforcement Strategies

This pillar includes laws and other interventions by agencies and recognises that plural regulatory strategies are often necessary for regulators to respond effectively to plural compliance motivations and characteristics (Parker & Nielsen, 2011, pp. 17–18). The two main challenges for regulatory agencies are: (a) *where* to intervene (allocating resources and targeting entities for inspection); and (b) *how* to intervene in the affairs of regulated enterprises to foster compliance (Black, 2010; Gunningham, 2011; Holley & Sinclair, 2015; May & Winter, 2000). There is now something approaching consensus that in terms of (a), the best way to allocate scarce regulatory resources is risk-based regulation (prioritising regulatory activities and deploying resources based on an assessment of the risks that groundwater users pose to a regulator's objectives e.g. sites of intense groundwater drawdown, or areas with significant groundwater dependent ecosystems) (Black, 2010 p. 187). This is certainty reflected in Australia, where the National Framework (and similar state-based compliance policies, NSW DPI, 2015) adopted a risk focused approach. As the National Framework notes: "This Framework is risk based, with increased compliance and enforcement with increased risk…A risk-based compliance strategy is one that identifies 'at risk' water resources and targets breaches of water resources legislation most likely to further stress the resource or which undermine the public's confidence in effective water resource management" (DSEWP, 2012, p. 5.)

In terms of (b), questions remain as to how a risk-based framework should best be applied with multiple theories competing in this domain (Black & Baldwin, 2010; Gunningham, 2011). This is especially the case with groundwater regulation where the challenges are numerous and complex, for example, large numbers of small farms, limited resources, geographical dispersion and historical factors that emphasise that agriculture is "special" and that private property is paramount (Bricknell, 2010; Holley & Sinclair, 2012; Robertson, 2014). Further, there is little consensus on what type of intervention strategy will best ensure compliance and facilitate enforcement (Gunningham, 2011). Various ideas have been identified, including deterrence (Kagan, Gunningham, & Thornton, 2003), advice/persuasion (Abbot, 2005), criteria based (Yeung, 2004), responsive (Ayres & Braithwaite, 1992), smart (Gunningham, Grabosky, & Sinclair, 1998), facilitative (Holley & Gunningham, 2006), risk-based (Sparrow, 2000) and meta (Parker, 2002) regulation, but none have demonstrated their applicability to all sectors or types of enterprises, particularly in the unique and often untested case of groundwater. For Australia, the preferred approach of states and the National Framework has largely been a responsive one, pursing a compliance pyramid (see e.g., Fig. 22.1) (DSEWP, 2012, p. 1). Inspired by the work of Ayres and Braithwaite (1992), "the pyramid is designed with most compliance action at the base…Further up the pyramid, actions are more concerned with directing compliance…The top, where generally there is the least activity, involves administrative remedies and criminal proceedings. For

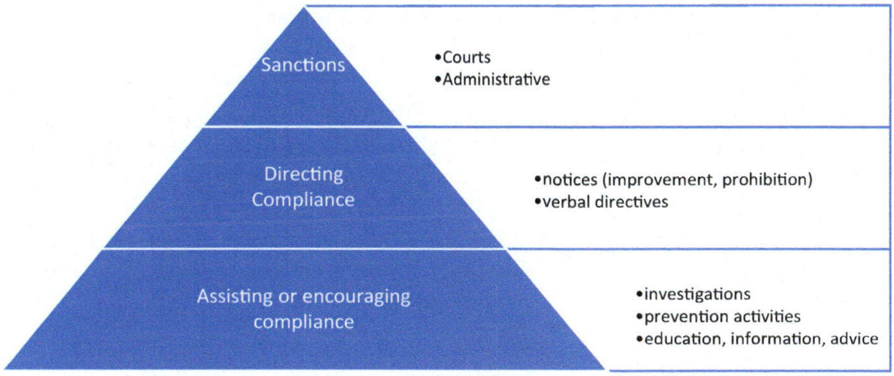

Fig. 22.1 Compliance pyramid. (Based on DSEWP, 2012, p. 1)

the pyramid to work effectively, each of the elements need to be effective and operate efficiently, to allow for the strategy's overall success" (DSEWP, 2012, p. 1).

22.2.4 Regulatory Pluralism

The regulation of groundwater involves a variety of parties (farmers, non-government actors, state, national and supra-national agencies) that are assembled via numerous regulatory approaches (e.g. self-management, co-management, cap and trade markets and other groundwater management arrangements). Achieving an optimal mix of parties and approaches is complex and context specific (Gunningham et al., 1998). Within this mix, studies have shown the importance of harnessing resources (e.g. peers, community groups, expert-driven civil society organisations or third parties) outside the public sector that can generate norms and harness others to influence compliance behaviour (March & Olsen, 1989; Ostrom, 1990). Even so, in Australia the majority of compliance and enforcement work is conducted by state government water regulators (COAG, 1994; NWI, 2004), who have spent the last two and half decades developing legislation, establishing compliance policies and advancing institutional capacity to encourage compliance and deliver enforcement. Certainly, there are practices involving mixed compliance roles. The most prominent of these are irrigation corporations, such as those in southern NSW where private companies hold bulk water entitlements on behalf of shareholders. While irrigation corporations are regulated by the NSW state government regulator, the companies monitor compliance of its water user shareholders, who themselves are regulated by a range of private contracts between the parties (ie. over-extraction by an individual member of an irrigation corporation does not per se trigger offence provisions in the *Water Management Act* 2000 (NSW)). There are also opportunities for governments to harness other parties in the compliance mix (e.g. groundwater bore drillers who may be required to ensure a farmer has a permit for the bore before drilling), however the vital task of ensuring non-urban water users comply with their obligations remains with each state and territory government.

In summary, the above four concepts offer a way to capture the complexity of compliance and enforcement in the regulation of non-urban groundwater use, and enable exploration of connections (and disconnections) between the plurality of motivations, capacities, regulation and enforcement of rules and other normative institutions (like third party actors) pertinent to compliance. This is apposite in the context of groundwater governance, where compliance motivations, capacities and strategies and third-party actors have been under-examined in practice.

22.3 Groundwater Compliance and Enforcement: Experiences from NSW

22.3.1 NSW Water Regulators – The Ongoing Evolution of Compliance and Enforcement Agencies

The centre piece of groundwater compliance and enforcement in New South Wales is the *Water Management Act* 2000 (NSW) (the Act). The Act was introduced in 2000, alongside, and to eventually replace, the *Water Act 1912* (NSW). A key objective was to transform water rights from a system where water licences were tied to land, to the separation of water licences from land to provide greater opportunities to trade water. The Act recognises the importance of protecting and providing water for the environmental health of groundwater systems (e.g. ss 3, 5, 324 *Water Management Act 2000*). It establishes mechanisms to allocate these protections in various ways including rules in statutory water sharing plans (e.g. Chapter 2, Part 3 *Water Management Act 2000*). These plans establish limits on the total extraction of water from water sources and rules for taking water.

Water access licences must be held to take water and they give the holder a share in the resource, which is realised in real terms through determinations made based on forecasts of water availability (e.g. s 56 *Water Management Act 2000)*. Approvals are required for the construction and operation of works used to take water, such as groundwater bores, and conditions imposed on these approvals aim to prevent negative impacts both to the aquifer and to other groundwater users (e.g. s 89–90 *Water Management Act 2000)*.

The Act makes provision for an approval to manage specific activities which interfere with groundwater systems (e.g. s 91F *Water Management Act 2000*). Although not commenced, the introduction of the Aquifer Interference policy (2012) provided greater clarity about water licensing requirements for developments which interfere with aquifers, such as mining and petroleum exploration, and strengthened the assessment of aquifer interference activities to help protect groundwater sources. Notably, a licence is now required under s60I of the *Water Management Act 2000* (NSW) for incidental take (i.e. take that occurs as a consequence of a mine cutting through an aquifer). However, this has resulted in cease-to-pump provisions being switched off in certain water sharing plans that apply to areas with high mining activity (i.e. Hunter Unregulated system), raising questions about equity and ultimately sustainable use (as modelling to estimate incidental take over the life of the mine and subsequent licensing requirements may not be adequate) (see e.g. Carmody, 2013).

While these rules were being developed, NSW, along with other parts of Australia, experienced a significant drought (ranging between the mid 1990s until about 2010). This drought saw reductions in surface water flows and put increasing pressure on groundwater resources. For instance, by July 2009, Burrinjuck dam, which feeds the Murrumbidgee regulated river, was at less than 30% of its storage capacity (MDBA, 2009) and data from monitoring bores in the Lower Murrumbidgee groundwater sources showed declining levels in the aquifer during this period (NSW Government, 2015). Inflows were so low that the Murrumbidgee regulated river water sharing plan was suspended from 2006 to 2011. Although a disputed point (see Grafton et al., 2014), it was argued that the suspension was necessary to enable necessary management interventions for the critically low water levels which had not been anticipated by the statutory plan.

During this period, access to water was a significant concern. Through government operational activities, it became clear the range of compliance and enforcement tools available needed to be strengthened and changes began to be introduced in 2008/2009.

Prior to this time, the regulation of groundwater in NSW initially occurred via shared compliance and enforcement functions amongst various departments. Indeed, as water reforms evolved, the administration of water management and regulation in NSW was restructured amongst different government agencies close to twenty times over the last two decades, including dividing departments, joining units and establishing new agencies (NSWO, 2017, pp. 5–6). This evolution in water agencies reflected in part the maturing nature of water governance in NSW (and Australia), with initial attention devoted to establishing caps and plans, before turning increasing attention to compliance and enforcement. Given compliance concerns arising in 2008/9, the NSW government established the Office of Water as the primary regulator. This would be followed by the Department of Primary Industries, Water in 2014, and most recently the 2018 establishment of the new Natural Resources Access Regulator (discussed below).

To assist the Office of Water's compliance activities, amendments to the Act commenced in 2009 to provide a range of new investigatory and enforcement powers and tiered offences analogous to the *NSW Protection of the Environment Operations Act* 1997 *(NSW),* which regulates point source pollution in NSW. The most relevant Tiers to this chapter are Tiers 1 and 2. Tier 1 is the most onerous of these, and is designed to deal with intentional, negligent and reckless acts by an individual or corporation (e.g. s60A(1) *Water Management Act 2000).* Tier 2 offences are designed to apply to remaining breaches of the law (such as, lacking appropriate approval) (e.g. s60A(2) *Water Management Act 2000).* This tiered system of offences has a range of penalties (that until recent increases in 2018), ranged up to $1.1 million and jail terms of up to 2 years for individuals, and up to $2.2 million for corporations for Tier 1 offences.

Additional offences were also introduced to reflect the range of possible unauthorised activities including taking water when not authorised by a licence (s60A *Water Management Act 2000*), constructing or using a work (such as a bore) when not authorised by an approval (s91B *Water Management Act 2000)*, and meter tampering (s91K *Water Management Act 2000)*.

A range of powers were included to enable authorised officers to investigate offences, including notices to require information (s338A *Water Management Act 2000*), and powers to enter land to conduct inspections (s339 *Water Management Act 2000*). Statutory powers provided for directions to be issued, for example, requiring a person to take measures to mitigate adverse effects on groundwater caused by the construction or use of a bore (Chapter 7, Part 1 *Water Management Act 2000)*. Enforcement powers included the ability to issue penalty notices, civil penalties for illegal water extraction, and prosecution (Chapter 7, Part 5, *Water Management Act 2000*). This suite of compliance and enforcement powers would come to apply to approximately 10,000 groundwater licences and about 90,000 work approvals for bores (data extracted from Water NSW, n.d.). Of these, approximately 87,000 are for bores used to take water for domestic and stock purposes (data extracted from Water NSW, n.d.).

Although significant institutional and policy advances were made with the introduction of these changes and the NSW Office of Water, Holley and Sinclair (2012) have argued elsewhere that the NSW Office of Water and, more particularly, its predecessors had struggled, historically at least, to deliver a comprehensive compliance and enforcement regime for groundwater (and surface water) across NSW. A range of reasons for this were identified (see Holley & Sinclair, 2012; NSWO, 2017 p. 6), including agency restructures (e.g. loss of staff expertise, corporate knowledge and continuity in strategy), historic cultural and institutional constraints (e.g. an early history of permissiveness and a reluctance to prosecute), limited sophistication in data management and analysis capabilities (e.g. limited records on the precise location of bores), challenges of internal and external coordination between different branches and departments relevant to compliance (e.g. water providers and regulators), as well as old, unreliable or non-existent metering technology. Augmenting these challenges at the time was perhaps the biggest and most consistent constraint on compliance and enforcement, namely limited resources. For example, for the first few years of its existence, the Office of Water had approximately 12 inspectors to enforce water allocations across all of NSW. Given that there are thousands of license holders, spread over such a vast geographical area (800,642 km^2), this posed a serious impediment to comprehensive enforcement (Holley & Sinclair, 2012). Of course, inspectors are not the only way in which potential breaches are identified — indeed, the public reportedly accounts for at least a third of all breach reports (see Holley & Sinclair, 2012). Even so, officer time (both desktop and inspection) is often needed to investigate and confirm publicly reported breaches, further stretching the capacity of regulatory staff. With limited inspectors on the ground, the scope of the regulator's ability to engage in the full range of enforcement activities, such as those envisaged in its Compliance Policy, was also constrained, including face-to-face communication and education (beyond a detailed website and posted bill flyers) and robust 'proactive' enforcement activities (e.g. targeted compliance audits). More broadly, ongoing resource limitations, as well as fluctuations in resources for prosecutions, led to illegal bores and the potential overuse of stock and domestic water, historically receiving minimal inspectoral attention (Holley & Sinclair, 2012; see also Matthews, 2017; NSWO, 2017).

The NSW Office of Water was not the only agency to confront such challenges, a fact recognised by Australia's Federal government and states and territories, who would develop and seek to implement a National Framework on Non-Urban Water Metering (2010) and the 5 year National Framework. The former (along with other programs) aimed to improve water metering technology, while the latter invested sixty million dollars in achieving, among other things, an appropriate and nationally consistent range of water offences, penalties and evidentiary requirements (DSEWP, 2012, p. 2).

An evaluation in 2012 (see e.g. NWC, 2014, pp. 352–354) found that NSW water legislation met all the Framework's minimum requirements which included having evidentiary presumptions, the concept of no-fault liability and evidence gathering tools. At the time NSW also had the highest penalties for water offences compared to all Australian states and territories and that is still the case in 2018. During the period of the National Framework and its additional resources, NSW compliance staff numbers increased, with the Office of Water and its later variant, DPI Water employing 8 monitoring officers, 11 strategic investigators and 50 water regulation officers (where compliance was about 20% of their role).

With the help of these compliance roles, over 17,000 audits were conducted between 2012 and 2015 on groundwater works, primarily authorised for domestic and stock purposes, in areas assessed as high risk (DPI, 2016). About 250 alleged breaches were detected, suggesting the incidence of breaches was low (DPI, 2016). However, although these audits did not detect significant numbers of alleged breaches a number of factors may have influenced these findings, including the time of year the audit was conducted, the prevailing climatic conditions, and officers' ability to determine compliance with conditions at the time of the audit. For example, at the time of the audit the work may not have been in 'use' so, it may not have been possible to determine some conditions, such as the purpose for which water was being used. Other operations based on alleged breaches reported by the public and other intelligence, have found breaches are occurring in groundwater systems, including over-extraction, unauthorised bores and people drilling bores without a driller's licence.

This work demonstrated the importance of having a range of methods to detect non-compliance, including the use of remote sensing methods which can be used across the significant areas that need to be regulated. Indeed, working to tackle the challenge of detecting non-compliance, research and trials into using satellite imagery were undertaken as part of the Framework. These trials evaluated the practicality of identifying irrigated areas and matching this with information on water entitlements and allocations. Initial evaluation showed successful results and further work is continuing.

Water bore drillers were also recognised as a significant actor in groundwater management and in NSW a water bore driller's licence is required to drill a bore (see e.g. s346 *Water Management Act 2000*). Partnering with a well-informed community of water bore drillers has significant potential benefits in both raising standards of groundwater protection and preventing and deterring non-compliance. In 2014, DPI Water consulted on a proposed new framework for the licensing of groundwater drillers. The new framework has been developed to strengthen protection of groundwater sources through ensuring drillers are appropriately trained in bore drilling techniques. To assist drillers in understanding their obligations and explaining rules to landholders, drillers will be required to pass a legislation knowledge test when they apply for a licence and at licence renewal. Additional enforcement powers will assist the regulator, and a register of licensed drillers will make it easier for landholders to find a driller with the appropriate class of licence.

Notwithstanding these positive developments in compliance activities, by the end of 2016 the National Framework (and its funding) was beginning to draw to a close. This was arguably representative of a broader national decline in water reform in Australia, including the dismantling of the National Water Commission in 2014, and water reform falling rapidly down the list of national priorities (Holley & Sinclair, 2018; Williams, 2017). Perhaps gripped by the hydro-illogical cycle (Wilhite, 2012), the absence of a Millennium scale drought saw national law and policy leadership wane (Wentworth Group of Concerned Scientists 2014, p. 0). Within this context, another shift occurred in NSW, with NSW Government altering some of the compliance staffing and DPI Water's functions by transferring them to a State-owned water corporation, Water NSW (NSWO, 2017 p. 6; Blair, 2017). During this transition, a number of allegations were made to the NSWO about DPI Water's performance of statutory functions, and the adequacy of enforcement actions (see NSWO, 2017 p. 7). This was closely followed by allegations of non-compliance aired on a national television current affairs program, particularly regarding surface water extraction in the Barwon-Darling River system in Northern NSW (see ABC, 2017; NSWO, 2017 p. 7).

These nationally broadcast claims sparked tangible concern over NSW and other states' water compliance and enforcement (PC, 2017). A public inquiry followed, which reported 'water related compliance and enforcement arrangements in New South Wales have been ineffectual and require significant and urgent improvement' (Matthews, 2017, p. 7). 'Considerable frustration' was also levelled at the intersecting role of the Murray Darling Basin (MDB) Authority and its response to alleged serious breaches (MDBA, 2017, p. 14). A subsequent MDB Authority inquiry found differences in Basin state compliance vigilance, with South Australia reportedly having a 'long commitment to a compliance culture', including extensively codified rules and comparatively higher resourcing, while Queensland, NSW and to a lesser extent Victoria evidenced a 'notable lack of transparency' and remain 'bedevilled by patchy metering … the lack of real-time, accurate water accounts …[and] a low level of compliance resourcing' (MDBA, 2017, p. 14; WG, 2017).

Following this recent series of inquiries, the NSW Government announced the creation of a new and independent Natural Resources Access Regulator (NRAR) to strengthen water compliance and enforcement in NSW. The NRAR independent board was established in January 2018 and the new agency is currently being created. One of the first actions of the NRAR has been to set up a governance framework, with a Regulatory Policy published setting out the Regulator's risk-based and outcomes-focused approach to compliance and enforcement. This policy will be supported by a range of strategic and operational documents to guide the Regulator in managing risks to surface and groundwater sources. Drawing on the last 10 years of experience, work has commenced on identifying the activities, entities and water sources that pose the highest risks in relation to non-compliance.

Ultimately, the continuing compliance and enforcement challenges and reforms evidenced in NSW should perhaps have been expected. Environmental regulation has taken decades to evolve and advance (Gunningham & Holley, 2016). But the luxury of time has not been afforded to water regulators. Urgent government interventions were arguably needed to offset a water crisis in Australia and those tasked with compliance and enforcement have had little time or resources (and arguably independence) to develop at their own pace (Bartel & Barclay, 2011, p. 168). This suggests that future groundwater policy reforms need to take account of the nascent culture of water regulatory institutions, and the limitations it can impose on compliance and enforcement (Holley & Sinclair, 2012; Torres, 1989, pp. 193–4). Without this, adequate resourcing and strategies to nurture desired change (e.g. training and policy guidance like that emerging in the case of NRAR), there is the risk that, as occurred for many of the early years in NSW, compliance and enforcement will fall well short of policy expectations.

22.3.2 NSW Water Users – Motivations, Characteristics and Third Parties

Throughout this period of experimentation and transition in water regulatory agencies, it was recognised that building knowledge of, and partnerships with, groundwater users were important to advancing effective programs to support them and achieve compliance (DSEWP, 2012, p. 7). This section draws on a survey (Holley & Sinclair, 2015, 2016, 2017) conducted in partnership with the NSW water regulator that aimed to enhance their understanding of water users' knowledge, behaviour, motivations and experiences with water compliance and enforcement. The survey was sent to approximately 4000 water users (22% response rate) across three NSW regions selected purposively to represent a diversity of water sources, locations, authorisations and risk levels between September 2012 and January 2013. The survey contained over 100 questions (tested and consulted with the NSW regulator, NSW Irrigators' Council and other industry associations) and asked respondents to score their level of agreement with a range of statements (typically using a 5-point scale e.g. 'Strongly Disagree' to 'Strongly Agree').

22.3.2.1 Motivations for Compliance and Non-compliance

Regarding drivers for non-compliance, the most commonly agreed justification for illegal water extraction was a desire for economic advantage (n = 596; 49%). There were also various motivations driving compliance. The highest percentage agreement included complying with water laws because it was the right thing to do (n = 617; 95%), fairness among other water users (n = 582; 93%) and reputation (reflects badly on all water users (n = 585: 84%) and peer reputation (n = 570; 81%). It was evident that users may also have complied because of the perceived legitimacy of laws, with respondents agreeing that water regulation was needed to sustainably manage water resources (n = 656; 90%), as well as protect the rights of water users (n = 657; 89%), the long-term viability of communities (n = 657; 86%), and the environment (n = 656; 72%). Notably, the threat of legal punishment (which is often identified as a major driver of compliance) (Kagan, 2004) had comparatively fewer (although still substantial) numbers of respondents agree that criminal records (n = 607; 62%) and penalties (n = 601; 35%) are a deterrent for illegal water extraction. This suggests that while traditional and sufficient legal punishment is an important motivator for compliance, respondents perceived softer sanctions (e.g. social and peer reputation) and morals (fairness) as more effective in driving compliance in NSW than administrative or criminal sanctions.

However, it is important to note that the impact of penalties and criminal records as a motivator of compliance depend on water users believing people illegally taking water will be caught and prosecuted. Just over one-third of respondents agreed they would be caught or prosecuted (n = 611; 33% and (n = 609) 35% respectively). Further, only 26% (n = 533) of respondents agreed that compliance officers (who enforce the laws and hand out penalties) regularly worked in their region. Given the survey was primarily conducted prior to securing the full suite of NSW compliance and enforcement staff under the National Framework, these findings are perhaps unsurprising. This finding was mirrored in a broader Murray Darling Basin survey in 2017 (following the ending of the National Framework), which reportedly found that 70% of respondents believed current compliance and enforcement was not a deterrent (MDBA, 2017, p. 29).

22.3.2.2 Characteristics of Water Users and Compliance

As noted above, compliance is influenced by various factors not least water users' understanding and knowledge of laws and regulation. The findings suggest that respondents were far more familiar and confident with topics of immediate interest to their on-property operations than compliance and enforcement generally. For instance, survey respondents indicated they had good or very good knowledge of options for trading water (n = 527; 28%), bore requirements (n = 603; 38%), the conditions of their licence or approval (n = 602; 47%), the requirements for water metering (n = 565; 47%) and stock and domestic requirements (n = 623; 50%).

In contrast, the survey respondents' knowledge of legislation, policy, compliance and enforcement was very low. In particular, few respondents indicated they had good or very good knowledge of key legislation (e.g. the *Water Management Act 2000* (n = 619; 8%), different enforcement actions under the *Water Management Act 2000* (n = 599; 7%), the regulator's compliance policy (n = 623; 7%), activities/works regulated by NSW legislation (n = 623; 9%), penalties for illegal water activities (n = 608; 9%), and national standards for water meters (n = 513; 12%). There was also a relatively high number of water users who reported water laws and regulations were too complex (n = 589; 47%) while a smaller number reported it was challenging to understand their licence or approval conditions (n = 544; 27%). Those who agreed laws were too complex or found it difficult to understand their licence conditions were more likely to state that tough economic conditions, high water costs, drought and flooding and a lack of awareness of the rules justify the illegal taking of water. They were also less likely to identify with various aims and content of the laws (and therefore may be more inclined to break them). As a result, this suggests that those who see water laws as difficult may be more willing to not comply (Holley & Sinclair, 2015).

The relationships between knowledge and views on penalties and criminal records as motivators of compliance was also worth noting. Those with good knowledge of the regulator's compliance policy, penalties for illegal water extraction, the *Water Management Act 2000*, *Water Act 1912* or different enforcement actions under the *Water Management Act 2000* were more likely to agree that penalties and a criminal record for illegal water extraction worked as a deterrent. The effectiveness of penalties and criminal record as a driver of compliance appeared to improve where there was good knowledge of a range of water laws and policies (Holley & Sinclair, 2015).

Of equal importance to knowledge is access and use of water monitoring technology. However, water users who were unfamiliar with or unsupportive of the monitoring technology and its maintenance posed a potential risk to the accurate monitoring of water extraction. According to survey respondents, 90% (n = 500) agreed on the importance of ensuring that meters are well maintained. The majority of respondents also responded positively towards the value and benefits of metering, with two-thirds of respondents agreeing that accurate measurement of water extraction by metering was necessary to sustainably manage water resources (n = 608; 66%). However, this support was greatest in regions with more meter experience (e.g. Murray Murrumbidgee and Central West (76–80% supportive (n = 394)), which had a history of irrigation schemes and metered river takes, whereas in the North Coast (45% supportive (n = 214)) many water users were far less likely to have metered takes).

22.3.2.3 The Role of Non-government Parties

Education and information about rules and penalties can help promote compliance with legislation and reduce the need for enforcement, but it was a sub-group of personal non-government sources that survey respondents ranked most highly as sources of information (rather than government). In particular, neighbours and family (n = 615; 63%) and bore drillers (n = 616; 56%) were the highest ranked as information sources. Neighbours or family (n = 616; 63%) and bore drillers (n = 616; 56%) were also ranked first and third respectively in terms of information usefulness. Industry peers were also ranked highly (fourth) as a source (n = 615; 55%) and in terms of usefulness (n = 615; 55%), closely followed by industry associations (sources – 49% (n = 616) and usefulness – 50% (n = 616)).

The importance of non-government actors was also evident in related research presented by Castilla-Rho et al. (2017), who developed an agent-based simulation (Epstein & Axtell, 1996) by behavioural metrics extracted from the most recent version of the World Values Survey (http://www.worldvaluessurvey.org). In their simulated world, water agencies place restrictions (in the form of caps, allocations or quotas) on groundwater pumping rates, but it is the social norms that emerge within each particular cultural setting that determine whether and to what extent farmers comply with those restrictions. The simulation builds on Axelrod's evolutionary model of social norms (Axelrod, 1986), but makes agent interactions spatially explicit by situating farmer agents within a landscape of farm properties. Illegal extractions generate groundwater depletion hotspots that increase operational costs to neighbouring agents (i.e., more energy is needed to pump groundwater at greater depths). As the tragedy of the commons predicts, if farmers only cared about their own revenue, unsurprisingly, the farming collective would end up overharvesting the groundwater resource. But in this case agents also consider cultural values in their decision-making, for which Castilla-Rho et al. rely on Grid-Group Cultural Theory (Douglas, 1966). The grid dimension ranges from the pursuit of independence and individuality (low grid), to the willingness to enforce social norms even if this comes at a cost to the individual (high grid). The group dimension ranges from a self-focused (low group) to a group-focused (high group) attitude. The cultural context of tolerance for illegal extractions (grid) and the importance of maintaining a good reputation (group) sway the agents' decisions and the social norms that emerge. To evaluate past decisions, agents not only rely on the economic outcomes of their farm enterprise, but they also consider the value of complying with the rules in their specific cultural context. To evaluate future decisions, agents apply the 'imitate the most successful' heuristic to make quick decisions under limitations of time and cognitive processing by exploiting the way information is structured in their environment (Goldstein & Gigerenzer, 2002). These fast and frugal heuristics enforce the social learning and ecological rationality view of decision making (Goldstein & Gigerenzer, 2002).

The simulations of Castilla-Rho et al. were statistically tested against data from the above empirical surveys in the Murray–Darling Basin. This stylized, yet empirically grounded Groundwater Commons Game model, suggested, inter alia, that

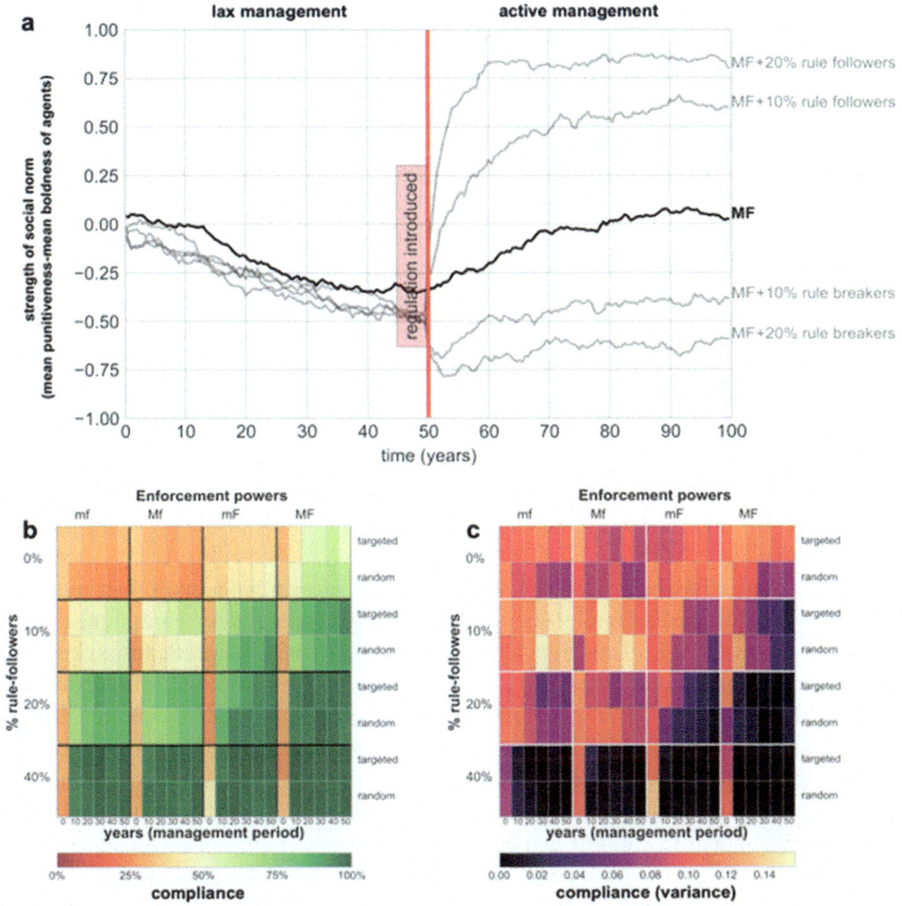

Fig. 22.2 Management landscapes derived from the Groundwater Commons Game (Castilla-Rho et al., 2017), a computer simulation capturing key social, economic, and institutional drivers of compliance. These management landscapes can help decision-makers better understand the full range of outcomes from groundwater conservation efforts, along with the social and management forces that are most likely to influence these outcomes. Water authorities may set specific objectives (e.g., >50% compliance within 10 years of management, blue line in panel b) and evaluate whether management action or inaction will meet those objectives (i.e., scenarios below the blue line) while considering the uncertainty of the outcomes (panel c). Panels b and c summarise the compliance response considering different proportions of rule followers and levels of regulatory enforcement (m = 10% users audited, M = 50% audited, f = low fines, F = high fines). Under targeted monitoring, the water authority allocates ground staff using a risk-based approach (audit highest and most frequent breaches). Under random monitoring, ground staff conduct audits randomly. Note how random monitoring triggers higher levels of compliance (for discussion, see Castilla-Rho, Rojas, Andersen, Holley, & Mariethoz, 2019)

increasing monitoring and fines would have very limited impact in the Australian context (although note the above challenges and limitations e.g. resourcing), and the most effective measure (across NSW as well as other cases studied by Castilla-Rho et al., 2017) was increasing the number of non-government leaders and rule-followers that attempt to influence others by their 'good example'. The impacts of individual groups of farmers either promoting or obstructing the consolidation of a social norm of compliance on achieving and maintaining sustainable pathways was investigated. Figure 22.2a shows the strength of social norms—defined as the mean punitiveness minus the mean boldness of the agent population—under a 50-year scenario of active management (monitoring of 50% of water user and imposition of heavy fines on breaches) and incremental efforts to deter rule breakers (20% to 0%) and engage rule-followers (0% to 20). Figure 22.2b–c summarise the levels of compliance (and its variability) that result at the end of the active management period under different management strategies (type and level of monitoring, level of fines). Importantly, results showed that beyond the point where rule-breakers are identified and dissuaded, engaging a minority of rule-followers exerts a strong, positive, non-linear effect in the rate of change and the activation of social norms (Castilla-Rho et al., 2017).

In summary, the above findings suggest strong in-principle support of the need for water regulation. Personal values and social reputation are key compliance motivators. Fear of legal enforcement does not appear to be as significant a driver of water users' compliance behaviour, most likely reflecting the lack of staff on the ground. The potential deterrent of enforcement powers may be compromised by a lack of knowledge of, in particular, enforcement actions and penalties. Similarly, a widespread lack of knowledge of broader water policy goals and mechanisms has the potential to undermine compliance initiatives in a range of areas, although most respondents appear to support and have a good understanding of metering technology and its use in compliance and on-property management. However, the survey found that knowledge and support of metering is greatest in those regions with more regular extraction/irrigation and more metering experience. Economic advantage remains a primary driver for non-compliance, suggesting compliance and enforcement tools need to target such benefits (e.g. using enforcement to leverage higher fines or recoup profits). Water users displayed a distinct preference for relying on information from trusted sources, in particular, family, neighbours and water user groups. This suggests working with third parties (given the limited public resources), in the provision of more detailed information may assist compliance activities to complement other information initiatives. Moreover, the bottom line from Castilla-Rho et al.'s (2017) computer modelling is that enforcement 'sticks' will not be sufficient to ensure compliance with groundwater regulations (especially in the absence of sufficient compliance and enforcement staff on the ground) and that using non-government actors (community leaders or unconditional rule-followers) to promote broader social norms will provide the true 'glue' that cements and holds cooperative compliance behaviours together (Janssen, 2017). With new reforms underway in NSW, there is an opportunity for responding to these insights and broader lessons on compliance and enforcement.

22.4 Conclusion

This chapter has proposed an analytical framework that connects studies in regulatory literature with aspirations in the Australian water compliance and enforcement approach. We invite future research to refine and interrogate this framework and its applicability and utility for other groundwater contexts, countries and literatures. The chapter has illustrated aspects of the framework through a case study of regulators and regulated actors in NSW, noting that effective groundwater management requires a combination of activities to support those who are motivated to comply, and deter those who are less so. Although we highlighted numerous challenges (e.g. resourcing, culture and institutional restructures), compliance officers in NSW and other states have a diverse range of tools available to investigate and manage non-compliance. We argued that it is important regulators continue to build their knowledge of, and partnerships with, groundwater users. We provided some important insights into water users' views, perceptions and experiences of compliance, enforcement and water extraction in NSW and proposed several policy implications for a compliance and enforcement regime that engages a broader range of parties and facilitates better management of groundwater extraction. Ultimately, this could lead to more effective and efficient groundwater regulation in NSW, Australia and potentially internationally.

References

Abbot, C. (2005). The regulatory enforcement of pollution control laws: The Australian experience. *Journal of Environmental Law, 17*(2), 161–180.

Allan, A. (2010). Does the law stymie the science? The role of law in achieving sustainable groundwater management [powerpoint presentation] (*EGU General Assembly,* 22–27 April 2012, Vienna, 2307). http://presentations.copernicus.org/EGU2012-12307_presentation.ppt

Anderson, K., & McCusker, R. (2005). Crime in the Australian fishing industry, 297 Crime and Criminal Justice 1.

Australian Broadcasting Corporation (ABC). (2017). *Four Corners* 'Pumped: Who's benefitting from the millions spent on the Murray-Darling?' https://iview.abc.net.au/show/four-corners/series/2017/video/NC1704H024S00

Australian Bureau of Statistics (ABS). (2017). *Water account Australia 2015–16*, AusGov.

Axelrod, R. (1986). An evolutionary approach to norms. *The American Political Science Review, 80*(4), 1095–1111.

Ayres, I., & Braithwaite, J. (1992). *Responsive regulation*. OUP.

Barclay, E., & Bartel, R. (2015). Defining environmental crime: The perspective of farmers. *Journal of Rural Studies, 39*, 188–198.

Bartel, R., & Barclay, E. (2011). Motivational postures and compliance with environmental law in Australian agriculture. *Journal of Rural Studies, 27*(2), 153–170.

Black, J. (2010). Risk-based regulation: Choices, practices and lessons being learnt. In *OECD, risk and regulatory policy: Improving the governance of risk* (pp. 185–224). Paris: OECD Publishing.

Black, J., & Baldwin, R. (2010). Really responsive Riskbased regulation. *Law & Policy, 32*(2), 181.

Blair, N. (2017). *NSW water reform package,* NSW Government.
Blomquist, W. (1992). *Dividing the waters: Governing groundwater in Southern California.* San Francisco, CA: Institute for Contemporary Studies Press.
Borck, J., & Coglianese, C. (2011). Beyond compliance: Explaining business participation in voluntary environmental programs. In C. Parker & V. L. Nielsen (Eds.), *Explaining compliance: Business responses to regulation* (pp. 12–16). Cheltenham, UK: Edward Elgar.
Braithwaite, V. (2009). *Defiance in taxation and governance.* Cheltenham, UK: Edward Elgar.
Bricknell, S. (2010). *Environmental crime in Australia.* Canberra, Australia: AIC.
Brown, V. F. (2017). *Water theft and water smuggling.* Washington, DC: Brookings.
Carmody, E. (2013). Exemptions from cease-to-pump rules in the hunter coal field: Mines 1, aquifers 0. *Australian Environment Review, 28*(4), 567–570.
Castilla-Rho, J., Rojas, R., Mariethoz, G., Andersen, M., & Holley, C. (2017). Social tipping points in global groundwater management. *Nature Human Behaviour, 1*(9), 640–649.
Castilla-Rho, J., Rojas, R., Andersen, M., Holley, C. & Mariethoz, G. (2019). Sustainable groundwater management: How long and what will it take? *Global Environmental Change, 58,* 101972.
Climate Council (CC). (2015). *Climate change 2015,* CC.
COAG. (1994). *Communiqué water reform,* Australian Government.
Department of Primary Industries (DPI). (2016). *National framework for compliance and enforcement systems for water resource management*; Final report to the commonwealth; New South Wales, ISBN 978-1-74256-903-1.
Douglas, M. (1966). *Purity and danger, an analysis of conceptions of pollution and taboo.* London: Routledge and Kegan Paul.
DSEWPC (Department of Sustainability, Environment, Water, Population and Communities). (2012). *National framework for compliance and enforcement systems for water resource management,* Australian Government.
Epstein, J. M., & Axtell, R. (1996). *Growing artificial societies: Social science from the bottom up.* Brookings Institution Press.
Fisher, D. (2011). A sustainable MDB: The legal challenges. In D. Connell & Q. Grafton (Eds.), *Basin futures.* Canberra, Australia: ANU EP.
Goldstein, D., & Gigerenzer, G. (2002). Models of ecological rationality: The recognition heuristic. *Psychological Review, 109*(1), 75–79.
Grafton, P., Pittock, J., Williams, J., et al. (2014). Water planning and hydro-climatic change in the Murray-Darling basin, Australia. *AMBIO: A Journal of the Human Environment, 43*(8), 1082–1092.
Gunningham, N. (2011). Enforcing environmental regulation. *Journal of Environmental Law, 23*(2), 169–201.
Gunningham, N., & Holley, C. (2016). Next-generation environmental regulation: Law, regulation, and governance. *Annual Review of Law and Social Science, 12,* 273–293.
Gunningham, N., Grabosky, P., & Sinclair, D. (1998). *Smart regulation.* Oxford, UK: OUP.
Gunningham, N., Kagan, R. A., & Thornton, D. (2003). *Shades of green: Business, regulation, and environment.* Stanford, CA: Stanford University Press.
Head, B. (2010). Water policy: Evidence, learning and the governance of uncertainty. *Policy & Society, 29,* 171–180.
Holley, C., & Gunningham, N. (2006). Environment improvement plans: Facilitative regulation in practice. *Environmental and Planning Law Journal, 23,* 448–464.
Holley, C., & Sinclair, D. (2012). Compliance and enforcement of water licences in NSW. *The Australasian Journal of Natural Resources Law and Policy, 15*(2), 149–189.
Holley, C., & Sinclair, D. (2013). Non-urban water metering policy: Water users' views on metering upgrades in New South Wales, Australia. *The Australasian Journal of Natural Resources Law and Policy, 16,* 101–131.
Holley, C., & Sinclair, D. (2015). *Water extraction in NSW: Stakeholder views and experience of compliance and enforcement,* CWI at http://www.water.nsw.gov.au/waterlicensing/compliance

Holley, C., & Sinclair, D. (2016). Governing water markets. *The Environmental and Planning Law Journal, 33*(4), 301–324.

Holley, C., & Sinclair, D. (2017). Enforcement strategies: Inspection, targeting and escalation. In D. Markell, L. Paddock, & Bryner (Eds.), *Environmental compliance and enforcement* (pp. 101–113). Edward Elgar.

Holley, C., & Sinclair, D. (Eds.). (2018). *Reforming water law and governance*. Springer.

Honneland, G. (1999). A model of compliance in fisheries: Theoretical foundations and practical application. *Ocean and Coastal Management, 42*, 699–716.

INTERPOL. (2016). *Strategic report: Environment, peace and security*. UN/Interpol.

Jagers, S., Berlin, D., & Jentoft, S. (2012). Why comply? Attitudes towards harvest regulations among Swedish fishers. *Marine Policy, 36*(5), 969–976.

Jakeman, T. (2010). Modelling and decision support for integrated groundwater management [powerpoint presentation] (National Groundwater Conference, *Groundwater* 2010*: The challenge of sustainable management*, Canberra, 31 October – 4 November 2010). http://www.groundwater2010.com/documents/HT3DTonyJakeman.pdf

Janssen, M. A. (2017). Modelling social norms of water conservation. *Nature Human Behaviour, 1*, 624–625.

Kagan, R. A. (2004). Regulators and regulatory processes. In *The Blackwell companion to law and society* (pp. 212–230). Austin: Sarat.

Kagan, R. A., Gunningham, N., & Thornton, D. (2003). Explaining corporate environmental performance: How does regulation matter? *Law and Society Review, 37*, 51–90.

March, J. G., & Olsen, J. P. (1989). *Rediscovering institutions*. New York: Free Press.

Martin, P., & Gunningham, N. (2011). Leading reform of natural resource management law: Core principles. *Environmental and Planning Law Journal, 28*(3), 137–158.

Matthews, K. (2017). *Independent investigations into NSW water management and compliance,* NSW Government.

May, P. J., & Winter, S. (2000). Reconsidering styles of regulatory enforcement: Patterns in Danish agro-environmental inspection. *Law & Policy, 22*, 143–173.

Murray Darling Basin Authority (MBDA). (2009). *Murray darling Basin government irrigation storages* – 1 July 2009 https://www.mdba.gov.au/sites/default/files/waterstorages/weekly-basinreports/MDB_Storage_2009-07-01.pdf

Murray Darling Basin Authority (MDBA). (2017). *The MDB water compliance review,* Aus Gov.

National Water Commission (NWC). (2014). *Australia's water blueprint: National reform assessment 2014,* Australian Gov.

National Water Initiative (NWI). (2004). *Intergovernmental agreement on a national water initiative*. Canberra, Australia: Australian Government.

NSW Department of Primary Industries, Office of Water (NSW DPI). (2015). *Compliance policy,* NSW Gov.

NSW Government. (2015). *Department of primary industries, Office of Water*. Lower Murrumbidgee Groundwater Sources Summary Report 2006–2014. https://www.water.nsw.gov.au/__data/assets/pdf_file/0017/555020/lower_murrumbidgee_groundwater_summary_report_2006_2014.pdf

NSW Ombudsman (NSWO). (2017). *Investigation into water compliance and enforcement 2007–17*. https://www.ombo.nsw.gov.au/news-and-publications/publications/reports/state-and-local-government/investigation-into-water-compliance-and-enforcement-2007-17

NSW Protection of the Environment Operations Act. (1997).

NSW Water Management Act. (2000).

Ostrom, E. (1990). *Governing the commons,* CUP.

Paddock, L., & Wentz, J. (2014). *Next generation environmental compliance & enforcement,* ELI.

Parker, C. (2002). *Is there a reliable way to evaluate organisational compliance programs?* Melbourne, Australia: Australian Institute of Criminology.

Parker, C., & Nielsen, V. L. (2011). Introduction. In C. Parker & V. L. Nielsen (Eds.), *Explaining compliance*. Edward Elgar.

Productivity Commission (PC). (2017). *National water reform draft report.* Australian Government.
Purdy, R. (2011). Attitudes of UK and Australian farmers towards monitoring activity with satellite technologies: Lessons to be learnt. *Space Policy, 27*, 202–212.
Rinaudo, J. D., Montginoul, M., Varanda, M., & Bento, S. (2012). Envisioning innovative groundwater regulation policies through scenario workshops in France and Portugal. *Irrigation and Drainage, 61*, 65–74.
Rittel, H. W. J., & Webber, M. M. (1973). Dilemmas in a general theory of planning. *Policy Sciences, 4*(2), 155–166.
Robertson, S. (2014). A regulatory framework for monitoring and enforcement of water access rights in Western Australia. *University of Western Australia Law Review, 37*(1), 215–242.
Rubenstein, N., Wallis, P., Ison, R., & Godden, L. (2009). *Strengthening water governance in Australia* (Briefing paper No. 1). Water Governance Research Initiative.
Sparrow, M. (2000). *The regulatory craft*. Washington, DC: Brookings.
Torres, G. (1989). Theoretical problems with the regulation of agriculture. *Virginia Environmental Law Journal* Vol. 8, No. 2, Symposium Issue: Agricultural Law and Policy, pp. 191–214.
Tyler, T. (2006). *Why people obey the law*. Princeton, NJ: Princeton University Press.
Tyran, J., & Feld, L. P. (2006). Achieving compliance when legal sanctions are non-deterrent. *Scandinavian Journal of Economics, 108*(1), 135–156.
Water Crimes (WC). (2017). *European Report on Water Crimes*, RECCEE.
Water NSW. (n.d.). *NSW water register*. https://waterregister.waternsw.com.au
Wentworth Group (WG). (2017). *Review of water reform in the MDB*, WG.
Wilhite, D. A. (2012). Breaking the hydro-illogical cycle: Changing the paradigm for drought management. *EARTH Magazine, 57*(7), 71–72.
Williams, J. (2017). Water reform in the Murray–Darling basin: A challenge in complexity in balancing social, economic and environmental perspectives. *Journal of Proceedings of the Royal Society of New South Wales, 150*(1), 68–92.
World Economic Forum (WEF). (2016). *The global risks report,* WEF.
Yeung, K. (2004). *Securing compliance. A principled approach.* Oxford, UK: Hart Publishing.
Zaelke, D., Stilwell, M., & Young, O. (2005). Compliance, rule of law and good governance. What reason demands: Making law work for sustainable development. In D. Zaelke, D. Kaniaru, E. Kruzikova (Eds.), *Making law work: Environmental compliance and sustainable development*. 1(1) Cameron May Ltd.

Dr. Cameron Holley is a Professor at University of New South Wales Law and is a member of the Global Water Institute and Connected Waters Initiative, University of New South Wales Sydney. Cameron has worked closely with Australian and international government and non-government organizations on a range of water and natural resource management research projects. He currently holds ARC Discovery Grants (DP170100281 DP190101584) on Non-Urban Water Regulation and Integrating the Governance of Water and Coal Seam Gas that supported this chapter. He is an Editorial Board member on the Environmental and Planning Law Journal, and in 2016 was the guest editor of a Special Issue (EPLJ Vol 33 Part 4), entitled Rethinking Water Law and Governance.

Dr. Tariro Mutongwizo is a Postdoctoral Fellow with UNSW Law, Australia. Tariro holds a PhD in Criminology from the University of Cape Town, South Africa. Her research interests include multidisciplinary approaches to exploring the non-state governance of security, the governance of contested spaces and the security of vulnerable and marginal-ized groups. Her work focuses broadly on regulation and governance particularly relating to human and environmental security. Tariro's work is centred on building resilience and promoting collaborative governance specifically where contests exist between diverse stakeholders. She has also published on the intersection between state and non-state actors in the governance of security in Africa.

Dr. Susan Pucci is a Principal Policy Officer in the NSW Department of Planning, Industry and Environment – Water. Susan has worked in the NSW Government for 19 years and started working as Manager Planning and Reform in the New South Wales Office of Water Licensing and Compliance Branch in February 2011. She moved into the NSW Natural Resources Access Regulator in 2018 as Manager Regulatory Practice, and commenced in a Principal Policy Officer role in the Water Group in January 2019. For about 10 years she led a team responsible for delivering a number of functions related to water regulation including: implementing regulatory policy frameworks; implementing procedures for operational regulatory staff and implementing operational policies to guide regulatory staff in the implementation of water policy and legislation. She was initially employed to apply the knowledge and skills developed during her research career to develop best practice field procedures for regulatory staff, and apply best practice and innovative techniques to natural resource regulation.

Dr. Juan Castilla-Rho is a transdisciplinary systems modeller working at the interface of natural resource management and computational social science. He combines numerical modelling, systems thinking and complexity science to better understand the dynamic interactions between people, water, infrastructure, and the environment. Juan is pioneering the use of interactive agent-based policy simulators to engage stakeholders in the exploration of future trajectories and improve policy decisions in groundwater- and other environmental-related dilemmas. His research has earned several keynote presentations, best early-career oral presentation award at the Australasian Groundwater Conference in 2015, the inaugural 2017 CSIRO Land & Water early-career award, and the cover of Nature Human Behaviour. His work on the Groundwater Commons Game was also recently featured and presented at the Research Institute of Humanity and Nature in Kyoto, Japan. His scientific and leadership skills have been recognised internationally, as a recipient of an AusAid Australian Leadership Award in 2010. Prior to pursuing a research career, Juan worked for six years in the Atacama Desert for the Chilean Water Authority, water utilities and also as consultant, dealing with complex groundwater issues in one of driest regions in the world.

Dr. Darren Sinclair is an Associate Professor at the Centre for Change Governance, University of Canberra. He has a BSc (Hons I) from the University of Sydney, and a Masters and PhD from the Australian National University. He represented Australia at the United Nations International Framework Convention on Climate Change (Geneva 1993 and 1994). He co-designed the self-regulatory and trading scheme to phase out the use of ozone-depleting gases under Australia's commitment to the Montreal Protocol. He currently holds a DECRA Fellowship and an ARC Discovery Grant (DP190101584) that supported this chapter. Prior to this, he was a Research Fellow at RegNet, the Australian Centre for Environmental Law (Faculty of Law) and the Fenner School of Environment and Society, all at the Australian National University. He has published widely on environmental governance, regulation and policy, and mining occupational health and safety regulation and policy. He has also worked as a consultant to government and industry.

Chapter 23
Compliance and Enforcement: The Achilles Heel of French Water Policy

Marielle Montginoul, Jean-Daniel Rinaudo, and Charlotte Alcouffe

Abstract This chapter examines the compliance and enforcement issues relating to groundwater policy in France. It is based on a review of existing grey and scientific literature and a series of interviews conducted by the authors with enforcement officers in 16 French counties. The chapter starts with a presentation of the existing regulations governing groundwater abstraction (Sect. 23.1), followed by a description of how the law enforcement agencies are organised (Sect. 23.2) and how they operate (Sect. 23.3). It then describes the infractions observed by regulators and analyses the factors that may explain compliance and non-compliance (Sect. 23.4). The problems that limit the effectiveness of enforcement are discussed.

Keywords Water crime · Criminal enforcement · Administrative sanction

23.1 Introduction

The water policy implemented in France since the 1960s is often presented as a model that has inspired European legislation, as well as legislation in other countries. Yet, its effectiveness is debatable: the environmental objectives set by European directives have not always been achieved. Many assessments have underlined that this situation could be due to problems of regulatory enforcement, which largely stem from the state's reluctance to prosecute violations of the law (Barone, 2018; Boutelet, 2014; Cour des comptes, 2010), as well as the difficulties of organising France's water police (Legrand et al., 2015; Simoni et al., 2005).

In the field of quantitative management, a volumetric system for managing water resources, as laid down by the 2006 law on water and aquatic environments (see

M. Montginoul (✉) · C. Alcouffe
INRAE – UMR G-EAU, Montpellier University, Montpellier, France
e-mail: marielle.montginoul@inrae.fr

J.-D. Rinaudo
BRGM, Montpellier University, Montpellier, France
e-mail: jd.rinaudo@brgm.fr

J.-D. Rinaudo et al. (eds.), *Sustainable Groundwater Management*,
Global Issues in Water Policy 24, https://doi.org/10.1007/978-3-030-32766-8_23

Chap. 3), can only function if the vast majority of users comply with the existing regulations. All abstraction points, wells or boreholes must be declared and authorised and users must comply with the abstraction limits allocated to them (in terms of both flow and volume). To achieve this, several counties are jointly responsible for water policing, which aims to prevent and punish regulatory non-compliance. This chapter describes how the water police are organised in France and analyses the difficulties they face. In particular, it focuses on the issues of groundwater management in regions where agricultural water use is dominant. Indeed, most of the problems of law enforcement have been identified in the agricultural sector and concern abstraction for crop irrigation (Legrand et al., 2015).

This chapter reviews several official publications on the subject and draws on a series of semi-structured interviews conducted with environmental inspectors operating at county levels (*Direction Départementale des Territoires*). Interviews focussed on compliance and enforcement issues specifically related to groundwater. The objective was to identify the most frequent infractions, to understand the factors explaining non-compliance and to highlight the main difficulties met by law enforcement officers. The survey was conducted by telephone in 2016. It focused on 17 French counties (see Fig. 23.1) characterised by the prevalence of irrigated agriculture (over 10,000 ha of irrigable land and 10% of the total land area irrigated), as well as the existence of major groundwater resources.

The chapter is organised as follows: the first section reviews the main regulatory obligations that apply to water users and focuses on the case of groundwater. Section 23.2 describes how the water police are organised. Section 23.3 attempts to describe the importance and nature of the infractions observed, based largely on the survey results. We then strive to highlight the factors that explain the scale of the problems of non-compliance. Section 23.5 describes the difficulties encountered by the water police and identifies possible courses of action to solve them.

23.2 Key Regulatory Provisions for Groundwater Abstraction

Groundwater users are subject to two main regulatory obligations relating to the construction of a well or borehole and to water abstraction.

23.2.1 Administrative Provisions for Constructing Wells and Boreholes

When installations are constructed, the mining code stipulates that any underground structure exceeding a depth of 10 m must be declared to the regional environment agency (the DREAL, the Regional Directorate for Environment, Planning and

Fig. 23.1 Counties studied in the framework of the survey of the water policing services

Housing). In the declaration, the applicant (owner) must include the geological data gathered during the construction (the drilling log). The underground installation must comply with the special building regulations set out in the environmental code.[1] These provisions apply irrespective of the intended use of the borehole (geothermics, water abstraction, exploratory drilling, etc.).

---

[1] Decree of 11th September 2003, which implements Decree n° 96–102 of 2nd February 1996 and lays down the general provisions applicable to drill holes, boreholes and the construction of wells or underground installations subject to a declaration pursuant to articles L. 214-1 to L. 214-3 of the environmental code and pertaining to category 1.1.1.0 of the nomenclature annexed in the modified Decree n° 93-743 of 29th March 1993.

If the underground installation is intended for abstracting water, the owner must obtain a permit (the right to access the resource), according to the 1992 Water Act.[2] A permit is only obligatory if the user intends to abstract over 10,000 m^3 per year. For volumes of between 10,000 and 200,000 m^3, the permit is automatically issued to the owner when they declare their installation to the government authority (*Direction Départementale des Territoires, DDT*). The procedure is simplified and rapid, but it gives the administration the option to refuse an application. For volumes over 200,000 m^3 per year, the permit is granted after a more complex authorisation procedure, involving an assessment of the potential impact of abstraction on third parties and the environment. This allows state services to prohibit the construction of these larger installations or limit their pumping capacities in certain zones: ecologically sensitive areas, zones reserved for drinking water and overexploited zones. The abstraction permit is tied to the installation and is not time-limited. It is automatically transferred with the installation in the event of a sale.

23.2.2 Administrative Provisions for Water Abstraction

Once the installation has been authorised, the user must apply for an abstraction permit. This specifies the restrictions applicable for the use of the installation. It indicates the flow rate and annual volume for abstraction. The annual volume may be expressed on a seasonal or monthly basis, depending on the existing local management systems (for further examples see Chaps. 13 and 18).

The authorisation procedure for abstraction also depends on the volume requested. When the volume abstracted is less than 10,000 m^3 per year, a declaration is not necessary, nor is an authorisation. For volumes between 10,000 and 200,000 m^3 per year, a simple declaration suffices. In excess of 200,000 m^3 per year, the state services must authorise the abstraction. This allows the state to ensure that the total authorised volume does not exceed the volume that can be abstracted (see Chaps 3 and 5). In restriction zones (ZRE in French), users must undergo the authorisation procedure if the pumping capacity (in flow rate) of the planned installation exceeds 8 m^3/h (Table 23.1).

Until 2017, abstraction permits were issued individually to the user (for example, the owner or tenant farmer). They were renewed each year and the state had the possibility of varying the allocated volume, depending on the state of the water resource and the total volume requested by all the users (with authorised boreholes). Since 2017, the state issues a single authorisation to the water user groups (collective management organisations or OUGC in French, see Chap. 3), which are responsible for distributing the volume between their members. The state approves the distribution and checks that there is no environmental impact.

[2] Article R214-1 of the environmental code.

Table 23.1 Regulatory provisions for groundwater use

	Type of abstraction	Administrative procedure governing boreholes or wells	Administrative procedure for abstraction
Outside water restriction zones (ZRE)	Annual abstraction <1000 m³/year (domestic use)	Local council declaration	Not applicable
	Annual abstraction between 1000 m³/year and 10,000 m³/year	Declaration to DDT	Not applicable
	Annual abstraction between 10,000 m³/year and 200,000 m³/year	Declaration to DDT	Declaration to DDT
	Annual abstraction >200,000 m³/year	Application for authorisation (DDT)	Application for authorisation (DDT)
In ZRE	Pumping capacity exceeding 8 m³/h	Same procedure as for outside the ZRE	Application for authorisation (DDT)

23.2.3 *Temporary Restrictions on Water Use*

When river flow rates or groundwater levels fall below a critical threshold, temporary restrictions on water use are introduced by prefectural decree. Abstraction can even be prohibited. Restrictions vary according to use. For domestic water use, a general ban may be announced for certain uses, for example, watering the garden, filling swimming pools and washing cars. In the case of agriculture, the restrictions affect the weekly duration of irrigation, which is gradually reduced from 7 days to 1 day per week and may even be totally banned.

23.3 The Organisation of Water Policing Services

23.3.1 *A Transversal Mission Involving Several Government Agencies*

In France, enforcing regulations is a strictly sovereign mission. Therefore, the government administrations alone are responsible for enforcement (no other actors are involved, e.g. user, NGO, local authorities). Since 1992, the prefect (the government representative at the county level) has been responsible for coordinating the water police.

For years, enforcement was carried out by several government agencies, with a different speciality field depending on the ministry they are affiliated to (industry, agriculture, environmental protection). This compartmentalisation impeded the effectiveness of enforcement because the different counties were simultaneously responsible for supporting economic development in a sector and managing its environmental impact. The 1992 Water Act unified the water policing service and placed it under the authority of the prefect. The coordination of the services greatly improved[3] in the early 2010s, partly in response to imperatives set by European directives.[4]

Three main counties are now involved in water policing activities. These are supervised by the prefect (for the administrative police) and by the public prosecutor (for the judiciary police), and supported by the police:

- At the county level, the County Directorate for Territories and the Sea (DDTM) supervises the operational coordination of the different water policing services. This is an inter-ministerial county supervised by the prefect at the local level and the prime minister at the national level. The DDTM is responsible for regional development planning (urban, transport, housing and environmental issues). Water policing is coordinated by an inter-service mission for water and the environment (the MISEN). The DDTM is directly responsible for monitoring all the activities that are likely to harm water resources and natural environments (excluding installations classified for environmental protection, ICPE). Monitoring deals with quantitative management (surface or groundwater abstraction), as well as pollution from point sources (sewage treatment plants, livestock effluent) or non-point sources (nitrates and pesticides of agricultural origin).
- At the regional level, the regional environment agency (DREAL, which represents the Ministry of the Environment) coordinates the plan of action for the different counties responsible for the water police. It also inspects the facilities that constitute industrial pollution hazards (known as installations classified for environmental protection, ICPE in French).
- Lastly, the French agency for biodiversity (AFB, ex-ONEMA) is responsible for monitoring aquatic environments and fishing, in particular. As an independent public body, AFB staff are not accountable to the prefect. The AFB coordinates its action with the national agency for hunting and wildlife (ONCFS).

The average number of staff dedicated to water and environmental (widelife, fishing, hunting, etc.) policing duties is 18.6 (full-time equivalent) per county (Legrand et al., 2015). However, only a fraction of them intervene in water management issues, often for part of their time.

[3] Decree n° 2012–34 of 11th January 2012 relating to the simplification, reform and harmonisation of the environmental code for the provisions pertaining to the administrative police and the judiciary police. JORF n°0010 of 12th January 2012, page 564.

[4] In particular, Directive 2008/99/CE regarding the application of criminal law to environmental protection, which led to an organisational reform of the police, procedures and sanctions.

23.3.2 A Dual Mission: Judicial and Administrative Enforcement

The counties referred to in the previous section have two complementary police missions. It is important to note that the term police refers to "*the power attributed to a person to restrict the liberty of individuals in order to prevent or repress disturbances to public order, public health and security, where necessary*" (Boutelet, Brun, & Van Bosterhaudt, 2012). Water policing involves administrative and judicial police, who are responsible for prevention and enforcement, respectively.

For environmental issues, the administrative police are accountable to the prefect (the executive representative). For matters of quantitative groundwater management, policing involves examining, monitoring and reviewing declarations and authorisations for installations and abstraction. The administrative police set out the pumping limits to prevent damage to the water resource and aquatic environments. In the field, they check whether the works/installations/activities comply with the permits obtained and recommend administrative sanctions if an infraction is observed. The administrative police's mission is primarily conducted by the DDTM and, to a lesser extent, the DREAL. Table 23.2 shows the administrative sanctions that can be imposed in the event of regulatory non-compliance.

The judiciary police are responsible for identifying infractions, gathering evidence and finding the offenders. The public prosecutor deploys the judiciary police to perform these tasks, which are generally performed by environmental investigators from the AFB. They, in turn, are accountable to the public prosecutor. Environmental investigators are sworn officers or engineers that have some of the prerogatives of judiciary police. Most breaches involve no more than a compliance notice, which is referred to the public prosecution service. However, investigators can also search, seize documents, take water samples or other measures and conduct hearings with witnesses or plaintiffs. The public prosecutor has full discretion to decide on the appropriate prosecution in view of the elements provided by the judiciary police.

Table 23.2 Administrative sanctions following an infraction (article L 216-1 of the environmental code)

Infraction	Sanctions
Use of an unauthorised installation	A compliance notice.
	Failure to comply: Installations are shut down or removed and the site is restored.
Failure to comply with regulatory provisions	An injunction demanding that work be undertaken within a specified time.
	If the work is not undertaken within the stated period, the administrative police can proceed as follows: (a) order the consignment of a sum to a public accountant, equal to the cost of the work to be undertaken; (b) order that the work be undertaken; (c) suspend the operations at the installation; (d) impose a fine of €15,000 maximum and a daily penalty of €1500 maximum; the fines and penalties are proportional to the severity of non-compliance.

Table 23.3 Judicial sanctions that can be applied following an infraction (article L 173-1 to 173-12 of the environmental code)

Infraction	Maximum sanctions
Use of an unauthorised installation	€75,000 and 1-year prison sentence (€100,000 and 2-year prison sentence in the case of a breach following a refusal to grant authorisation)
Failure to respect a compliance notice relating to an abstraction point	€100,000 and 2-year prison sentence
Use of a facility without complying with the provisions, causing substantial degradation to flora, fauna or water	€300,000 and 5-year prison sentence
Obstructing the regulators	€15,000 and 6-month prison sentence
In the event of a conviction for a breach, the court can impose the following additional sanctions	Suspension of authorisation to use facilities for the duration of 1 year maximum. An injunction to restore the site and repair the environmental damage caused; a fine of €3000 per day after the deadline set for restoration is reached
Additional possible sentences for natural persons guilty of infractions	The decision is published in medias. The material involved in the regulatory breach is confiscated. The operator's activities are shut down (5 years)

When the judiciary police identify a breach during their operations, a compliance notice is referred to the public prosecutor, who can proceed in several ways. If the evidence provided is judged to be insufficient, the case can be closed with no follow-up. The second possible option consists of a criminal fine, whereby the offender is fined and ordered to repair the damage caused by their breach.[5] The third option involves bringing the offender before a magistrate's court, in the case of an offence, or before a police tribunal, in the case of a contravention. If the infraction represents a serious threat to the environment or public health, the public prosecutor can suspend the associated activity for a maximum duration of 3 months (interim measure) (Table 23.3).

23.4 The Water Police's Methods of Intervention

23.4.1 Limited Monitoring Pressure

The administrative services responsible for water policing put little pressure on users. The national objectives stipulate that the services should devote 20% of their time to monitoring (preparation, execution, follow-up) and conduct a minimum of 400 inspections per year and per county (on- and off-site inspections), with an average national target of 600 per county. The investigations must focus on the

[5] Although this approach should be limited to minor cases, reports suggest that it is used far too frequently (Court of Auditors annual report, Cour des comptes, 2010).

geographic sectors and water management issues defined in the inter-institutional control plans.

These goals may appear even less ambitious given that they cover all areas of environmental and water policy. Thus, in 2014, only 8% of the 25,000 inspections conducted by ONEMA's 600 agents concerned quantitative resource management (surface and groundwater). Most of these inspections focused on monitoring aquatic environments (42%) and water quality and pollution (38%). Monitoring fishing represented 6% of inspections and monitoring species and natural habitats represented 5%.

23.4.2 An Inter-institutional Control Plan

As the state administration has limited means, the available resources are allocated to investigating priority issues and regions, according to a plan drawn up in each French county. The plan is established under the joint direction and supervision of the prefect (administrative police) and the public prosecution (judiciary police). It involves all the services and institutions that perform policing duties related to water and nature. It identifies the most effective investigative actions for meeting the objectives of protecting aquatic environments, habitats and species. In particular, the plan identifies the activities or installations, which exert major pressure on resources and natural environments, and which are generating a risk that the objectives of EU directives are not achieved. It determines the operational goals, namely, the number of investigations to be performed per administrative service, per theme and per sector, by specifying the orientation for each type of investigation (administrative police, judiciary police). The plan takes account of case history. Thus, controls are intensified where past activities have revealed frequent non-compliance depending on observed non-compliance. Lastly, random checks of installations or activities are performed (across all sectors and categories of person), to ensure that no one a priori escapes the control policy. One of the regulatory agencies is designated to organise and coordinate each type of inspection with associate services, if necessary.

Inspections related to abstraction are conducted at approximately 1% of abstraction points, all of which are located in sectors where there is pressure on the resource. The inspection usually involves checking the following points: (i) the presence of a meter; (ii) the existence of a record of meter readings dating back to when the authorisation was granted; (iii) whether the installation (well or borehole and drill head) complies with current regulations; (iv) the instantaneous pumping rate; (v) the period and periodicity of pumping; and (vi) the records of all the declarations of the volume abstracted that have been transmitted to the water police.

23.4.3 How Controls Are Conducted

The 2012–34 ruling of 11th January 2012[6] governs and defines the methods of inspection. Several reports published in the 2010s, as well as people that we surveyed in this study, stated that conducting inspections was difficult, especially in the agricultural sector (Legrand et al., 2015). Paradoxically, the counties where tension is the most acute are those with the fewest inspections. The regulators may be met with a very hostile reception. The people being inspected may express animosity or rudeness and verbal abuse, in some cases, or even death threats (Boizard, Garcin, Menager, & Tosi, 2016). The AFB recorded 96 incidents (insults, threats, intimidation) that occurred during inspections in 2015. The situation is extremely difficult in south-west France, where agricultural representatives (unions) encourage farmers to refuse to cooperate during the inspections or even to prevent the regulator from carrying out their inspection. Thus, Boizard reports that farmers in the Lot-et-Garonne organise "reception committees" comprised of about 40 farmers, who support their colleague (the one being inspected) and put almost unbearable psychological pressure on the regulators (Boizard et al., 2016). For security reasons, inspections are systematically conducted by several officers. The regulator is generally unarmed (except the AFB personnel) but he may request the presence of the gendarmerie.

These tense situations can be explained, in part, by the fact that the farming profession does not understand the regulations or why they are justified, which undermines their legitimacy (Legrand et al., 2015). Farmers view water regulation as illegitimate because it pays disproportionate attention to the protection of environment, while totally neglecting what they think their mission is: to produce food to feed the world (Boutelet, 2014: p 149). Legrand et al. (2015) also suggest that the difficulty of applying the regulations may be because they are recent and are being applied to historically accepted uses. In this way, "*the people inspected believe that they are within their rights, living on the land or farming it, to apply the law as they see fit*".

Sometimes, farmers' hostility to controls can be explained by the high number of inspections imposed on their profession. Indeed, agriculture is subject to numerous and diverse environmental, labour and agro-food regulations, etc. This observation has led state services to coordinate the controls they perform in all fields relating to agriculture.

[6] Published in the JORF on 12th January 2012.

23.4.4 Coordination Between the Water Police and the Judicial System

In the early 2010s, several reports showed that many of the compliance notices served by the water police were not followed up by the public prosecutor (91% in 2008, for example). Several reasons have been put forward to explain this situation (Barone, 2016, 2018; Boutelet, 2014): court congestion; the magistrates' lack of technical expertise, since they generally have little training in environmental law;[7] and the water police officers may not prepare the cases properly. Nonetheless, the lack of human resources in the judicial system seems to be the primary factor, as Marguerite Boutelet illustrates when she quotes a magistrate, who explains: "*We are overloaded with work. When young girls are being raped, when drug dealers must be prosecuted, fish survival can wait*" (Boutelet, 2014: p. 150).

Between 2012 and 2015, the state took steps to improve the coordination between the actions performed by the water police and the public prosecutor (MEDDE 2012; Ministère de la Justice 2015; Barone, 2018). At the prosecution level, referral magistrates were appointed in the environmental field. They were informed about the economic issues associated with environmental protection and given the responsibility of coordinating the actions performed by the water police and the public prosecutor. Coordination was formalised by establishing a protocol, which defines the operational arrangements for legal action, from the field to the tribunal. In 2015, 78% of counties had a memorandum of understanding signed by the water policing services and the public prosecution service (MEEM, 2016).[8]

The established protocols have several goals. First, they strive to clearly define how to conduct inspections and open an enforcement case to minimise the risk of procedural defect. They also seek to limit the number of cases presented to the public prosecutor, by selecting the most serious. Therefore, the water police are invited to apply sanctions progressively. The gradual approach (shown in Fig. 23.2) involves providing an administrative response to all the regulatory breaches that have not caused environmental damage (❶ in the figure). This response should be incremental: (i) compliance notice, specifying the actions to be undertaken in order to comply within a specified deadline; (ii) administrative sanctions in the event of non-execution (fine, suspension of activity, etc.); and referral to the public prosecutor for criminal prosecution, as a last resort. In the event that the breach caused damage that is reparable ❷, the ministry recommends recourse to a criminal fine.[9]

[7] In France, there are no specialised judges, offences that have an impact on nature and the environment are dealt with by generalist magistrates and jurisdictions.

[8] Barone (2018) has shown that these memorandum are only efficient if there is a true commitment from their signatory to implement them, which is not systematically the case.

[9] The Court of Auditors criticised the fact that recourse to criminal fines was too systematic. According to Barone (to be published) and van Bosterhaudt (2014), recourse to criminal fines encourages economic actors to commit environmental crimes because the financial profit generated by non-compliance with the regulatory constraint is far greater than the fine imposed in the event of a criminal fine.

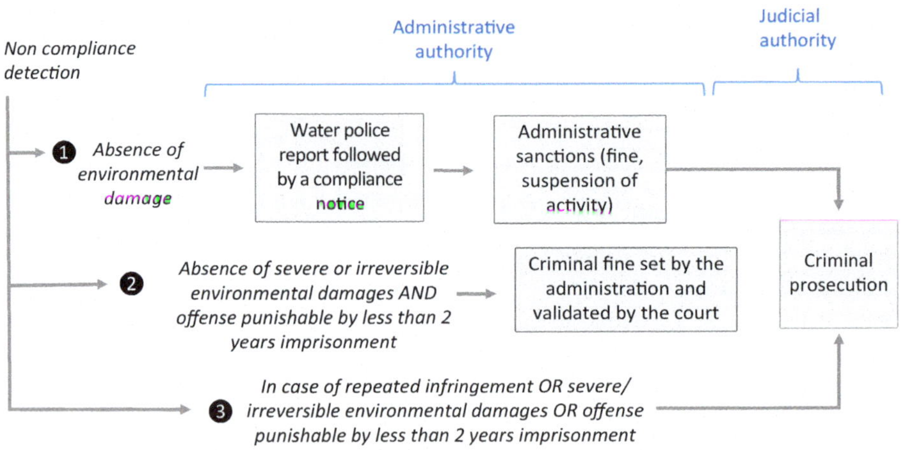

Fig. 23.2 The gradual approach to sanctions recommended by the Ministry of Ecology (Executive Order of 20th October 2014)

This option is requested and approved by the prosecutor. The administrative authority prepares the settlement. Criminal fines can only be used in cases where the potential sanction is less than a 2-year prison sentence. When the law calls for sanctions exceeding a 2-year prison sentence, criminal sanctions must be applied, for example, in the event of a repeat offence or when the breach is deliberate ❸.

23.5 Level of Compliance: Observations and Interpretation

23.5.1 Statistical Information on Compliance

Detailed data relating to the inspections carried out by the water police, as well as the nature of the infractions observed, are not made public by the state services. This rules out the possibility of statistical analyses. The only available data, produced separately by the Ministry of the Environment and the French agency for biodiversity, are shown in Tables 23.4 and 23.5. These data show that between 20% and 37% of inspections detect non-compliance. A substantial number of cases of non-compliance are infractions. They are sanctioned by compliance notice and referred to the public prosecutor. The proportion of violations observed by the AFB is higher than that observed by the DDT. This reflects the fact that the AFB primarily conducts field visits as part of its judiciary police duties. These figures should be considered with caution because they concern all the inspections undertaken relating to the management of water resources, aquatic environments and nature. There is no way of identifying which inspections concern quantitative water management, let alone groundwater management.

Table 23.4 Evolution in the number of inspections where non-compliance is detected and infractions observed by AFB agents (named ONEMA before 2016)

	Inspections undertaken	Non-compliance[a]	Infractions[b]
2016	19,500	7215 (37%)	2500 (34%)
2015	22,833	7535 (33%)	3000 (40%)
2014	25,500	8415 (33%)	7405 (88%)
2013	25,200	8316 (33%)	5112 (61%)
2012	22,932	8485 (37%)	5618 (66%)

Source: ONEMA (2010, 2011, 2012, 2013, 2014, 2015) and AFB (2017)
[a]Number of inspections where non-compliance is detected and global level of non-compliance (%)
[b]Number of infractions and infractions weight as a percentage of the number of cases of non-compliance

23.5.2 *Type of Infractions Observed*

The results of the survey conducted by the authors helped identify the characteristics of the main infractions observed relating to the management of groundwater abstraction.

Overall, compliance with the regulations is considered to be a genuine problem in 11 out of 17 counties. As far as volumetric management is concerned, the two main problems are: the absence of flow meters (four counties) and the lack of meter readings or unsatisfactory records (six counties, see Fig. 23.3). Compliance with the authorised volumes is considered problematic in three counties. The inspectors claim not to have sufficient information to determine how often the meters may be tampered with. The use of wells and boreholes for irrigation, which are not declared for that purpose, is considered significant in three counties. Other cases of non-compliance have also been described relating to borehole construction, for example (no drill head cover). Compliance with the temporary restrictions on water use is not perceived as a problem.

The situation varies depending on the counties. In order to visualise the differences, we devised an indicator of compliance, incorporating an assessment of the frequency of the seven infractions described in Fig. 23.3. The indicator has a value of 1, if all seven of the infractions are considered very frequent; and 0, if no infractions are reported. Figure 23.4 is a simplified representation of the diversity of situations. The first group of counties (eight counties) is characterised by the relative absence of problems (the indicator = 0.07 and no infraction is considered "frequent or very frequent"). In the second group (four counties), the indicator has an average value of 0.21, with 1.25 infractions considered very frequent. Lastly, the third group (five counties) includes the counties where almost three out the seven infractions are considered very frequent and the indicator has a value of 0.41. These results (Fig. 23.4) illustrate the diversity of situations as perceived by the water police. It is only a qualitative indicator, which is not based on a precise measure of the number of infractions.

Table 23.5 Evolution in the number of inspections involving the administrative and judiciary police undertaken by the DDTM, the level of compliance and infractions reported

	Administrative police						Judiciary police
	Inspections undertaken		Inspections that detect non-compliance			Administrative procedures	Infractions
	Off-site	On-site	Off-site	On-site	Total		
2015	25,247	19,273	3947	5129	20%	4180 (46%)	1190 (3%)
2014	33,361	22,043	4263	5889	18%	5050 (50%)	1330 (2%)
2013	33,694	37,355	4208	9135	19%	4693 (35%)	6169 (9%)
2012	32,142	29,188	3404	9540	21%	6147 (47%)	7534 (12%)
2011	41,737	38,783	12,537	10,361	28%	10,493 (46%)	8157 (10%)

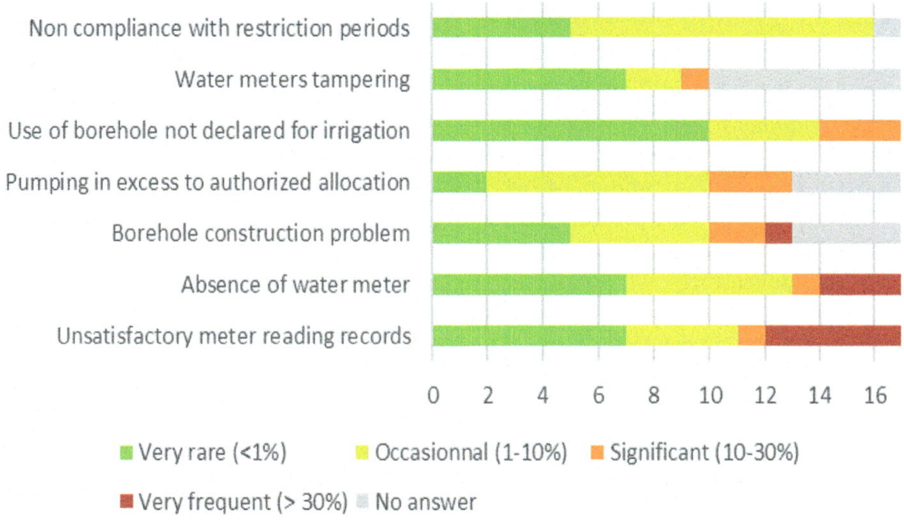

Fig. 23.3 Frequency of non-compliance problems as perceived by regulators interviewed in 17 counties

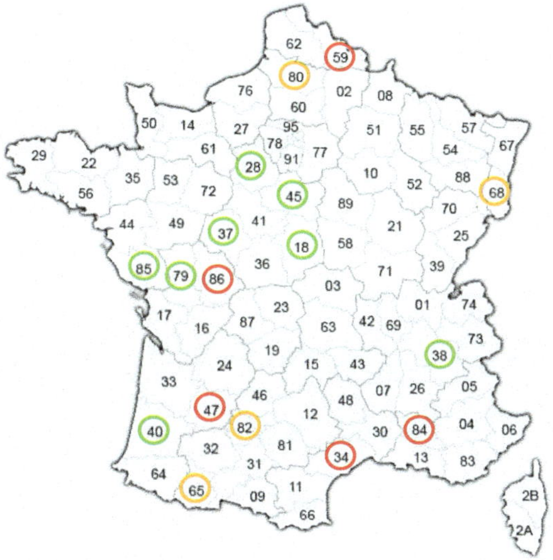

	○	○	○
Average value of the compliance indicator	0.07	0.21	041
Number of infractions considered as frequent	0	1.25	2.8
Number of counties	8	4	5

Fig. 23.4 Classification of counties into three groups according to the intensity of problems of compliance (results of the survey)

23.5.3 *Factors that Determine Non-compliance*

The survey also shed light on the factors that determine regulatory compliance or non-compliance in the different counties surveyed (see Tables 23.4 and 23.5). In the case of non-compliance, the water police officers consulted in the 17 counties studied agree on the observations below.

First, the economic pressure facing farmers can explain the majority of breaches related to water abstraction (12 out of 15 counties, see Fig. 23.5). This is particularly pertinent for farmers, who produce high added value crops such as seeds, fruit and vegetables under contract for the industry. It is also the case for dairy producers whose production depends on irrigated fodder (11/13).

In addition, the level of financial sanctions recommended by law is too low to be an incentive (11/17). The probability of an inspection remains very low (12/17). The fact that criminal sanctions are not systematically applied when a breach is reported is also perceived as a factor that may explain the regulatory non-compliance observed in 5 out of 15 counties.

Interviewees also report that, in a minority of counties, the social climate in the agricultural sector does not encourage regulatory compliance. Farmers who breach the regulations are not stigmatised by their peers when they receive a compliance

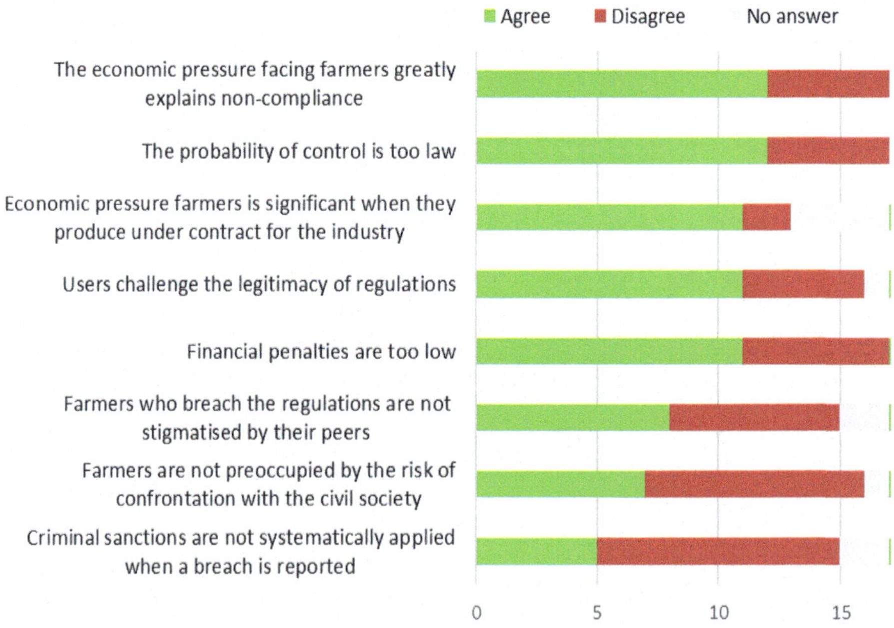

Fig. 23.5 The main factors that explain situations of regulatory non-compliance (results of the survey with the enforcement officers from 17 counties)

notice (8/15). In most counties, there is no social pressure within the agricultural sector to encourage self-monitoring. A water police officer working in a county in the south-west explained that offenders derive a certain pride or social prestige from being booked. *"It's a bit like in the suburbs, they are proud to be against the water police. People used to hide it, but now they show it"*. Farmers seem unperturbed by the risk of a confrontation with the civil society (7/16).

Water users challenge the legitimacy of the regulations (11 out of 16 counties). The regulations are perceived as giving too much priority to environmental protection, to the detriment of the productive use of water resources (farming, in particular). The agricultural sector is reluctant to let a water resource "*flow to the sea*", when it could be used to produce food to feed the world. Farmers do not always understand how the regulations are applied, particularly, when restrictions of use are applied incrementally within the same county. This observation reflects M. Boutelet's analysis (2014).

23.5.4 Factors That Facilitate Compliance

The agents surveyed agree with the following observations regarding factors that facilitate compliance (see Fig. 23.6). Overall, farmers are aware of the regulations that they must comply with (12/15), notably, because the professional farming organisations and state services make an effort to inform them. Generally, farmers have the necessary available resources (time, money and advice) to comply (9/16)

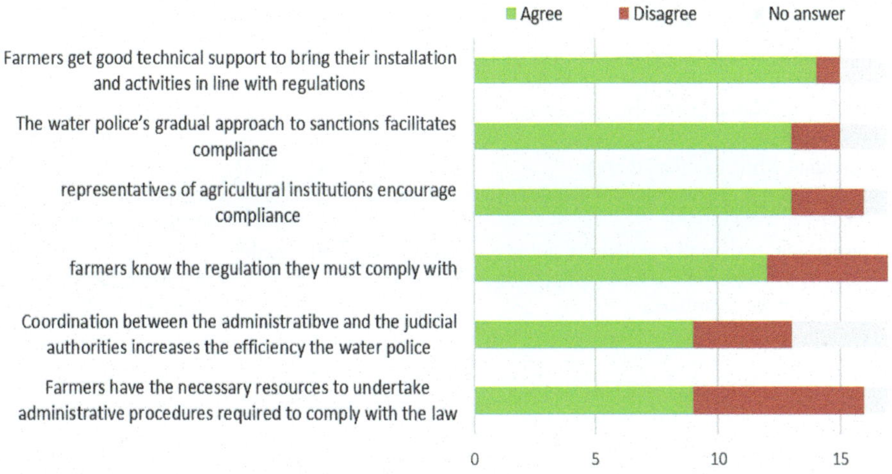

Fig. 23.6 Main factors that facilitate compliance according to the enforcement officers surveyed

because the professional organisations (Chamber of Agriculture, collective management organisations or OUGC) and the state services can provide support (14/15).

In addition, farming representatives encourage farmers to comply with regulations (13/15), except in a few counties, where the profession has a more ambiguous position and may even overtly encourage disobedience.[10]

The water police's gradual approach to sanctions also facilitates regulatory compliance (13/15). Indeed, the aim of prioritising pedagogy over punishment is to improve the compliance rate. The coordination between the water police and the public prosecutor is also helpful (9/13).

23.5.5 A Behavioural Typology

The survey also examined whether there are types of users (profiles), who may or may not comply with regulations. Generally, the survey suggests that it is difficult to establish a typology of offenders, even if non-compliance may reflect different types of rationale (Table 23.6).

Table 23.6 Different types of rationale that could lead to regulatory non-compliance

Type of rationale	Determinant	Trigger for action
Ill-informed small farmer, lacking the resources to comply	Lack of information & support	Information about the procedures
Farmer involved in professional organisations, who has political support to defend his case in the event of a conflict with the administration	Political power and capacity to influence administrations	Impartial application of sanctions prescribed by the state
Small farm, whose economic survival would be at risk in the event of a water restriction: Market gardening, livestock production	Economic pressure	Diagnosis to reduce economic vulnerability to water shortage
Individual who considers that the regulations are illegitimate and decides to resist by not complying	Ideological and political motivation	Extended information about water management issues
Farmer who resists all regulations and is prepared to contravene if non-compliance costs less than compliance	Maximum profit seeking	Strict application of sanctions, including criminal sanctions

[10] In 2013, ONEMA reported 19 public statements expressing offensive comments about the water police, most of which came from professional farming organisations (Legrand et al., 2015).

23.6 Improving the Effectiveness of Law Enforcement Agencies

23.6.1 The Main Difficulties Reported by the Law Enforcement Officers

The survey also focused on the main difficulties that the enforcement officers encounter while performing their duties (Fig. 23.7). Some difficulties are mentioned in all the counties, while others are more specific to certain regions.

The main difficulty common to all the counties is the lack of human resources for carrying out inspections. It is considered as a major obstacle in 8 out of 17 counties, especially in counties with large areas of agricultural land.[11] As the probability of inspection is very low, a sense of impunity has developed among users who do not comply with the regulations, while those who do comply feel that the situation is unfair or unjust.

The lack of modern technology for inspection is also considered to be a limiting factor (8/15) or even a major obstacle (2/15) for effective water policing. Thus, during a field inspection, the use of tablets (with an Internet connection) would provide regulators with access to all the information relating to the abstraction point, the

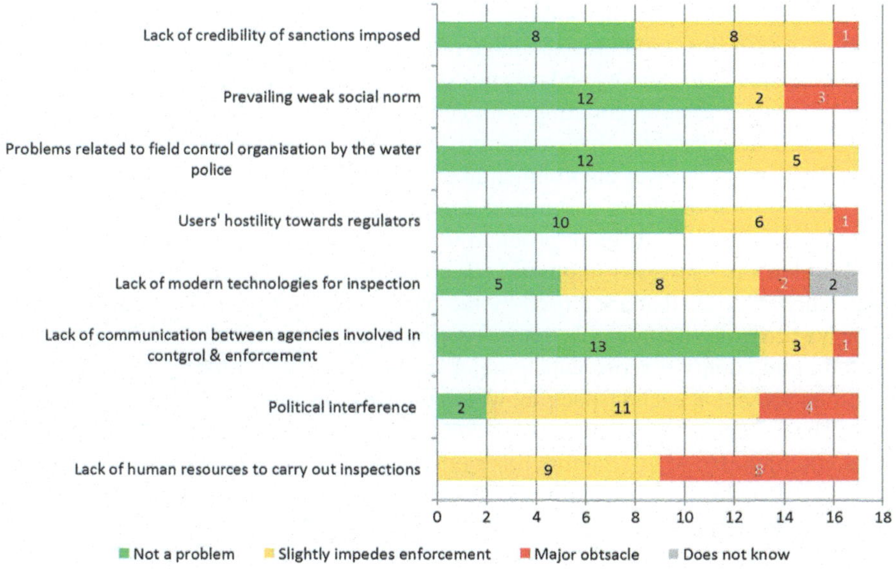

Fig. 23.7 The main difficulties encountered by the water police during operations

[11] According to the Court of Auditors, the number of inspections undertaken by the water and environmental police was reduced by half between 2011 and 2015 (Cour des comptes, 2017).

history of use, crop data, etc. The use of airborne methods, such as ultra-light aircraft or drones would mean that compliance during temporary restrictions on water use could be checked. Similarly, when the manager of an irrigation system installs smart meters (see Chap. 18), the water police should be able to access data in real time, which would make their action more effective. Lastly, some of the enforcement officers surveyed suggest using transparency as an incentive. This would entail making water abstraction data available on the Internet to encourage self-monitoring among users.

The majority of police services also mention that political interference prevents them from doing their job properly. This interference is considered to limit the water police's effectiveness in 11 counties. It is seen as a major obstacle in four counties. The nature of the political interference is roughly as follows: the Chambers of Agriculture and the farming unions ask the prefect to ensure that when the water police perform their duties, they take economic issues into account and do not sanction infractions committed in years that are difficult for climatic reasons. The prefect may be sensitive to this kind of request if the local social climate is tense and the agricultural sector could implicitly threaten to disturb public order by organising demonstrations. The prefect can ask the water police to refrain from serving a compliance notice, except in cases where there is evident abuse. Therefore, the application of the regulations is subject to negotiation between the various social groups. Interference of this type has an adverse impact in the long term. Farmers (users) expect support from their representatives and do not comply scrupulously with the regulations (especially the authorised volumes). This is detrimental to the water police's credibility and may encourage other users to flout the regulations. This was observed during the survey and reflects Sylvain Barone's work (2018), which shows that the state prefers negotiating with economic actors, rather than applying sanctions. This demonstrates that economic development and the preservation of social peace have priority over environmental protection.

Other difficulties were mentioned during the survey, although they did not concern all the counties. A major problem in one county (only mentioned in eight others) is that the sanctions imposed lack credibility. This is largely due to the high incidence of cases that are closed by the public prosecution with no further action. The hostility of users towards the enforcement officers is mentioned in seven counties. In one county it represents a major obstacle because it means several officers must be present at each inspection, which reduces the number of inspections carried out. The existence of a dominant lax culture (or weak social norm) is mentioned in five counties. However, dysfunctions involving the water police are seldom mentioned, be it the way inspections are planned and undertaken or the coordination between the different police services.

23.6.2 Past and Future Evolution

In over half of the counties (eight), the officers surveyed estimate that the situation has improved significantly in the last 10 years. This is primarily due to improvements in the services' internal procedures: the different services have pooled their resources; intervention protocols have been established; communication with the public prosecution has improved. The improvement is also the result of pedagogical action that targets different audiences: users, the Chambers of Agriculture, the prefect and political decision makers. Clearly, the fact that pedagogy was preferred to sanctions means that the police are regarded as partners, who can help the users comply.

On the contrary, two counties consider that the situation is worse due to a combination of three factors. The first relates to the state's capacity of action, which seems to be eroding: fewer resources are allocated to sovereign missions; political interference has increased; and certain responsibilities have been delegated to the users, with the creation of the collective management organisations, the OUGC (see Chap. 3). The second factor is of an economic order: the agricultural sector is in crisis and farms' economic survival takes precedence over environmental protection. Lastly, climate change exacerbates the problems of shortage, especially during the summer period.

The vision of the future is more contrasted. The majority of officers surveyed are confident about the future and in the capacity of the state services to adapt to the current changes, particularly, climate change. Several factors are mentioned: (1) the increase in the frequency of crises (drought) is seen as a positive factor because it helps raise awareness of the issues among users and other actors in civil society, such as environmental protection organisations. (2) Setting up the OUGC (delegated by the state to share the resource), is also seen as a factor that encourages the farming profession to take responsibility. (3) Overall, the institutions responsible for managing water at the level of river or groundwater basins, especially the organisations in charge of the local water management plans (SAGE), will have more technical, financial and legal resources to put pressure on users that fail to comply with regulations. (4) The state is constantly improving service coordination, which could compensate for the small number of staff.[12] (5) The adoption of new technology (e.g. smart meters) should also facilitate inspections and make it easier to apply regulations. (6) The users are gradually improving the efficiency of their irrigation techniques, which means that the volume required per hectare can be reduced. (7) Lastly, the construction of reservoirs for storing excess water in the winter will help reduce conflicts in the summer (see Chap. 18).

Several agents surveyed are more pessimistic about how compliance will evolve in the future. In their view, the sustainable water resource management policy is bound to fail because of the reduction in the human resources allocated to monitor-

[12] See Jevakhoff, Barthod, Cartier, Delaunay, and Lavarde (2018) for a discussion on the need to rethink the spatial distribution of police staff at the AFB and ONEMA across French territory.

ing. In addition, they consider that the volumes allocated for abstraction are over-optimistic, often as a result of pressure from the agricultural profession (see Chap. 11). This will inevitably cause environmental crisis situations, which are likely to be more frequent as a result of climate change. The state will no longer have the wherewithal to enforce the rules for crisis management (temporary restrictions of use); the crises will be too frequent and applying the regulations would threaten the survival of a large number of farms. Society may well accept giving priority to economic activities and employment, to the detriment of protecting water resources.

23.7 Conclusion

The implementation of a quantitative management policy for water resources implies that the state has the capacity to enforce the often complex regulatory provisions on thousands of users: users that are scattered over vast regions and whose behaviour is not easy to discern. This can only be achieved if the state allocates considerable human, technical and financial resources to water policing. The results of the survey presented in this chapter show that this is not the case, even in a country like France, which has sufficient economic resources and a multisecular tradition of state intervention in water management.

The primary lesson drawn from this survey is that it will take years or even decades to make groundwater users comply with regulations. This is because regulations restrict the use of a resource, which users have considered to be freely accessible for years. Users will only comply with regulations if they fully understand their justification and are convinced that they are genuinely in the public interest. This requires political discussion, which has not necessarily occurred in all the French counties analysed in this survey. Nevertheless, the survey shows that there are less serious problems of compliance in regions where quantitative management was set up over 20 years ago (Clain basin, Beauce aquifer, Poitevin marshlands[13]) in comparison to the south-west, where it is relatively recent. When regulations are enforced, the users should also be trained in the procedures to apply. French law states that ignorance of the law is no excuse. Yet, it is common sense to suggest that the state should be responsible for training users. Therefore, the water police's primary mission is pedagogical. The punitive approach is only applied gradually over time.

The second lesson is that it is harder to convince people of the value of environmental protection compared to other issues, which are considered more fundamental by society as a whole. To quote M. Boutelet (2014, p. 150), local actors "*are vaguely aware of the need to protect the environment but not to the point of regarding environmental damage as the violation of a fundamental value, such as damage to private property, for example, theft*". This raises the question: are the recently

[13] Departments (counties) 18, 28, 37, 45, 79, 85 and 86 in Fig. 23.4.

established user groups (OUGC, see Chap. 3) capable of implementing the regulations? The transfer of certain state responsibilities (for example, resource allocation) begs the question: would it not be better to strengthen the state's capacity to control and apply sanctions in parallel?

The third lesson drawn from the survey is related to how the water police are organised and how their action is coordinated with that of the judicial system. The French experience demonstrates the advantage of this type of coordination when it comes to: targeting the main local water management issues; developing a common culture shared by magistrates and water police officers; and defining the methods for field intervention and case preparation. Administrative sanctions and criminal fines are applied more frequently than other penalties in order to improve user compliance. Criminal sanctions, which have a moral dimension, are only applied in exceptional cases. They should be exemplary. Indeed, the credibility of public action depends on it. This would also help avoid overloading the legal system, which is already struggling in France, as is the case in many other democracies.

The challenges of compliance and enforcement are by no means specific to France, as shown in Chaps. 22, 27 and 29, which deal with the issue in relation to Australia, Chile and Morocco.

References

AFB (Agence Française de la Biodiversité). (2017, October). *10 ans d'activité et bilan 2016*, 40 p. https://www.afbiodiversite.fr/actualites/le-rapport-dactivite-2016-de-lagence-est-publie

Barone, S. (2016). *Comment la Justice traite les dommages écologiques: une sociologie des acteurs judiciaires face aux atteintes à l'eau et aux milieux aquatiques* (Final report). Montpellier, France: Irstea. 46 p.

Barone, S. (2018). L'impunité environnementale: l'Etat entre gestion différencielle des inégalismes et désinvestissement global. *Champs Pénal, XV*, 25. https://journals.openedition.org/champpenal/9947

Boizard, P., Garcin, M.-L., Ménager, T., & Tosi, J.-C. (2016). *Audit de la mise en oeuvre des politiques de l'eau et de la biodiversité dans le département de Lot-et-Garonne* (139 p). Paris: Ministère de l'Environnement, de l'Energie et de la mer et Ministère de l'Agriculture, de l'AgroAlimentaire et de la Forêt.

Boutelet, M. (2014). Polices de l'eau et ordre public écologique, valeurs exprimées, valeurs protégées. *Droits et Cultures, 68*, 143–163.

Boutelet, M., Brun, A, & Van Bosterhaudt, P. (2012). La police de l'eau. L'autre versant de la politique de l'eau en France. In: Brun and Lasserre (dir.), *Gestion de l'eau, approche territoriale et institutionnelle* (pp. 95–121). Presse de l'Université du Québec.

Cour des Comptes. (2010). *Les instruments de la gestion durable de l'eau.* Paris: Cour des comptes. 39 p.

Cour des comptes. (2017). *Les services déconcentrés de l'Etat: clarifier leurs missions, adapter leur organisation, leur faire confiance*, Paris, 269 p.

Jevakhoff, A., Barthod, C., Cartier, J., Delaunay, A., & Lavarde, P. (2018). *L'avenir des opérateurs de l'eau et de la biodiversité.* Inspection Générale des Finances and CGEDD, 27 p.

Legrand, H., Rébeillé-Borgella, E., Chabrol, D., Flam, G., Marchal, Y., & Soulié, M.-C. (2015). *Evaluation de la police de l'environnement* (160 p). Paris: rapport interministériel, Ministères chargés de l'Ecologie, de la Justice, de l'Intérieur et de l'Agriculture.

MEDDE (Ministère de l'Ecologie, du Développement Durable et de l'Energie). (2012). *Rapport d'activité des services déconcentrés – Gestion des ressources naturelles*, 23 p.
Ministère de l'Environnement, de l'Energie et de la Mer. (2016). *Rapport d'activité des services déconcentrés 2015* (39 p). Paris: Gestion des ressources naturelles.
Ministère de la Justice. (2015). Circular n°2015-04 of 21st April 2015, relating to: les orientations de politique pénale en matière d'atteinte à l'environnement. *Journal Officiel du Ministère de la Justice*, 30 April 2015, 30 p.
ONEMA. (2010). *Les temps forts 2010 de l'Onema*. 48 p.
ONEMA. (2011). *Les temps forts 2011 de l'Onema*. 48 p. http://www.onema.fr/sites/default/files/images/temps-forts-2011.pdf. Accessed 14 Sept 2018.
ONEMA. (2012). *Les temps forts 2012 de l'Onema*. 25 p. http://www.onema.fr/sites/default/files/pdf/Onema-Temps-forts.pdf. Accessed 14 Sept 2018.
ONEMA. (2013). *Les temps forts 2013 de l'Onema*. 52 p. http://www.onema.fr/sites/default/files/images/les-temps-forts-2013.pdf. Accessed 14 Sept 2018.
ONEMA. (2014). *Les temps forts 2014 de l'Onema*. 52 p. http://www.onema.fr/sites/default/files/images/temps-forts2014.pdf. Accessed 14 Sept 2018.
ONEMA. (2015). *Les temps forts 2015 de l'Onema*. 36 p.
Simoni, M-L., Perriez, F., Berriat A., Delbos, V., Maziere B., Tetart S., Bourgau J-M., Guth M-O., & Verrel J-L. (2005). *Renforcement et structuration des polices de l'environnement*. Paris: rapport interministériel des ministères chargés de l'Intérieur, de l'Equipement et de l'Aménagement du Territoire, de l'Agriculture, de la Justice et de l'Ecologie, 70 p + annexes.
Van Bosterhaudt, P. (2014). Les valeurs des polices de l'eau. Efficacité du droit pénal de l'environnement à l'épreuve de la protection de l'eau et des milieux aquatiques? *Droit et cultures. Revue internationale interdisciplinaire, 68*, 81–142.

Dr. Marielle Montginoul is senior researcher in Economics at the National Research Institute for Agriculture, Food and Environment (INRAE – previously IRSTEA) and she is a member of the Joint Research Unit G-Eau. Her work focuses on understanding and modeling farmers and households' water consumption behaviors. She more specifically studies instruments that can be used to reveal these behaviors when information is incomplete. Her research also focuses on economic tools to manage water withdrawals, with a focus on water pricing. She mobilizes a wide range of methodologies including surveys, experimental economics, and scenarios workshops. Marielle is member of the scientific council of the Rhone Méditerranée and Corsica Water agency. She coordinates a Master in Social sciences applied to water management in Montpellier University

Dr. Jean-Daniel Rinaudo is researcher at Brgm, Montpellier University, where he coordinates the scientific program on environmental and risk economics Initially trained as an agricultural engineer (Montpellier SupAgro 1994), he specialized in agricultural and resource economics (PhD University of Auvergne, 2000). Prior to joining Brgm, he worked for the International Water Management Institute in Pakistan where his research focused on the political economy of irrigation management reforms. His current research mainly focuses on the institutional economic dimension of groundwater management. Most of his research is conducted in France but he also works in Morocco and Chile. He is currently developing new research activities in the field of natural disasters economics, focusing on the methods to assess economic vulnerability and resilience. Dr. Jean Daniel Rinaudo is also member of the Scientific Council of the Adour Garonne River basin agency.

Charlotte Alcouffe has recently obtained a Master's degree in Water Sciences and Agriculture at Agropritech School in Montpellier, France. During her Master's Degree studies, she did an internship in 2017 related to water governance, water policy and water management. The research was on the key factors that explained the compliance of the regulation on groundwater management in

France. She read, analyzed and synthesized some annual reports of the French Environmental Ministry; she conducted interviews with the French services of « Police de l'Eau ». In 2018, during her internship of her last year of Master's degree, she conducted a study at the International Office of Water (France) throughout Europe. She collaborated in the preparation of the report 'Agricultural phosphorus regulation in Europe: experience-sharing for 4 European countries' that has been published in November 2018. Charlotte is mother tongue French and Italian and fluently speaks English, Spanish and Greek.

Chapter 24
Tracing the Impact of Agricultural Policies on Irrigation Water Demand and Groundwater Extraction in France

Josselin Rouillard

Abstract Sustainable groundwater quantitative management does not only depend on implementing the right water policy instruments. It also relies on enabling sectoral policies that work in synergy with water policy objectives. To explore this link, this chapter presents the evolution of European agricultural policies, their level of support to irrigated farming, and consequences for groundwater abstraction in France. Three phases are identified. Until 1992, the French government encouraged the deployment of irrigated farming through price support mechanisms, market measures, subsidies for agricultural modernisation, and large scale supply infrastructure projects. The second phase, from 1992 and 2003, is a transitional period during which agricultural policies maintained an explicit support to irrigated farming, while the first agro-environmental schemes were established. The third and on-going phase (2003–2020) is associated with the progressive removal of direct payments for irrigated crops, while rural development funding offers mixed incentives. The chapter then presents current policy instruments contributing to reduce structural water deficits due to agricultural abstraction. To date, most projects to achieve groundwater quantitative targets focus on improvements in irrigation efficiency and the building of "compensatory" water storage schemes. To meet the challenges of climate change and increased scarcity, future initiatives should focus on water savings through the diversification of agricultural and food systems.

Keywords Groundwater pumping · Irrigation · Common agricultural policy · Mainstreaming · Demand-side management · Agri-food systems

J. Rouillard (✉)
BRGM, Montpellier University, Montpellier, France
e-mail: J.rouillard@brgm.fr

J.-D. Rinaudo et al. (eds.), *Sustainable Groundwater Management*,
Global Issues in Water Policy 24, https://doi.org/10.1007/978-3-030-32766-8_24

24.1 Introduction

In France, the agricultural sector is responsible for around 20% of groundwater use and 30% of surface water use, but it is the largest annual net water consumer (50%) and can represent up to 80% of total consumptive use in summer in some regions. About 40% of water abstracted annually for agriculture in the country is from groundwater (AFB, 2017). In central and western regions, groundwater represents the majority if not the only source of water (e.g. Beauce, see Chap. 5).

The progressive tightening of abstraction controls in 1990s have posed significant challenges to irrigated agriculture in France. Farmers have invested in more efficient irrigation techniques as well as reservoirs to store winter flow for consumptive use during the summer period. Conflicts have nevertheless become frequent, especially in regions where irrigation is mainly used for intensive cereal production. Proponents of irrigated agriculture emphasise its role in enhancing crop productivity and the competitiveness of the sector as well as in reducing exposure to drought risks and stabilising farm income. Critics emphasise the impacts of abstraction on environmental flows, ecological continuity and natural habitats, the appropriation of water by an intensive form of agriculture, and the high cost of building water supply infrastructure.

To reduce conflicts and align water demand with available resources, French authorities have set quantitative targets for priority aquifers and catchments (Erdlenbruch, Loubier, Montginoul, Morardet, & Lefebvre, 2013). They also require a reduction of agricultural water allocations which must be mutualised and allocated annually to irrigators (see Chap. 3). This approach largely assumes that farmers would adapt their choice of crop production and irrigation management according to their allocations. However, this perspective does not account for the sectoral incentives that work against water policy objectives and contribute to increase agricultural water demand.

This chapter posits that successful groundwater quantitative management is not only dependent on water policy instruments such as those reducing water allocations but also on enabling sectoral policies. It presents how, historically, agricultural price support mechanisms, market measures and subsidies for agricultural modernisation have largely contributed to promote irrigation and increase groundwater abstraction (Fig. 24.1, Table 24.1). It also examines how reformed agricultural and rural development policies can contribute to reduce abstraction pressure and help reach quantitative management targets by encouraging changes in farm and irrigation management.

The chapter is organised in the following way. Section 24.2 presents a historical narrative (1950s–2010s) of how different agricultural and rural development policy phases have influenced the development of irrigation and consequently groundwater abstraction. Table 24.1 provides an overview of the policy instruments reviewed, their relationship with the development of irrigation, and their impact on groundwater. Section 24.3 presents current strategies used to reduce structural water deficits due to agricultural abstraction. The conclusion highlights opportunities and

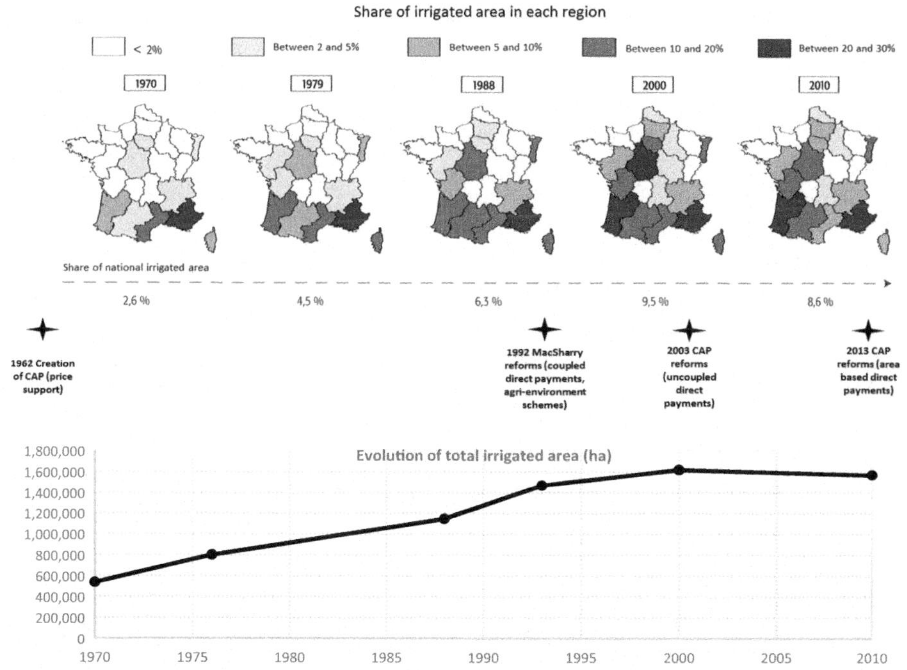

Fig. 24.1 Key dates in the evolution of agricultural policies and irrigated areas in France. (Source: modified from Lerbourg, 2012; Loubier, Campardon, & Morardet, 2013)

challenges for integrating groundwater quantitative management and agricultural policies in France.

24.2 Irrigation Development and Groundwater Use in France

24.2.1 Increasing Food Production Through Irrigation: 1945–1992

Irrigation has long been an essential element of agriculture in the southern and drier Mediterranean regions of France. It was traditionally based on the diversion of surface river water into canals and applied through gravity to orchards, vegetable crops, and rice fields (in the Rhône delta). In other regions of France, surface irrigation was limited to areas with specific climatic, topographic or soil characteristics. Irrigated agricultural land increased significantly in the second half of the twentieth century, from 402,000 ha in 1955 to 539,000 ha in 1970 (+25% in 15 years). By this time, irrigation remained concentrated in southern France and could still be generally

Table 24.1 Policy mechanisms influencing investment and maintenance of irrigation systems in France

Key policy mechanism	Impact on the development of irrigation	Likely impact on groundwater abstraction
1945–1992		
EU market and price interventions (CAP guarantee fund)	Minimum commodity price and stable income offer favourable grounds for private investments in irrigation infrastructure	+
National and European structural funding for the modernisation of agricultural holdings	Subsidies for targeted investments in irrigation infrastructure	++
1992–2003		
Direct coupled payments and irrigation premium	Direct income support strengthens farm-level investment capacity (e.g. to develop new irrigation infrastructure) and cash flow to maintain existing infrastructure	++
Rural development plans –investments	Subsidies for targeted investments in irrigation infrastructure	+
Rural development plans (agri-environment payments)	Compensation of income loss and additional costs for the uptake of less water intensive land use and management practice	–
2003–2020		
Direct uncoupled payments	Direct income support maintains investment capacity and cash flow, but does not directly encourage increased agricultural production	0
Greening	Payments rewarding crop diversification, permanent grasslands and ecological focus areas contribute to reduce incentive to farm water intensive crops	–
Rural development plans (investments)	Subsidies for targeted investments in irrigation infrastructure (additional environmental conditionality and funding of compensatory reservoirs)	+/–
Rural development plans (agri-environment payments)	Compensation of income loss and additional costs for the uptake of less water intensive land use and management practice	–

Legend: ++ actively supports abstraction, + contributes to encourage abstraction, 0 neutral, –: reduces incentive to abstract

characterised as a form of "structural" irrigation, whereby irrigated water represents most water supply to the crop in the dry season. Yet, 18 years later in 1988, irrigation reached 1,147,000 ha (+112%) (Janin, 1996), expanding northwards and westwards (Table 24.1) into regions where irrigation can be characterised as "complementary" because irrigated water mainly serves to control the timing and quality of crops.

The vast expansion of irrigation in France is associated with a pro-active policy to increase food production and the competitiveness of French agriculture on international markets (Brun, Lasserre, & Bureau, 2006; Rieu & Arlot, 1992). The EU Common Agricultural Policy (CAP) was launched in 1962, guaranteeing minimum commodity prices to European farmers. In parallel, a vigorous policy to modernise agriculture was initiated via subsidies for agricultural equipment and infrastructure such as irrigation schemes (Dechambre, 2007; Perrin et al., 2003). Thanks to CAP incentives, private investments and technical progress, food production increased by 64% between 1960 and 1980 and France became the second net exporter of food product worldwide in 1981 (Brun et al., 2006). CAP price support mechanisms particularly favoured cereal and maize production. By 1988, irrigated maize represented 48% of irrigated areas in France (Janin, 1996).

The development of irrigation took three main forms (see also Amigues et al., 2006):

- *Large schemes based on surface irrigation and managed by regional development agencies.* Starting in the 1950s, these multi-purpose schemes aimed to supply domestic (including tourism), agricultural and industrial uses. The objective was to develop rural areas and reduce poverty in the southern and central regions of France. Large reservoirs, heavily subsidised, were constructed in the Alps, Pyrenees and Massif Central mountains as well as long distance canals and water transfers. Large water storage also helped maintain river flows during the low flow season and ensure sufficient water supply in downstream areas of the river basin.
- *Collective irrigation schemes created and managed by irrigation associations.* These schemes involved the derivation of surface water through canals and reservoirs, although some example existed of collective groundwater schemes[1]. Although collective irrigation schemes represent an old form of partnerships between farm businesses, public authorities took an active part in supporting their development in the second half of the twentieth century with up to 60% subvention rate in the 1960s.
- *Private investments consisting of farm-level irrigation infrastructure and material.* These investments included individual pumping units in surface water bodies or boreholes. Up until the early 1970s, these initiatives were mostly found in northern France where individual farm businesses had the financial means to support large investments individually (Martin, 1972). The 1980s saw the vast expansion of individual irrigation schemes, sometimes supported by public subsidies.

While collective schemes led to the greatest increase in irrigation between 1955 and 1965, individual initiatives became more popular from 1966 onwards (Martin, 1972). Between 1970 and 1988, individual initiatives represented two-third of the

[1] One such example is located in the Rhône county. See http://www.smhar.fr/presentation/historique-du-smhar/

additional 400,000 ha of irrigated areas (Loubier et al., 2013). During this period, groundwater pumping became a major source of agricultural water (see Chap. 3).

Structural water deficits in the 1980s in many catchments and aquifers such as in the Beauce region (see Chap. 5) or in the Marais Poitevin (see Chap. 18) led to a change in water policy in France in 1992 (see Chap. 3). However, reforms in agricultural policy did not remove the incentive for intensive cereal and maize production until the 2000s, leading to an increase in irrigated areas throughout the 1990s.

24.2.2 *The Difficult Reform of a Productivist Model: 1992–2003*

24.2.2.1 A System of Direct Payments Incentivising Irrigation Use

In the 1980s, the CAP was increasingly criticised for its increasing burden on European public finance, for causing environmental degradation and for generating large food surpluses and international market distortions it created. The CAP underwent a first major transformation in 1992 under the MacSharry Reforms. The reforms led to a requirement on establishing set-asides, initially 15% of land on each farm[2]. Most importantly, a system of direct payments per hectare of farmed area would replace price support mechanisms. From 1996, payments were based on reference values set for each country at European level, then adjusted by a reference yield for each crop. In France, this reference yield was set at county level, i.e. higher payments were provided in counties with higher yield reference values. While the transition from price support to direct payments resulted in a net loss of income for farmers, a form of "coupling" on production (yield) was thus retained.

To compensate income loss to farmers who had invested in irrigation infrastructure and material in the 1980s and 1990s, French authorities created an additional premium on irrigated crops by mean of a regional reference yield value which accounted for the greater yield usually obtained from irrigated crops (Brun et al., 2006). As a result, direct payments for irrigated could be as high as 80% above the payments for dry cereal crops (Martin, 1996, see also Table 24.2 for examples of

Table 24.2 Income potential from cereal crops in the Midi-Pyrénées region in 2000/2001

Crop	Average yield	Price (€/q)	Product (€/ha)	Direct payment (€/ha)
Sunflower (dry)	28	26	734	337
Maize (dry)	75	12	994	296
Maize (irrigated)	110	12	1286	441
Sorgho (irrigated)	85	9	805	441
Soja (irrigated)	33	23	1219	531

Modified from Amigues et al. (2006)

[2] Set asides were abandoned at EU level in 2008.

direct payments on dry and irrigated crops). Overall, the MacSharry reforms in France did not lead to the removal of all incentives to grow irrigated maize and cereals, which still benefited from higher CAP payments and better market conditions (Hurand, 1998; Simon, 1998).

By 2000, the extent of irrigated area reached 1.57 million ha (around +40% from 1988) while the area equipped for irrigation covered 2.6 million ha (around +50% from 1988). At this time, the development of irrigation occurred through a mix of collective and individual, farm-level schemes (Loubier et al., 2013). A significant development of groundwater abstraction can be observed during that period (see Chap. 3).

24.2.2.2 The Growth of Agri-Environmental Measures in Rural Development Policy

The development of rural areas has been a priority for the French government since the immediate post Second World War period. Initially focused on modernising agriculture, rural development policy progressively broadened its scope to promote the development of rural infrastructure, the economic diversification of rural areas and, more recently, the management of natural resources and environmental protection (Dechambre, 2007; Perrin et al., 2003; Vandenbroucke, 2013).

At European level, a support scheme to farms in less favoured areas and regional structural funds to develop poorer regions were created in 1975, while the LEADER programme was created in 1991. Agri-environment schemes, created in 1988 and made compulsory for all member states in 1992, started to provide payments covering income loss and additional costs for measures increasing the environmental performance of farms. The 1999 reforms consolidated these multiple EU policy instruments around a coherent EU-wide rural development policy, also called the "second pillar" of the CAP[3]. EU rural development policy offered the possibility to Member States to define their own priorities and select from 22 types of measures from which countries or regions could design their rural development programs according to their needs. Some measures aimed to modernise the agricultural or forestry sector while others aimed to strengthen rural economies at large or preserve natural resources (i.e. agri-environment schemes).

Building on localised experiments since 1989, French authorities established their first national agri-environment scheme in 1993 (Couvreur, Mitteault, & Michel, 1999). This first scheme did not include any measures to reduce agricultural irrigation or tackle abstraction pressures. Nevertheless, several measures protecting biodiversity and reducing nitrogen or pesticide pollution could indirectly reduce abstraction pressures, such as crop diversification, crop rotation, and conversion of cereal crops into permanent grasslands. Measures preventing the conversion of

[3] The "first pillar" refers, since 1991, to the system of direct payments and other market interventions to stabilise the agricultural sector, see previous section.

grasslands into cropland also helped indirectly by reducing the incentive to convert to more water-intensive crops.

In contrast to the first scheme, the second agri-environment scheme (2000–2006) explicitly integrated the issues of quantitative water management. One measure specifically aimed to reduce irrigated areas by replacing irrigated cereal crops with non-irrigated crops and another to reduce irrigation intensity by rotating cereal crops with non-irrigated leguminous crops. As a condition for payment, the farmer had to give up the associated abstraction allocation. However, their uptake was nearly non-existent (CNASEA, 2008); thus, their impact on reducing agricultural abstraction from groundwater and surface water was very limited.

24.2.3 Tackling Production Incentives, Promoting Good Practice: 2003 Onwards

24.2.3.1 Des-incentivising Irrigation in Direct Payments

As discussed previously, CAP reforms in the 1990s maintained an incentive to increase agricultural production by coupling payments to crop yields. A subsequent reform in 2003 decoupled most direct payments. Two options were available to Member States: payments to farmers on an area basis (i.e. a uniform payment for all farms based on ha and type of crops produced) or on an historical basis (i.e. farms would receive the average of payments received in the period 2000–2002). In addition to direct decoupled payments, Member States could offer (limited) additional direct coupled payments to support specific crop production.

Because France opted for the historical approach, decoupled direct payments during the 2006–2013 CAP programming period remained higher for farmers who irrigated during the 2000–2002 reference period than for those farmers who did not. In addition, France opted to maintain some coupled payments, together with the irrigation premium, on cereal, oleaginous and protein crops (Boulanger, 2007).

As observed by Loubier et al. (2013), total irrigated area in France did not change significantly between 2000 and 2010 while the area equipped with irrigation has reduced by 12%. Furthermore, during the same time, the total area of irrigated maize reduced by 8% while other irrigated cereal crops increased by 11% (Fig. 24.2).

The latest CAP 2014–2020 aims to phase-out decoupled payments based on historical references towards fixed area-based payments. In France, decoupled payments should converge towards a unit value of 132€/ha by 2020 (MAAF, 2017b). In addition, a compulsory "greening" top-up to the basic decoupled payments rewards crop diversification, the maintenance of permanent grasslands, and ecologically focus areas (e.g. field margins, buffer strips along rivers, hedges, N-fixing crops, green cover, landscape elements). Most coupled payments target livestock farming rather than crop production. Overall, the convergence in area payments between irrigators and non-irrigators and greening should further reduce historical policy incentives for irrigation.

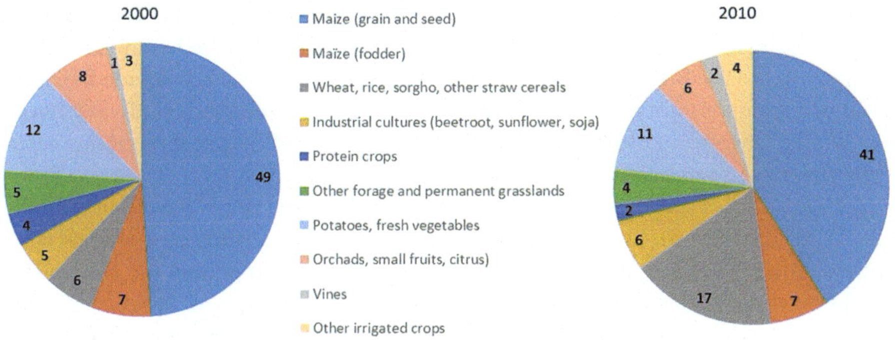

Fig. 24.2 Main irrigated crops in France in 2000 and 2010. (Source: modified from Loubier et al., 2013)

In addition to the change in direct payments, the 2003 CAP reform attached cross compliance requirements to all direct payments to ensure farm compliance with a range of existing environmental and sanitary regulations and best practice. One conditionality, still applied to direct payments nowadays, relates to irrigation and requires that irrigators benefiting from CAP payments install water meters on their irrigation equipment and have a water abstraction authorisation from the authorities (or have made the relevant declaration)[4].

24.2.3.2 Promoting Good Practice in Irrigation Development and Management

The third French agri-environment scheme 2007–2013 offered similar measures to the 2000–2006 French agri-environment scheme (MAP, 2007). The uptake of agri-environment measures to reduce abstraction from irrigation was also very limited during the 2007–2013 scheme (<5000 ha) (MAAF, 2017a). The on-going agri-environment scheme 2014–2020 (MAAF, 2016) provides payments for introducing rotations of leguminous crops in irrigated cereal farming systems (i.e. from 78€/ha/year to 215€/ha/year depending on the amount of land targeted) (MAAF, 2016). A recent assessment of agri-environment schemes encouraging water savings in agriculture in five countries (including France) showed that the measures have not been effective and faced significant implementation barriers (Oréade-Brèche, 2018).

Subsidies to construct, renovate and expand irrigation infrastructure for individual farms or collective schemes have long been available. Several conditions were attached to these subsidies in the 1990s and 2000s to account for stricter environmental requirements (e.g. impact assessments) and additional abstraction controls (see Chap. 3). In particular, funding started to differentiate traditional reservoirs,

[4]This conditionality has been applied since 2000 in France for irrigated maize.

which aim is to increase water supply to agriculture during dry periods, and "compensatory" reservoirs (in French, "réserves de substitution") which aim is to reduce the environmental impact of agricultural abstraction. Constructed outside the minor river bed, compensatory reservoirs are filled with surface and/or groundwater during the winter season and used in summer by farmers instead of direct pumping in surface water or groundwater bodies.

Under rules established for measure 125C of the rural development plan 2007–2013, irrigation projects leading to an increase in abstracted volumes could only be funded (1) if the river basin or aquifer had no water deficit and (2) where it could not impact the good status of the water body. In areas with water deficits, subsidies were only available to reduce abstraction pressure by temporally or spatially redistributing abstraction through water transfers or compensatory reservoirs. Overall, between 2006 and 2013, measure 125C was associated with 116 million € worth of investment for 185 projects across France.

Subsidies for developing irrigation are still available in many of the on-going regional rural development plans 2014–2020. However, additional rules were attached, in particular to avoid an increase in irrigated areas following the construction of water storage or of more efficient irrigation scheme. Specifically, EU rural development plans must meet Article 46 of Regulation (EU) No 1305/2013 which requires that any investment in irrigation subsidised with EU funds, whether in new or existing irrigated areas, should meet a number of criteria which include:

- A river basin management plan is in place in the irrigated area and water metering is carried out on all abstraction points;
- Investments into existing installation result in potential water savings of at least 5–25% according to the technical parameters of the existing installation.

Any investment in areas where ground or surface waters are in less than good status for reasons related to water quantity (according to the river basin management plans under the EU Water Framework Directive) has to ensure an effective reduction in water use of at least 50% of the potential water saving. Net increases in irrigation area are only possible in areas where water bodies are not failing good status for reasons related to water quantity.

An evaluation made by Rouillard and Berglund (2017) demonstrated compliance with these requirements in several French rural development plans. Some plans are more ambitious. For example, the Poitou Charentes region, which partly administered the Marais Poitevin (see also Chap. 18), only funded compensatory reservoirs, thereby aiming that new irrigation infrastructure funded via rural development policy did not lead to more abstraction in groundwater bodies.

24.3 Collective Approaches for Quantitative Groundwater Management in Agricultural Areas

24.3.1 Establishing Volumetric Management of Agricultural Abstraction

The favourable public policy support for irrigation in the second half of the twentieth century has led to a vast increase in agricultural irrigation and the creation of multiple abstraction points. More specifically, the importance of individual abstraction points (1.16 million ha in 2010 compared to 410,000 ha in collective or mixed irrigation schemes, see Loubier et al., 2013) represents a significant challenge to regulators whose monitoring and enforcement capacities are limited.

To increase the capacity to regulate agricultural abstraction, the 1992 Water Law established a requirement on irrigators to install water meters and request yearly abstraction licences (see Chap. 3). The 1992 Water Law created two additional mechanisms for managing irrigation abstraction: temporary restrictions on abstractions to reduce abstraction pressures during drought situations and volumetric management of abstraction authorisations to tackle structural water deficits.

These instruments formed the basis for greater control on irrigation abstraction with the implementation of low flow and groundwater level targets, and spring and summer abstraction caps. Volumetric management in particular helped initiate some first schemes to control irrigation abstraction at aquifer level (e.g. Beauce) or catchment level. However, the instrument was voluntary and was thus first implemented for catchments and aquifers with intense water conflicts (e.g. Beauce, see Chap. 5).

The 2006 Water Law made allocation caps compulsory in priority basins and aquifers, and requires agricultural water user associations ("Organisme Unique de Gestion Collective" or OUGC) which role is to facilitate the fair and equitable distribution of water allocated to irrigation between irrigators (see Chap. 3).

As the implementation of the 2006 Law progressed, large mismatches were confirmed in several catchments and aquifers between irrigation water demand and allocations available for irrigation water use. In the late 2000s, several studies examined the economic impacts of reducing water allocations to the agricultural sector (Bouarfa et al., 2011; Danel, 2011; Hébert et al., 2012; Lejars et al., 2012). These studies suggested that the new restrictions could have a significant impact on farm businesses and agro-food chain that are highly dependent on irrigation.

24.3.2 *Emerging "Territorial" Contracts Between Agricultural Water Users and the State*

The response to reduce large structural water deficits while avoiding severe economic impact has usually been to promote the construction of compensatory water storage. However, building reservoirs is costly and is not economically viable without public subsidies in most regions in France (Loubier, Poussin, Gleyses, Le Mat, & Garin, 2011). The agricultural sector thus negotiated public support to build water storage where irrigation abstraction led to structural water deficits. In 2011, the main agricultural union requested € 1 billion for the creation of 100 Mm^3 of water storage through compensatory reservoirs across France. The French government promised in return 100 million Euros to subsidise the creation of 40 Mm^3 of water storage (DGALN, 2011). In the Adour-Garonne river basin, a regional agreement was signed the same year to allow the subsidised construction of up to 69 Mm^3 in 59 reservoirs; no quantitative objective for water savings were set (Aypahssorho, Caude, & Etaix, 2016).

Plans to build reservoirs were met with strong resistance by non-agricultural actors, citing the visual and environmental (i.e. affecting winter flow dynamics) impact of reservoirs, high public costs, and the explicit support to an intensive form of agriculture (see also Granjou & Garin, 2006). Following national elections in 2012, the French government imposed a ban on state funding for compensatory reservoirs. A parliamentary investigation was set up to identify "*a new vision for quantitative water management in agriculture*".

The resulting report (Martin, 2013) and stakeholder consultation led to a first ministerial decree in 2015 (RF, 2015) setting out new conditions on water agencies funding of compensatory reservoirs. More specifically, funding for compensatory reservoirs should be justified through local "projets de territoire" (i.e. "territorial" projects), presenting a comprehensive strategy to meeting quantitative management targets for a given groundwater body or catchment strategy through a balanced combination of supply and demand management measures.

24.3.3 *The "projet de territoire": An Integration of Agricultural and Water Policies, and Regional Development?*

As defined in the governmental decree of 2015, the "projet de territoire" must:

- Aim for a balanced quantitative management of water resources, considering the impacts of climate change; objectives must include quantified targets on reducing total abstraction;
- Take into account the chemical and ecological status of water bodies, notably by promoting the development of agro-ecological systems and crop diversification;

- Implement a variety of measures, including water demand management, changes in crop production and rotation, development of alternative agro-food value chains, improvement of water efficiency (e.g. drip irrigation, irrigation management) and infrastructure modernisation; water supply options should not only include new reservoirs and transfers but also water reuse;
- Be an outcome of a dialogue between all local actors.

The initial intention of the "projet de territoire" was to encourage an integrated approach to planning water storage for agriculture, taking into account the environmental and socio-economic impacts of the proposed infrastructure. The project should not only have an environmental objective, but also add "value" to the area in social and economic terms. For example, the planning should shed light on the economic benefits of irrigation and on the value for the local economy of non-irrigated agri-food value chains (RF, 2015). Thus, in theory, the "projet de territoire" is at the crossroads of several planning processes, including those related to river basin and catchment plans, agricultural and rural development policies, and regional development plans (Fig. 24.3).

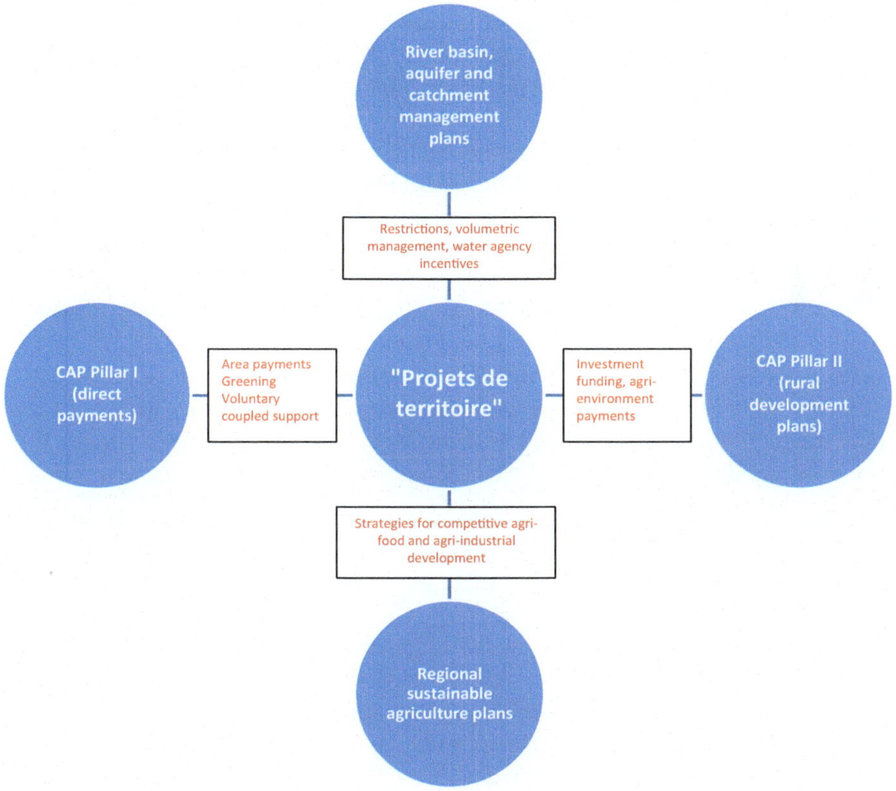

Fig. 24.3 The "projet de territoire" at the crossroad of water, agricultural, rural development and regional development policies

A recent ministerial communication (Bisch, Hubert, Mailleau, Denier-Pasquier, & Servant, 2018) and governmental decree (RF, 2019) broadens the scope of the "projet de territoire". It should consider quantitative targets for all sectors (e.g. agriculture, drinking water, industrial) and propose a balanced combination of measures across those. It should also integrate qualitative issues in water management (e.g. diffuse and point source pollution). Regarding agriculture, the communication reinforces the need to adapt agricultural production systems. Water savings should take priority over water storage or transfers. Finding synergies between sectors are encouraged (e.g. reuse of wastewater in agriculture, creation of multi-use water storage). The building of water storage for irrigation purposes is justified if it contributes more widely to regional development.

The "projet de territoire" builds on previous experiences in the Loire-Bretagne and the Rhone-Mediterranean-Corsica river basins where quantitative management contracts were used to manage structural water deficits in priority catchments and aquifers, i.e. respectively "Contrat Territorial de Gestion Quantitative" (CTGQ) and "Plan de Gestion des Ressources en Eau" (PGRE). Evaluations of the implementation of the CTGQ and PGRE have highlighted their value in raising awareness of the issues linked to intensive agricultural water use (Epices and AScA, 2015; Epices et al., 2017). However, several limitations were also found, one of which being that projects tend to focus on creating reservoirs (in the case of CTGQ) and water transfers (in the case of PGRE) rather than securing real water savings.

More recently, Bisch et al. (2018) showed that the "projet de territoire" concept did not help in overcoming conflicts around the management of irrigation water: out of 60 existing projects (including CTGQ and PGRE), only five have been validated and implemented. More significantly, most measures proposed in those projects focused on building water storage or water transfers, rather than securing water savings.

For example, in the Marais Poitevin area where over-abstraction of groundwater has been a recurring problem since the 1980s (see also Chap. 18), three CTGQs were adopted in 2012. 18.48 Mm^3 of new compensatory storage (e.g. filled with groundwater abstracted in winter and used for spring and summer irrigation) were planned in 2012. This compared to 1.56 Mm^3 of planned water savings[5] to be achieved through:

- Carry out farm-level audits to determine the adequate modifications to irrigation techniques and crop rotations to reduce overall water consumption;
- Develop tensiometer-based irrigation and more water efficient irrigation (e.g. drip irrigation);
- Encourage the uptake of more water efficient crops during the spring and summer season (e.g. spring crops, irrigated grasslands, sorghum) and adapt sowing dates depending on spring hydrological situation;
- Improve communication on the quantitative status of the water resources and offer training in irrigation management.

[5] Total calculated from the Sèvre Niortaise, Lay and Vendée CTGQs

Table 24.3 Planned and estimated water savings in the Lay and Vendée sub-basin of the Marais Poitevin

Measure	Planned saving in CTGQ (m^3)	Estimated saving 2017 (m^3)
Tensiometer-based irrigation	427,000	862,500
Agri-environment measure on irrigation	300,000	300,000
Earlier sowing date	305,780	173,000
Spring crop variety, water stress resistant crops	305,420	147,500
Crop diversification	162,000	48,000
Total	**1,500,200**	**1,405,500**

Source: SMMP & CA Vendée (2012a, b); CA Vendée (2017)

Table 24.3 presents planned and estimated water savings for two sub-basins of the Marais Poitevin. It shows that most savings were achieved through the application of tensiometer-based irrigation to optimise water use. The implementation of the agri-environmental measure to stop irrigation, which offered better payment conditions in the 2007–2013 than the 2000–2006 (see above), also contributed to reduce water abstraction allocations by 1.4 Mm^3.

Measures modifying cropping patterns and types are taking longer to implement, in part due to unwillingness to modify existing farm management practices, market demand, and requirements from agro-food chains. Although agricultural diversification has been promoted since the 1990s for biodiversity reasons (Simon, 1998), few farmers have taken up these options (Aypahssorho et al., 2016) and this approach does not appear prominently in the CTGQs.

Recent research evaluating four alternatives to solving water imbalance suggest strong trade-offs between environmental effectiveness and economic impacts on agriculture (Allain, Obiang Ndong, Lardy, & Leenhardt, 2018). The four alternatives included: reducing irrigated areas, assisting irrigation with decision-support tools, implementing crop rotations and merging water storage into later reservoirs. The study showed that crop rotations (in this case study, switching maize monoculture into a sunflower-straw cereal-oilseed rape and maize rotation) had the greatest potential for long term environmental preservation but the highest impact on farm economies. The study however focused on changes to gross margin in a similar production system. It did not explore the possible cushioning long term effect of a transformation of the production system towards higher value crops and value chains.

24.4 Conclusion

The chapter presented the evolution of French and EU agricultural policies, their role in irrigation management, and the consequences on groundwater and surface water use. Three phases were identified.

The first phase, from the immediate post-war to the early 1990s, is associated with a vast development of irrigation across France. The increase in groundwater use was associated with the uptake of complementary irrigation in central and northern regions of France to increase the productivity of cereal and maize farming. The second phase, mainly in the 1990s, is a transitional period, during which agricultural policies maintained an explicit support to irrigated farming. Issues of overexploitation resulted in the adoption of the first major policies to monitor and regulate groundwater abstraction and install water metering on irrigation equipment. The current phase (2000s–2010s) is associated with the progressive removal of incentives for irrigated crops. Rural development policies remain ambiguous: on the one hand, some agri-environmental measures tackling agricultural pressures on water quantity are proposed; on the other, funding is available for the construction of reservoirs to support irrigated farming. Conflicts around the funding of compensatory reservoirs show that this approach is not devoid of controversy.

More broadly, the chapter highlighted that French authorities have opted for a decentralised form of governance on agricultural water use though the use of local "contracts" and "projets de territoire" to achieve a balance between irrigation water demand and availability. The policy framework thus sets out a comprehensive strategy to manage irrigation water. However, in practice, water quantity targets are mostly met via the building of compensatory reservoirs and via efficiency gains (e.g. improvements in irrigation techniques). Less emphasis is given to modifying production types and optimising their commercialisation in order to enhance economic sustainability (i.e. increasing the value added per water consumed).

The liberalisation of agricultural markets and increased meteorological variability due to climate change are likely to increase the demand for irrigated water and water storage to secure fodder production and high quality crops. To achieve future quantitative groundwater targets, it is essential to ensure a coherent set of agricultural and water policy instruments. Policy action should not solely focus on water supply management (e.g. compensatory reservoirs, water allocations) or the efficiency of irrigation systems, but also on production choices made by farmers. Future work should thus explore in more detail the potential of agricultural diversification, alternative value chains, agro-food systems and rural development trajectories in meeting quantitative groundwater management targets.

Acknowledgement This work is part of the RURECO project ("Institutions for groundwater dependent rural economies"), funded by the Marie Skłodowska-Curie Action Individual Fellowship of the EU's Horizon 2020 Programme for Research, Technological Development and Demonstration under Grant Agreement no. 750553.

References

AFB (Agence Francaise pour la Biodiversité). (2017). Bulletin n°2: prélèvements quantitatifs sur la resource en eau. Edition Mars 2017.

Allain, S., Obiang Ndong, G., Lardy, R., & Leenhardt, D. (2018). Integrated assessment of four strategies for solving water imbalance in an agricultural landscape. *Agronomy for Sustainable Development, 38*(6), 12.

Amigues, J. P., P. Debaeke, B. Itier, G. Lemaire, B. Seguin, F. Tardieu, & Thomas, A. (éditeurs). (2006). *Sécheresse et agriculture. Réduire la vulnérabilité de l'agriculture à un risque accru de manque d'eau.* Expertise scientifique collective, Rapport, INRA, France, 380pp.

Aypahssorho, H., Caude, G., & Etaix, C. (2016). *Le Marais poitevin: état des lieux actualisé des actions gouvernemental 2003–2013 et orientations.* Ministère de l'Environnement, de l'Energie et de la Mer, Paris, France, 236pp.

Bisch, P. E., Hubert, L., Mailleau, C., Denier-Pasquier, F., & Servant, L. (2018). *Cellule d'expertise relative à la gestion quantitative de l'eau pour faire face aux épisodes de sècheresses.* Ministère de la Transition Ecologique et Solidaire et Ministère de l'Agriculture et de l'Alimentation, Paris, France, 128pp.

Bouarfa, S., Brunel, L., Granier, J., Mailhol, J.C., Morardet, S., & Ruelle, P. (2011). Evaluation en partenariat des stratégies d'irrigation en cas de restriction des prélèvements dans la nappe de Beauce (France). Cahier de l'Agriculture, 20, n° 1–2, 124–129.

Boulanger, P. (2007). *Subventions directes agricoles et gestion quantitative des ressources en eau.* Policy brief GEMPB-2007-4. Groupe d'économie mondiale, Septembre 2007.

Brun, A., Lasserre, F., & Bureau, J. C. (2006). Mise en perspective comparée du développement de l'irrigation aux Etats-Unis et en France. *Géocarrefour, 81*(1), 5–14. https://doi.org/10.4000/geocarrefour.1701

CA (Chambre d'Agriculture) Vendée. (2017). Bilan CTGQ du Bassin Lay et Vendée. Présentation.

CNASEA (Centre National pour l'Aménagement des Structures des Exploitations Agricoles). (2008). *Evaluation ex post du Plan de Développement Rural National.* Soutien à l'agroenvironnement. Synthèse de l'évaluation. Rapport final définitif. Paris, France, Décembre 2008.

Couvreur, M., Mitteault, F., & Michel, P. (1999). Les mesures agri-environnementales mises en œuvre en France. *Economie rurale, 249*, 6–10. https://doi.org/10.3406/ecoru.1999.5053

Danel, J.B. (2011). *Conséquences sur les filières agricoles et agroalimentaires de l'atteinte des objectifs quantitatifs de la Directive cadre sur l'eau et du SDAGE dans le bassin Adour-Garonne.* Conseil général de l'alimentation, agriculture et des espaces ruraux, Paris, France, 38pp.

Dechambre, B. (2007). La PAC et le développement rural. L'agriculture: de nouveaux défis. Editions 2007, INSEE. Available at https://www.insee.fr/fr/statistiques/1372302?sommaire=1372309. Accessed 1 Apr 2018.

DGALN (Direction Générale de l'Aménagement, du Logement et de la Nature). (2011). *Plan d'adaptation de la gestion d'eau: soutien à la création de retenues d'eau et à l'adaptation des cultures.* Ministère de l'Ecologie et Ministère de l'Agriculture, Paris, France.

Epices and AScA. (2015). *Evaluation des politiques publiques de l'agene de l'eau Loire-Bretagne.* Rapport final – Evaluation du contrat territorial. 164pp.

Epices, AScA, Artélia. (2017). *Evaluation de la politique de réduction des déséquilibres quantitatifs de la ressource en eau.* Synthèse. Bassins Rhône-Méditerranée et de Corse. 9pp.

Erdlenbruch, K., Loubier, S., Montginoul, M., Morardet, S., & Lefebvre, M. (2013). La gestion du manque d'eau structurel et des sécheresses en France. *Sciences Eaux et Territoires, 11*, 78–85.

Granjou, C., & Garin, P. (2006). Organiser la proximité entre usagers de l'eau: le cas de la gestion volumétrique dans le bassin de la Charente. *Développement durable et territoires*, dossier 7.

Hébert, N., Grandmougin, B., Loubier, S., Graveline, N., Marsac, S., Amen, J.-F., et al. (2012). Réforme des autorisations de prélèvement dans le bassin Adour-Garonne: impacts sur l'économie agricole. *Agronomie, Environnement & Sociétés, 2*(2), 113–126.

Hurand, P. (1998). La PAC et l'irrigation des grandes cultures. *Aménagement et Nature, 130*, 35–48.
Janin, J. L. (1996). L'irrigation en France depuis 1988. *La Houille Blanche, 8*, 27–34. https://doi.org/10.1051/lhb/1996083.
Lejars, C., Fusillier, J. L., Bouarfa, S., Brunel, G., Rucheton, G., Girard, X., et al. (2012). Impacts de restrictions en eau d'irrigation sur les exploitations et les filières agricoles en Beauce. *Agronomie, Environnement et Sociétés, 2*(2), 139–154.
Lerbourg, L. (2012). Des surfaces irrigables en baisse à partir de 2000. Ministère de l'Agriculture, de l'Agroalimentaire et de la Forêt. Agreste Primeur, numéro 292, Novembre 2012, Montreuil, France.
Loubier, S., Campardon, M., & Morardet, S. (2013). L'irrigation diminue-t-elle en France ? Premiers enseignements tirés du recensement agricole de 2010. *Science eaux et territoires, 11*, 12–19.
Loubiet, S., Poussin, J. C., Gleyses, G., Le Mat, O., & Garin, P. (2011). Faut-il subventionner la création de réserves d'eau pour l'irrigation ? Une approche par la modélisation microéconomique. *Cahier de l'Agriculture, 20*, 157–164.
MAAF (Ministère de l'Agriculture, de l'Alimentation et de la Forêt). (2016). *Plan de Développement Rural Français - Cadre National.* Version 03 Aug 2016. Paris, France, 950pp.
MAAF (Ministère de l'Agriculture, de l'Alimentation et de la Forêt). (2017a). *Evaluation ex post du programme de développement rural hexagonal (PDRH) Programme FEADER 2007–2013.* Rapport final, Tome 3. Paris, France, 163pp.
MAAF (Ministère de l'Agriculture, de l'Alimentation et de la Forêt). (2017b). *La Politique Agricole Commune, Cap sur la PAC 2015–2020.* Version révisée de Mars 2017. Available on http://agriculture.gouv.fr/la-pac-en-un-coup-doeil. Accessed 20 Apr 2018.
MAP (Ministère de l'Agriculture et de la Pêche). (2007). *Plan stratégique national de développement rural 2007–2013.* Version 16 avril 2007. Paris, France, 178pp.
Martin, J. (1972). L'évolution de l'irrigation en France. *Revue de géographie alpine, 60*(3), 419–443. https://doi.org/10.3406/rga.1972.1276
Martin, P. (2013). *La gestion quantitative de l'eau en agriculture : une nouvelle vision, pour un meilleur partage.* Parlemantaire en mission auprès du gouvernement, Paris, France, 87pp.
Martin, Y. (1996). *Rapport sur la gestion durable des eaux souterraines.* Ministère de l'Industrie, de la Poste et des Télécommunications. Conseil Général des Mines, Paris, France, 74pp.
Oréade-Brèche. (2018). *Economiser l'eau pour l'irrigation par les changements de pratiques agricoles : analyse comparée de politiques publiques et pistes d'amélioration en France.* Etude n°15.14., Paris, France, 86pp.
Perrin, J., Malet, J., Brun, G., Guérin, M., Ruas, J.F., Templé, P., et al. (2003). *Rapport de l'instance d'évaluation des politiques de développement rural.* Conseil National de l'Evaluation, Commissariat Général du Plan. Paris, France, 479pp.
RF (République Française). (2015). *Instruction du Gouvernement du 4 juin 2015 relative au financement par les agences de l'eau des retenues de substitution.* NOR : DEVL1508139J.
RF (République Française). (2019). *Instruction du Gouvernement du 7 mai 2019 relative au projet de territoire pour la gestion de l'eau.* NOR: TREL1904750J.
Rieu, T., & Arlot, M. P. (1992). Le poids de l'agriculture dans la répartition et l'usage de l'eau. *Economie rurale, 208*, 57–60. https://doi.org/10.3406/ecoru.1992.4452
Rouillard, J. J., & Berglund, M. (2017). *European level report: Key descriptive statistics on the consideration of water issues in the Rural Development Programmes 2014–2020.* Brussels, Belgium: European Commission.
Simon, G. (1998). *Le Marais Poitevin.* Conseil Général des Ponts et Chaussées. Paris, France, 22pp.
SMMP (Syndicat Mixte Marais Poitevin) & CA Vendée. (2012a). *Contrat Territorial Gestion Quantitative Secteur Lay.* Période 2013–2017.
SMMP (Syndicat Mixte Marais Poitevin) & CA Vendée. (2012b). *Contrat Territorial Gestion Quantitative Secteur Vendée.* Période 2013–2017.

Vandenbroucke, P. (2013). *Transformations de l'unité de production agricole : d'une exploitation sectorielle à une exploitation agricole territoriale. Exploitations agricoles, agriculteurs et territoires dans les Monts du Lyonnais et en Flandre intérieure de 1970 à 2010*. Thèse de Doctorat en Géographie, Aménagement et Urbanisme. Université Lumière Lyon 2, Lyon, France, 607pp.

Josselin Rouillard is currently a recipient of a Marie-Skłodowska Individual Fellowship working at the French geological survey (Brgm) on quantitative groundwater management and agriculture. His work in the past 12 years has focused on the assessments of governance arrangements and economic instruments in water management, focusing on the EU with some experience in Asia, North Africa and South America. His research work draws on institutional economics, collective action theories and the policy sciences. He holds a PhD (Dundee/Edinburgh) in the management of environmental systems, a MSc. (Oxford) in environmental change and management, and BSc. (Reading) in environmental sciences of the earth and atmosphere.

Chapter 25
Groundwater Management Lessons from Chile

Guillermo Donoso, Elisabeth Lictevout, and Jean-Daniel Rinaudo

Abstract Groundwater has increasingly become a water supply source in Chile. In the future this trend is expected to grow as a consequence of the increased water use due to economic growth, together with population growth, urbanization, water contamination and pollution, as well as the projected climate change impacts. The Water Code of 1981, as well as previous water codes, were in essence designed for surface water and, thus, contained only few references to groundwater. This regulatory absence has been covered with groundwater guidelines established through internal administrative acts. As it stands, the legal and institutional context considers the required instruments and mechanisms to balance growing demand and the need to protect and preserve groundwater resources. This chapter investigates whether this framework has been effective to ensure that groundwater is managed sustainably, through the analysis of two cases located in an arid region of northern Chile: the Copiapó Valley and the Pampa del Tamarugal Aquifer.

Keywords Groundwater governance · Groundwater management · Collective groundwater management · Groundwater communities · Chile

---

G. Donoso (✉)
Deparment of Agricultural Economics, Pontificia Universidad Catolica de Chile, Santiago, Chile

Water Law and Management Center, Pontificia Universidad Catolica de Chile, Santiago, Chile
e-mail: gdonosoh@uc.cl

E. Lictevout
Universidad de Concepcion, Concepción, Chile

Carpe Science, San Pedro de la Paz, Chile

J.-D. Rinaudo
BRGM, Montpellier University, Montpellier, France
e-mail: jd.rinaudo@brgm.fr

J.-D. Rinaudo et al. (eds.), *Sustainable Groundwater Management*,
Global Issues in Water Policy 24, https://doi.org/10.1007/978-3-030-32766-8_25

25.1 Introduction

Water has always been a strategic resource for economic development, particularly in the arid and semi-arid regions of the north of Chile. Water supply infrastructures, in particular irrigation systems, were developed by Native Americans even before the Spanish colonial period, allowing for the development of a prosperous agricultural economy in temperate, semi-arid and even arid regions. From the end of the XIXth century to the 1970's, the State has heavily invested in the development of large reservoirs which contributed to increase water supply security. As construction costs increased and public funds became relatively scarcer, private sectors increasingly invested in the development of a new series of reservoirs between 1980–1990. In parallel, many users turned towards groundwater, which was relatively cheap and easy to access both in technical and regulatory terms. As in many other countries, the development of groundwater exploitation has taken place in a relatively weak regulatory framework. And it is only when problems began to appear that the State timidly set up a groundwater management policy.

Thus, over decades, groundwater has gradually become an essential water supply source particularly in northern Chile (Peña et al., 2011). The number of granted groundwater rights increased 4350% between 2001 and 2017, while surface water rights grew 207% during the same period. Today, the importance of groundwater as a water source is particularly evident in the North and Center Macroregions (Table 25.1). In the future this trend is expected to grow as a consequence of the increased water use due to economic growth, together with population growth, urbanization, water contamination and pollution, as well as the projected climate change impacts on the availability of surface water (Vicuña, Garreaud, & McPhee, 2011; Vicuña, Meza, Bustos, & Poblete, 2012).

The rapid development of groundwater use has generated a number of problems threatening the sustainability of the resources. Groundwater levels have been declining in a number of regions, revealing that aquifers were exploited beyond sustainable limits. In many Andean valleys, river base flows significantly decreased, with rivers drying-up in extreme cases, threatening traditional irrigation systems. Dependent ecosystems were also impacted by groundwater depletion. Groundwater over-allocation has increased water conflicts. At first, conflicts concerning groundwater were typically in the North and Center Macroregions of the country. However, the intensification of its use has expanded the territorial extent of such conflicts

Table 25.1 Granted surface and groundwater rights

	Surface Water		Groundwater	
Macroregion	N°	l/s	N°	l/s
North	5826	161,145	9097	78,536
Center	8380	1,231,989	24,078	290,664
South	25,381	1,209,944	13,136	82,517
Extreme south	3359	355,848	651	1123

DGA (2016a)

(Rivera et al., 2016), and everything presumes that their number and complexity will continue to expand.

From the 1990's onwards, there was growing awareness that Chilean water policy and management practices were inadapted to ensure a sustainable use of groundwater resources. Various problems associated with groundwater management have been identified (Salazar, 2003; World Bank, 2011). A major concern is the general lack of information about groundwater and insufficient knowledge about its dynamics, in particular its interaction with surface waters. There are significant gaps in the registry of wells, extraction and quality measurements, recharge balances, and identification of pollution sources. Furthermore, there is no collective management, due to the lack of effective groundwater user associations. An additional challenge for a sustainable groundwater management is the fact that at present ground and surface waters are managed independently despite their recognized interrelations and even though the 2005 reform of the Water Code of 1981 (WC81) establishes that surface and groundwater must be jointly managed (Briscoe, Anguita, & Peña, 1996; Vergara & Rivera, 2018). Finally, the effects on the recharge of aquifers of the direct subsidies for the modernisation of irrigation have not been analyzed.

In spite of several policy responses and changes in the legal framework, evidences of groundwater over-allocation have been growing, increasing concerns over the sustainability of actual groundwater use. The sustainability of northern aquifers is compromised due to the over-allocation of groundwater rights (GWR). For example, McPhee et al. (2012) points out that in the North Macroegion of Chile, where aquifers have a significant role as a source of water resources, mainly for mining and agricultural activities, annual estimated recharge is 10 m^3/s while average discharge ranges between 10 m^3/s to 20 m^3/s. A similar situation occurs in the Center Macroregion where annual estimated recharge ranges from 50 m^3/s to 100 m^3/s, while annual discharge fluctuates between 54 m^3/s and 120 m^3/s (DGA, 2016a).

In the following sections, we illustrate the legal and institutional framework for groundwater management in Chile. We then investigate the effectiveness of this framework through the analysis of two case studies. The chapter is organized as follows. It starts with a broad description of the legal and institutional groundwater management scheme in Chile. The chapter then goes on with a presentation of groundwater management in two cases located in an arid region of northern Chile: the Copiapó Valley and the Pampa del Tamarugal Aquifer. The chapter ends with a discussion of the lessons learnt and implications for groundwater management.

25.2 Legal and Institutional Framework

The historical evolution of groundwater development and management can be broken down into four major phases. The first phase corresponds to the 1960–1990 period during which groundwater use significantly developed. During those three decades, landowners could freely appropriate the water located beneath their land, obtaining permanent water rights that were granted by the State under different conditions as the law changed in 1951, 1967 and 1981. The end of this phase is marked by the introduction of legal rules allowing to restrict or prohibit new groundwater use in the WC81. The second phase corresponds to the 10 years of transition period that followed, during which users resisted the implementation of these procedures, and managed to obtain that new rights be allocated, leading to increased overexploitation and groundwater depletion. Phase 3 covers the period from the mid-2000's to today. It has been marked by the development of an increasingly sophisticated groundwater regulation, and the promotion of a form of co-management implying both the State and groundwater users' associations. The remainder of this section presents this evolution of the Chilean groundwater policy in more detail, looking at five main issues: (i) the definition of groundwater property rights; (ii) the initial allocation of those rights; (iii) procedures implemented to restrict water use when groundwater is overexploited; (iv) the establishment of groundwater users' associations; and (v) GWR reallocation through markets.

Water laws and institutions that emerged in Chile during the colonial time, were largely based on those that existed in Spain. Water law in Chile during the pre-colonial period was strongly influenced by the following two principles of Spanish water law: (1) the principle of Roman law that held that waters were common to all men and therefore could not be part of anyone's private property and, (2) that the use of these same waters could be reclaimed as part of the private property of certain feudal lords (Ugarte Araya, 2003). Thus, under the Laws of the Indies, through an express concession (called Merced), a private person was allowed to acquire property over water rights. Hence, water concessions were expressly granted only for the use of water, and in no way referred to the domain of the water resource (Ugarte Araya, 2003).

Water management in Chile has therefore, been governed throughout its history by water rights (WR) granted by the State. The nature of these rights changed under different legislations. For example, the Supreme Decree of November 18, 1819, the first legal provision with respect to water of the Government of the Republic of Chile, established tradable private water rights. The Water Code of 1951 (Gobierno de Chile, 1951) also granted private tradable water rights. The 1967 code (Gobierno de Chile, 1969), on the other hand, established that water rights were administrative rights that could expire (Hearne & Donoso, 2005). Finally, the Water Code of 1981 (Gobierno de Chile, 1981) maintained water as a national property for public use, granting private tradable water rights and reduced the participation of the State (Montginoul, Rinaudo, Brozović, & Donoso, 2016; Vergara & Rivera, 2018); the user is the owner of the right in perpetuity, ownership that is protected constitutionally (Vergara, Arévalo, Muñoz, Rivera Bravo, & Vergara, 2011; Vergara & Rivera, 2018). Additionally, WR are not sector specific and can be transferred between sectors as well as within economic sectors.

The WC81 specifies that WR can be exercised in a permanent or contingent manner and in a continuous, discontinuous or alternating mode. Permanent WR are rights that authorize the extraction of a specified water flow, unless water supply is insufficient to satisfy all permanent WR and they are recognized as shares of water flows (Vergara & Rivera, 2018; World Bank, 2011). Contingent rights are specified as a volume per unit of time and only authorize the user to extract water once permanent rights have been satisfied. Continuous WR allow users to extract water continually over time, discontinuous rights only permit water extraction at given periods, and, lastly, alternating WR distribute water among two or more persons.

It is important to point out that the WC81, as well as previous water codes, were in essence designed for surface water and, thus, contains only few references to groundwater. Thus, groundwater development has taken place in an institutional setting that put no or few limits on groundwater use (Rivera, 2015; Vergara & Rivera, 2018). This regulatory absence has been covered with groundwater guidelines established by the DGA through internal administrative acts (Rivera, 2015). While this trend has experienced some variations in recent years, the precariousness of the treatment of groundwater remains, in general terms, a characteristic feature of the WC81 which thus, contains insufficient rules to effectively regulate groundwater resources (Rivera, 2018; Vergara & Rivera, 2018).

25.2.1 Initial Allocation of Groundwater Use Rights

Historically, the Civil Code and all Water Codes defined groundwater as "water that is hidden within the core of the earth and has not been found". Interested parties must then explore for the existence of groundwater. Any person can explore in order to find groundwater on their own property but requires an authorization by the Directorate General of Water (Dirección General de Aguas – DGA) to do so on public lands (Zañartu Rosselot, 2001). Should two or more petitions for exploration be presented for the same geographic area, the DGA will define who receives the exploration right based on an auction[1].

Decree N° 203 of 2014 (Gobierno de Chile, 2014) sets the legal and technical regulations for groundwater exploitation and exploration. It is only possible to grant groundwater rights (GWR) once it has been verified that groundwater is available (Rivera, 2018; Zañartu Rosselot, 2001), and that its extraction would not affect GWRs of third parties. Therefore, it is not enough to only prove the physical existence of groundwater to obtain a GWR. By strict legal mandate, the DGA must also consider an additional element when studying the availability of groundwater; the exploitation of the aquifer should be appropriate for its conservation and protection in the long term, according to the recharge estimates and to existing and foreseeable uses. Thus, the analysis should not only focus on the present situation of the aquifer when assessing the request, but it should consider its projected use, in order to ensure its sustainability.

[1] Evidence has shown that auctions have not been frequently used to allocate exploration, as well as GWR requests.

Thus, in the case that the exploration efforts are successful, and groundwater is found, the user can petition the DGA for a new GWR presenting pumping tests that certify the existence of the requested water flow. This petition must:

(a) Identify the aquifer from which the water is to be extracted;
(b) Define the quantity of water to be extracted, expressed in liters per second;
(c) Specify the yield and depth of the extraction well;
(d) Specify the water extraction points and the method of extraction; and
(e) Define whether the right is permanent or contingent, continuous, discontinuous or alternating.

The GWR petition must be made known to other potential interested parties by being published in the Diario Oficial, in a daily Santiago newspaper, and in a regional newspaper, where applicable. Previous to the WC81 reform of 2005, the DGA could not refuse to grant new GWR without infringing a constitutional guarantee, provided there was technical evidence of the availability of water resources and that the new use would not harm existent rights holders. At present, the DGA can refuse to grant the solicited GWR if the petition is perceived to be for speculative reasons. To assess this, the petitioner must present a brief technical description of the project that requires the GRW. If there is competition for solicited water rights, they are to be allocated through an auction granting the GRW to the highest bidder.

The GWR is specified as a flow rate expressed in liters per second, that is granted in perpetuity and allows its holders total freedom to use the allocated water for the purpose they wish (Hearne & Donoso, 2005; Lictevout & Faysse, 2018; Vergara & Rivera, 2018); thus, GWR are not sector specific (Donoso, 2015). In 2005 the reform of the WC81 changed the required characterization of groundwater rights, specifying both the maximum instantaneous flow and maximum pumped volume per year (Donoso, 2015).

Due to the concern about the lack of effective water use, Law No. 20,017 of 2005, which amended the WC81, introduced a non-use tariff (patente de no-uso). The non-use tariff is applied to all consumptive permanent GWR that do not count with the required pumps and equipment to extract the granted water flow (Gobierno de Chile, 2005).

25.2.2 *Procedure to Restrict Groundwater Use to Ensure Sustainable Groundwater Exploitation*

Despite the deregulation that prevails in groundwater, the recently approved Law N° 21,064 (Gobierno de Chile, 2018) allows for a greater control and administrative intervention of the DGA focused on groundwater availability. Within a framework of sustainability, the Supreme Decree N° 203 recognizes the need to regulate groundwater extraction, endowing the DGA with powers to limit the extraction of groundwater when there is evidence that extraction rates have had a direct impact on groundwater levels or existing GWR, declaring a temporary reduction of the exercise of groundwater rights or declaring the aquifer under restriction or under prohibition. GWR remain secure and are not threatened under these limitations (Mechlem, 2016).

In situations where there was proof that extraction rates had a direct impact on groundwater levels or existing GWR, the DGA could limit groundwater use by temporally reducing the allocated groundwater flow, only at the request of an interested party. The possibility of limiting withdrawals has been contemplated since 1983 (Res DGA 207 of 1983). However, this measure has never been implemented in practice; users have never petitioned the DGA to temporarily limit the exercise of their own GWR (Rivera, 2015). Law N° 21,064 takes this into account and now allows the DGA to take the initiative to temporally reduce the allocated groundwater flow imposing a prorrata *ex officio*, i.e. in the absence of any request of an interested party.

The declaration of a groundwater resource under restriction is appropriate, as a generic cause, when there is a serious risk of diminishing water levels, with the consequent damage to constituted or recognized GWR. To declare a restriction area, the DGA must present evidence of at least one of the following conditions:

(i) there has been a general decline in groundwater levels, affecting groundwater uses;
(ii) groundwater extractions exceed the recharge rate, reducing groundwater levels and the volume of water stored in the aquifer by more than 5% of the total volume over a period of 50 years;
(iii) average low flow of springs and surface water have decreased 10% or more, affecting existing WR;
(iv) exploitation generates a risk of groundwater contamination from polluted water or saline intrusion;
(v) exploitation induces environmental risks in protected dependent ecosystems due to reduced water flows and groundwater levels.

The declaration of a restricted area can be requested by any water user or the DGA itself as of 2018 (Law N° 21,064 of 2018). This declaration implies that only provisional GWR can be granted. The allocation of, albeit provisional, rights to an aquifer considered to be in a fragile state may be understood as a way to account for uncertainties in hydrogeological studies before a final decision is taken on the appropriate pumping rate (Lictevout & Faysse, 2018). Provisional rights can, in principle, become permanent water rights if they are used continuously for at least 5 years, on condition that they do not affect other users[2]. Should negative impacts be identified in these areas, these provisional WR can be annulled by the DGA; this was the case in La Ligua-Petorca where the DGA, based on hydrogeological models of the basins, determined that the decreases in the water levels of the aquifers correspond to a permanent deficit and not to a temporary effect. As a result, the totality

[2] No provisional GWR have become permanent, even though they have been in use for more than 5 years.

Table 25.2 Number of aquifers or hydrogeological sectors of common use declared under restriction or prohibition

Macroregion	Restriction	Prohibition
North	47	5
Center	103	1
South	3	0
Extreme south	0	0
Total	153	6

DGA (2016a)

of the provisional GWRs in these basins were revoked (Res 1703, 2014). The prohibition area arises from the need for greater protection of the aquifer (Rivera, 2018), and, hence, no further rights can be granted.

Approximately 70% of Chilean territory presents no restrictions for groundwater exploitation. Between 1997 and 2015, 153 restriction areas have been declared in the Central and Northern Macroregions. There are only 5 prohibition areas in the Northern Macroregion and 1 the Central Macroregion (Table 25.2).

Additionally, Law 21,064 establishes that whenever a restriction or prohibition area is declared, GWR holders must install and maintain a water flow extraction and volume measurement system and send this information to the DGA. For the fulfillment of this requirement, the DGA may *ex officio* or based on the complaint of any groundwater user initiate a procedure to sanction non-compliance with this norm. This has not yet been implemented since the DGA has yet to promulgate the required rules of operation.

25.2.3 Groundwater User Associations or Aquifer Management Organizations

Similar to the legislation of a number of countries, Decree N° 203 of 2014 provides for the establishment of groundwater user community (comunidad de aguas subterráneas – GUC) or aquifer management organizations. These associations are responsible for the management of the aquifer; more specifically they have the authority to:

(i) set limits for each user's pumping rate whenever necessary to avoid a decrease of the water table;
(ii) control extractions;
(iii) monitor the quality and quantity of groundwater;
(iv) report to the DGA extraction levels.

When it comes to decision-making, the number of votes allocated to groundwater users in the GUC is proportional to their groundwater rights (Lictevout & Faysse, 2018).

According to the Water Code and Decree N° 203 of 2014, a GUC has to be constituted when an aquifer has been declared a restricted or prohibited area. However, even though we should find at least 159 GUC[3], only 13 have been in fact established, mostly in the regions of Copiapó and Ligua-Petorca – Valparaíso (DGA, 2016b). The majority of these were created during the last 5 years, and thus are still in an implementation stage.

The fact that users have not yet organized themselves in GUCs to take over the management of groundwater reflects the lack of understanding of a large proportion of users of the long-term effects that uncontrolled exploitation of aquifers may cause. A large part of the users are not aware of the legal possibility of setting up a restriction of use and they have difficulty understanding how this mechanism will be put in place by the GUC.

Additionally, groundwater users are relatively reluctant to create and participate in GUCs, since most of them still consider GUCs as more restrictive in terms of water abstraction than surface associations and they tend to view the process as state controlled (Rinaudo & Donoso, 2018). Moreover, user participation is poor, and this is particularly true of small-scale users since votes are proportional to their allocation of WR. The possibility of exercising the power to resolve conflicts by the administration of a GUC is affected by the low participation and legitimacy of the directory before its users. Thus, in general, the performance of GUCs is poor (Rinaudo & Donoso, 2018; Vergara & Rivera, 2018). This can be explained by the fact that GUCs do not fully satisfy Ostrom's 8 principles for an effective collective groundwater management (Ostrom, 2000). The main difficulties that limit GUC's effective water management are (Rinaudo & Donoso, 2018; Vergara, Donoso, Rivera Bravo, Blanco, & Moyano, 2013):

(i) Legal and administrative obstacles in the determination of their statutes and rules of operation.
(ii) Lack of adequate professional management.
(iii) Insufficient budgets for an effective water management.
(iv) Administrative presence and intervention in some aquifers.
(v) Aquifer sections with autonomous and independent GUC, limiting an integrated water management.
(vi) Lack of effective participation of all water users.
(vii) Lack of a complete registry of GWR.

Due to these concerns, the DGA, Comisión Nacional de Riego (National Irrigation Committee – CNR), and Dirección de Obras Hidráulicas (Public Works Directorate – DOH) have implemented programs to create and strengthen GUC (Donoso, Blanco, Vergara, & Rivera, 2014; Fuster et al., 2016; Ravanal Salinas, 2011).

In the absence of GUCs, the WC81 establishes that the DGA is responsible for controlling and monitoring groundwater withdrawals. Evidence has shown, however, that the DGA has not had the necessary resources (human, technical, and financial) to monitor all groundwater extractions (World Bank, 2013).

[3] One for each aquifer or hydrogeological sector of common use declared under restriction or prohibition.

25.2.4 Reallocation of Groundwater Use Rights Through Markets

As previously pointed out, the WC81 established that WR are transferable so as to facilitate and achieve the efficiencies of market reallocation of water. Although private WR existed in Chile prior to 1981, the previous water codes restricted the creation and operation of efficient water markets (Hearne & Donoso, 2005). Thus, the WC81 was designed to protect traditional and customary WR and to foster economically beneficial reallocation through market transfers (Montginoul et al., 2016).

The existence of groundwater markets has been documented (Cristi, Melo, & Donoso, 2014; Hearne, 2018). The regions with the greatest GWR market transactions have been the Metropolitan region, with 73%, and Araucanía, with 10% of total transactions. GWR markets have also been active in river basins in northern Chile, allowing expanding mines and growing cities to purchase water rights from farmers (Hearne, 2018). Notwithstanding, the majority of transactions have been between agricultural users, moving GWRs towards high valued agricultural export sector with resulting efficiency gains (Cristi et al., 2014; Donoso, 2013; Hearne & Donoso, 2014).

A key conclusion is that GWR markets are driven by demand from relatively high-valued water uses and facilitated by low transactions costs and are more active in those aquifers that the DGA has declared as restricted or protected and where there are GUCs present that assist in the transfer of water. For example, as Fig. 25.1 shows, in the Copiapó basin, the volume of water and number of GWR traded began to significantly increase as of 1994, when the DGA declared the aquifer under protection (Donoso et al., 2014). There is a second increase as of 2002 when the DGA maintained the prohibition for Sectors 1–4 and declared restriction for Sectors 5 and 6. This resolution reinforced the signal to water users that new GWR were not available for the Copiapó aquifer and, thus, new water demands must be satisfied through the market for GWR. In the absence of these conditions, trading has been rare and water markets have not become institutionalized in most aquifers.

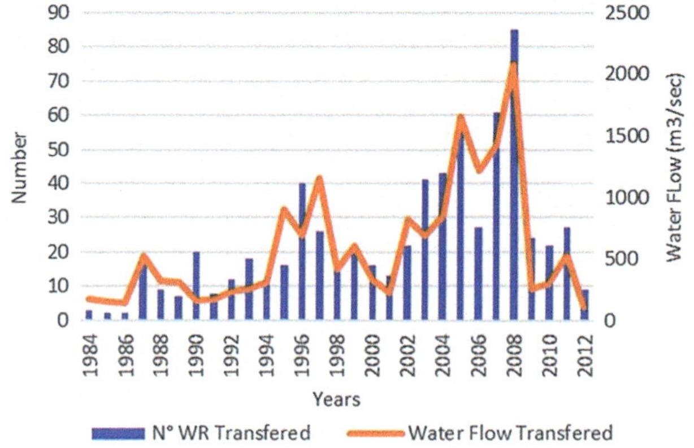

Fig. 25.1 Water flow and number of WR traded in the Copiapó Aquifer. (Donoso et al., 2014; Montginoul et al., 2016)

25.2.5 Summary

Over the years there has been an acknowledgement that legal and institutional frameworks play a crucial role for effective groundwater governance and, thus, the precariousness of the treatment of groundwater in the WC81 has been covered with groundwater guidelines established by the DGA through internal administrative acts. As it stands, it considers the required instruments and mechanisms to balance growing demand and the need to protect and preserve groundwater resources.

This governance framework will in principle be effective and lead to a sustainable groundwater management to the extent that the implementation requirements are met. That is, that the state and GUCs have the technical and financial capacity to perform the required tasks to effectively implement the legal and institutional framework. Has Chile been able to apply its groundwater management policy in practice? We will investigate this through the analysis of two cases located in the arid region of northern Chile: the Copiapó Valley and the Pamapa del Tamarugal Aquifer.

25.3 Groundwater Management in the Copiapo Valley

25.3.1 Case Study Presentation

The Copiapó Valley is located in the Atacama Region, in the semi-desert region of northern Chile (Fig. 25.2). The watershed covers an area of 18,000 km^2. It extends from the Andean summits, at elevations exceeding 6000 m, as far as the sea; the river is 160 km in length.

The groundwater resource lies in the valley bottom, over a width ranging from 1 to 5 km. It constitutes a major quaternary alluvial aquifer, averaging 100 m in thickness and divided into six large sectors (Fig. 25.2). The aquifer's yearly groundwater recharge of approximately 4000 l/s, is derived almost exclusively from melting snow and ice in the Andes and by the Lautaro Reservoir, located at the head of the river, which has an intra-annual water regulation capacity of 25,4 million m^3.

The total authorized pumping flow increased from about 160 l/sec in 1965 to 13,200 l/sec in 1993, an increase of nearly 8200%. Noting the fall in piezometric levels in Sectors 3 and 4 due to the excess pumping, respect to recharge rates, the DGA classified all 6 sectors of the Copiapó Basin as a prohibition area in 1993 (Resolución 193); thus, no new pumping permits could be granted. However, in 1994, DGA established Resolución 232 reducing the prohibition area, and allowing new GWR allocation to areas located more than 35 km from the river. As can be seen in Fig. 25.3, DGA continued granting GWR after the declaration of prohibition. Rinaudo and Donoso (2018) point out that this occurred, on one hand, due to a judicial sentence that required the DGA to process user's applications that had been

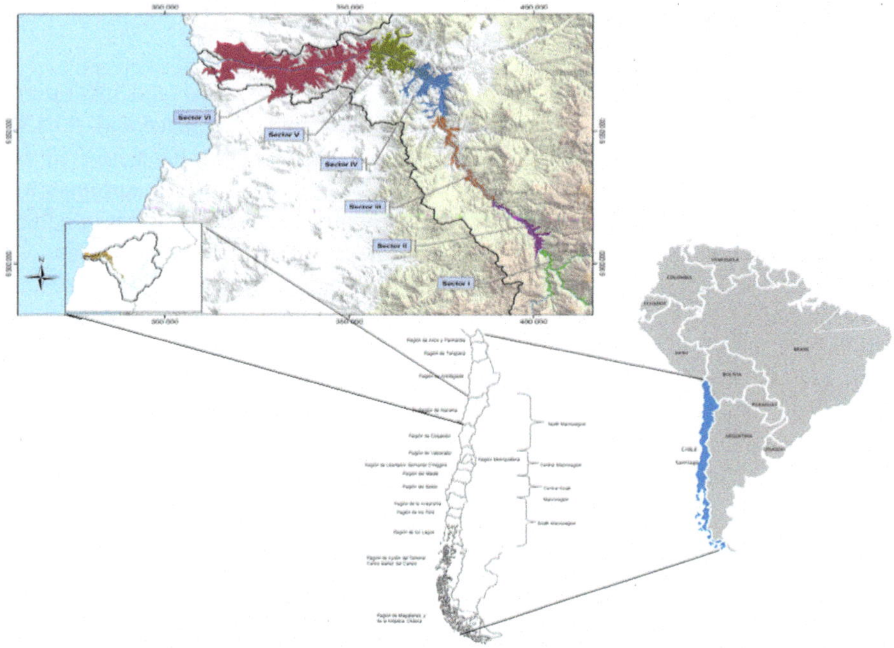

Fig. 25.2 Location of the Copiapó Valley and division of the alluvial water table into management sectors. (Rinaudo & Donoso, 2018)

Fig. 25.3 Accumulative Granted GWR l/s. (Adapted from Rinaudo & Donoso, 2018)

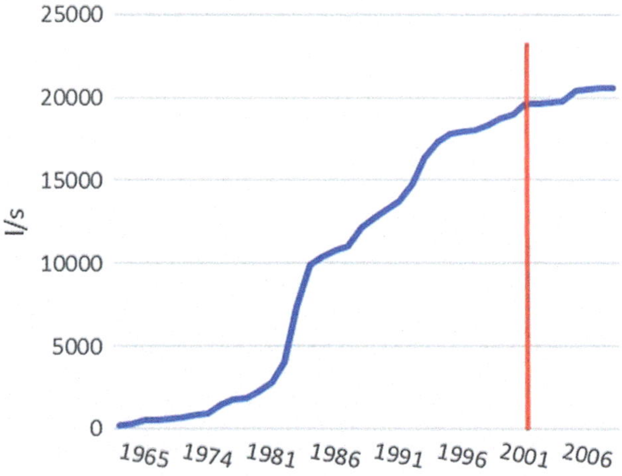

submitted prior to the publication of the prohibition. On the other hand, expecting the DGA to declare prohibition, recognized customary groundwater users (historic usages) requested the regularization of their GWR following the procedures of the transitory second article of the WC81. Thus, total authorized pumping flow increased to about 18,500 l/sec in 2001, an increase of nearly 30% (Fig. 25.3).

In 2001 the DGA lifted the prohibition for sectors 5 & 6 classifying them as a restricted zone and granting 57 provisional GWR for a total of about 450 l/s, which represent 1.8% of the total authorized pumping flow rate of 25,300 l/s (Fuster et al., 2016). At present, agriculture is the main GWR user representing 73% of total granted GWR flows. Mining is the second most important GWR user, with 12%, while urban water supply presents the lowest participation with 7%.

Since the 1980s, a number of studies have shown that withdrawals exceed by far the average recharge of the alluvial aquifer, estimated at 4 m^3/s (Troncoso et al., 2012). Thus, the total available water reserve fell by about 38 million m^3 per year between 1990 and 2012, the shortfall then accelerated sharply between 2005 and 2012 to 55 million m^3 per year (Hidromás CEF limitada, 2013). The total estimated reserve lost is approximately 830 million m^3.

25.3.2 Factors Explaining Overexploitation

Rinaudo and Donoso (2018) researched the question of how this over-extraction situation came to be when in theory the State has the regulatory tools to avoid it. They identify the following five explanatory factors: limited knowledge of the groundwater, legal complexity and political pressures, poorly-defined water permits, compliance and enforcement problems, and inconsistency between the management of surface water and groundwater. In this section we will briefly present these factors.

25.3.2.1 Limited Knowledge of the Groundwater

Over the past 20 years, several studies have been conducted of the Copiapó aquifer, most of which have alerted the authorities to the danger of overexploitation (DICTUC, 2010; Hidromás CEF limitada, 2013; Troncoso et al., 2012). However, these studies offered different conclusions with respect to the situation of the aquifer. For example, Uri Hammer y Asociados' (1980) study showed that extraction exceeded recharge by 18% and, thus, groundwater reserve decreased in 59 million m^3 between 1970 and 1979. Just a few years later, Alamos y Peralta (1987) presented contradictory results, indicating that the aquifer could continue to be exploited without a significant reduction in piezometric levels. Given that the declaration of a groundwater resource under restriction or prohibition requires proof that there exists a serious risk of diminishing water levels, the lack of reliable technical information on Copiapó's aquifer characteristics led to a late establishment of protection areas and, thus, overallocation.

25.3.2.2 Legal Complexity and Political Pressures

When significant reductions in piezometric levels in Sectors 3 and 4 became evident, the DGA declared the Copiapó Basin as a prohibition area in 1993. However, a year later, the declaration was modified, reducing the prohibition areas located more than 35 km from the river, thus allowing a number of mining projects to obtain GWR. This, together with the judicial obligation that the DGA process GWR petitions that had been submitted prior to the publication of the prohibition and the regularization of recognized customary GWR, led to an increase in the total authorized pumping flow to 18,500 l/sec in 2001, an increase of nearly 30% with respect to authorized extractions in 1993. Rinaudo and Donoso (2018) also suggest that the local authority may have been subjected to pressures from influential users, who had relations in the central government and in Santiago's political circles, to continue granting GWR. Additionally, the *Instituto Nacional de Desarrollo Agropecuario* (INDAP), the agency of the Ministry of Agriculture whose objective is to promote the development of small agriculture, reached an agreement with the DGA to lift the prohibition declaration for sectors 5 and 6 and classify them as a restricted zone. This allowed the DGA to grant provisional GWR in these sectors to small farmers. However, not only small farmers received provisional GWR; some mining companies were also benefited.

25.3.2.3 Poorly-Defined Water Permits

Granting GWR based on a foreseeable use factor is a third issue that helps explain the overexploitation. The number of GWR the DGA could grant considering the foreseeable use that the beneficiary intended to make of the permit is higher given that water availability is less restrictive. Consider that a permanent and continuous GWR of 1 l/s gives the right to pump 31,500 m^3/year, if this permit is used 365 days a year, 24 h a day. Since agricultural producers only pump water between the end of spring and the end of summer, the DGA assumed a use factor of 20% and thus a water-permit of 1 l/s for agriculture theoretically implied a total demanded volume of 6307 m^3/year. This theoretical use factor was 75% for drinking water and the mining industry. Therefore, considering the GWR allocated to the different economic sectors, the total theoretical granted volume was 239 million m^3/year.

However, actual water use has increased due to improvements in water use efficiency and intersectoral market trades. Due to these factors, actual volume extractions reached 421 million m^3/year, which represents a 56% increase in volumetric extraction without increasing the number of granted GWR.

25.3.2.4 Compliance and Enforcement Problems

The WC81 establishes that water users, organized in GUC are responsible for monitoring and enforcing water extraction limit compliance. However, they have not fulfilled this obligation and, thus, the DGA has unsuccessfully tried to fill this gap. In order to implement an effective extraction monitoring system, Decree 203 of 2014 required that all groundwater users install measuring equipment and inform the DGA of their water extraction levels. However, this was not effective as only a minority of users up to date have installed measuring equipment (mainly in the downstream section of the valley) and very few inform the DGA of their extraction levels. Additionally, the random monitoring conducted by the DGA only led to seven official police reports in a 12-year period. Therefore, this policy instrument does not incentivize stakeholders to comply since the cost of non-compliance is low; some farmers openly acknowledge that they do not comply with the terms of their water rights.

25.3.3 The Challenge of Collective Action

Although the Copiapó aquifer was declared under prohibition in 1993, no GUC were formed until 10 years later. The Comunidad de Aguas Subterráneas Copiapó – Piedra Colgada; Piedra Colgada – Desembocadura (CASUB), that pulls together all GWR holders of sections 5 and 6, was finally judicially constituted in 2004. In the other sectors of the aquifer, the GUCs were established judicially in 2015. CASUB is composed of users that have 424 GWR with a total water flow of 5745 l/s. The participation of the different economic sectors is similar to that of the total aquifer; agriculture holding most of the GWR (representing 81% of total granted flows, according to National Water Right registry of 2017), mining is the second most important GWR holder, with 13%, while water supply presents the lowest participation with 6%.

CASUB is managed by of a board of 7 elected directors. In order to guarantee that all users are represented in the board, the agricultural, mining, industrial and water supply company must have, at least, one representative in this collegiate body[4]. The Board presents proposals to the general assembly for approval, where the vote of each user is proportional to the number of liters per second he/she holds. It is financed by the users via a contribution proportional to the size of of the GWR (in l/s per). The main objectives of CASUB are to (i) represent interests of its users, (ii) manage and protect its water source, (iii) regulate uses so as to insure a sustainable exploitation of the hydrogeological sectors of the aquifer, and (iv) control groundwater extractions and apply sanctions. The community offers technical and administrative support to its users and helps resolve conflicts between users and users vis-à-vis the state.

[4]This is not a common characteristic in GUC.

To control groundwater extractions, CASUB monitors pumping flow-rates as well as aquifer static and dynamic levels in real time through telemetry. In the case of non-compliance with the authorized flow the GUC can theoretically apply sanctions, such as to remotely cut the power supply to the pump. However, users find that this monitoring system is not fully exploited to suppress illegal wells and ensure that users respect their extraction limits.

CASUB only became operational in 2008 and its management capacities have been limited due to the lack of rules of operation, which were only approved in 2013. However, a majority of its members do not clearly understand which are the attributions of CASUB and the DGA. This is particularly evident with respect to the delivery of information on wells, GWR, and extraction levels. There is a belief that since CASUB already collects information (e.g. telemetry), users are not obliged to give additional information to CASUB or DGA, thus not complying with Law 21,064. To support users help generate a sense of belonging to the community and of its duties and responsibilities, CASUB has implemented several activities such as a website which will soon enable users to obtain information on the levels of the water table and their dynamic trends and support tools to help manage irrigation at the farm level, which will allow individual farmers to identify faulty operations in their irrigation systems. However, most members have yet to be convinced. Rinaudo and Donoso (2018) found that only a small group of members are now fully aware of the community's significance and are actively involved in the association.

Additionally, a large proportion of CASUB's users are not aware of the legal possibility of applying a temporary extraction limit so as to reduce the groundwater reductions and tend to a sustainable extraction level. They have difficulty understanding how this mechanism would be put in place.

25.4 Groundwater Management in the Pampa de Tamarugal

25.4.1 The Pampa del Tamarugal Aquifer

The Pampa del Tamarugal aquifer is a major aquifer located in the far north of Chile (Atacama Desert), in the region of Tarapacá (Fig. 25.4). It lies within the Pampa del Tamarugal basin – a hyper-arid and relatively flat sedimentary basin (Central Depression) bounded in the east by the Precordillera Mountains and the Chilean Altiplano and in the west by the Coastal Range (Fig. 25.4). The precipitation in the area is almost nil, so the aquifer recharge comes from lateral groundwater flow within the piedmont area which originates from the precipitations that occur on the western flank of the Andes (Scheihing, Moya, Struck, Lictevout, & Tröger, 2017; Viguier et al., 2018), where the average annual precipitation ranges between 150 and 180 mm (Lictevout, Maass, Córdoba, Herrera, & Payano, 2013). Until recently, little was known about the aquifer's limits, structure and recharge (Lictevout et al., 2013). The exploited aquifer has a total estimated saturated thickness of 100 to

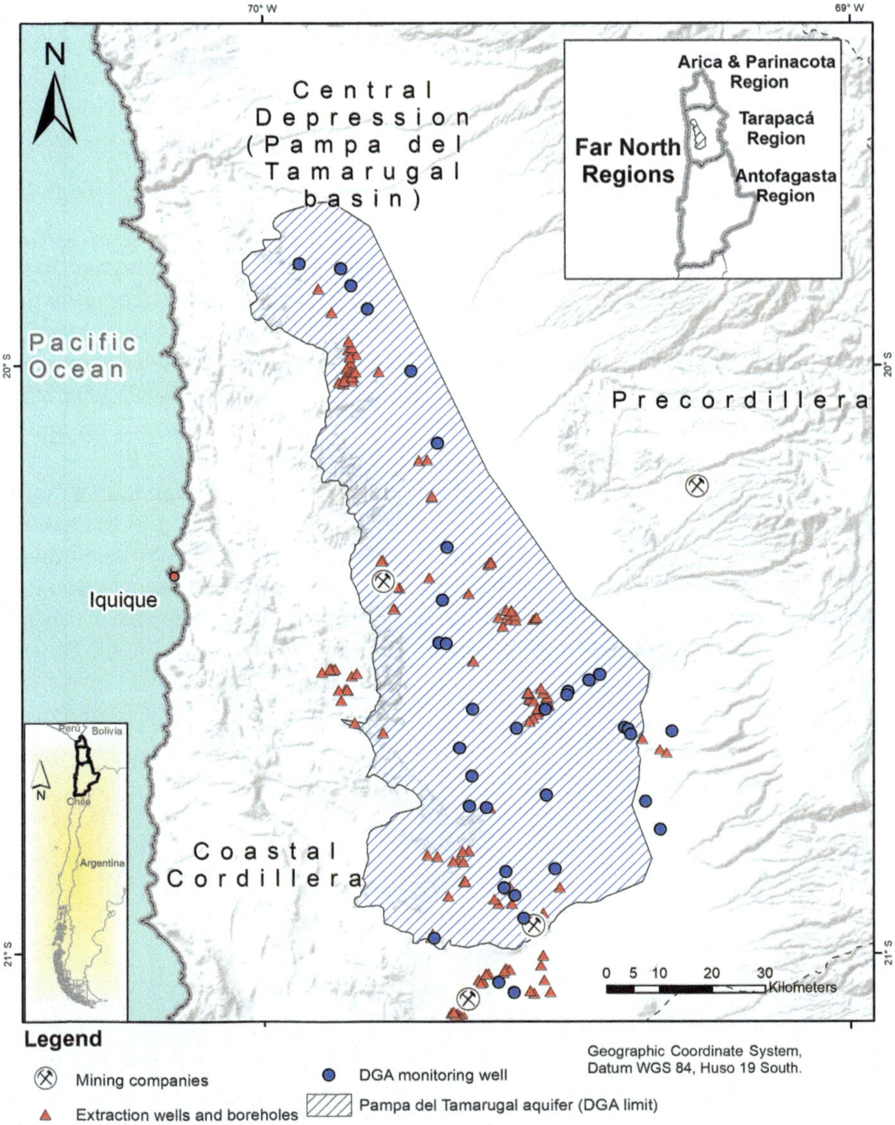

Fig. 25.4 Location of the Pampa del Tamarugal Aquifer

300 m (Rojas & Dassargues, 2007). The recharge was estimated to be approximately 1180 l/s. The main natural discharge occurs through evaporation of the western and southern parts of the aquifer (145 l/s) and evapotranspiration of the tamarugo trees (900 l/s) (JICA, 1995). There is also an outflow to other aquifers estimated at 135 l/s (DICTUC, 2008).

The Pampa del Tamarugal (hereafter PdT) aquifer plays an important role in providing drinking water, as well as water for mining companies and agricultural use.

25.4.2 Groundwater Use and Management

The PdT basin harbours the world's biggest deposit of natural nitrate and iodine which exploitation requires water for the transformation of nitrate. The growth in mining activities in the Region of Tarapacá since the 1990's also triggered a 51% population increase between 1992 and 2017. The corresponding rise in demand for drinking water was met with groundwater from the PdT Aquifer. Small-scale farmers also pump water for irrigation purposes. In 2009, the DGA deemed the aquifer to be overused and declared it a restricted area at the beginning of 2010.

Figure 25.5 shows the temporal evolution of the granted extractions yields (groundwater rights) in the PdT Aquifer since the 1980s according to the national Water Rights Registry. The total cumulative granted water flow is 3758 l/s. Similarly to what happened in Copiapo, the GWR granted for agricultural use increased from 529 l/s to 1021 l/s in 2009, following the regularisation of groundwater use as permitted by the 2005 reform of the Water Code.

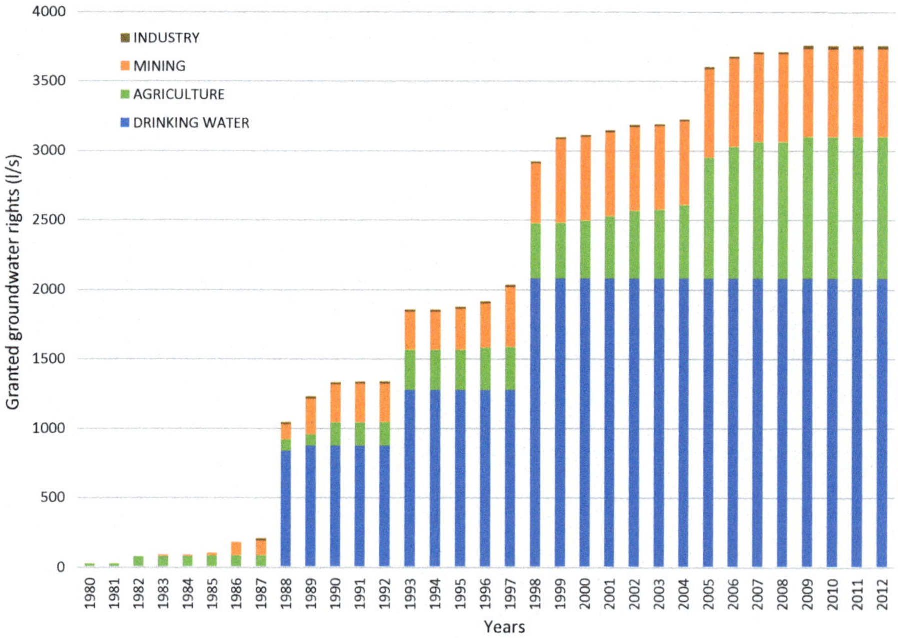

Fig. 25.5 Evolution of granted groundwater rights (as extraction yield) in the Pampa del Tamarugal aquifer since 1980. (DGA, 2016a)

However, these figures may not reflect the true situation because the registry of water rights is not exhaustive (World Bank, 2011). This is due to the fact that: (i) WR obtained prior to the WC81 may not have been regularised; (ii) users, law courts and particularly notaries (*conservadores de bienes raíces*), who legally register water rights, do not always inform the DGA, which is responsible for updating the registry, of the final resolutions for the allocation of water rights; and (iii) WR transactions between users which are not required to be informed to the DGA.

Lictevout and Faysse (2018) analyzed the groundwater management of the PdT aquifer, paying particular attention to the links between: (i) how information relating to groundwater resources and its uses is applied to groundwater management; and (ii) the actors' strategies and discourses regarding groundwater management.

25.4.3 The Declaration of a Restricted Area

The declaration of the PdT aquifer as restricted area was first requested in 2004 by a farmer with GWR in the PdT Aquifer based on the arguments that (i) groundwater levels were decreasing at a rate of 12–20 cm/year in the wells monitored by the DGA; and (ii) the granting of further groundwater rights would increase the imbalance between the recharge and use, thus representing a threat to existing water rights and protected areas. Four groundwater users objected the request: one farmer and three mining companies. No follow up was given to the request and, in 2006, the major mining company withdrew its opposition.

In 2009 the DGA issued a report (DGA, 2009), which confirmed the arguments expressed in the 2004 request and rejected all those upheld by its opponents. The analysis was based on a previous report (DGA, 1996), which used data produced by a regional study (JICA, 1995). The DGA's, 2009 report demonstrated that the existing extraction rates would cause a continuous decline in groundwater levels and that the PdT Aquifer was at serious risk of depletion. The report also described the evolution of groundwater levels in the 11 wells monitored by the DGA in the PdT Aquifer. The data showed decreasing levels in seven wells, four of which showed a constant decrease since 1997 of 0.05 cm/year to 0.15 cm/year. This led the DGA to declare the PdT Aquifer a restricted area at the beginning of 2010.

However, the report was heavily biased in order to 'force' a calculation that would lead to the establishment of a restricted area while maintaining existing water use and granting new entitlements to those who had already submitted their requests. First, the 1996 and 2009 DGA reports calculated the sustainable extraction flow based on 5% of the stored volume over a 20-year period rather than a 50-year period as officially required. If a 50-year period had been taken into account, no additional water rights should have been granted in 1996. Moreover, the DGA did not consider recharge and natural discharge. Finally, a decrease of 0.05–0.15 cm/year of the groundwater level in a few wells may be considered limited for an aquifer whose total saturated thickness is estimated at between 100 and 300 m. In her study of La Ligua Valley, Budds (2009) pointed out that extremely tenuous hydrogeological calculations had been used to justify the declaration of a restriction on the aquifer.

25.4.4 Groundwater Monitoring and Control

Every year users must declare the water flow of GWRs left unused, for which a fee is paid to the DGA. Figure 25.6 shows that the total GWRs declared as unused decreased between 2010 and 2016. Although this flow is officially not pumped, the DGA considers it to be pumped in the groundwater balance. In addition, a total of 2080 l/s was allocated for drinking water (Fig. 25.6), but the used extraction flow for this purpose was 973 l/s in 2011 (Superintendencia de Servicios Sanitarios, 2013). In the same year the company declared only 145 l/s of unused rights (Fig. 25.6). However, no official information is available on the remaining flow allocated, i.e. whether it was extracted and, if so, for what purpose. So overall, official estimates of groundwater use are over-estimated.

In the Taparacá Region, users with GWR exceeding 20 l/s are required to install measuring equipment and report the volume pumped, average flow and groundwater level to the DGA (DGA, 2011). Between 2014 and 2017 the DGA conducted 58 checks to determine the amount of water actually pumped from the PdT Aquifer. The majority of these were carried out following denunciations, although only one led to a penalty for illegal extraction (DGA, 2018). In practice, small-scale users are not subject to control. However, the flow officially granted to agriculture (1021 l/s in 2009) does not correspond to the effective irrigated land in the PdT, which is much smaller. Some of those GWR may have been transferred to other uses, such as mining, or left unused (for example, waiting to sell for a profit) but this information is not available to the DGA and used for decision-making.

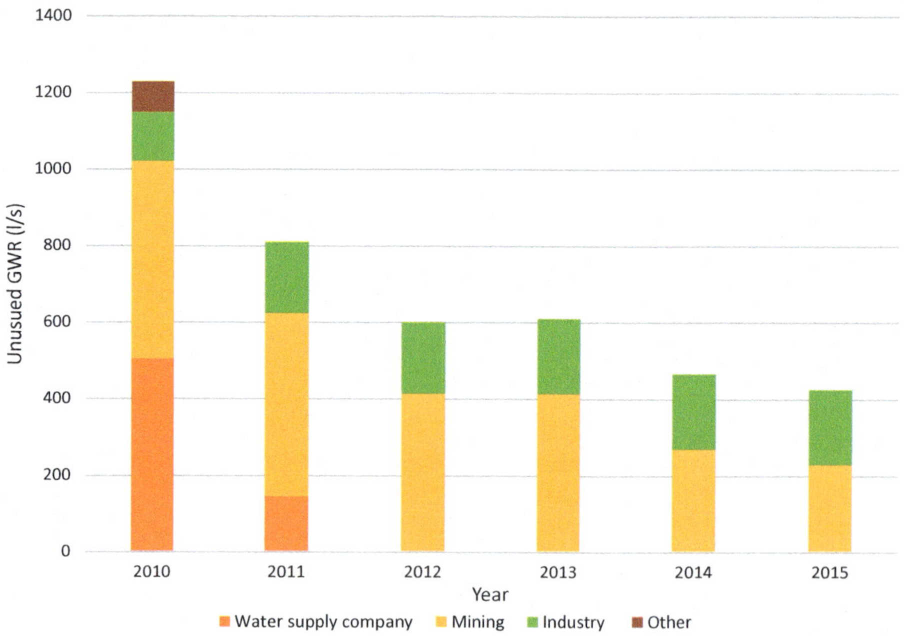

Fig. 25.6 Unused GWR. (Adapted from DGA, 2018)

Large-scale users, such as mining companies, need to be more cautious, since it is possible that their actual pumping rates are monitored. In 2008 the SQM company brought a lawsuit against its main rival, Cosayach, accusing the company of using illegal wells in the PdT Aquifer. In 2011 a court ordered the closure of the illegal wells used by Cosayach. Subsequently, the mine was forbidden from using any more than the volume specified in its water rights (28 l/s were legally granted before the closure of the aquifer).

25.4.5 Groundwater User Community

As previously pointed out, once an aquifer has been declared a restricted area, a GUC has to be established. The process was initiated in 2015 by the DGA but did not succeed and is currently at a dead end. The reason is that farmers are firmly opposed to the GUC since, as stipulated in the Water Code, the number of votes allocated to groundwater users in the community is proportional to their groundwater rights. This results in that there would be no farmers within the board of directors. The main concern of the farmers is that the community would be controlled by mining companies and the water supply company as both of them own the majority of the GWR in the aquifer.

25.4.6 Creation of Scientific Knowledge on the Aquifer

In 2010, in light of the concern over excessive water use for mining activities, the regional government in Tarapacá, the Regional University of Arturo Prat and the National Science and Technology Research Commission created a regional research and development centre for water resources (CIDERH) as part of a national programme to decentralise science. Its primary objectives were: i) to study the surface and groundwater resources in arid zones for the development of an integrated approach to water resource management; and ii) to develop technological innovations in water processes in order to increase water supply from natural sources and promote water reuse.

The research conducted by the CIDERH improved the understanding of the PdT aquifer's structure, its recharge processes and its evolution over several decades (Lictevout, Amaro, Córdoba, & Rodriguez, 2014; Moya et al., 2015; Scheihing et al., 2017; Viguier, 2016; Viguier et al., 2018). The CIDERH redefined the aquifer limits on its eastern margin far beyond the previously estimated limits (Viguier et al., 2018), involving a much larger volume of groundwater – around two times that reported by the DGA in 2009. Piezometric measurements taken in 2012, 2013 and 2014 showed that groundwater levels had decreased by an average of 5.6 cm/year since the 1980s, reaching 12–17 cm/year in some places. Thus, water levels dropped by 1–2 m over two decades (1993–2014), reaching a maximum of 3–5 m in some areas (Lictevout et al., 2014). In other areas the level remained stable, while in several boreholes groundwater levels actually rose over the same period. Although the CIDERH reported all of its results to the regional and national offices of the DGA, these never officially endorsed the findings. In 2015 the regional government withdrew its support and funding for the centre for no official reason. As a result, the centre was forced to halt its work on surface and groundwater resources, including studying the PdT aquifer.

25.4.7 Actors' Actions and Discourses Relating to Groundwater Management

Although the restriction on the PdT Aquifer is considered a correct decision by actors, the decision-making process that led to the declaration of a restricted area or of the data used to justify it is unknown. In 2017 a group of PdT farmers and native american representatives filed a lawsuit against the DGA in an attempt to reverse the restriction. They claimed that the decision denied their ancestral access to groundwater, but the claim was ultimately rejected (Court of appeal of Iquique, 2017). The farmers complained that, during the process of declaration of restriction area, the remaining water rights had been divided between the mining companies, the water supply company and some well-connected farmers, who, they claimed, obtained GWR for speculation rather than agricultural use. In addition, many of the actors interviewed expressed the opinion that the restriction failed to improve the situation, for two reasons: (i) public organisations lacked the capacity to control extractions, and (ii) GWR already allocated exceeded the aquifer's capacity, which meant that continued over-extraction was inevitable.

The community leaders interviewed suggested that the problem went further and originated with governmental bias towards private companies. In their view, people at the grass roots lose out in the process. The community leaders also claimed that insufficient water rights limited agricultural activities and the possibility of obtaining public subsidies.

All the actors interviewed said that they had never discussed the question of how much water should actually be pumped from the aquifer and that far greater capacity was required to monitor actual use.

25.4.8 A "Status Quo" Situation

The poor groundwater management in the PdT Aquifer stems from serious groundwater management policy implementation deficiencies. The high level of uncertainty in all terms of groundwater balance (in particular, outflows and actual water pumping) and the lack of an updated registry of water rights seriously compromise the implementation of a management system based on WR. This also opens the way for opportunistic use of data. This high level of uncertainty arises both from the legal setting and from insufficient financial and human resources dedicated to water resource management, in particular in State agencies.

The key decision taken on aquifer management was to restrict use in 2010, which has led to a situation of status quo. The calculations presented in support of the decision meant that existing water use, and any pending request, were unaffected.

Neither public nor private actors discussed the current or desired status of the PdT Aquifer, despite being informed of the CIDERH's research results. The CIDERH's analyses challenged the calculations made by the DGA in 2010 and 2011. For example, the results of the CIDERH analysis of the volume of water stored in the PdT Aquifer differed from the DGA's figures. Yet the dissemination of its research findings failed to break the status quo, which emerged as a result of the governance framework and actor's strategies.

Actors had very different views on water resources and their use but lacked a forum for discussion. This is a recurrent problem in Chile for issues relating to groundwater (Rinaudo & Donoso, 2018; Usón, Henríquez, & Dame, 2017) and surface water (Palomino-Schalscha et al., 2016).

With regard to the PdT Aquifer, some actors avoided discussion for strategic reasons. Several were driven by an ulterior motive and not by groundwater management itself. Usón et al. (2017) describe a case where key economic actors were in favour of restricting groundwater use. Although the groundwater level drawdown of the PdT Aquifer was limited, the main mining company in the PdT area supported a restriction. In addition, local community leaders raised the issue of groundwater overuse in order to express their opposition to the general development model designed to promote mining, complain about the failure to share the benefits derived from mining activities and improve their visibility with regard to the local authorities. The actors did not appear to want a genuine discussion on the actual and desired status of the aquifer. Together, these strategies served to reinforce the status quo.

25.5 Conclusions

In Chile, groundwater use has increased at an exponential rate in the last decades. Even though concern for the long-term sustainability of groundwater resources has gained prominence in the agenda, groundwater policy to address the common-pool resource losses remains in its precarious stage. The evolution to date, the level of sophistication and complexity attained have not been effective to reconcile sustainable groundwater management with changing hydrogeological, technological, economic, environmental and political circumstances. Hence, Chile was not prepared, from a legal and technical point of view, for this increase in the exploitation of groundwater in the magnitude that occurred. This is evidenced by the fact that, at present, many important aquifers in Chile are under considerable stress as withdrawals, predominantly for agricultural use, outpace recharge.

An examination of the Copiapó Valley and Pampa del Tamarugal cases have shown a number of common problems:

(i) Limited knowledge of the groundwater as the result of a high level of uncertainty about groundwater balance calculation (evaluation of the recharge and groundwater stock), due to the complexity of the aquifers and insufficient state groundwater monitoring networks. It is difficult to obtain reliable data that is accepted by all parties to support the decision to limit groundwater extraction. Moreover, technical studies can be manipulated to favor certain decisions, which also opens the way for opportunistic use of data, as was presented in both case studies. Given that the declaration of a groundwater resource under restriction or prohibition requires proof that there exists a serious risk of diminishing water levels, the lack of reliable technical information on Copiapó's and PdT aquifer characteristics led to a late establishment of protection areas and, thus, overallocation.

(ii) Lack of forum where diverse stakeholders could gather to talk and debate about water issues. The natural instance where different actors could gather to discuss the current or desired status of groundwater resources are GUC. However, only a small number of GUC have been constituted, and there is a significant low participation in these; thus, they do not represent a valid and effective forum. Neither public nor private actors discussed the current or desired status of the Copiapo and PdT aquifers.

(iii) Enforcement has clearly been a problem. Water users organized in GUC are responsible for monitoring and enforcing water extraction limit compliance. However, they have not fulfilled this obligation; for example, in Copiapo the data from telemetry are not used. Thus, the DGA has tried to fill this gap. Evidence has shown, however, that the DGA has not been effective due to the lack of an accurate GWR registry and the the necessary resources (human, technical, and financial) to monitor all groundwater extractions This is a prerequisite for an effective quantitative management which is missing in Chile (like in other countries, see Chap. 23 for France and Chap. 22 for Australia).

(iv) Over-allocation has been an inevitable problem. The regulator (and users) are usually not concerned by overexploitation until it happens, so the declaration of prohibition or restriction often comes too late. In the Copiapó case, since the 1980s, a number of studies showed that withdrawals exceeded by far the average recharge of the alluvial aquifer, however the DGA only classified the Copiapó Basin as a prohibition area in 1993, but by then the total available water reserve had significantly fallen. Moreover, users will try to obstruct the process of declaring restriction/prohibition zones when this becomes an issue, in order to obtain "last minute" water rights. For example, in Copiapó a year after, the prohibition declaration was modified, reducing the prohibition areas located more than 35 km from the river, thus allowing a number of mining projects to obtain GWR. The declaration of the PdT aquifer as restricted area was first requested in 2004 but was objected by one farmer and three mining companies. Only 5 years later the DGA declared the aquifer as a restriction area.

(v) Groundwater allocation is a political issue, in PdT as in Copiapo. A number of accounts suggested that the local authority may have been under pressure from influential users, to increase the total authorized pumping flow in the Copiapó aquifer after the prohibition declaration. Furthermore, the DGA declared the PdT aquifer a restricted area at the beginning of 2010 based on a biased report maintaining existing water use and granting new entitlements to those who had already submitted their requests. Even though users suffering from insufficient water supply contested the legitimacy of the initial allocation, legal complexities and political pressures led to the allocation of GWR that exceeded the established abstraction limits

(vi) Overall, the State (acting through the DGA) lacks the financial, technical and human resources to implement all the provisions of the WC81. This code is very sophisticated "on the paper" but many of its dispositions are left unimplemented. For example, GUC are responsible for monitoring and enforcing water extraction limit compliance but they have not fulfilled this obligation and, thus, the DGA has tried to fill this gap. However, evidence has shown that the DGA has not had the necessary resources to monitor all groundwater extractions. The State has not had sufficient power to impose action to GUC, in particular in terms of data collection, and for designing rules to reduce abstraction.

(vii) The legal framework established that GUC have to be established to collectively manage aquifers, however this has not happened in practice. Difficulties arise from the stage of creation of the GUC and the design of the governance. The two examples are quite contrasted, as one GUC was successfully created in Copiapo and continuously increases the range of its activity, while it could not be established in PdT. More research is needed to identify the conditions facilitating the emergence of collective action in the Chilean context.

Therefore, the analysis of the trajectory of Chile's groundwater legislation and state of its aquifers, suggests that establishing a solid and well developed legal and institutional framework is a necessary but insufficient condition to ensure that groundwater is managed sustainably. It is also important to strengthen the state and GUC's capacities and abilities to perform the key missions and tasks that the established groundwater governance framework requires.

References

Alamos y Peralta. (1987). Analisis y evaluacion de los recursos hidrogeologicos Valle del Copiapo - III Región D.-. MOP. Available at: http://documentos.dga.cl/REH591anexo3.pdf

Briscoe, J., Anguita, P. A., & Peña, H. (1996). *Managing water as an economic resource: Reflections on the Chilean experience* (Environmental Department Papers N° 62). The World Bank. http://documents.worldbank.org/curated/en/535281468768723989/pdf/multi-page.pdf

Budds, J. (2009). Contested H2O: Science, policy and politics in water resources management in Chile. *Geoforum, 40*(3), 418–430.

Court of Appeal of Iquique. (2017). Decision n° 534-2017 to refuse the call of M. Godoy Villalobos et al., Iquique, unpublished document.
Cristi, O., Melo, O., & Donoso, G. (2014). *Análisis Estimación Del Precio Privado De Los Derechos De Aprovechamiento De Aguas*. Santiago, Chile: Comisión Nacional de Riego (CNR).
DGA. (1996). *Determinación de la Disponibilidad de Recursos Hídricos para constituir nuevos derechos de aprovechamiento de aguas subterráneas en el sector del acuífero de la Pampa del Tamarugal* (Minuta técnica 68. S.D.T.). Santiago de Chile: Dirección General de Aguas (DGA).
DGA. (2009). *Declaración Área Restricción Sector Hidrogeológico Pampa del Tamarugal* (Informe técnico 607). Santiago de Chile: Dirección General de Aguas (DGA).
DGA. (2011). *Resolución n°395: Ordena a los titulares de derechos de aprovechamiento de agua subterránea que se indica instalar los dispositivos necesarios que permitan controlar las extracciones que realizan*. Santiago de Chile: DGA, Ministerio de Obras Públicas, Gobierno de Chile.
DGA. (2016a). *Atlas del Agua Chile 2016*. Santiago, Chile: Dirección General de Aguas, Ministerio Obras Púublicas. Available at: http://www.dga.cl/DGADocumentos/Atlas2016parte1-17marzo2016b.pdf
DGA. (2016b). *Constituidas comunidades de aguas subterráneas de La Ligua y Petorca*. www.dga.cl/noticias/Paginas/DetalledeNoticias.aspx?item=402 (9 May 2016).
DGA. (2018). *Listado de derechos afectos al pago de patente por no uso*. www.dga.cl/fiscalizacion/Paginas/patentes.aspx
DICTUC. (2008). *Estudio de impacto ambiental proyecto Pampa Hermosa*. Santiago, Chile: Estudio de impacto ambiental. Sociedad Química y Minera de Chile (SQM).
DICTUC. (2010). *Análisis integrado de gestión en cuenca del Río Copiapó*. DGA – MOP.
Donoso, G. (2013). The evolution of water Markets in Chile. In J. Maetsu (Ed.), *Water trading and global water scarcity: International perspectives* (pp. 111–129). Washington, DC: RFF Press.
Donoso, G. (2015). Chilean water rights markets as a water allocation mechanism. In M. Lago, J. Mysiak, C. M. Gómez, G. Delacámara, & A. Maziotis (Eds.), *Use of economic instruments in water management – Insights from international experience* (p. 423). Cham, Switzerland: Springer. https://doi.org/10.1007/978-3-319-18287-2
Donoso, G., Blanco, E., Vergara, A., & Rivera, D. (2014). *Capacitación y apoyo a comunidades de aguas subterráneas en el valle de Copiapó, Región de Atacama. Informe Final*. Comisión Nacional de Riego, Ministerio Agricultura.
Fuster, R., Kremer, C., Palacios, A. K., Haberland, J., Gálvez, R., Moya, H., et al. (2016). *Diagnóstico Para Desarrollar Plan De Riego En Cuenca De Copiapó*. Santiago, Chile: CNR, Ministerio de Agricultura. Available at: https://www.cnr.gob.cl/DivisionDeEstudios/Documents/4_%20IF%20PGR%20Cuenca%20Copiap%C3%B3%20Tomo%20I.pdf
Gobierno de Chile. (1951). *Ley 9909, Decreto 2310, Fija los Textos Definitivos del Codigo de Aguas*. Santiago, Chile: Ministerio de Justicia. Available at: https://www.leychile.cl/Navegar?idNorma=125984
Gobierno de Chile. (1969). *Decreto 162: Fija los Textos Definitivos del Codigo de Aguas*. Santiago, Chile: Ministerio de Justicia. Available at: http://documentos.dga.cl/LEG4441.pdf
Gobierno de Chile. (1981). *DFL 1122: Fija Texto del Codigo de Aguas*. Santiago, Chile: Ministerio de Justicia. Available at: https://www.leychile.cl/Navegar?idNorma=5605
Gobierno de Chile. (2005). *Ley 20.017 que modifica el Código de Aguas*. Valparaíso, Chile: Ministerio de Obras Públicas.
Gobierno de Chile. (2014). *Decreto 203, Reglamento y Normas de Exploración y Explotación de Aguas Subterráneas*. Santiago, Chile: Ministerio de Obras Públicas. Dirección General de Aguas. Available at: http://www.leychile.cl/Navegar?idNorma=1060095
Gobierno de Chile. (2018). *Ley N° 21.064 Introduce modificaciones al marco normativo que rige las aguas en materia de fiscalización y sanciones*. Santiago Chile: Ministerio de Obras

Públicas. Dirección General de Aguas. Available at: http://www.dga.cl/Documents/Ley21064-2018.pdf

Hearne, R. (2018). Water markets. In G. Donoso (Ed.), *Water policy in Chile* (pp. 117–127). Cham, Switzerland: Springer. https://doi.org/10.1007/978-3-319-76702-4-8

Hearne, R., & Donoso, G. (2005). Water institutional reforms in Chile. *Water Policy, 7*(1), 53–69.

Hearne, R., & Donoso, G. (2014). Water markets in Chile: Are they meeting needs? In W. Easter & Q. Qiuqiong (Eds.), *Water markets for the 21st century: What have we learned?* (Global issues in water policy) (Vol. 11, pp. 103–126). Dordrecht, The Netherlands: Springer.

Hidromás CEF limitada. (2013). *Actualización de la modelación integrada y subterránea del acuífero de la cuenca del río Copiapó*. Copiapó, Chile: DGA.

JICA. (1995). *The study of the development of water resources in Northern Chile*. Santiago de Chile: JICA (Japon Internacional Cooperation Agency) – DGA (Dirección General de Aguas).

Lictevout, E., Amaro, S., Córdoba, D., & Rodriguez, J. (2014). Estudio histórico de la variación del nivel piezometrico del acuífero Pampa Del Tamarugal, Norte de Chile. In *XII Congreso Latinoamericano de Hidrogeología*, Santiago de Chile, Chile.

Lictevout, E., & Faysse, N. (2018). A doubly invisible aquifer: Hydrogeological studies and Actors' strategies in the Pampa del Tamarugal aquifer, northern Chile. *Water Alternatives, 11*(3), 592–606.

Lictevout, E., Maass, C., Córdoba, D., Herrera, V., & Payano, R. (2013). *Recursos hídricos de la región de Tarapacá: Diagnostico y sistematización de la información*. Iquique, Chile: UNAP.

McPhee, J., de la Fuente, A., Herrera, P., Niño, Y., Olivares, M., Sancha, M. A., et al. (2012). *Diagnóstico del Agua en las Américas*. Mexico Distrito Federal: Foro Consultivo Científico y Tecnológico, AC. Available at: http://www.ianas.org/water/book/diagnostico_del_agua_en_las_americas.pdf

Mechlem, K. (2016). Groundwater governance: The role of legal frameworks at the local and national level—Established practice and emerging trends. *Water, 8*(8), 347.

Montginoul, M., Rinaudo, J.-D., Brozović, N., & Donoso, G. (2016). Controlling groundwater exploitation through economic instruments: Current practices, challenges and innovative approaches. In A. J. Jakeman, O. Barreteau, R. J. Hunt, J.-D. Rinaudo, & A. Ross (Eds.), *Integrated groundwater management: Concepts, approaches and challenges* (pp. 551–581). Cham, Switzerland: Springer. https://doi.org/10.1007/978-3-319-23576-9_22

Moya, C.E., Lictevout, E., Amaro, S., Viguier, B., Lira, E., Maringue, J., et al. (2015). Three-dimensional hydrogeological modelling of the Pampa del Tamarugal aquifer. Modelling with limited well-log data. In *XIV Congreso Geologico Chileno*, La Serena, Chile.

Ostrom, E. (2000). Collective action and the evolution of social norms. *The Journal of Economic Perspectives, 14*(3), 137–158.

Palomino-Schalscha, M., Leaman-Constanzo, C., & Bond, S. (2016). Contested water, contested development: Unpacking the hydro-social cycle of the Ñuble River, Chile. *Third World Quarterly, 37*(5), 883–901.

Peña, H., Brown, E., Ahumada, G., Berroeta, C., Carvallo, J., Contreras, M., et al. (2011) *Temas Prioritarios para una Política Nacional de Recursos Hídricos*. I. d. I. d. C. Comisión de Aguas. Available at: http://www.iing.cl/images/iing/pdf/Informe_Tem_prior_rec_hid.pdf

Ravanal Salinas, N. V. (2011). *Las Comunidades de Aguas Subterraneas en Chile: Regulación, Problemas y Prospectivas*. Santiago, Chile: Universidad de Chile.

Rinaudo, J.-D., & Donoso, G. (2018). State, market or community failure? Untangling the determinants of groundwater depletion in Copiapó (Chile). *International Journal of Water Resources Development, 35*, 1–21. https://doi.org/10.1080/07900627.2017.1417116

Rivera, D. (2015). Diagnóstico jurídico de las aguas subterráneas. *Ius et Praxis, 21*(2), 225–266.

Rivera, D. (2018). Alumbrando conflictos: disponibilidad y asignación de derechos de aguas subterráneas en la jurisprudencia chilena. *Revista de Derecho, 31*(1), 159–183.

Rivera, D., Godoy-Faúndez, A., Lillo, M., Alvez, A., Delgado, V., Gonzalo-Martín, C., et al. (2016). Legal disputes as a proxy for regional conflicts over water rights in Chile. *Journal of Hydrology, 535*(Supplement C), 36–45. https://doi.org/10.1016/j.jhydrol.2016.01.057

Rojas, R., & Dassargues, A. (2007). Groundwater flow modelling of the regional aquifer of the Pampa del Tamarugal, northern Chile. *Hydrogeology Journal, 15*(3), 537–551.
Salazar, C. (2003). *Situación de los recursos hídricos en Chile. Con el apoyo de la Fundación Nippon*. México: Centro del Tercer Mundo para el Manejo del Agua, A. C..
Scheihing, K., Moya, C., Struck, U., Lictevout, E., & Tröger, U. (2017). Reassessing hydrological processes that control stable isotope tracers in groundwater of the Atacama Desert (northern Chile). *Hydrology, 5*(1), 3. https://doi.org/10.3390/hydrology5010003
Superintendencia de Servicios Sanitarios. (2013). *Estudio de determinación de tarifas 2013–2018 Aguas del Altiplano S.A. Tomo I. Informe final*. SISS: Santiago de Chile.
Troncoso, R., Pérez, Y., Castro, R., Lorca, M. E., Vega, N., et al. (2012). *Evaluación hidrogeológica de la cuenca del río Copiapó, con énfasis en cuantificación, dinámica y calidad química de los recursos hídricos superficiales y subterráneos*. Sernageomin. Available at: http://biblioteca.sernageomin.cl/opac/index.asp?param=o%AD%88%92bl%99%8Cq%5E&Op=3.
Ugarte Araya, P. (2003). *Derecho de Aprovechamiento de Aguas. Análisis Histórico, Extensión y Alcance en la Legislación Vigente*. Universidad de Chile.
Uri Hammer y Asociados. (1980). *Plan Maestro de Acción Inmediata para el Desarrollo de los Recursos de Agua y Suelo del Valle de Copiapó*. Dirección de Riego - MOP.
Usón, T. J., Henríquez, C., & Dame, J. (2017). Disputed water: Competing knowledge and power asymmetries in the Yali Alto basin, Chile. *Geoforum, 85*, 247–258.
Vergara, A., Arévalo, G., Muñoz, G., Rivera Bravo, D., & Vergara, C. (2011). *Codigo de Aguas Comentado*. Santiago, Chile: Abeledo Perrot Legal Publishing Chile.
Vergara, A., Donoso, G., Rivera Bravo, D., Blanco, E., & Moyano, V. (2013). Aguas y energía: propuestas para su autogobierno y resolución especializada de conflictos. In I. Irarrázaval, M. Morandé, & M. Letelier (Eds.), *Concurso Políticas Públicas: Propuestas para Chile Santiago*. Santiago, Chile: Salesianos Impresores.
Vergara, A., & Rivera, D. (2018). Legal and institutional framework of water resources. In G. Donoso (Ed.), *Water policy in Chile* (pp. 67–85). Cham, Switzerland: Springer.
Vicuña, S., Garreaud, R. D., & McPhee, J. (2011). Climate change impacts on the hydrology of a snowmelt driven basin in semiarid Chile. *Climatic Change, 105*(3), 469–488. https://doi.org/10.1007/s10584-010-9888-4.
Vicuña, S., Meza, F., Bustos, E., & Poblete, D. (2012). *Los nuevos desafíos a la gestión de los recursos hídricos en Chile en el marco del Cambio Global*. (Paper presented at the Desafíos del recurso hídrico en Chile, Santiago, Chile).
Viguier, B. (2016). *Caractérisation des facteurs de contrôle de la recharge et des écoulements souterrains a différentes échelles de temps en zone de piedmont aride a hyper-aride*. PhD thesis, University of Montpellier, France.
Viguier, B., Jourde, H., Lira, E., Yáñez, G., Leonardi, V., Moya, C., et al. (2018). Assessment of the geometry, boundaries and preferential recharge zones of a regional aquifer in 1 dryland piedmonts by geophysical and hydro(geo)logical surveys: Case of the Pampa del Tamarugal Aquifer (Northern Chile). *Journal of South American Earth Sciences, 86*, 366–383.
World Bank. (2011). *Diagnóstico de la gestión de los recursos hídricos*. Santiago, Chile: Dirección General de Aguas, Ministerio de Obras Públicas, Chile. http://www.dga.cl/eventos/Diagnostico%20gestion%20de%20recursos%20hidricos%20en%20Chile_Banco%20Mundial.pdf
World Bank. (2013). *Estudio para el mejoramiento del marco institucional para la gestión del agua*. Santiago, Chile: Dirección General de Aguas, Ministerio de Obras Públicas de Chile. http://reformacodigodeaguas.carey.cl/wp-content/uploads/2014/09/Informe-Banco-Mundial-Estudio-para-el-mejoramiento-del-marco-institucional.pdf
Zañartu Rosselot, J. H. (2001). Aguas Subterráneas: ¿Necesidad De Un Nuevo Enfoque? *Revista de Derecho Administrativo Económico de Recursos Naturales, III*(2), 539–545.

Guillermo Donoso is Agricultural and Natural Resource Economist, specializing in Water Economics. Full Professor at the Water Law and Economics Center of the Pontificia Universidad Católica de Chile (PUC), he holds a Ph.D. in Agricultural and Natural Resource Economics from the University of Maryland, College Park. He has researched water economics, management, and water allocation mechanisms with an emphasis on water markets the past 20 years. He has over 16 years of experience as grants researcher in Water Economics and Management in projects for Latin America

Dr. Elisabeth Lictevout is senior hydrogeologist in Chile, with a Ph.D in Water Sciences from the University of Montpellier. She is associate researcher at the University of Concepcion and founder and director of Carpes Science. From 2011 to 2015, she was director of CIDERH (Centro de Investigación y Desarrollo en Recursos Hídricos) where she developed extensive interdisciplinary research on groundwater management.

Dr. Jean-Daniel Rinaudo is researcher at Brgm, Montpellier University, where he coordinates the scientific program on environmental and risk economics Initially trained as an agricultural engineer (Montpellier SupAgro 1994), he specialized in agricultural and resource economics (PhD University of Auvergne, 2000). Prior to joining Brgm, he worked for the International Water Management Institute in Pakistan where his research focused on the political economy of irrigation management reforms. His current research mainly focuses on the institutional economic dimension of groundwater management. Most of his research is conducted in France but he also works in Morocco and Chile. He is currently developing new research activities in the field of natural disasters economics, focusing on the methods to assess economic vulnerability and resilience. Dr. Jean Daniel Rinaudo is also member of the Scientific Council of the Adour Garonne River basin agency.

Chapter 26
California's 2014 Sustainable Groundwater Management Act – From the Back Seat to the Driver Seat in the (Inter)National Groundwater Sustainability Movement

Thomas Harter

Abstract California's geography and Mediterranean, semi-arid climate has attracted both a burgeoning population and one of the largest irrigated agricultural developments in the world. Water resources are important to the livelihood of the state. With dry summers and highly variable annual winter precipitation, groundwater is a critical resource, drought buffer, and long-term storage reservoir for the state. Only during the most recent five-year drought, California adopted statutory control of groundwater resources: in 2014, the legislature passed the Sustainable Groundwater Management Act (SGMA). The law is the most significant California water law reform since the legislature took statutory control of surface water rights in 1914 and of water quality in 1969. This chapter provides an overview of groundwater management during the state's 150-year history, with often uncontrolled groundwater development, with conflict resolution and groundwater adjudications through the courts in some areas, and continued groundwater overdraft in others. Where courts have set limits on groundwater extraction, the objective has been to ensure stable and reliable groundwater level dynamics to avoid well outages, land subsidence, and seawater intrusion. Shortages are shared in sometimes complicated arrangements among overlying users and prior appropriators of groundwater. Under SGMA, groundwater management decisions will be made at the local level, with state oversight, to achieve long-term sustainability. We explore SGMA's vision for sustainability, stakeholder engagement, technical-scientific assessment, planning, and infrastructure practices. We also describe the role of state enforcement as a key driver for successful implementation of local groundwater sustainability plans. Importantly, local groundwater management, for the first time, will also need to consider groundwater pumping effects on surface water, on groundwater-dependent ecosystems, and on water quality. Hence, key

T. Harter (✉)
Department of Land, Air, and Water Resources, University of California, Davis, CA, USA
e-mail: thharter@ucdavis.edu

J.-D. Rinaudo et al. (eds.), *Sustainable Groundwater Management*,
Global Issues in Water Policy 24, https://doi.org/10.1007/978-3-030-32766-8_26

challenges facing local groundwater management agencies also arise from needing to address overlapping and potentially competing and less well-defined legal doctrines and federal and state laws pertaining surface water rights, ecosystems, and water quality.

Keywords Groundwater management · Groundwater law · Sustainability · California · Sustainable Groundwater Management Act (SGMA) · Overdraft · Land subsidence · Seawater intrusion · Water quality · Groundwater-dependent ecosystems

26.1 Introduction

California's semi-arid Mediterranean climate and geography share some similarities with southern Australia and southern France. Precipitation occurs almost exclusively in winter and early spring. The majority of winter precipitation is associated with just a handful of powerful storms. Hence, precipitation totals vary widely from year to year. California's landscape is dominated by the contrast between large mountain ranges (Coastal Range, Sierra Nevada-Cascade Ranges, and the Ranges of the southeast desert province) and predominantly flat, alluvial basins. The Central Valley is California's most prominent valley (47,000 km^2). Like many of California's coastal basins, it is endowed with fertile agricultural soils. Rapid population and industrial growth has also occurred primarily in these central and southern California basins.

Precipitation is most abundant in California's central and northern mountain ranges. In contrast, water users are located in Central and Southern California's valleys and basins: about 4 million hectare of irrigated agricultural lands use 40 km^3 of water each year. Annual water consumption in urban areas – the San Francisco Bay Area (10 million people), the Los Angeles – San Diego Southern California Area (20 million people), and cities in the Central Valley (5 million people[1]) – amounts to 10 km^3. Currently protected environmental water uses account for another 50 km^3 in an average year (CDWR, 2014).

California engaged from its early gold mining days in the 1850s through the 1970s in building massive water infrastructure consisting of an elaborate network of surface water storage reservoirs (40 km^3 combined storage) and thousands of kilometers of canals now spanning across the state. The infrastructure allows for winter precipitation (with over 15 km^3 as snow, Wrzesien et al., 2017) to be captured for summer water use and for water to be transferred from the rainfall-rich north to the

[1] https://en.wikipedia.org/wiki/Combined_statistical_area#List_of_Combined_Statistical_Areas, accessed 18 December 2018.

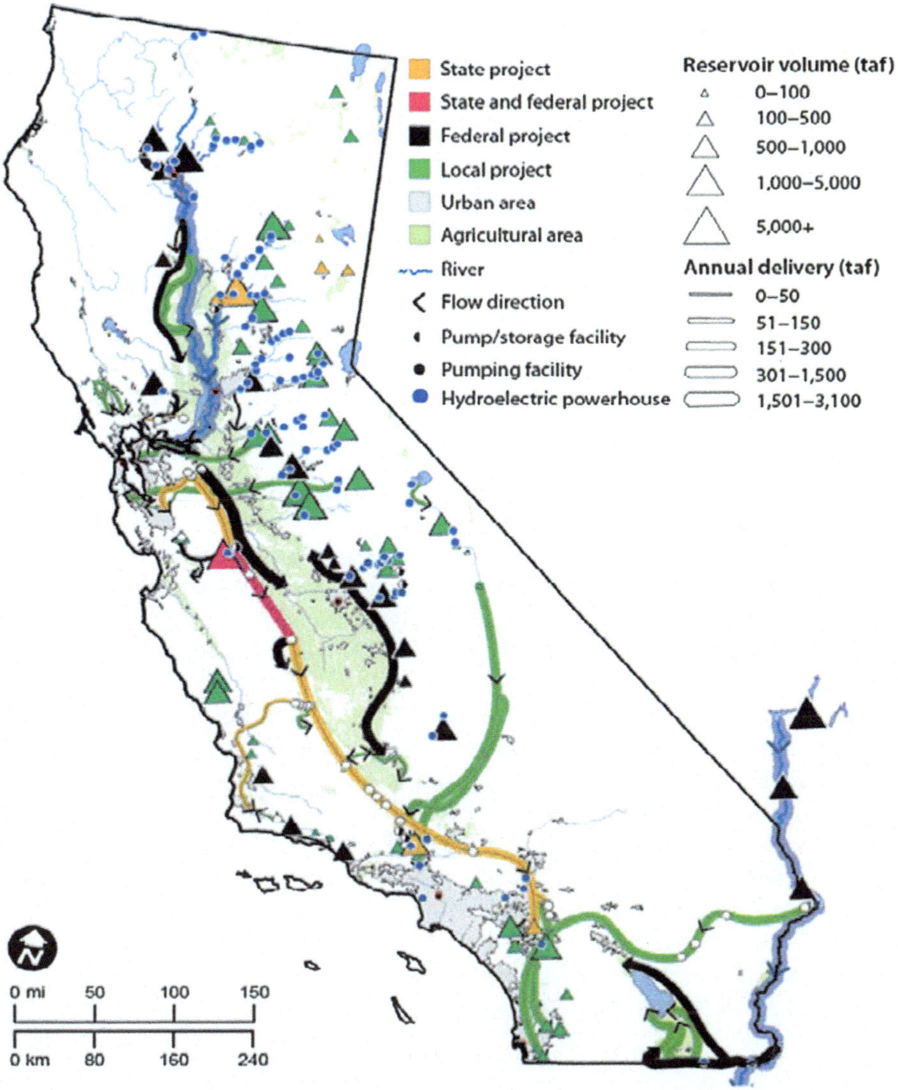

Fig. 26.1 California's surface water infrastructure and major landuse regions. "Project" refers to surface water reservoirs and associated canals. From: Hanak and Stryjewski (2012), http://www.ppic.org/content/pubs/report/R_1112EHR.pdf, accessed 18 December 2018

center and southern parts of the state (Fig. 26.1). Since the 1930s, this infrastructure has been supplemented by large wells equipped with turbine pumps that tap into the alluvial aquifers underlying California's urban areas and irrigated agricultural lands. Groundwater accounts for one-third (in dry years over one-half) of California's water use (CDWR, 2014). Partially depleted aquifers represent the state's largest potential storage capacity (Brush, Dogrul, & Kadir, 2013). Aquifers also function as a ubiquitous, natural local water treatment and delivery system.

Conflicts between urban, agricultural, and environmental water uses have always played an important role in California's economic and political development given the limited renewable water supplies (e.g., Reisner, 1993). While early water use focused on streams, competing uses of groundwater resources led to litigation and court trials already by the late 1800s. By 1904, the fundamental legal doctrine guiding California's groundwater resource management was well-defined.[2] The rapid expansion of groundwater use in the early and mid-1900s expanded the scope of groundwater conflicts, leading to far-reaching court decisions and the construction of massive surface water projects to address groundwater overdraft in California's coastal urban centers and its irrigated agricultural regions in the 1940s–1960s. Comprehensive statutory control of groundwater resources, while being considered for over a century, would not be realized until 2014. In the meantime, economic growth and increasing environmental awareness about in-stream flows continued to increase water demands beyond the capacity of existing surface water infrastructure and aquifer recharge, depleting groundwater supplies.

This chapter explores the history of and current developments in California's approaches to managing groundwater resources. An overview of the historical development of California's groundwater rights framework (Sect. 26.2) sets the background for exploring the development of groundwater management policy in the state (Sect. 26.3), which culminated in the enactment of the Sustainable Groundwater Management Act of 2014 (SGMA, Sect. 26.4). Subsequent sections explore stakeholder engagement (Sect. 26.5), technical-scientific assessment (Sect. 26.6), and infrastructure (Sect. 26.7) practices envisioned under the new legislation. Enforcement as the key driver for successful implementation of sustainable groundwater management is also briefly explored (Sect. 26.8). Finally, the chapter concludes (Sect. 26.9) by discussing key challenges facing local groundwater management agencies, as their work is not only subject to the provisions of SGMA, but also to overlapping and potentially competing and less well-defined legal doctrines and federal and state laws pertaining to the management of surface water resources, ecosystems, and water quality.

26.2 California's Water Rights Framework

In the United States, individual states rather than the federal government control water property and water use rights, except on federally owned lands (national parks, national forests, native American lands, etc.). California achieved statehood only in 1850, shortly after the beginning of the 1848 gold rush. Absent a long-standing body of state or federal law, California's people elected to use English

[2] *Katz v. Walkinshaw*, 141 Cal. 116 (1903).

common law as the foundation for deciding on many legal matters including the use and sharing of water resources.

Part of Mexico until 1848, early California settlements had already established community rights to surface water for irrigation, domestic, and other needs that were formally recognized by Congress as "pueblo water rights" when California became a state.[3] Lands under federal administration (45% of California's land area) hold "federal reserved rights" to surface water[4] and to groundwater[5] to meet the water needs of these federal lands.

Early on, the state adopted common law riparian rights, which assigns a right to use water on lands adjacent to a stream. The water right only extends to the natural flows in the stream, not to additional flows that may be released, e.g., by upstream reservoirs. The right also does not permit storage of water for more than 60 days. Riparian water rights do not expire due to non-use. Shortages in natural stream flows are shared among users proportional to their water right.

Given the semi-arid nature of the state, most urban, industrial, and agricultural development in need of water occurred in areas not adjacent to streams. For those water users, the state adopted the principle of prior appropriation, commonly applied in other Western states with semi-arid and arid climates: water rights for non-riparian water users would be defined by the date of first diversion, the point of diversion, the diversion amount, and the use and place of use of the water. This may include storage for later use. Within the appropriative system, shortages are not shared among users. Instead, seniority decides who will have either full access or no access to their water right in case of water shortages ("first in time, first in right"). Non-use for more than 5 years terminates an appropriative water right, whereas riparian rights do not expire due to non-use. Pueblo water rights and federal reserved water rights are senior to most appropriative water rights.

Until 1914, water rights were declared by individual notification. In 1914, the state established statutory control over surface water rights. Since then, water rights applications have formally been submitted to and decided by the State Water Board (and its predecessor agencies). For groundwater, the state did not assume similar statutory control although the possibility was strongly considered in the development of the Water Commission Act of 1914 (Sax, 2002). Hence, groundwater users have not been regulated and do not need to apply for a water use permits. The only permit required for a new well is a county-issued well construction (drilling) permit.

Absent statutory control and explicit state policy, water conflicts over groundwater use have historically been deferred to and decided by the courts. Court decisions in turn were subject to interpretation of established legal doctrine and case law. Groundwater rights work somewhat analogous to surface water: under California's

[3] *City of Los Angeles v. City of San Fernando* et al., No. 650079, Superior Court of the State of California for the County of Los Angeles, 26 January 1979.

[4] *Winters v. United States*, 207 U.S. 564 (1908); *Arizona v. California*, 373 U.S. 546 (1963); *U.S. v. New Mexico*, 438 U.S. 696 (1978).

[5] *Cappaert v. United States*, 426 U.S. 128 (1976); *Agua Caliente v. Coachella Valley Water Dist.*, 849 F. 3d 1262 (2017).

correlative rights doctrine, landowners overlying an aquifer system have a right to use groundwater on their land. Similar to riparian rights, these overlying rights cannot be extinguished by non-use, i.e., they may exist as dormant water rights. Among overlying landowners (agricultural, private, and industrial water users), the water right is shared in relation to use by others, to the characteristics of the land parcel, and to the characteristics of the aquifer[6] (hence, "correlative" right). Groundwater that is not claimed by overlying landowners can be claimed by other pumpers under the prior appropriation doctrine. The most important appropriators in many groundwater basins are public water agencies (cities, water districts, etc.) that serve overlying landowners or export water to neighboring basins. Their right is generally considered junior to overlying, correlative rights.

Some groundwater rights can be obtained by "mutual prescription", that is, by use of water that is "actual, open and notorious, hostile and adverse to the original owner" for at least 5 years.[7] Prescriptive water rights have become an important element to protect some water rights of cities and public water agencies overlying a groundwater basin against the (more senior) overlying rights of industrial and agricultural landowners and against dormant overlying rights.

Groundwater rights – like surface water rights – are a right to the use of water ("usufruct") rather than outright ownership, which remains with the people of the state (Matthews, 2003). California's constitution[8] dictates that water can only be pumped or diverted for reasonable and beneficial uses, a central element to all water rights decisions. Furthermore, the total amount of water pumped must not exceed the safe yield of a groundwater basin. In other words, groundwater rights do not extend to all groundwater physically present in a groundwater basin, only to the renewable amount of groundwater. In California, courts have never recognized a right to outright mining of groundwater.

Historically, groundwater conflicts among users over each party's volumetric water right have occurred mostly in Southern California's comparatively smaller (and more arid) groundwater basins. These are most susceptible to overdraft and – along coastal groundwater basins – suffer from seawater intrusion. The need to involve the courts for remedy meant the pursuit of costly lawsuits. Such efforts would most likely be extended only where overlying and appropriative users include influential and economically powerful parties – cities, industrial landowners, and large agricultural landowners and their associations. There, adjudications – court decreed allocations of groundwater rights among multiple users – have been performed, now including all or parts of 27 groundwater basins (Blomquist, 1992; Langridge, Brown, Rudestam, & Conrad, 2016a; Ostrom, 1990).

Adjudications have been initiated for widely differing reasons, may have involved few or many parties, and have led to a diversity of water management arrangements

[6] *Katz v. Walkinshaw*, 141 Cal. 116 (1903).

[7] *City of Pasadena v. City of Alhambra*, 33 Cal.2d 908 (1949).

[8] California Constitution Article 10 Section 2, enacted in 1928.

(Langridge et al., 2016a). Adjudications typically involve consideration of existing water rights (pueblo rights, federally reserved rights, overlying rights, appropriative rights, prescriptive rights, etc.), historic water use by individuals or groups of water users, groundwater basin conditions and safe yield, and, in few cases, seawater intrusion or other water quality issues. Adjudications have sometimes employed a "physical solution", settling on a negotiated allocation of groundwater rights that seeks a pragmatic balance between historic water use and various, sometimes conflicting water rights, including dormant rights under California' constitutional directive "that the water resources of the State be put to beneficial use to the fullest extent of which they are capable".[9] The court appoints a Watermaster to execute and oversee the adjudication. The Watermaster is most often a local entity representing basin water interests, although small pumpers and disadvantaged communities are rarely participating in the adjudication or subsequent Watermaster activities (Langridge et al., 2016a). The Watermaster is in charge of monitoring and annual reporting, but may also engage in other water management activities – facilitating an increase in water supplies, additional groundwater storage, and water trading, or overseeing the reduction in groundwater pumping. Watermaster activities, if any, vary widely between adjudications as does the outcome with respect to addressing overdraft conditions (Langridge et al., 2016a).

A steady, central historical tenet of court decisions throughout the past century, including many of the adjudications, has been that groundwater rights are separate from and unrelated to surface water rights, despite their obvious hydrologic connectivity. This has left surface water right holders and environmental interests largely without legal tools to address negative impacts of groundwater pumping on stream flow and groundwater-dependent ecosystems. A recent case involving the public trust doctrine, a legal concept going back to Roman law, may change that. First employed in a 1980s court decision, it allowed the state to limit diversions of surface water and modify existing surface water rights permits.[9] In a 2018 decision, courts have – for the first time – affirmed the application of the public trust doctrine to groundwater pumping.[10] The public trust doctrine protects flows in navigable waters to the extent feasible and reasonable. Importantly, it may override existing water rights to tributary streams of navigable waters and to groundwater pumping that harms flow in navigable streams.

[9] *National Audobon Society vs. Superior Court*, 33 Cal. 3d 419 (1983); Bay-Delta Plan Update: Lower San Joaquin River and Southern Delta https://www.waterboards.ca.gov/waterrights/water_issues/programs/bay_delta/bay_delta_plan/water_quality_control_planning/, accessed 18 December 2018.

[10] *Environmental Law Foundation v. State Water Board*, 3rd District Court of Appeals, No. C083239, August 29, 2018.

26.3 History of California's Groundwater Management Policy

California has a diverse and expansive institutional landscape that includes hundreds of irrigation and water related special acts districts (LAO, 2002) – quasi-governmental institutions with elected boards, taxing powers, and powers of eminent domain. First authorized by the state through the Wright Act of 1887, these subdivisions of the State government manage and distribute surface water among their landowners (including cities and communities). In contrast, groundwater is typically developed and used by landowners directly, not subject to scheduling requirements and shortages. Only 15 special acts districts have some groundwater management related authority, with large urban districts (Orange County Water District, Santa Clara Valley Water District) having been among the most active, since about the middle of the twentieth century (Langridge, Sepaniak, & Conrad, 2016b). Special acts districts with groundwater management responsibilities are mostly focused on water supply augmentation to address potential groundwater overdraft and, along the coast, looming seawater intrusion. They employ financial incentives, replenishments fees, and conservation measures to control groundwater extraction, where needed.

Groundwater management has been part of California's political discussion since the early twentieth century. As early as 1912, legislative proposals existed to address groundwater extraction. But already at that time agricultural interests appear to have been most influential in nudging the legislature against taking statutory control over groundwater as part of the controversial 1914 water rights reform (Sax, 2002, p. 296). Subsequently, groundwater management was most actively pursued in urbanizing areas of southern California and the San Francisco Bay Area. There, groundwater resources quickly became overdrafted, land subsidence threatened and damaged infrastructure, most prominently the surface water canal infrastructure, and seawater intrusion forced water supply wells of large cities to be closed (Hanak et al., 2011; Lipson, 1979; Schneider, 1977). Adjudication of groundwater rights led to Watermaster appointments that variously oversaw groundwater extraction.

Early to mid-twentieth century court-decreed limitations to groundwater extractions reinforced the period's efforts to build out surface water infrastructure in California, transferring surface water from the north and from surrounding mountains to Central and Southern California's urban and agricultural basins. Already by the middle of the twentieth century, the concept of "conjunctive use" of groundwater and surface water began to take hold.[11] Urban water districts took advantage of storage capacity in groundwater basins for seasonal or long-term transfer of surplus surface water; or to trade groundwater (that would otherwise be pumped) for surface water deliveries in exchange for title to the unpumped groundwater volume ('in lieu recharge'). Courts confirmed the security of surface water stored underground

[11] *Los Angeles v. Glendale*, 23 Cal.2d 6 (1943).

against claim by nearby groundwater pumpers, within the correlative rights of overlying owners.[12] Very few, mostly urban special acts district (e.g., Orange County Water District) used their tax authority to levy extraction fees from landowners (including cities) within the district boundaries that would pay for the replenishment of the aquifer through a portfolio of water management measures (Blomquist, 1992; Langridge et al, 2016b).

Groundwater replenishment would take numerous forms through the second half of the twentieth century: A major approach in the 1950s through 1970s was to recover lowering water tables through less groundwater pumping that was made possible by developing additional surface water supplies, locally, regionally, or from across the state (Colorado River Project, Central Valley Project, State Water Project). Groundwater recharge basins were built to supply aquifers with additional recharge that would balance extraction by groundwater pumpers. In some agricultural regions, irrigation districts actively or inadvertently replenished groundwater during the wet winter and the spring runoff season, by filling unlined canals or by using landscape depressions as natural flooding basins (e.g., Consolidated Irrigation District, 2009).

By the 1980s, environmental concerns and nearly full build-out of the surface water infrastructure stopped the expansion of water supplies. Importantly, new environmental legislation (the Endangered Species Act 1973 (ESA)[13] and the Clean Water Act 1972 (CWA)[14] and court decisions (the Public Trust doctrine[15]) began to limit or reduce the amount of surface water being diverted or transferred through California's water grid.

Constraints on surface water development critically widened groundwater management portfolios over the recent three decades, adding some creative solutions (Nelson, 2011): urban areas engaging in local stormwater capture for groundwater recharge; urban-agricultural exchanges of treated wastewater for use in irrigation and in turn traded for agricultural groundwater that remained unpumped for groundwater protection; treatment of poorer quality native groundwater; groundwater replenishment with highly treated urban wastewater; use of aquifer storage and recovery (ASR) schemes (Dahlke et al., 2018b); development of water markets (Hanak & Stryjewski, 2012); and some water conservation measures that culminated in a statewide voluntary urban conservation cutback of 25% during the 2012–2016 drought (Palazzo et al., 2017). Large groundwater banks were being developed beginning in the 1980s, holding over 2 km^3 of water for long-term storage and tied into the statewide water transfer grid (Hanak & Stryjewski, 2012).

Despite these efforts and significant success in restoring groundwater storage in some of the most severely affected urban regions of southern California and the San Francisco Bay Area, groundwater overdraft remained an issue, particularly in

---

[12] Alameda County Water District vs. Niles Sand and Gravel Co., 37 Cal. App. 3d 924 (1974).

[13] 16 U.S.C. § 1531 et seq.

[14] 33 U.S.C. §1251 et seq.

[15] *National Audubon Society v. Superior Court*, 33 Cal.3d 419 (1983).

irrigated agricultural regions, the largest of which is California's Central Valley. The Central Valley encompasses about 30,000 km^2 of irrigated lands, representing 75% of California's irrigated agriculture (CDWR, 2014).

Concern over continued over-allocation of groundwater would be particularly palpable during California's frequent drought periods, inevitably leading to heated political discussions and triggering calls for action from constituents. Over the past half century, each major drought period would see gubernatorial or legislative actions attempting to address the largely unchecked groundwater overdraft (Cannon Leahy, 2016). The 1959–1962 drought yielded legislation that funded extensive groundwater investigations to assess the state's groundwater resources. An Interim Committee on Water, in 1962, concluded with a sobering assessment of groundwater conditions. Almost 20 years later, following a record dry year in 1977–1978, then-governor Jerry Brown called for a Water Rights Commission to review water rights and groundwater management in California. The Commission recommended that the state take statutory control of groundwater management and outlined a governance structure that emphasised local control under state oversight.

Wet years followed, and the legislature had little appetite for creating the proposed legislation. However unlike the 1950s, 1960s, and 1970s, the next 40 years would bring drought conditions to California at an accelerating pace, intensifying discussions over state control of groundwater. Following the 1988–1992 drought, the legislature passed Assembly Bill (AB) 3030, providing a wide range of local agencies the authority to develop local groundwater management plans[16]. The legislation provided neither binding requirements to do so nor specific guidance on the implementation of local groundwater management. The latter shortcoming was addressed 10 years later, during another drought. In 2002, Senate Bill (SB) 1938[17] was enacted to provide a stiff financial incentive to local water agencies for developing groundwater management plans (GMPs): state support for water projects would not only be contingent on local agencies having in place a GMP, but significant minimum requirements for the content of such GMPs were put in place. Still, there was no requirement to go beyond "planning a plan". The 2007–2009 drought added to the discussion of groundwater reform law, but yielded little beyond additional legislatively required water level monitoring[18] to supplement already existing, long-term groundwater level monitoring programs.

Water, irrigation, and special acts districts throughout California, representing both urban and agricultural water users, would continue to be the key lobby against additional regulations and bureaucracy for groundwater use, which remained the by far least expensive and simplest water resource to tap into and manage at the discretion of individual landowners and local authorities, without state oversight.

---

[16] California Water Code §10750–10755

[17] Senate Bill 1938, 2002, https://leginfo.legislature.ca.gov/faces/billNavClient.xhtml?bill_id=200120020SB1938, accessed 18 December 2018

[18] Senate Bill X7-6, 2009, http://www.leginfo.ca.gov/pub/09-10/bill/sen/sb_0001-0050/sbx7_6_bill_20091106_chaptered.html, accessed 18 December 2018

With groundwater conditions becoming increasingly critical in the early 2000s, and with many urban districts advancing extensive groundwater management efforts, agricultural districts not only found themselves more and more isolated in resisting groundwater reform, but voices grew louder from within the agricultural community to more seriously begin to address groundwater management in the irrigated regions of California. In 2009, a third drought year in succession, the Association of California Water Agencies, the state's largest affiliation of water agencies representing both urban and agricultural water agencies and their water user members, developed a set of policy principles on groundwater management.[19] The policy principles supported strong local groundwater management, integrated with surface water management, and demanded significantly more accountability and transparency than currently practiced, while rejecting outright state control.

Only 3 years later, the State was facing yet another drought, one that would last 5 years, from 2012 to 2016. Already by 2013, Governor Jerry Brown, re-elected in 2011 and having initiated the earlier Water Rights Commission during his 1975–1983 term, made a public call for new legislation to be developed. Two major proposals emerged by spring of 2014 including proposed legislation from the Association of California Water Agencies. By fall of 2014, the legislature passed and the governor signed the Sustainable Groundwater Management Act (SGMA),[20] setting in motion the largest water management reform in California history since the 1914 Water Commission Act and the 1969 Porter Cologne Water Quality Control Act. SGMA represents the first comprehensive statutory law governing the management of groundwater in California.

26.4 Principles of Sustainable Groundwater Management in California

The principles of the Sustainable Groundwater Management Act follow in the footsteps of the Water Rights Commission's 1978 final report recommendation: that groundwater be managed locally by local agencies; and that "groundwater resources be managed sustainably for long-term reliability and multiple economic, social, and environmental benefits for current and future beneficial uses".[21] These are the two founding pillars for twenty-first century California groundwater management. Importantly, and with substantial foresight and experience, SGMA provides an extensive and detailed definition of sustainability that establishes unequivocal

[19] https://www.acwa.com/wp-content/uploads/2017/03/groundwatermanage_policy_3.pdf, accessed 18 December 2018.

[20] http://opr.ca.gov/docs/2014_Sustainable_Groundwater_Management_Legislation_092914.pdf, accessed 18 December 2018.

[21] California Water Code §113.

frontiers in future California groundwater management. Sustainability is defined as the absence of six specific "undesirable results" which are defined as:[22]

1. Chronic lowering of groundwater levels indicating a significant and unreasonable depletion of supply if continued over the planning and implementation horizon. Overdraft during a period of drought is not sufficient to establish a chronic lowering of groundwater levels if extractions and groundwater recharge are managed as necessary to ensure that reductions in groundwater levels or storage during a period of drought are offset by increases in groundwater levels or storage during other periods.
2. Significant and unreasonable reduction of groundwater storage.
3. Significant and unreasonable seawater intrusion.
4. Significant and unreasonable degraded water quality, including the migration of contaminant plumes that impair water supplies.
5. Significant and unreasonable land subsidence that substantially interferes with surface land uses.
6. Depletions of interconnected surface water that have significant and unreasonable adverse impacts on beneficial uses of the surface water.

The significance of these definitions and principles cannot be overstated. In 2014, about 2 months prior to the passage of SGMA, but at a time when much of the statute had been developed and undergone broad discussion, Paul Wenger, then-president of the California Farm Bureau Federation, wrote an opinion article headlined "Groundwater legislation could be checkmate".[23] While perhaps meant to draw attention to a perceived looming defeat for agriculture, the analogy was, perhaps inadvertently, but also implicitly rather foresightful: all the pieces on the chess board would finally be in complete and full relation to each other. For the first time, California's statutes would recognize the physical linkage between groundwater and surface water, between groundwater and ecosystems, between water supply and water quality, and between water use and land use. Importantly, the legislation also mandated that sustainable management of groundwater be substantively considerate of these linkages.

It is too early to begin to assess the full impact of the broad, integrated water management perspective that SGMA takes on sustainability. Local stakeholders will need to balance conflicting interests of groundwater users, surface water users, environmental interests, environmental justice concerns, land use planners, and others. But already, SGMA is beginning to become a catalyst for more holistic, integrated thinking in water and land management in regions that have historically been reluctant to engage in groundwater management. While SGMA and the focus on achieving groundwater sustainability may temporarily be distracting from nearly 20 years of statewide efforts in integrated regional water management planning

[22] California Water Code §10721(x))

[23] http://agalert.com/story/?id=6829, accessed 18 December 2018

(IRWMP[24]), SGMA may become a critical catalyst to bring IRWMPs to full maturity, across all of California's important groundwater basins.

The SGMA legislation[25] required the establishment of local groundwater sustainability agencies (GSAs) by 2017 and the development of groundwater sustainability plans (GSPs) – by 2020 for critically overdrafted basins and by 2022 for all other basins. GSPs must be designed to achieve sustainable groundwater conditions within 20 years of the initial GSP completion (2040 or 2042, respectively). The state has substantial oversight of the local process with requirements for regular, 5 year review of GSPs and their implementation. Failure to form a local GSA, to develop a GSP, or to implement a GSP leads to mandated state take-over of local government in managing groundwater, at cost to affected groundwater users.

The legislation designated two existing state agencies with the oversight and implementation of new regulations: The Department of Water Resources (DWR, within the California Natural Resources Agency), primarily a state planning and technical support agency (but also the operator of the State Water Project), was designated to develop the detailed regulations within 2 years of passing of the act, to provide technical guidance and assistance, and also to administer state financial support for local GSAs. DWR will be the agency in charge of reviewing GSAs and GSPs on a regular 5-year basis. DWR is the agency that determines compliance with SGMA. In case of non-compliance, DWR turns matters over to the State Water Resources Control Board (within the California Environmental Protection Agency), an enforcement agency currently overseeing surface water rights, water quality (through its nine regional member agencies, the Regional Water Boards), and implementation of drinking water regulations.

While SGMA applies to all California groundwater and to addressing all past, ongoing, and future undesirable results, regulatory and therefore most practical implementation is limited both, in space and in time: Spatially, SGMA requires GSA formation and GSP development and implementation only in areas overlying medium and high priority groundwater basins, but not in low and very-low priority groundwater basins. Priority is set by the state (DWR) based on technical-scientific criteria related to groundwater use, population density, water use, and existing groundwater conditions. The prioritization criteria and basin status will be reviewed by DWR every 5 years. California is divided into 515 groundwater basins (Fig. 26.2). Some of these basins are delineated using hydrologic boundaries, many are delineated by a combination of hydrologic and political boundaries, particularly in the Central Valley and other large hydrologic groundwater basins. Only alluvial (unconsolidated sedimentary) groundwater basins are currently designated as groundwater basins. DWR has instituted a basin boundary adjustment process that will occur every few years, as needed or requested by local agencies. California also has some smaller volcanic aquifer systems. Historically, DWR has not designated those as

---

[24] Senate Bill 1672, 2002, http://www.leginfo.ca.gov/pub/01-02/bill/sen/sb_1651-1700/sb_1672_bill_20020921_chaptered.html, accessed 18 December 2018.

[25] http://groundwater.ucdavis.edu/sgma/, accessed 18 December 2018.

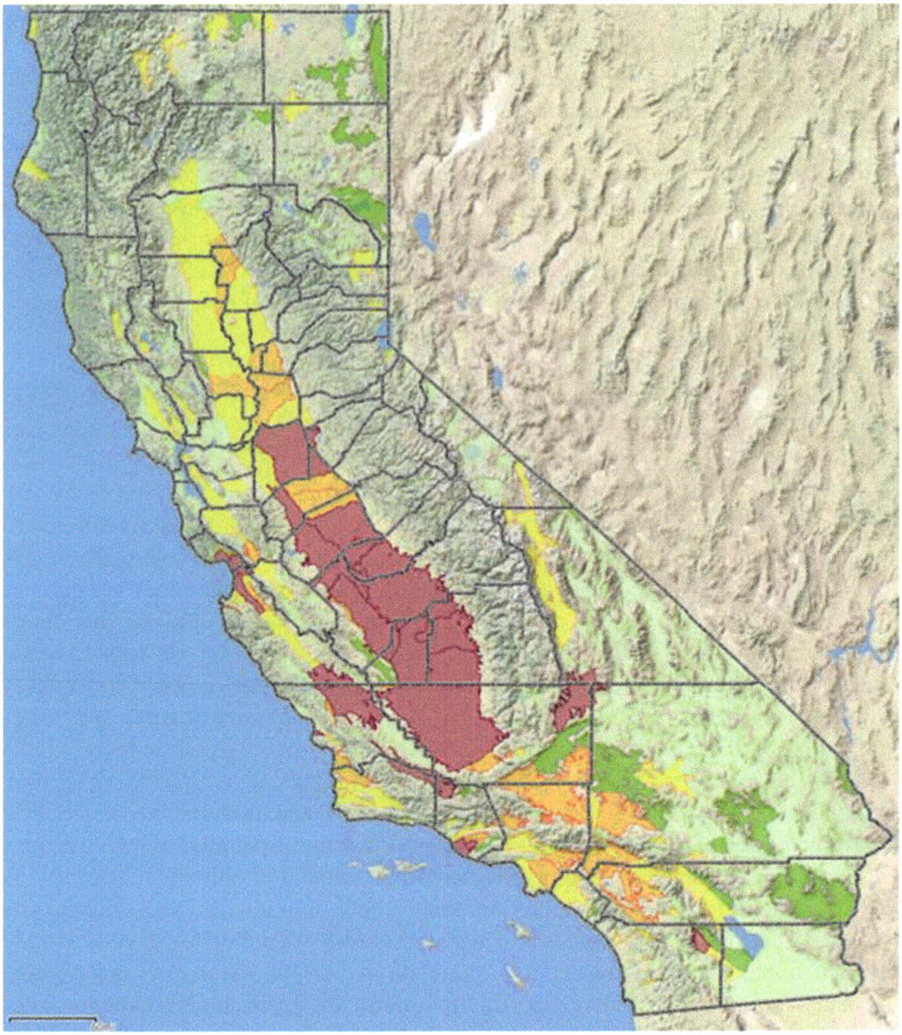

Fig. 26.2 California's 515 groundwater basin: very low and low priority (light and dark green), medium priority (yellow), high priority (orange), critically overdrafted high priority (burgundy). Status: 2017. From: https://gis.water.ca.gov/app/boundaries/, accessed 15 September 2018)

groundwater basins, but has recently accepted the first volcanic aquifer basin boundary adjustment and may add more in the future. Over 120 of the 515 groundwater basins are currently classified as medium- and high priority basins with 21 basins classified as "critically overdrafted" (Fig. 26.2). These medium and high priority basins encompass over 95% of current groundwater usage.

Temporally – with respect to the "age" of undesirable results – SGMA limits requirements for addressing undesirable results to those undesirable results that occurred on or after January 1, 2015. In other words, undesirable results such as water level decline, seawater intrusion, land subsidence, or groundwater capture of surface water must be addressed by a GSP only if they began to occur or to the degree that they degraded after 2014. Undesirable results that already existed at the time the legislation was enacted may be addressed by a GSA, but the GSA is not required to do so. In other words, further decline in water table, additional land subsidence, the advancement of a seawater intrusion front, or additional depletion of streams must be avoided. But an already existing low water table need not be reversed to higher levels that existed perhaps decades ago, except to control seawater intrusion. By focusing the regulatory efforts on those basins with the highest usage and on continued expansion of undesirable results rather than recovery of historic conditions, the legislation cast a practicable management framework that would also have broad support from constituents.

Local agencies that were given eligibility to form a GSA include cities, counties, irrigation, water, and other special acts districts, resources conservation districts, flood control districts, and others. By 2017 – within 3 years of the legislation – over 300 new agencies had formed, overlying the over 120 medium- and high priority groundwater basins. Agency formation required a minimum amount of notification and public hearing. Where multiple agencies provided notification to become a GSA, agencies needed to work out among themselves how to proceed before finalizing the GSA(s). The path to creating this many GSAs in such a short period of time was a perhaps unique experience in administrative practice. The process diverged widely between GSAs, some based on broad engagement of stakeholders and the public, others with only minimal public participation. SGMA does not prescribe the governance structure of GSAs.

The resulting GSAs are equally diverse. Some are single agencies partially or fully overlying a groundwater basin. Some single agency GSAs may even overly multiple groundwater basins (e.g., a county agency with multiple subbasins within its boundaries). Some GSAs are contractual arrangements (through a memorandum of understanding or through a joint powers agreement) among multiple local agencies with agreed-upon governance structures and representation. Some of these GSAs will write a single GSP, some will write multiple GSPs (if overlying multiple basins), some GSAs will collaborate with other GSAs to write a GSP that applies across multiple GSAs. Analogously, some GSPs will cover only a part of a groundwater basin, some GSPs will extend over the entire groundwater basins, some GSPs may partially or fully overlap with multiple groundwater basins. The legislation requires very close coordination if multiple GSPs are written within a single designated groundwater basin. Some coordination is also required by law between GSPs in adjacent groundwater basins.

The development and content of groundwater sustainability plans was the subject of detailed regulations developed by the Department of Water Resources. The department sought substantial input from stakeholders, regulated entities, and technical experts during the drafting of the regulations in 2015 and 2016. The regulations set up the requirements for GSP elements and reflect on the criteria by which the GSPs will be evaluated in DWR's initial and five-yearly reviews.

- Groundwater Sustainability Plans cover three broad areas:
- governance, process, stakeholder engagement, learning, and communication;
- technical and scientific assessment, monitoring, modeling, data management and reporting; and
- project development and implementation to reduce groundwater demand or enhance groundwater supplies.

These areas are discussed in more detail in the following sections.[26]

26.5 People: Communication and Engagement in the GSP Process

Water users and other stakeholders in California's groundwater basins are diverse and include farmers (agricultural landowners), cities and communities, environmental interests and NGOs, environmental justice representatives, representatives of domestic well users, land use zoning agencies, water agencies, representatives of minority communities, economically disadvantaged communities, and Native American tribes. Water users and stakeholders are given participatory roles in the GSP process. GSAs are required to provide public notification and opportunities for public participation in the GSP development. The governance of the GSA may reflect some of that diversity; many GSAs have advisory committees that meet regularly to assure a broader participatory approach across all stakeholder groups.

An emerging challenge is capacity: With over 300 GSAs, some user and stakeholder groups find themselves limited in their (personnel or people) capacity to attend the numerous meetings. Particularly public members of advisory committees and those representing smaller NGOs have found themselves stretched thin by the large number of meetings in areas with many GSAs. Some participation is also limited by the ability to provide funding toward the cost of operating a GSA, providing travel cost, or finding in-kind volunteer contributions.

26.6 Creating the Knowledge Base: GSP Monitoring, Assessing, Reporting

A thorough understanding of the groundwater system within the governance area and – more broadly – within the groundwater basin (where GSAs share governance of a basin) is a critical basis for sustainable management. Education and information of stakeholders, assessment of the groundwater sustainability status, and evalu-

[26] Further supporting information on SGMA and its implementation is available at http://groundwater.ucdavis.edu/sgma, accessed 18 December 2018

ation of any actions needed is based on sufficiently detailed groundwater characterization. All GSPs are therefore required to include the following elements[27]:

- development of a hydrogeologic conceptual model
- description of groundwater conditions with respect to potential undesirable results
- development of a water budget
- development of sustainable management framework with goals, desired outcomes, and thresholds for actions
- development of a monitoring network
- descriptions of project and management actions
- reporting and data management

For each element, the state laid out broad minimum requirements in the regulations, but those will leave significant flexibility to individual GSAs as they are developing their GSPs. DWR has also developed non-binding best management practices and guidelines[28] that are intended to provide some basic technical education and guidance, but also to articulate DWR's expectations when reviewing GSPs.

While details in the GSP development are anticipated to vary widely, DWR personnel will likely be engaged in the local process as observer and in an advisory role to ensure that there is ongoing feedback between the regulator and the GSA. The longer-term process is designed to encourage adaptive management, whereby new information will update the conceptual models, water budgets, numerical models where used, and inform decisions on projects and actions, and planning of additional monitoring. GSPs will need to be updated by GSAs every 5 years for formal review by DWR.

The six undesirable results will play a central role in the development of the GSP. The regulations have coined "sustainability indicator" as an operational term for speaking and articulating the linkage between undesirable results, monitoring systems, management goals, and thresholds (Fig. 26.3). Sustainability indicator

Fig. 26.3 The six sustainability indicators to be considered in the GSPs (CDWR, 2017)

[27] California Code of Regulations Title 23(Div.2) §350 -§358, 2016, https://water.ca.gov/LegacyFiles/groundwater/sgm/pdfs/GSP_Emergency_Regulations.pdf, accessed 18 December 2018.

[28] https://water.ca.gov/Programs/Groundwater-Management/SGMA-Groundwater-Management/Best-Management-Practices-and-Guidance-Documents, accessed 18 December 2018.

refers to "any of the effects caused by groundwater conditions occurring throughout the basin that, when significant and unreasonable, cause undesirable results".[29]

Each sustainability indicator – if at all relevant to the basin – requires a monitoring system that can provide metrics on the status of the sustainability indicator. Monitoring systems need not be separate between the sustainability indicators. The overall sustainability goal, identified by the GSA, must be translated into "measurable objectives" and "minimum thresholds" for each sustainability indicator. Measurable objectives are defined in SGMA and represent the range of values in the sustainability indicator metrics that reflect a desirable, perhaps optimal sustainability indicator status, e.g., an acceptable range of desirable water levels, groundwater storage, or concentrations in water quality indicators. Minimum thresholds, conceptually defined in SGMA, operationally are values beyond which the status of the sustainability indicator becomes undesirable. The minimum threshold is a numeric value within the same metric (measurement or aggregated measurement obtained from the monitoring system) used for measurable objectives.

With few exceptions explicitly contained in SGMA, the state provides some guidance, but does not prescribe what these metrics should be or what values they must take on – this will be left to the GSA to decide. However, in the state's review of the GSP, the state has a duty to evaluate whether measurable objectives and minimum thresholds are defined appropriately, especially when compared to those defined by other GSAs that share the same groundwater basin or are located in adjacent groundwater (sub-)basins.

Equipped with a hydrogeologic conceptual model, the water budget – perhaps from a groundwater model, and knowledge of the relevant sustainability indicators, GSA managers and stakeholders decide on monitoring networks needed, set measurable objectives and minimum thresholds, and possibly develop trigger thresholds for actions and projects.

26.7 Aligning Water Use with Abstraction Limits: GSP Projects

Actions and projects will need to be identified in the GSP to demonstrate to the state that the GSA has the capacity to address undesirable results when they occur. GSAs will initially focus on increasing groundwater availability through additional groundwater recharge before turning to the politically more painful, challenging task of reducing groundwater demand. Funding for GSA activities will partly be provided through (competitive) grants offered by the state, but a significant portion will come from taxes and fees locally generated by the GSAs, which have been given authority through SGMA to raise such fees and taxes.

[29] California Code of Regulations Title 23 (Div. 2) §351(ah)

Many areas have a long history of groundwater enhancement projects, through managed aquifer recharge of local stormwater runoff, wastewater recycling, and others as outlined above. However, some agricultural regions are losing historic groundwater recharge from surplus irrigation, as farmers adopt more efficient irrigation practices to address water shortages and to control pollutants (especially nitrate) leaching to groundwater.

Recently, high flood flows, typically uncaptured, have been identified as an additional source of water (Kocis & Dahlke, 2017). Managing these high flow requires additional storage, not currently available. A new form of groundwater enhancement is emerging in form of agricultural managed aquifer recharge during the winter ("Ag-MAR", Harter & Dahlke, 2015; Niswonger, Morway, Triana, & Huntington, 2017). Ag-MAR would take advantage of the existing agricultural landscape, at a large scale and using existing infrastructure, during a period when crops are dormant and when the risk for leaching agricultural chemicals can be minimized. Ag-MAR would provide potentially multiple benefits. Besides enhancing groundwater storage, winter flooding and recharge in agricultural landscapes offers opportunities to improve groundwater quality, but also to enhance ecosystem services. Current pilot projects are implemented in permanent crops – almond orchards, grape vineyards, and alfalfa (Dahlke, Brown, Orloff, Putnam, & O'Geen, 2018a). To take full advantage of available stormwater flows, additional action needs have been identified, including re-operation of surface water reservoirs for conjunctive storage in groundwater and surface water, additional infrastructure investment for conveyance of stormwater runoff, agronomic research to investigate feasibility of off-season recharge in a variety of crops, and clarification of water rights (CDWR, 2018; Fogg & Bernacchi, 2018).

Controlling groundwater demands will be necessary where enhancement of groundwater recharge is insufficient to meet sustainability goals. In urban areas, water conservation has played an important role in adjusting to limited water resources. During the 2012–2016 drought, urban areas achieved a statewide conservation goal of 25% water use reduction (Palazzo et al., 2017).

Agricultural regions will bear the most significant economic impact where groundwater pumping restrictions need to be put in place. There, less groundwater pumping translates into immediate economic losses, as other sources of water are unavailable. In the San Joaquin Valley (the southern and central part of the Central Valley), annual groundwater overdraft is estimated to be on the order of 2 km^3 (Schneider, 1977). Additional surface water supplies that may be developed for increasing groundwater recharge (including in lieu recharge) are limited by the ability of farms to pay. Current estimates suggest that additional surface and groundwater projects in this region may address only about one-third of the overdraft (Hanak et al., 2019).

The remaining overdraft will need to be achieved by fallowing at least 200,000 ha or more of currently irrigated agricultural production, a reduction of 5–10% of the current irrigated agricultural footprint. Ways to lessen the impact of this landuse change are currently under discussion but have yet to yield substantial changes in governance or local planning decisions. Key elements being proposed include (Hanak et al., 2019):

- creation of water markets that provide growers flexibility to sell or buy groundwater across their region and would allow for water transfers from areas with larger water endowments to areas with limited or no ability to enhance groundwater recharge.
- integration of groundwater planning with landuse planning to develop economically viable alternative land uses that do not depend on groundwater or much less so (e.g. development of natural dry-land conservation areas, especially where markets or government payments are available for conversion to conservation habitat; development of solar photovoltaic parks or other low water-impact industries).
- integration of planning activities between local surface water, groundwater, water quality, and landuse planning agencies.

For the San Joaquin Valley, current estimates show economic impacts to agriculture can be limited to about 5% of current production, with water markets and other measures in place (Hanak et al., 2019). Agricultural production within a reduced spatial footprint will likely increase the market share of high value crops, with low value crops disappearing. GSAs will need to each make their own choice in electing from this portfolio of options to achieve local groundwater pumping reductions.

26.8 Designing an Effective Enforcement System

Enforcement occurs at three levels: at the GSA-level, through DWR, and through SWB. GSAs are given statutory authority to ensure compliance by individual groundwater users and may impose civil penalties on individual parties that do not comply with a GSP. DWR was given the authority to develop GSP requirements and criteria and will review individual GSPs on a five-yearly schedule. GSPs not found in compliance will be designated "probationary", which puts the management of the area into the hands of the State Water Board (SWB).

The SWB is not only the designated enforcement agency for SGMA, but already administers surface water rights and oversees groundwater quality regulations through its nine regional water boards. Hence, the agency – unlike DWR – will be able to draw from decades of experience in law enforcement on water matters. Perhaps most importantly, the success of SGMA hinges largely on the motivation of local stakeholders to implement painful and costly SGMA measures because they feel sufficiently threatened by the prospect of SWB taking over groundwater management, if local GSAs fail to form, fail to develop an appropriate GSP, or fail to implement a GSP properly.

The role of the SWB is defined in Chapter 11 "State Intervention" of SGMA.[30] Not inadvertently, "chapter 11" makes open and notorious reference to the colloquial term "filing for chapter 11", that is, filing bankruptcy. The expression refers to

[30] California Water Code §10735 and §10736

Chapter 11, Title 11 of the United States Code, commonly known as the "Bankruptcy Code", which allows for reorganization under U.S. bankruptcy laws. SGMA's chapter 11 allows a local agency to remedy deficiencies within 6 months. Otherwise the SWB will be responsible for developing an interim plan that emphasises reduction of groundwater extraction over other groundwater management tools. Local control may be re-established under qualifying conditions at a later time.

Chapter 11 and the concept of state control of groundwater resources was designed to create strong motivation to local groundwater stakeholders and agencies to comply with SGMA regulations rather than leaving matters to a central state agency. In the short term, the state agency would likely be overwhelmed if a large number of basins had refused to form GSAs or failed to provide adequate GSPs. But for the intermediate and long-term, the economic, social, and local political cost of state control needed to be setup in ways that sends a clear signal to local agencies that local control and compliance would be preferable over control under chapter 11. To that end, SWB reacted to SGMA by immediately creating a SGMA enforcement unit with substantial funding to organize and prepare for enforcement actions. An early component of those developments was publication of a fee schedule that would be imposed on individual groundwater pumpers in areas managed under chapter 11. These fees would only cover the cost of state management. These costs are in addition to the pumpers' financial responsibility for planning costs and implementation of projects and actions, costs that incur even under local management. But under local management, these costs can be partially recovered through state grants or funded through local markets (Hanak & Stryjewski, 2012).

26.9 SGMA and GSPs at the Intersection with Other Laws and Rights

The framework outlined in SGMA extends well beyond the groundwater management efforts that California has historically engaged in, e.g., under adjudications or through special acts districts, and beyond efforts that will be the focus of critically overdrafted basins – addressing groundwater overdraft, land subsidence, and possibly seawater intrusion. SGMA requires groundwater management agencies to also consider the water quality implications of their activities (through the water quality sustainability indicator), and the connectivity between groundwater, surface water, and groundwater-dependent ecosystems (through the groundwater-surface water sustainability indicator). These mandates close the loop on many unintended consequences of groundwater pumping. But they represent largely unexplored frontiers in California and the Western U.S. water management landscape.

GSAs responsibility for groundwater quality and for pumping impacts on surface water and groundwater-dependent ecosystems overlap and interact substantially with other regulatory efforts and legal doctrines under federal and state law (Cantor, Owen, Harter, Green Nylen, & Kiparsky, 2018). This adds significant uncertainty to the development of GSPs as science and new technical approaches will need to be developed to appropriately address these two sustainability indicators, while the relationship to other regulatory programs and potential legal liabilities remains without much state or legal guidance (Cantor et al., 2018).

The mandate for management of the groundwater-surface water interaction raises a number of questions that GSAs will need to consider, with little guidance from the state or the courts beyond basic requirements for quantifying the amount of historic and current depletion of surface water due to groundwater pumping. These questions include (Cantor et al., 2018):

- the interaction of surface water law with groundwater law
- the definition of "significant and unreasonable" adverse impacts to beneficial uses and users of surface water
- the allocation of responsibilities to various entities and parties in addressing groundwater – surface water interactions
- finding processes to effectively resolve conflicts among parties
- the deployment of a variety of monitoring and assessment tools to quantify the groundwater – surface water interactions at various spatial and temporal scales

Table 26.1 provides an overview of key legal doctrines, regulations, and laws that will need to be considered by GSAs – their boards, advisory committees, and stakeholders. Among those, several have significant potential to interfere or conflict with or supersede basic SGMA requirements:

- SGMA explicitly protects existing surface water and groundwater rights. However, the requirement to bring groundwater use into harmony with surface water use may directly conflict with existing surface water rights or with existing groundwater rights. A GSA will need to set an ambitious agenda to address these conflicts (Owen, Cantor, Green-Nylen, Harter, & Kiparsky, 2019).
- instream flow requirements, in addition to surface water rights, have been adjudicated by SWB on only a few streams. SWB is in the process of expanding instream flow rights, which may require curtailments by both surface water and groundwater users.[31] The distribution of these curtailments between existing surface water and groundwater rights holders is highly uncertain, providing both, opportunities and risks for GSAs to play an active, perhaps central role in facilitating a resolution to such conflicts.
- SGMA explicitly requires GSPs to comply with existing laws. The federal (and state version of the) ESA and CWA have proven to bear significant importance on the management of surface water to protect ecosystems and water quality

[31] https://www.waterboards.ca.gov/waterrights/water_issues/programs/applications/instream_flow_dedication/, accessed 18 December 2018; https://www.waterboards.ca.gov/waterrights/water_issues/programs/bay_delta/bay_delta_plan/water_quality_control_planning/, accessed 18 December 2018; https://www.waterboards.ca.gov/waterrights/water_issues/programs/bay_delta/docs/sed/sac_delta_framework_070618%20.pdf, accessed 18 December 2018

Table 26.1 Summary of important laws, regulations, and legal doctrines that may affect and overlap with the development of the groundwater-surface water interaction sustainability indicator in a groundwater sustainability plan in California

Area of law or regulation	Key intersections between SGMA and other laws in the context of groundwater-surface water interactions
Reasonable Use Doctrine	Groundwater use, like all water use in California, is subject to the reasonable use doctrine. But the practical implications of the doctrine are not entirely clear. Reasonable use is, by nature, a flexible and highly context-dependent concept that is based in part on value judgments.
Water rights	SGMA explicitly does not alter surface water or groundwater rights. However, the implications of bringing a groundwater basin's water budget into sustainable balance may bear directly on both. SGMA does not provide a formula for resolving conflicts between surface water and groundwater rights, but it does provide opportunity and a potential forum for doing so—if GSAs are ambitious.
Regulatory takings	Water rights in California are property rights, and surface or groundwater users may bring takings claims if they believe regulatory restrictions on use have effectively taken their property. However, inherent in those rights is susceptibility to reasonable regulation. GSAs can reduce the risk of takings liability by managing groundwater in a manner generally consistent with California water rights.
Public Trust Doctrine	If groundwater pumping within a GSA's jurisdiction draws water from aquifers that are tributary to surface waterways, the public trust doctrine is likely to be relevant.
Federal and State Endangered Species Acts CESAs)	Endangered species laws apply to groundwater allocation decisions that may impact listed species. GSAs seeking to avoid consequences under the ESA should be aware of these species within the basin and explicitly address their needs when developing GSPs.
California Environmental Quality Act (CEQA)	The preparation and adoption of GSPs is specifically exempt from CEQA. However, implementation actions taken by a GSA under a GSP would remain subject to CEQA. Compliance with CEQA would include analyzing and mitigating potential negative impacts on interconnected surface waters.
Clean Water Act and Porter-Cologne Act	Although water quality is also addressed separately within SGMA, it is relevant to groundwater-surface water interactions, including through effects on streamflow volume and temperature.
Instream flow requirements	To avoid significant and unreasonable adverse impacts on surface water, minimize risk of litigation^ and maximize their GSPs' defensibility. GSAs will need to be aware of instream flow requirements set by the State Water Resources Control Board and consider them when developing and implementing GSPs.

Source: Cantor et al. (2018)

consistent with a stream's beneficial uses. SGMA effectively extends those responsibilities to GSAs to the degree that surface water depletion by groundwater pumping may affect water quality (including temperature) and species protected under ESA.

- the public trust doctrine has recently been affirmed by California courts to apply not only to surface water diversions, but also to groundwater pumping that reduces instream flows in navigable rivers. The state (and, by extension, GSAs) must consider their public trust duty, which is not pre-empted by SGMA. Most significantly, perhaps, the public trust doctrine, ESA, and CWA do not waive the need for addressing pre-existing undesirable conditions – conditions that existed prior to 2014. While SGMA itself only requires that new post-2014 undesirable results be avoided, SGMA explicitly does not exempt or supercede any existing law.[32] The latter may mean that GSPs need to consider groundwater pumping impacts on surface water that have existed already for decades.

The role of GSAs in managing the water quality sustainability indicator similarly remains uncertain. Regional Water Boards (RWBs) already have extensive regulatory powers to protect both surface water and groundwater quality. At a minimum, GSAs will be required to take an active role in understanding existing water quality and the potential impacts of groundwater management projects and actions on future water quality. This will necessitate significant data collection and data management, assessment, and monitoring, possibly in collaboration with RWBs. Some GSA projects may require permits from the respective RWB, particularly for some recharge projects.

It remains to be seen how much technical assistance and legal guidance GSAs will receive from the state to support these efforts. But the water quality and groundwater-surface water interaction sustainability indicators, for the first time, will greatly expand the scope of groundwater management and require active engagement of the GSAs with a wide range of local, regional, state, and possibly federal agencies. Thus, GSP development, at its best, offers an opportunity to be the catalyst for comprehensive integrated regional water management planning and implementation across groundwater and surface water, across water supply and water quality, and across water and land use management.

References

Blomquist, W. (1992). *Dividing the waters: Governing ground water in Southern California*. San Francisco, 415 p.

Brush, C. F., Dogrul, C., & Kadir, T. N. (2013). *Development and calibration of the California Central Valley groundwater-surface water simulation model (C2VSIM), Version 3.02-CG*. Sacramento, CA: DWR Technical Memorandum, California Department of Water Resources. https://water.ca.gov/-/media/DWR-Website/Web-Pages/Library/Modeling-And-Analysis/C2VSim/Files/Coarse-Grid-C2VSim-Model-Version-R374_rev-C2VSim-CG_R374_rev-Released-April-27-2018/C2VSim_Model_Report_2016-03_vR374.pdf?la=en&hash=532C0F1BE59C6FE587D81B9250FC18D845FC38C6. Accessed 18 Dec 2018

California Department of Water Resources (CDWR). (2014). California Water Plan Update 2013, Volume 1, Chapter 3 (California Water Today), 115 p. https://water.ca.gov/-/media/DWR-Website/Web-Pages/Programs/California-Water-Plan/Water-Plan-Updates/Files/Update-2013/Water-Plan-Update-2013-Volume-1.pdf. Accessed 18 Dec 2018.

---

[32] California Water Code §10726.8(a)

California Department of Water Resources (CDWR). (2017). *Draft sustainable management criteria BMP*, 38 p. https://water.ca.gov/-/media/DWR-Website/Web-Pages/Programs/Groundwater-Management/Sustainable-Groundwater-Management/Best-Management-Practices-and-Guidance-Documents/Files/BMP-6-Sustainable-Management-Criteria-DRAFT.pdf. Accessed 18 Dec 2018.

California Department of Water Resources (CDWR). (2018). *Flood-MAR: Using flood water for managed aquifer recharge to support sustainable water resources*. Sacramento, 56p. https://water.ca.gov/-/media/DWR-Website/Web-Pages/Programs/All-Programs/Flood-MAR/DWR_FloodMAR-White-Paper_06_2018_updated.pdf?la=en&hash=350DBD68452230C5CF1706C3E8EB1E3E3E613C25. Accessed 18 Dec 2018.

Cannon Leahy, T. (2016). Desperate Times Call for Sensible Measures: The Making of the California Sustainable Groundwater Management Act, 9 Golden Gate U. Envtl. L.J. 5. http://digitalcommons.law.ggu.edu/gguelj/vol9/iss1/4. Accessed 18 Dec 2018.

Cantor, A., Owen, D., Harter, T., Green Nylen, N., & Kiparsky, M. (2018). *Navigating groundwater-surface water interactions under the Sustainable Groundwater Management Act*. Berkeley, CA: Center for Law, Energy & the Environment, UC Berkeley School of Law, 50 p. https://doi.org/10.15779/J23P87. Accessed 18 Dec 2018

Consolidated Irrigation District. (2009). Groundwater Management Plan, 137 p. http://kings-groundwater.info/_documents/GWMPs/CID_SB1938_GWMP.pdf. Accessed 18 Dec 2018.

Dahlke, H. E., Brown, A., Orloff, S., Putnam, D., & O'Geen, T. (2018a). Managed winter flooding of alfalfa recharges groundwater with minimal crop damage. *California Agriculture, 72*(1), 65–75.

Dahlke, H. E., LaHue, G. T., Mautner, M. R. L., Murphy, N. P., Patterson, N. K., Waterhouse, H., Yang, F., & Foglia, L. (2018b). Managed aquifer recharge as a tool to enhance sustainable groundwater management in California: examples from field and modeling studies. In J. Friesen & Rodriguez-Sinobas (Eds.), *Advanced tools for integrated water resources management volume 3: Advances in chemical pollution, environmental management and protection* (pp. 215–276).

Fogg, G. E., & Bernacchi, L. (Eds.) (2018). *Recharge roundtable call to action: Key steps for replenshing California groundwater*, 27p. http://ucwater.org/recharge. Accessed 18 Dec 2018.

Hanak, E., Escriva-Bou, A., Gray, B., Green, S., Harter, T., Jezdimirovic, J., Lund, J., Medellin-Azuara, J, Moyle, P., & Seavy, N. (2019). *Water and the future of the San Joaquin Valley*. San Francisco, CA: Public Policy Institute of California, 100 p. https://www.ppic.org/publication/water-and-the-future-of-the-san-joaquin-valley/. Accessed 22 Nov 2019

Hanak, E., Lund, J., Dinar, A., Gray, B., Howitt, R., Mount, J., et al. (2011). *Managing California's water: From conflict to reconciliation*. San Francisco, CA: Public Policy Institute of California. https://www.ppic.org/content/pubs/report/R_211EHR.pdf. Accessed 18 Dec 2018

Hanak, E., & Stryjewski, E. (2012). *California's water market, By the numbers: Update 2012*, 48 p. http://www.ppic.org/content/pubs/report/R_1112EHR.pdf. Accessed 18 Dec 2018.

Harter, T., & Dahlke, H. E. (2015). Out of sight but not out of mind: California refocuses on groundwater. *California Agriculture, 68*(3), 54–55.

Kocis, T. N., & Dahlke, H. E. (2017). Availability of high-magnitude streamflow for groundwater banking in the Central Valley, California. *Environmental Research Letters, 12*(8), 084009.

Langridge, R., Brown, A., Rudestam, K., & Conrad, E. (2016a). *An evaluation of California's adjudicated groundwater basins*, Report to the State Water Resources Control Board, 272 p. https://www.waterboards.ca.gov/water_issues/programs/gmp/docs/resources/swrcb_012816.pdf. Accessed 18 Dec 2018.

Langridge, R., Sepaniak, S., & Conrad, E. (2016b). *An evaluation of California's special act districts*, Report to the State Water Resources Control Board, 183 p. https://escholarship.org/uc/item/3cr8k66v. Accessed 18 Dec 2018.

Legislative Analyst's Office (LAO). (2002). https://lao.ca.gov/2002/water_districts/special_water_districts.html. Accessed 18 Dec 2018.

Lipson, A. (1979). *Efficient water use in California: The evolution of groundwater mangaement in Southern California*, Rand Report R-2387/2-CSA/RF, ISBN 0-8330-0135-3. https://www.rand.org/content/dam/rand/pubs/reports/2009/R2387.2.pdf. Accessed 18 Dec 2018.

Matthews, O. P. (2003). Simplifying western water rights to facilitate water marketing. *UCOWR Water Resources Update, 126*, 40–44.

Nelson, R. (2011). *Uncommon innovation: Developments in groundwater management planning in California*. Stanford University. 43 p. http://waterinthewest.stanford.edu/resources/publications-directory/uncommon-innovation-developments-groundwater-management-planning. Accessed 18 Dec 2018.

Niswonger, R. G., Morway, E. D., Triana, E., & Huntington, J. L. (2017). Managed aquifer recharge through off-season irrigation in agricultural regions. *Water Resources Research, 53*, 6970–6992. https://doi.org/10.1002/2017WR020458

Ostrom, E. (1990). *Governing the commons: The evolution of institutions for collective action*. New York, NY: Cambridge University Press.

Owen, D., Cantor, A., Green-Nylen, N., Harter, T., & Kiparsky, M. (2019). California groundwater management, science-policy interfaces, and the legacies of artificial legal distrinctions. *Environmental Research Letters, 14*(4). https://iopscience.iop. https://doi.org/10.1088/1748-9326/ab0751

Palazzo, J., Liu, O. R., Stillinger, T., Song, R., Wang, Y., Hiroyasu, E. H. T., et al. (2017). Urban responses to restrictiveconservation policy during drought. *Water Resources Research, 53*, 4459–4475. https://doi.org/10.1002/2016WR020136

Reisner, M. (1993). *Cadillac desert*. Penguin USA. 582 p.

Sax, J. L. (2002). We Don't do groundwater: A morsel of California legal history. *University of Denver Water Law Review, 6*, 269. https://scholarship.law.berkeley.edu/cgi/viewcontent.cgi?referer=https://www.google.com/&httpsredir=1&article=2394&context=facpubs. Accessed 18 Dec 2018

Schneider, A. (1977). Groundwater Rights in California: Background and Issues, Governor's Commission to Review California Water Rights Law Staff Paper No. 2. https://ia600501.us.archive.org/32/items/GovernorsCommissionMemos/Govcomm-GwRights_text.pdf. Accessed 18 Dec 2018.

Wrzesien, M. L., Durand, M. T., Pavelsky, T. M., Howat, I. M., Margulis, S. A., & Huning, L. S. (2017). Comparison of methods to estimate snow water equivalent at the mountain range scale: A case study of the California Sierra Nevada. *Journal of Hydrometeorology, 18*(4), 1101–1119.

Thomas Harter is the Robert M. Hagan Endowed Chair in Water Management and Policy within the hydrology program in the Department of Land, Air, and Water Resources. Thomas Harter has a B.S. in hydrology from the University of Freiburg, Germany and a M.S. in hydrology from the University of Stuttgart, Germany. He received his Ph.D. in hydrology (with emphasis on subsurface hydrology) at the University of Arizona. In 1995, he joined the faculty at the Department of Land, Air, and Water Resources, University of California, Davis. His research focuses on nonpoint-source pollution of groundwater, sustainable groundwater management, groundwater modeling, groundwater resources evaluation under uncertainty, groundwater-surface water interaction, and on contaminant transport. Dr. Harter's research group has done extensive modeling, laboratory, and field work to evaluate the impacts of agriculture and human activity on groundwater flow and contaminant transport in complex aquifer and soil systems. In 2007, Dr. Harter was appointed Robert M. Hagan Endowed Chair in Water Management and Policy. Also in 2008, Dr. Harter's research and extension program received the Kevin J. Neese Award in recognition of its efforts to engage scientists, regulators, farm advisors, dairy industry representatives, and dairy farmers to better understand the effects of dairy operations on water quality.

Chapter 27
Changing from Unrestricted Access to Sustainable Abstraction Management Regimes: Lessons Learnt from France and Australia

Jean-Daniel Rinaudo, Steve Barnett, and Cameron Holley

Abstract This concluding chapter compares the important features of the groundwater policy and management approaches that have been implemented in France and Australia and draws lessons that may be relevant to other countries who are implementing groundwater management regimes. To support the comparison, the chapter looks at six main stages of the policy development process: (1) political awareness raising; (2) increasing the groundwater knowledge base; (3) defining and allocating water use rights; (4) defining sustainability objectives and setting extraction limits; (5) returning over-allocated and overused ground-water systems to sustainable levels of extraction; and (6) enforcement policies.

Keywords Allocation · Enforcement · Extraction limits · Groundwater knowledge base · Political awareness raising · Water use rights

J.-D. Rinaudo (✉)
BRGM, Montpellier University, Montpellier, France
e-mail: jd.rinaudo@brgm.fr

S. Barnett
Adelaide, SA, Australia
e-mail: Steve.Barnett@sa.gov.au

C. Holley
Faculty of Law, University of New South Wales Sydney, Sydney, NSW, Australia
e-mail: c.holley@unsw.edu.au

J.-D. Rinaudo et al. (eds.), *Sustainable Groundwater Management*,
Global Issues in Water Policy 24, https://doi.org/10.1007/978-3-030-32766-8_27

27.1 Introduction

This concluding chapter compares the important features of the groundwater policy and management approaches that have been implemented in France and Australia and draws lessons that may be relevant to other countries who are implementing groundwater management regimes. It shows that, in spite of spite of huge climatic, environmental, socio-economic and legal differences (Fig. 27.1), there are many similarities in the groundwater management approaches developed and implemented in France and Australia. Key differences are also highlighted. The comparative analysis is based on the case studies presented in the preceding chapters. To support the comparison, the chapter looks at the main stages of the policy development process which are listed below. Jakeman, Barreteau, Hunt, Rinaudo, and Ross (2016) also provides insights into this process.

- Stage 1 is the process through which groundwater management is brought on the political agenda and becomes a public policy issue.
- Stage 2 involves increasing the knowledge base and understanding of the groundwater systems, which is required to underpin the foundations of a groundwater management regime. In most of the case studies covered in this book, management has been initiated with very limited information, which later improved over several decades.
- Stage 3 consists of defining and allocating water use rights, a policy issue where France and Australia have taken very different pathways.
- Stage 4 requires the definition of sustainability objectives. This chapter compares the approaches taken in France and Australia. While the theoretical approaches apparently differ, in practice the definition of sustainability objec-

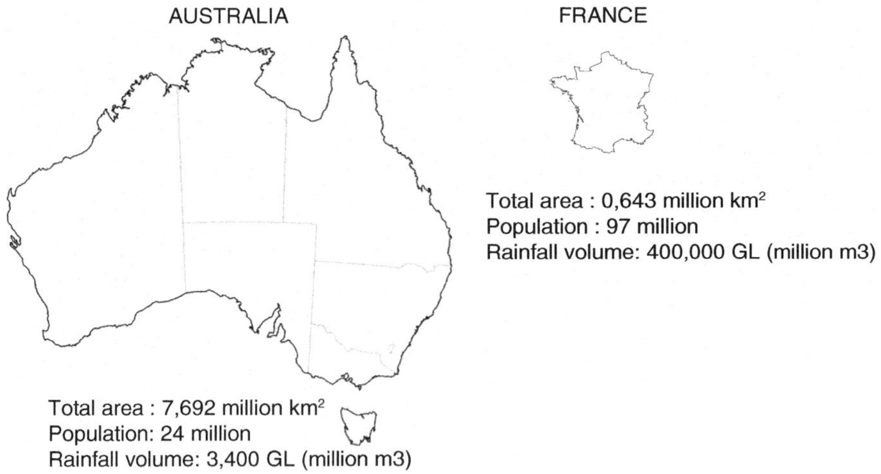

Fig. 27.1 Area, population and yearly precipitation in France and Australia

tives results in some cases more from a negotiation with water users than a scientific objective evaluation.
- Stage 5 corresponds to the process where the State aligns water use entitlements to the extraction limits which correspond to the sustainable objectives. The comparison of the case studies covered in the book shows the diversity of strategies chosen by water managers. It also highlights the political challenges associated with this crucial stage.
- The chapter finally compares enforcement policies in Stage 6 as well as their capacity to adapt to a rapidly changing social, economic and climatic context.

The chapter concludes by summing up key insights that may be relevant for other countries and regions on the journey to develop sustainable groundwater management policies.

27.2 Bringing Groundwater Management on the Political Agenda (Stage 1)

Overall, bringing groundwater management on the political agenda, engaging stakeholders and developing dedicated policies has been a very gradual process which has developed over the last 80 years and is still evolving. One of the main reasons for this is that, in France as in Australia, political attention has long been focused on the more visible surface water resources through the construction of reservoirs and canals, and then the management of those resources through the development of sophisticated rules to allocate surface water amongst competing demands, supported by a legal framework. Meanwhile, groundwater remained unrecognized and unmanaged and was long considered a resource that could be freely accessed without controls imposed by the State.

Deep confined artesian aquifers were the first groundwater resources to be intensively exploited. From the beginning of the nineteenth century, the progress of drilling technologies led to the development of thousands of wells tapping these artesian aquifers for industrial uses (e.g. the Parisian Basin in France) or for agricultural purposes (e.g. the Great Artesian Basin in Australia). Although these supplies must have seemed endless, the States progressively realized after several decades that some important groundwater resources were being used beyond sustainable limits. Declining water pressure levels resulted in impacts such as loss of artesian conditions in France and the drying of springs in Australia. This raised awareness of the need to regulate groundwater extraction.

When the State finally decided to intervene through the development of groundwater laws or regulation, problems due to over-exploitation were already being experienced in some areas. For instance, when the French government passed the 1935 groundwater decree, groundwater pressure levels had already dropped significantly in deep confined aquifers of the Parisian Basin, and most wells were no longer artesian. In Australia, Interstate Conferences on Artesian Water were held between 1912 and 1928, long after pressure levels started to decline due to more than 1500 deep wells having been drilled in less than 30 years.

It is worth noting that those State interventions in both France and Australia were motivated by the desire to protect the "common good", and were not a response to pressures exerted by third parties impacted by groundwater overdraft. In France in particular, the 1935 State decree was a response to concerns that private industries were overexploiting aquifers which would put high quality drinking water supplies for cities at risk in the future.

Between 1960 and the end of the 1980s, improvements in drilling and pumping technology led to an exponential increase of water extraction from both deep confined aquifers and shallow unconfined aquifers. Exploitation of those resources was further accelerated by increased restriction imposed on surface water resources, which had the unintentional result of shifting the demand for extraction to nearby groundwater resources. A second wave of environmental problems ensued, including aquifer depletion, declining baseflows discharging into rivers, drying up of wetlands, seawater intrusion from the sea or from adjacent saline groundwater, etc. A significant number of stakeholders were affected by these impacts leading to a number of conflicts that compelled governments to take action and initiate the development of new regulations. In areas of over-extraction in both countries, legislation and policies were introduced to protect the resource for sustainable use into the future.

By the early 1990s, there was widespread acceptance of the need for "ecologically sustainable development" resulting in a second wave of legislation and agreements (1992 Water law in France, 1994 CoAG Agreement in Australia) that recognized the environment needed an allocation of water and that the impacts of groundwater extraction on dependent ecosystems should be managed.

The analysis of the French and Australian policy developments shows several similarities that potentially have implications to other countries. First, based on the French and Australian experiences, it seems very unlikely that policy makers will anticipate groundwater management problems and pre-emptively establish a constraining or precautionary legal and regulatory framework. Like the management of many environmental issues, groundwater management may only appear on the political agenda once problems and conflicts emerge, because politicians and decision makers are generally reactive with awareness, funding and reform only occurring during droughts and water resource decline, which then seem to evaporate when water is plentiful or the crisis has been averted. This process, called the "hydro-illogical cycle" by Wilhite (2012), is also described in the Californian and Chilean case studies presented in Chaps. 25 and 26. The implication for other countries is that their policy makers and managers should be aware of this cycle and attempt to initiate management intervention before crisis situations develop.

Because groundwater resources are hidden from view (Chap. 10), the effects of groundwater depletion are often only recognised when there are visible impacts on surface water resources and dependent ecosystems such as springs and wetlands. In response to these impacts, a management regime should include an appropriate environmental protection legislative framework which requires a proper scientific understanding of the relationships between surface and groundwater.

Perhaps a reason why it took so long to bring groundwater on the political agenda in France and Australia was the lack of understanding of important groundwater resources and their relationships with surface water resources and ecosystem. Accessing this knowledge is much easier today than it was 50–80 years ago. Modern societies now have access to incomparable knowledge (theoretical frameworks, data, measurement technologies, modeling capacities) as well as to a wide range of communication tools that can improve awareness raising, facilitate stakeholders' engagement and training. Bringing groundwater on the political agenda should therefore be facilitated. This however requires that public institutions invest the required resources to develop and transfer that knowledge, since private actors do not have sufficient incentives to do it.

27.3 Increasing the Knowledge Base (Stage 2)

The comparison of the Australian and French case studies presented in the previous chapters of this book highlights that the development of the knowledge base needed to establish sustainable groundwater management consists of six main consecutive steps:

1. the definition of aquifer and management unit boundaries;
2. the establishment of groundwater monitoring network;
3. the development of an information system to locate and quantify groundwater extraction;
4. the development of a conceptual model describing the aquifer flow system;
5. the development of a numerical model that allows simulating alternative management approaches;
6. the identification of groundwater dependent ecosystems and assessing their water requirements

The first step consists of defining aquifers boundaries (extent and geometry) and management units, both of which are essential for the development of groundwater laws and policies (Cuadrado-Quesada, Holley, & Gupta, 2018; Nelson & Quevauviller, 2016). When there is no central regulator or agency to carry out this work, a methodology has to be developed and its use made mandatory to ensure that all the aquifers are delineated in a consistent manner. This is particularly important in sedimentary basins which can contain numerous superimposed aquifer layers which need to be grouped for efficient management, based on connectivity and common aquifer characteristics. This is a key issue in large basins that extend across several regions or states that have their own water management agency and policies (e.g. Murray Darling Basin). The same issue applies to Europe in general (and France in particular), where groundwater aquifers were delineated by several River Basin District Authorities. In addition to aquifer definition, it is essential to define the boundaries of management units (a unit typically covers the whole aquifer or basin extent). Finally, smaller management zones (including areas of similar hydrogeological characteristics or similar sustainability issues) can be defined within a management unit to allow for targeted management policies that may not need to apply to the whole management unit.

The second step consist in establishing groundwater monitoring networks that allow assessing long term trends in groundwater levels and quality (Chap. 9). Long-term groundwater monitoring not only indicates whether water levels or salinities are rising or falling, but also helps determine the major drivers causing the changes (e.g. rainfall or extraction). Because of the expense for drilling new observation wells, existing privately owned wells should be used where possible. However, the French and the Australian case studies show that coordinating the data collection from existing but often heterogeneous monitoring networks is challenging, in particular where they have been developed independently over time by several different institutions for different purposes (municipal water supply, government agencies, local or regional governments, etc.).

The third step consists of locating all groundwater abstraction points and developing a database that accounts for all significant extractions. This information is not only a prerequisite for the calibration of groundwater models, but is essential for the implementation of a compliance and enforcement regime (see Stage 6 below). A State agency usually performs this task, based on existing regulations defining the need for groundwater permits or licenses. The French and Australian experiences suggest implementing a progressive approach is desirable, focusing first on the identification of abstractions points, then developing estimates of the volume of groundwater abstraction. While French law requires that all uses be metered in volume (which sometimes results in problems of meter tampering), Australia historically implemented a more pragmatic approach, accepting to use indirect measurement approaches (use estimate based on crop type and crop area). It is worth noting that the use of more sophisticated technologies such as smart volumetric meters or connected pumping flow meters may not always be efficient if the relevant agencies or regulators does not invest sufficient effort in the analysis of the huge amount of data produced by those technologies (Holley & Sinclair, 2016; Rinaudo and Donoso, 2018).

Based on the previous information, hydrogeologist can then develop a simple conceptual model of how the aquifer system works (step 4). This model describes groundwater flow directions (based on water level elevation maps), that can identify recharge and discharge areas as well as interactions with surface water. Estimates of hydrogeological parameters (using pumping tests where possible) may allow the calculation of order of magnitude estimates of the groundwater storage volumes and recharge rates to unconfined aquifers. Any such conceptual model will almost certainly be characterized by significant uncertainty in some components (Chap. 11). The French and Australian groundwater experience suggest that areas of uncertainty should be communicated to stakeholders, even if it undermines the argument for a reduction in extraction. Failing to provide this transparency might later result in a break down in trust between water users on the one hand (Daniell, 2011), and scientific experts and government officials on the other, which could result in significant compliance and enforcement problems in the future (Holley & Sinclair, 2012).

The fifth step consist in developing numerical groundwater models that can predict the impacts of changes in extraction, climate or land use. Numerical models are also useful to verify conceptual models as well as to highlight data gaps that require further investigation. The development of a numerical model should not be considered as an end in itself but rather as a key component of the knowledge development process. This is well illustrated in the case of Bordeaux (Chap. 12) where a numerical model has been progressively developed and continuously improved over more than five decades. The French experience also shows that the choice of a modeling technique (lumped vs fully distributed model) needs to be adapted to the resource characteristics, existing knowledge, management issue and budget constraints (Chap. 11).

The sixth and last step consists of identifying groundwater dependent ecosystems and assessing their water requirements. In France and Australia, these are frequently expressed as critical groundwater levels or the maintenance of base-flow discharge to streams. In France, the definition of these trigger levels is mainly based on scientific knowledge of aquatic ecosystems while Australian managers also consider stakeholders' needs.

The process of increasing the knowledge and understanding of groundwater systems is ongoing and is one of 'continuous improvement' over decades. Despite this, a key message from several case studies presented in this book is that policy makers should not wait to have perfect knowledge in order to initiate groundwater management. Early intervention with minimal information can be beneficial, provided there is a commitment to refine policies over time as new knowledge is purposefully and progressively acquired (Cosens, 2018).

Another key message from Australia and France is that significant State investment is needed to develop the knowledge base. This is mainly because of the absence of direct benefits that can be derived by private users from understanding of groundwater systems. They are therefore highly unlikely to directly invest in the acquisition of such knowledge. However, the cost of groundwater knowledge production, planning and management can be recovered from users through water abstraction fees or taxes. The cost recovery principle is generally implemented in France, with the River Basin District Agencies playing a key role in levying water abstraction charges and subsidizing actors developing water resource knowledge and planning. Cost recovery explicitly for groundwater management is only carried out to a limited extent in Australia.

Last but not least, the Australian experience suggests that other non-State organisations may generate information useful for increasing the knowledge base. These include mining and petroleum companies, universities and private companies that use groundwater. The main challenge in incorporating this knowledge lies in the design of interoperable information systems, an issue which is discussed in Chap. 9.

27.4 Defining and Allocating Groundwater Use Rights (Stage 3)

In France and Australia, as in many other countries, groundwater management fundamentally relies on water use rights (WUR), defined as legal rights to abstract a specified quantity of water from the ground. While the history of establishing those rights in France and Australia is quite similar, the evolution of their characteristics over time show interesting differences concerning:

(i) their property status
(ii) the conditions under which WUR are required;
(iii) the specification of WUR;
(iv) the procedure used to allocate WUR;
(v) the potential to transfer those rights.

These differences are discussed in the following paragraphs.

27.4.1 Property Status

In the beginning of the nineteenth century, Australian States' and French legislators adopted the "rule of capture" which gives landowners the right to take all the water they can capture from under their property. This can be explained by the fact that at that time, groundwater was then considered as a permanent resource, whose origins and movement were poorly understood by scientists, courts and governments (Margat et al., 2013). At that time, the legislators would also not consider groundwater to be an economic resource since it was not widely used because drilling and pumping technologies were still in their infancy. Groundwater was thus largely treated as a private property, maintained outside the realm of state intervention.

While this unregulated groundwater exploitation regime was not problematic during an era of limited drilling and pumping capabilities, technical advances that took place in the 1930s and 1940s lead to an increased exploitation of groundwater resources, in particular of confined artesian aquifers. As these resources were being increasingly depleted and threatening local public water supplies, French and Australian States progressively implemented a system water use rights, based on a combination of drilling permits and groundwater use licenses. In France, this first took place in 1935, with the introduction of a system of drilling permits applying to deep confined aquifers. The 1992 water law extended State control to shallow aquifers, thereby confirming the key role of the State as guardian of all groundwater. Interestingly, the State increasingly regulated groundwater abstraction without incorporating groundwater in the public domain: it remains the property of landowners, although its exploitation became subject to permits granted by the State. In Australia, it was only in the 1970s that some States imposed a system of abstraction licenses in areas where groundwater resources were considered as at risk due to over-extraction for irrigation, while maintaining the common law rights to a certain degree in other areas where extractions were minimal. However a permit was required to drill a borehole or construct a well in all areas. Overall in Australia, the control of all waters is now vested with the State.

The progressive strengthening of State control over groundwater extraction has been observed in many other countries, whether water is considered a private or public good. However, the rule of capture still prevails in other contexts like Texas. In such locations, the State has traditionally not interfered in groundwater management and allocation (although this is beginning to change with the introduction of the Sustainable Groundwater Management Act of California, see Chap. 24).

27.4.2 Conditions Under Which a WUR Is Required

Since defining, allocating and managing WUR requires significant human and financial resources, some States may only use this instrument in areas where groundwater resources are at risk. Where groundwater is of poor quality (e.g. highly saline) or where the resource capacity far exceeds demand, groundwater use often remains unregulated. This risk-based approach is implemented in the Australian states of South Australia and Queensland for instance. By contrast, France and other Australian states such as Victoria, have adopted a universal licensing approach where all uses must be authorized, independent of the local water demand or quality.

In both universal and risk based approaches, a number of exemptions may apply, allowing certain categories of users to abstract groundwater without holding a WUR; e.g. wells used for domestic supplies or stock watering. The same type of exemptions were reported in other countries studied in this book (Chile, California). A justification of these exemptions is that these users abstract a limited amount of groundwater[1]. At the other extreme, some activities which do not extract groundwater but which modify the recharge can be subject to the obligation to hold a licence. This is for instance the case of forest plantations in South Australia which requires a license corresponding to the quantity of precipitation intercepted by the forest. In areas of shallow water tables, the uptake of groundwater by forests also requires a licence (Avey & Harvey, 2014).

27.4.3 Specification of Groundwater Use Rights

In France, Australia and other countries studied in this book, groundwater use rights have been defined in very diverse ways:

[1] This is however not the case for mining activities which are surprisingly exempted from groundwater regulation in several Australian States despite having large extractions. However some controls may occur under mining legislation (Productivity Commission, 2018).

(i) individual versus collective rights;
(ii) specification in pumping flow rate, or an annual or seasonal volume;
(iii) a nominal value (fixed over time) or as a share of the available resource which may fluctuate over time.

Individual groundwater use rights have historically been used both France and Australia. This approach was consistent with the private property status granted to groundwater in the initial policy stages. In France, this approach prevailed until the enactment of 2006 Water Law which radically reformed WUR, by establishing a collective WUR to cover all pre-existing individual rights (Chap. 3). This drastic measure only applies to agricultural users in restricted areas, where all users are compelled to establish an association, which has the responsibility of crafting rules for sharing the water they are entitled to take within the collective WUR. This approach is at odds with the Australian approach which considers that water use rights should be treated as individual property which can be freely traded in a water market subject to conditions that minimise impacts on the resource, other water users and the environment. It is still too early to judge the environmental, economic and social performance of this new approach of WUR in France. However, its mere existence should help policy makers in other countries to broaden the range of options they consider when designing their own groundwater management policies.

WURs can be specified in terms of a pumping flow rate, or a volume or area over which water can be used. The simplest approach is to define WURs in the form of an area that can be irrigated. While it facilitates enforcement (the State only has to monitor the irrigated area), it does not allow for a precise limitation of groundwater abstraction, as the water use per unit of area may greatly vary depending on crops cultivated and irrigation technologies used. Most Australian States initially issued area-based licences, but these have mostly now been converted to volumetric allocations (exceptions include the Northern Territory and Western Australia; Productivity Commission, 2018).

An alternative approach consists in specifying WUR in pumping flow rates (as occurred in France up until 2003, and in Chile up until now). Enforcement of such WUR only requires checking the capacity of the pump when the well/borehole is constructed and occasionally in random surveys after that. The disadvantage of this approach is that again, it does not allow for precisely controlling the volume of water extracted, as the duration during which the pump is used is not defined in the WUR. Chapter 25 highlights the problems associated with this approach in Chile.

A third approach consists in specifying water use rights as a volume that can be abstracted over a year or irrigation season. This approach theoretically allows for the better control of water abstraction, provided that volumetric meters are installed and regularly monitored (to avoid inaccuracies or tampering problems). Another advantage of the volumetric approach is that it provides incentives for users to increase water use efficiency – because each cubic meter that can be saved can later on be used in production, or can be sold or leased if the trading of WUR is allowed.

Historically, licenses or permits granted by the French and Australian States included different conditions controlling how the WUR could be used: the location where water can be abstracted; the characteristics of the well or borehole; and the quantity of water that can be abstracted annually from that groundwater source. More recently, there has been a tendency to unbundle the different components of the rights. In France, site use and well/borehole construction approvals are managed with a system of permits granted by the State, while allocation of water is managed with a system of annual authorizations granted for agricultural purposes by Water Users' Associations in restriction zones and by the State elsewhere. Allocations are usually granted for long periods (about 15 years) without annual adjustment, but use remains subject to seasonal restrictions if groundwater levels decline below threshold levels. In some French groundwater basins, stakeholders have agreed to unbundle water access entitlements from water allocation, with the entitlements specifying a share of the available resource, while the allocation defines the specific volume of water that can be abstracted from the resource in a given year or season depending on the resource availability. Chapters 5 and 13 respectively describe how entitlements and allocation have been unbundled in the Beauce and Tarn et Garonne aquifers. The main advantage of separating entitlements from allocation is that it allows flexibility to manage resources where sustainable extraction can vary with the climate.

In Australian States, the various legislative instruments also increasingly facilitate the unbundling of existing water licenses, but the process has not yet been widely implemented due to legal and administrative complexities and the longevity of the current management plans. In those areas where it has been introduced, licensed water users are provided with an opening allocation for each category at the start of each new water use year on 1 July. This may be anywhere between zero and 100% of their full entitlement, as illustrated in Table 27.1 below. In Australia, another motivation for unbundling WUR through the separation of water rights from a specific piece of land, was to facilitate water trading, in particular the transfer of seasonal volume of water (allocation) independently of the water access entitlement (Chap. 21).

Table 27.1 Groundwater allocations in % of entitlement for 2018–2019 in New South Wales

Groundwater resource	Allocation (%)
Eastern recharge (NSW Great Artesian Basin)	50
Peel alluvium	69.4
Murrumbidgee alluvial	98.4
All other groundwater sources	100

Source: https://www.industry.nsw.gov.au/water/allocations-availability/allocations/summary

27.4.4 Duration and Transferability

Duration and transferability of WURs are two issues on which the French and Australian policies are radically different. In Australian states' legislation, water use rights (including volumetric licences and water access entitlements) and water allocations seek to approximate a personal property right[2] that in areas that have been unbundled, can be transferred independently of land. In most States, entitlements are granted in perpetuity (with the exception of Victoria and Western Australia and specific types of licenses in several states[3]), although the states retain the right to make changes to these entitlements.

In France, volumetric water use licenses are not considered as personal property and they cannot be transferred. Such authorizations are generally renewed annually. The State is theoretically authorized to modify or even cancel authorizations without any compensation, provided the decision is taken for the general public interest. Such changes are generally limited and in practice, most licenses can remain unchanged for decades. And in case of severe reduction, the State often offers compensation through public subsidies granted by the Water Agency to develop alternative water resources that can substitute groundwater. However, compensation is only partial and the beneficiaries have to comply with a set of environmental rules (see Chap. 18).

The French and Australian views on property rights illustrate two opposing policy approaches. However between them lies a continuum where variations of each WUR approach could be applied to suit different circumstances. For instance, water use licenses (or concessions) are granted for respectively 40, 50 and 75 years in South Africa, Mexico and Spain, while they are granted in perpetuity in Chile. They are tradable in Spain and Chile, but not in the two other countries. This illustrates the challenge of striking an appropriate balance between the security needed to encourage investment and the need for flexibility to adapt to climate change, societal needs, environmental requirements and to take into account increases in understanding of groundwater systems.

27.4.5 Use Priorities

An important characteristic of WUR is the existence of priorities for their allocation. In both France and Australia, environmental water requirements are given the first priority, since sustainable extraction limits are calculated in such a way that they aim to prevent severe environmental impacts from groundwater extraction. The

[2]The characteristic of the right as property is debatable (as a matter of law). For further see, e.g. *ICM Agriculture Pty Ltd v Common-wealth* [2009] HCA 51.

[3]In several States where different types of licenses coexists, WUR may not be granted in perpetuity (for instance, area-based licenses issued under the NSW Water Act of 1912).

maximum permissible volume is then shared between economic sectors, giving priority to public water supplies (although in Australia, this is not explicitly stated in legislation). In France, industrial users are generally prioritised next, with the remaining volume of water allocated to farmers. In Australia, where groundwater use for irrigation is far greater than any other demand, the remaining users are generally treated equally.

27.4.6 Allocation Rules

France and Australia have also adopted different policy approaches to the way in which water allocations are issued to groundwater users. In France, the allocation policy has evolved over time. Before 2006, if a basin was designated as a restricted zone, existing users would receive a water use right proportional to their average past use, estimated over a reference period of 3–10 years depending on the local context. New applicants could only obtain a water use right if some of the historical users no longer exercised their right. The 2006 water law drastically changed this practice by requiring all potential users be given access to water resources, including within restriction zones. This was not really an issue for the drinking water sector, since there are no "new users". In the agricultural sector, Water Users Associations were asked to craft rules allowing the entry of new water users. Although each WUA has developed different rules, the most frequent approach relies on the following main principles:

(i) water use can be transferred to a new owner when a farmer retires and sells his farm; WUR only follows the farm is the activity planned by the new owners makes beneficial use of water; and if not, the WUR is reverted to a WUR reserve managed by the association;
(ii) for each transfer, a portion of the right (up to 20%) can be reverted to the WUR reserve;
(iii) rights (or portions of rights) which are not used over a number of years (typically five) also return to the association reserve;
(iv) the associations develop a rule to redistribute WUR held in reserve to farmers willing to expand their activities or to new users; the criteria used to rank competing applicants cover a number of factors – economic (added value, employment, strengthening of existing value chains), social (young applicants favored) and the environment (organic farming, crops with limited impacts favored).
(v) This allocation decision is taken by the association only; the State only verifies that the rules do not involve any discrimination and are correctly applied in practice.

Traditionally in Australia, the allocation of groundwater depended on a case-by-case assessment of individual applications to take water. However, since the significant reform instigated by the National Water Initiative in 2004, the different States have adopted broadly similar approaches to the allocation of water rights which are usually issued in the broader context of creating a management plan and the determination of a sustainable extraction limit. Two case studies are presented in Chap. 7. In general, the following principles are applied.

(i) Allocations are made by the State to meet the reasonable requirements of existing users.
(ii) This is based on use during a specified qualifying period which usually extends over several years. Those who can demonstrate a financial commitment to develop water use may be considered as existing users.
(iii) If there is no meter information available to quantify the reasonable requirements, the theoretical crop irrigation requirement can be used.
(iv) If the volume of existing user allocations is less than the sustainable extraction limit, the State may issue new allocations using a variety of methods (by application, ballot, auction).
(v) If the volume of existing user allocations exceeds the sustainable extraction limit, the State may reduce allocations (often through the management plan which requires extensive consultation).
(vi) If an area is fully allocated up to the sustainable extraction limit, the entry of new water users can only occur through trading of existing allocations on the water market.

27.5 Defining Sustainable Objectives and Setting Extraction Limits (Stage 4)

France and Australia both consider the establishment of a limit for extraction to be a fundamental requirement for the long-term sustainable development of groundwater resources. Both countries generally define this limit as the level of extraction from a particular groundwater management zone which, if exceeded, would compromise key groundwater dependent ecosystems and cause adverse impacts on the productive base of the resource. Although this concept was developed and implemented between the 1990s and the 2000s by both countries, there have been some differences in how it has been applied.

27.5.1 French Approach

In France, the transition from unrestricted access to a management regime incorporating sustainable extraction limits in groundwater basins has been a complex process that has generally been conducted progressively, using a methodology that involved a number of steps over a significant time frame.

- Existing extraction is first capped at the current level (with no new users allowed) in order to prevent degradation of the resource and inform water users that the resource is not unlimited and needs to be regulated. The cap can be defined as a seasonal or yearly maximum volume or a borehole extraction rate. The cap is usually based on the number of hectares of agricultural land and the type of crops that have been irrigated over a reference period.
- At catchment or groundwater basin level, Local Water Commissions (composed of representative of users, local communities and government agencies) establish monitoring networks and conduct investigations to inform the determination of a sustainable extraction limit. This often involves the construction of groundwater flow models (Chap. 11). The sustainable extraction is then specified in Local Water Management plans, which give them legal force. The extraction limit specified in the plan may differ from scientific recommendation, reflecting negotiations that take place within Local Water Commissions (Chap. 4).
- Because of political and economic considerations, the implementation of management actions is phased in over time to give water users time to adjust their operations.

Chapters 5, 12, and 13 give examples of how this approach was followed in three groundwater basins. In France, 581 groundwater aquifers have been identified for management purposes and of these, only 10% are considered to be in a "poor quantitative status" with sustainability issues and are consequently being managed with volumetric limits.

Australian Approach

While there is little difference between Australia and France in the fundamental approach to setting extraction limits, the implementation is different. The steps generally adopted by the States for establishing a groundwater management regime are described below. It is important to note that this process has frequently been implemented pre-emptively before over-extraction has occurred.

- Like France, the extraction is capped at the current level of pumping with no new development allowed.
- The existing knowledge of the groundwater systems is assessed and additional investigations are carried out if necessary, including groundwater modelling where appropriate.
- In parallel with these investigations, agencies begin the process of preparing a groundwater management plan which involves extensive community consultation to increase their understanding of the groundwater system and to work through various management options. The preparation of the plan may take up to 5 years.
- The sustainable extraction limit is determined for the plan, and in the States where a universal requirement for licensing does not exist (as explained in Chap. 7), allocations for existing users are granted. The limit can be calculated a number of ways (recharge estimation, modelling the impacts of extractions or resource condition limits) and is usually expressed as an annual volume.

- If the total volume of allocations is below the extraction limit, the management plan may define how new allocations can be issued and conversely, if the total volume of allocations exceeds the extraction limit, the plan may contain a process to reduce the allocations.
- Any reduction in allocations is generally phased in over several years to give water users time to adjust.
- The management plans are generally reviewed every 10 years to take into account any new understanding of the resource, changes in demand and impacts of variations in climate e.g. declining water levels due to lower rainfall.

In Australia, 288 groundwater management areas have been created. Of these, 136 have volumetric limits for extraction, with 25% of these classified as Over-allocated and only 2% considered to be overused (Chap. 6).

27.5.2 Why the Difference?

There are a number of factors which have resulted in a different style of implementing a sustainable extraction regime. The population of Australia (24 million) is much lower than France (65 million), and is highly concentrated in large cities on the coast. About 70% of Australia's groundwater extraction is for agricultural purposes which occurs in sparsely populated areas. Compared to France, there are far fewer groundwater users, less stakeholders to involve in consultation and fewer layers of bureaucracy involved in administration. In addition, there was significant investment into the investigation of major groundwater resources by the Federal and State governments during the 1970s and 80s.

This means that the transition from unrestricted access to a sustainable management regime is much less difficult in Australia than in France, and can be achieved more quickly because of the existing knowledge base and relatively small number of users. This has allowed the establishment of management regimes in many resources before over-extraction has occurred which at the time, avoided the potentially painful and difficult process of reducing allocations.

27.5.3 Common Challenges

One of the key on-going challenges encountered by French and Australian managers was to establish criteria which can be used to define what sustainable extraction means. There are two main aspects that need to be considered when establishing these criteria. The first is of a technical nature and should be considered fundamental. Sustainable limits should be set to prevent resource depletion, salt water intrusion and unacceptable impacts on streamflow and ecosystems. French and Australian managers have used a number of methods to determine these 'technical' limits which have been covered in earlier chapters e.g. recharge estimates, groundwater modelling and groundwater level thresholds. The second aspect is the consideration of social/economic factors which requires consultation with stakeholders and groundwater users and may lead to other criteria for determining an extraction limit e.g. timing of extraction, location of new wells, critical water levels for existing well completion depths etc. (See the case of the Barossa valley in Chap. 16). The consultation process should fully explore any trade-offs that may occur if there are differences between the ‘technical’ and ‘social/economic’ limits.

The definition of such criteria is more complex concerning confined aquifers. Indeed, the volume stored in those aquifers as well as water levels (or pressure) are doomed to decrease in those aquifers, as soon as they become exploited. It may take a few years, decades or even centuries to reach a new steady state equilibrium, in which the water storage and pressure will stabilize, at a new level. Meanwhile, it is extremely difficult to assess if extraction level is excessive and endangers the aquifer or not. A possible approach then consist in assuming that a storage decrease does not endanger the sustainability of the resource if it does not result in (i) permanent and extensive dewatering of the reservoir; (ii) flow directions and patterns causing the inflow of extraneous water (inland saline or sea water); (iii) insufficient outflow into dependent ecosystems which would threaten their ecological status.

Another key challenge faced when determining extraction limits in France and Australia lies in properly accounting for groundwater – stream interactions. This requires an understanding of how the volume, timing and location of groundwater pumping will affect baseflow to streams and interactions with ecosystems such as wetlands. These are complex processes which require sophisticated management tools such as well-calibrated groundwater flow models.

Experience from France and Australia has shown that because all aquifer systems are unique with different complexity and different levels of data availability and understanding, managers should use fit for purpose hydrogeological approaches to determine extraction limits – it is not always possible or desirable to construct a well calibrated multi-layered groundwater flow model which could cost millions of dollars/euros. In some cases, a simple spread sheet analytical model using representative hydraulic parameters may suffice (Chaps. 11 and 14).

Finally, the authors strongly emphasise the need to engage all stakeholders and water users in the debate that leads to the definition of extraction limits, and recognise the challenges involved with this process. Where there are significant technical and scientific uncertainties, they should be explained to stakeholders. Any assumptions made should be shared and if possible, be accepted to ensure that the final outcome is supported. If the scientific approach is not transparent and understood, there is a high probability that extraction limits will be challenged.

27.6 Returning Over-Allocated and Overused Groundwater Systems to Sustainable Levels of Extraction (Stage 5)

Having gone through the four stages previously described in this chapter (political awareness, increased understanding of groundwater systems, allocated water rights and set sustainable extraction limits), one could assume the journey to a sustainable abstraction management regime was virtually complete. However as governments, decision makers and academic researchers across the globe have come to realise, it is the remaining two steps –returning Over-allocated systems to sustainable levels and ensuring compliance and enforcement – that are arguably the most complex and difficult groundwater management challenges. Indeed in some areas, the impacts of a new sustainable management regime on established political, economic and social interests can become so complex and difficult, that these final stages have often been hampered by sluggish progress, or remained an afterthought for policymakers. In short, significant work is still needed to deliver on these steps.

There is no better illustration of this fact than Chaps. 16 and 20 documenting Australia's attempts to return over-allocated and overused systems to sustainable levels. Challenges such as a perceived "top down" unilateral approach adopted by some state governments (Chap. 20), different criteria and interpretations of the terms "over-allocation" and "over-use", the development of "short term" responses, and lengthy contentious debates about the economic and social trade-offs associated with re-allocating water away from agriculture to the environment, have all made it difficult to identify and evaluate the steps taken to deal with over-allocation and over-use. While some progress has been made, including in areas such as Tintinara in South Australia (Chap. 19), recent national assessments suggest there is still more work to do (Productivity Commission, 2018). Compared to Australia, France has faced far few challenges in this regard, partly because of the greater involvement of affected users in developing rules for the allocation reductions. Each Agricultural users' association (OUGC) has made use of the power given to them to develop their own rules, with considerable differences in the choices made by different OUGCs. Even so, experiences in places such as Poitou Marshes (Chap. 18) reveal important insights on the fragility of pathways for returning over-allocated or overused systems to sustainable levels.

For both France and Australia, methods to reduce the permanent share/entitlement to groundwater resources typically saw the use of some form of compensation (although this was often not legally required). This included financial payments (buy back programs in Australia, Chap. 17), and infrastructure or substitution water reservoirs (Australia and France, Chaps. 17 and 18), to lessen the social and economic impact of reductions. In some areas in Australia, this was often the most difficult process in the journey to sustainable groundwater management, featuring conflict, moratoriums and court challenges (Chaps. 17 and 20).

Temporary adjustments to account for seasonal variation in available groundwater resources (e.g. times where aquifer levels are lower), were comparatively easier to implement for unconfined aquifers which have limited or highly variable storage volumes that are controlled by rainfall variability. In Australia, adjustments to the volume that can be pumped from these types of aquifer arise from periodic changes made to the 'available resource' (which is the sustainable extraction limit). Whilst a water user's entitlement may be a permanent percentage share of the 'available resource' (e.g. 1.0%), the 'available resource' can change periodically depending on the aquifer levels (e.g. from 20,000 ML to 15,000 ML). The water user's annual allocation will consequently reduce from 200 ML to 150 ML. If the water user requires more water, additional allocations or entitlements can be purchased through water trading. A similar approach to seasonal adjustments occurs in France. If the State imposes a reduction in allocations in a management area due to lower groundwater levels, the OUGC then decides how to share the seasonal reduction amongst their members, which may not necessarily be a universal reduction (Chap. 18).

Regardless of the implementation process, it was clear from the case studies that there are numerous disputes about the precise rules that should be used to determine a reduction in allocations. Much of this controversy arises from the fact that governments in both France and Australia are the ultimate decision makers when it comes to determining global reductions for a given management area. In contrast in places such as Chile, Texas or other locations, the State does not impose this decision, it is up to users to decide if they want to reduce allocations and how that process might occur.

Regardless of who imposes the initial reductions, tensions clearly arise as to how the reduction effort is shared among the various groups of users. This was more prominent in Australia where conflicts and tradeoffs between environment and agricultural interests occurred, as well as amongst agricultural users themselves. While some advocated universal reductions, others called for differentiated approaches, with higher reduction imposed on recent users than historical ones or giving priority to certain activities. These processes reveal diverging concepts of social justice and alternative visions of how to reconcile impacts on agricultural communities with the needs of the environment and economic efficiency.

So how should such conflicts be solved? In France, users are asked to agree on principles, define rules and then apply them at catchment or groundwater basin level. This devolved and context specific approach has arguably had some success in France, partly because farmers are given a level of agency and autonomy to decide how the cuts should be shared and what rules will be chosen to implement them. It is however too early to tell if this approach will deliver the desired outcomes. Australia aspired to a similar approach by involving communities in water planning processes. This consultative process regarding allocation reductions was sometimes a relatively smooth (e.g. Tintinara in Chap. 19), but it sometimes was not (e.g. Lower Murrumbidgee in Chap. 20), producing ongoing distrust from affected agricultural users, calls for compensation, the threat of court action and calls for improved procedural justice (Daniell, 2011).

Ultimately, the above insights from both France and Australia on possible pathways to returning over-allocated groundwater systems to sustainable levels remain important for policy makers as the world continues to confront the uncertain impacts of climate change and growing demands on water from increasing population and energy use.

27.7 Stage 6 – Implementing a Cost-Effective Enforcement System (Stage 6)

The final stage confronting policy makers is to implement cost effective compliance and enforcement. Although the last in the policy making process laid out in this chapter, it is arguably the foundation on which all other elements within the system come to rest. If people do not comply with rules, and rule breakers are not identified and brought back into compliance, the entire system of groundwater management can be undermined, producing aggravated effects to humans, the environment and future generations (Interpol, 2016; Segato, Mattioli, Capello, & Migliorini, 2017). Yet despite the importance of compliance and enforcement, agencies or groups responsible for groundwater management have devoted comparatively little time or effort on this final stage. This is perhaps most clearly illustrated by the estimates of Interpol (2016), that suggests that not only is up to 50% of the global water supply illegally purchased, but that there are millions of unregulated wells worldwide (including more than 20 million in Africa alone).

With global consumption of water doubling every 20 years, demand for water from agriculture projected to increase by 50%, and up to 85% more water projected to be consumed by the energy sector over the next 15 years (Segato et al., 2017), improvements in compliance and enforcement will be vital to prevent further development of illegal water use that could lead to major degradation of groundwater resources, particularly in areas of depleting surface water resources due to climate change (Brown, 2017; Interpol, 2016).

As Segato et al. (2017) note: "*Groundwater reserves are depleting in many places, leaving current and future generations with close to no buffer against increased climate variability, and without effective regulation and suppression of water crimes, the sustainability, long-term viability, and inclusive and equitable use of water can-not easily be achieved*".

Notwithstanding the importance of implementing a cost-effective compliance and enforcement regime, studies of compliance and enforcement in quantitative groundwater contexts in France, Australia and indeed other parts of the world, are quite rare. Drawing on Chaps. 22 and 23, at least two broad sets of insights emerge relating to: (1) common factors explaining compliance and non-compliance; (2) common factors explaining enforcement success/problems.

27.7.1 Common Factors Explaining Compliance and Non-compliance

There were at least three similar drivers of compliance in both France and Australia. Firstly it was clear that compliance was facilitated by relatively strong groundwater user understanding of the core compliance requirements. This particularly related to the obligations that were required of them on their farm and in their day to day operations.

Secondly, recognising that education and information about rules and penalties can help promote compliance, both France and Australia demonstrated an important educating role by non-government professional farming organisations/industry associations. These bodies were seen to provide useful sources of information for farmers.

Thirdly, compliance and enforcement government officials in both countries also took a graduated punitive approach (Ayres & Braithwaite, 1992) to interacting with farmers when a breach was suspected. This was seen to reduce risks of conflict and helped to facilitate improved compliance over time.

In terms of differences between France and Australia, one notable distinction was the different motivations for agricultural water users to follow rules. Chapter 22 found that farmers in NSW were more likely to follow rules because of their desire to do the right thing, to ensure fairness amongst other water users, because of social and peer reputation, and the perceived legitimacy of laws (e.g. protecting water resources, user rights, viability of communities and the environment). Interestingly, these motivations did not appear to be strongly echoed in France. In some areas, where the tension between government agency and farmers is maximal, it was even suggested that not following the rules may in fact lead a farmer to have a better reputation (rather than worse) with their peers. To what extent this difference can be generalised across both nations, and/or reflects different values between the two nations peoples (see e.g. World Values Survey discussed in Chap. 22) remains an open issue worth exploring.

Turning to trends in non-compliance between France and Australia, it is notable that precise levels of non-compliance were difficult to obtain in either nation. However both Chaps. 22 and 23 suggest non-compliance remains a fundamental issue, including being subject to public inquiries. Even so, both countries revealed a common justification for illegal water extraction, namely economic pressure on farmers and a desire for economic advantage. Both also showed a lack of deterrence from enforcement practices (discussed below), including perceptions of a low probability of an inspection on farm or users being caught for illegal activities.

27.7.2 Common Features of Enforcement

France and Australia have undergone steady improvement in groundwater compliance and enforcement. As detailed in Chaps. 22 and 23, there has been a general trend of moving away from having a diverse set of government agencies with mandates to support agriculture development and enforce compliance. Both countries have accordingly consolidated enforcement policies and produced more structured and separate organisations (e.g. NSW Natural Resource Access Regulator).

Despite these improvements, four common challenges were identified in both countries. First and foremost were resourcing barriers, including low numbers of permanent groundwater focused staff, combined with the lack of human resources in the judicial system.

A second challenge was the uneasy relationships between government and agricultural interests. As Chap. 23 explains in France, a significant barrier to enforcement in the agricultural sector is a tendency for government to avoid areas with significant agricultural conflicts or concerns. This issue is echoed in Australia, albeit in a wider context of inquiries pointing to ineffectual processes applied to agriculture, and a wider climate of concern arising from the murder of an environmental compliance officer during a visit to a farm[4]. Third, and intertwined with this challenge, was a perception of political interference in compliance and enforcement activities, that prevented regulators from doing their job properly in France, and led to allegations of corruption and recent inquiries in Australia.

Fourth and finally, a lack of modern technology for inspection was reported to be a significant limiting factor in both nations, including metering challenges, access to data and the need for greater use of new technology like aerial drones.

Ultimately, both Chaps. 22 and 23 suggest a need for regulators and water users to devote more resources and effort to build on early successes and to fix problems that becoming evident.

27.8 Conclusion

27.8.1 Similar Approaches to Groundwater Management

The material presented in that book and summarized in this chapter was presented during a French-Australian workshop held in September 2015, bringing together 30 French and 13 Australian attendees who were all involved in groundwater management at different administrative and geographic levels. Overall, the main conclusion of that workshop was that, in spite of huge climatic, environmental, socio-economic and legal differences "*there is more that unites us than divides us*". Essentially, the philosophy and approaches to groundwater management are very similar:

- The State plays a key role in water governance, although significant differences exist concerning water users' participation and the role of water markets.
- The State and users jointly contribute to financing the human, technical and financial resources dedicated to the management of water.
- There is a hierarchy of plans generally based on surface water catchments. Planning is supported by groundwater monitoring, including the metering of extractions, and sophisticated groundwater information systems.

[4]ABC. 2016. Moree shooting: Farmer Ian Turnbull jailed for 35 years for murdering environmental officer https://www.abc.net.au/news/2016-06-23/moree-shooting-ian-turnbull-sentenced-over-murder/7535808

- There is a clear recognition of the importance of providing water for environment, in particular groundwater dependent ecosystems.
- Policies recognizes the importance of community consultation to achieve satisfactory outcomes, and the continual need to educate decision makers and users about how our groundwater systems work.
- Policy makers, managers and users recognize that science-based decision making is the basis of responsible and sustainable groundwater management. However, other factors influence actual decision making that reflect existing power relationships and political balances within society.

27.8.2 A Different Visio of Water Use Rights

There is however a key issue on which the French and Australian philosophies diverge quite significantly: the approach to water use rights.

Australian aspires to treat water use rights on an individual and private basis. They promote the development of water markets which have proved to be very efficient tools to minimise the economic impacts of drought in the recent past. Australia is internationally recognised for the success of this approach, it has and will continue to inspire other countries.

France is promoting an alternative approach, based on common property regime. This approach is based on the creation of hybrid institutions, holding collective resource use rights, and bringing together representatives of the State and users to define management rules. These hybrid institutions are responsible for setting extraction limits that ensure the long-term sustainability of the resource and keep environmental impacts at an acceptable level. They also define how to share among members the allocation hold in common. These rules exclude any permanent individual appropriation of natural resources.

The contrast between these two approaches is an invitation for the reader to reconsider the issue of water property. Historically, there has been a systematic attempt to eradicate community ownership regimes over natural resources in the Western world following the Enclosures movement in the seventeenth and eighteenth centuries. We then witnessed the emergence of a polarized vision opposing public and private ownerships of natural resources. Private regimes have gradually become a dominant model, leading many countries to implement policies based on private individual tradable use rights. Australian water policy aspires to be a perfect illustration of that model (albeit with mixed experiences between its different states and territories). But the French example (and similar approaches tested in New Zealand) shows that common property regime could rise like a phoenix from the ashes.

While the Australian model has been adopted by many countries (USA, Chile, Spain) and is being popularized by scholars (mainly economists) and international institutions, the French model is a rather unique experiment which is worth being considered as an alternative to the market approach by countries engaging in the development of a new groundwater allocation and management policy.

This controversial property issue should be addressed by all countries entering into groundwater management reform. A constructive debate should be organised, bringing together researchers from various disciplines (economists, lawyers, political scientist), policy makers and water resources managers. The confrontation of different points of view should help thinking outside the box and it could lead to the identification of highly innovative approaches, combining elements of two apparently exclusive models.

27.8.3 Common Challenges

The two countries also face similar challenges, in particular related to the following technical and institutional issues:

- The development of the knowledge base is a long and costly process which should be intensively supported by the State, with a contribution from users. Uncertainty will always remain but that should not prevent policy makers and water resources managers to make decisions. A key challenge consists in understanding and modelling groundwater-surface water interaction and developing management and planning procedures that integrate both resources.
- Involvement of users in the development of groundwater water management rules and plans is more difficult than that for surface water resources, considering the hidden nature of that resource, the absence of collective infrastructure, and the fact that it has often been considered an open access resource for decades.
- Compliance is a key challenge, even in developed countries which can dedicate significant resources to enforcement policies. Countries initiating groundwater management reforms should treat this issue as a high priority, to prevent the installation of a weak social norm where deviant behaviours become the rule, and which will be extremely difficult to reverse.

To end this book, we would like to stress again the benefits of creating a dialogue between practitioners from different countries, and between practitioners and scholars. Confronting visions of experts having very diverse backgrounds helps reconsidering assumptions each take for granted for historical, legal or regulatory reasons and it is source of creativity. We hope that readers of that book will have been inspired by the cases studies presented and experiences shared in this book and that it will help them developing innovative groundwater management approaches, adapted to the specific technical, economic, social and institutional characteristics of their context.

References

Avey, S., & Harvey, D. (2014). How water scientists and lawyers can work together: A 'down under' solution to a water resource management problem. *Journal of Water Law, 24*, 45–61.

Ayres, I., & Braithwaite, J. (1992). *Responsive regulation*. Oxford, UK: Oxford University Press.

Brown, V. F. (2017). *Water theft and water smuggling*. Washington, DC: Brookings.

Cosens, B. (2018). Governing the freshwater commons: Lessons from application of the trilogy of governance tools in Australia and the Western United States. In C. Holley & D. Sinclair (Eds.), *Re-forming water law and governance*. Singapore, Singapore: Springer.

Cuadrado-Quesada, G., Holley, C., & Gupta, J. (2018). Groundwater governance in the Anthropocene: A close look at Costa Rica. *Water Policy, 20*(3), 475–489.

Daniell, K. A. (2011). Enhancing collaborative management in the basin, chapter 26. In D. Connell & R. Q. Grafton (Eds.), *Basin futures: Water reform in the Murray-Darling basin* (pp. 413–438). Canberra, Australia: ANU E-Press.

Holley, C., & Sinclair, D. (2012). Compliance and enforcement of water licences in NSW: Limitations in law, policy and institutions. *The Australasian Journal of Natural Resources Law and Policy, 15*, 149–189.

Holley, C., & Sinclair, D. (2016). Regulation, technology, and water: "Buy-in" as a precondition for effective real- time advanced monitoring, compliance, and enforcement. *George Washington Journal of Energy & Environmental Law., 7*(1), 52–66.

INTERPOL. (2016). *Strategic report: Environment, peace and security*. UN/Interpol.

Jakeman, A., Barreteau, O., Hunt, R., Rinaudo, J. D., & Ross, A. (Eds.). (2016). *Integrated groundwater management: Concepts, approaches and challenges*. Cham, Switzerland: Springer.

Margat, J., Pennequin, D, & Roux, J.-P. (2013). *Histoire de l'hydrogéologie française*. Paris: Comité français d'hydrogéologie, Association internationale des hydrogéologues.

Nelson, R., & Quevauviller, P. (2016). Groundwater law. In A. J. Jakeman, O. Barreteau, R. J. Hunt, J. D. Rinaudo, & A. Ross (Eds.), *Integrated groundwater management* (pp. 173–196). Cham, Switzerland: Springer.

Productivity Commission. (2018). *National water reform* (Report no. 87), Canberra, Australia.

Rinaudo, J. D., & Donoso, G. (2018). State, market or community failure? Untangling the determinants of groundwater depletion in Copiapó (Chile). *International Journal of Water Resources Development*, 1–21.

Segato, L., Mattioli, W., Capello, N., & Migliorini, M. (2017). Environmental crimes in the water sector. *European Water, 60*, 373–379.

Wilhite, D. A. (2012). Breaking the hydro-illogical cycle: Changing the paradigm for drought management. *EARTH Magazine, 57*(7), 71–72.

Dr. Jean-Daniel Rinaudo is researcher at Brgm, Montpellier University, where he coordinates the scientific program on environmental and risk economics Initially trained as an agricultural engineer (Montpellier SupAgro), he specialized in agricultural and resource economics (PhD University of Auvergne). Prior to joining Brgm, he worked for the International Water Management Institute in Pakistan where his research focused on the political economy of irrigation management reforms. His current research mainly focuses on the institutional economic dimension of groundwater management. Most of his research is conducted in France but he also works in Morocco and Chile. He is currently developing new research activities in the field of natural disasters economics, focusing on the methods to assess economic vulnerability and resilience. Dr. Jean Daniel Rinaudo is also member of the Scientific Council of the Adour Garonne River basin agency.

Steve Barnett is Principal Hydrogeologist at the Water Science and Monitoring Branch of the Department for Environment and Water in South Australia. He has been involved in the investigation, monitoring and management of groundwater resources in SA for over 40 years, and has contributed technical and policy input into ten groundwater management plans which incorporate a variety of different aquifers and management issues. He is a past-president of Australian Chapter of the International Association of Hydrogeologists.

Dr. Cameron Holley is a Professor at University of New South Wales Law and is a member of the Global Water Institute and Connected Waters Initiative, University of New South Wales Sydney. Cameron has worked closely with Australian and international government and non-government organizations on a range of water and natural resource management research projects. He currently holds ARC Discovery Grants (DP170100281 DP190101584) on Non-Urban Water Regulation and Integrating the Governance of Water and Coal Seam Gas that supported this chapter. He is an Editorial Board member on the Environmental and Planning Law Journal, and in 2016 was the guest editor of a Special Issue (EPLJ Vol 33 Part 4), entitled Rethinking Water Law and Governance.

Printed by Printforce, the Netherlands